Nitrogen-Rich Energetic Materials

Nitrogen-Rich Energetic Materials

Edited by Michael Gozin and Leonid L. Fershtat

WILEY-VCH

Editors

Prof. Michael Gozin
School of Chemistry
Faculty of Exact Sciences
Tel Aviv University
Tel Aviv 6997801
Israel

Center for Nanoscience and Nanotechnology
Tel Aviv University
Tel Aviv 6997801
Israel

Center for Advanced Combustion Science
Tel Aviv University
Tel Aviv 6997801
Israel

Prof. Leonid L. Fershtat
Russian Academy of Sciences
N.D. Zelinsky Institute of Organic Chemistry
Leninsky prosp. 47
119991 Moscow
Russia

Cover Image:
© videoduck/Shutterstock

Library of Congress Card No.: applied for

British Library Cataloguing-in-Publication Data
A catalogue record for this book is available from the British Library.

Bibliographic information published by the Deutsche Nationalbibliothek
The Deutsche Nationalbibliothek lists this publication in the Deutsche Nationalbibliografie; detailed bibliographic data are available on the Internet at <http://dnb.d-nb.de>.

Print ISBN: 978-3-527-34929-6
ePDF ISBN: 978-3-527-83262-0
ePub ISBN: 978-3-527-83263-7
oBook ISBN: 978-3-527-83264-4

Cover Design: Wiley
Typesetting Straive, Chennai, India
Printing and Binding CPI Group (UK) Ltd, Croydon, CR0 4YY

C114244_011222

Contents

Preface

Energetic materials research is currently undergoing a global renaissance. At no point in history has the field possessed the breadth and geographic diversity of contributors that it does today. From Germany, to Russia, to the United States, to China, to Israel, numerous nations are currently involved in intense research in the field, and several nations are extensively funding work in this area in order to compete with others. This field has also grown to be incredibly diverse, and no single book could cover the entire field. Thus, for this book, we have chosen to focus on several recent and popular areas in the field of energetics. We have aimed to focus on the topics of energetic materials including all of the propellants, explosives, and pyrotechnics and to ensure the relevance of the topics discussed; the following chapters have been solicited from leading experts in the field from around the world. We hope that this book serves to be a reference to energetic materials chemists in the industry, academia, and government as well as for the interested individuals.

Chapter 1 is by Ming Lu and coauthors from the Nanjing University of Science and Technology in China and concerns the chemistry of pentazoles. We chose to place this chapter first as the all-nitrogen pentazole ring is one of the most important recent advances in energetic materials as higher and higher performances are being sought. While the thermal stabilities and densities of pentazoles known to date remain too low for practical use, if this can be overcome with further synthetic tailoring, the pentazole structural motif promises to allow the creation of energetic materials of unprecedented performance.

Chapter 2 by Qinghua Zhang and coauthors from the China Academy of Engineering Physics continues our discussion on energetic materials to aromatic fused-ring energetic systems. This structural motif for compounds is highly important as many members of this class of compounds exhibit high densities, high stabilities, and high performances – all very important for practical materials. Additionally, this chapter covers annulated 1,2,3- and 1,2,4-triazines that remain relatively rare in the energetics field, and the compounds covered give insights into the possibilities that may be afforded by these unique heterocyclic systems.

In Chapter 3, Lei Zhang and coauthors at the Laboratory for Computational Physics take a detour from the intense focus on the synthesis and properties as found in other chapters to focus on computational aspects in energetic materials.

Computation remains an important tool for guiding synthesis toward practical materials of tailored properties as well as pushing the limits of energetic performances to new levels through the determination of the possibility of stability of newly conceived chemical structures.

In Chapter 4, Jörg Stierstorfer and coauthors at the University of Munich discuss an emerging and promising subset of energetic materials, that of laser-ignitable compounds. These types of materials serve as prospective lead azide replacements, whose replacement is mandated as a result of environmental concerns, with the additional advantage of initiation from laser stimulation, which promises to offer new and advanced possibilities for firing systems. The energetic coordination compounds (ECCs) discussed in this chapter are one class of laser-ignitable compounds.

Chapter 5 continues with a discussion of one of the most promising heterocyclic systems in energetic materials design, the 1,2,3,4-tetrazine-1,3-dioxides by Churakov and Tartakovsky and coauthors at the Zelinsky Institute for Organic Chemistry, the original location of this heterocycle's discovery. The 1,2,3,4-tetrazine-1,3-dioxides are interesting from both a fundamental and practical perspective as their discovery and study led to fundamental insights into the factors affecting the stability of high-nitrogen systems, and also several members of this class of compounds are predicted to have very high energetic performances, exceeding the performance of the current energetic materials.

In Chapter 6, we discuss recent advances in energetic polymer materials. When energetic materials are considered, they are often not used as pure compounds and instead are formulated into actual compositions for use. Unfortunately, in many current formulations, the polymers used alongside the energetic material are completely nonenergetic, and one way to improve the performance of the energetic system is to improve the energy content of the binder polymer.

In Chapter 7 Jian-Guo Zhang and coauthors focused on the preparation of a wide series of tetrazole-based energetic salts bearing various explosophoric moieties at the tetrazole ring. Although energetic tetrazoles are one of the classic heterocyclic high-energy materials, such substances have good functional properties and, based on their molecular composition, they are of interest as either primary or secondary explosives.

Chapter 8 by Fengqi Zhao and coauthors from Xi'an Modern Chemistry Research Institute continues the discussion on polynitrogen heterocycle-based energetic materials and covers the synthesis, physicochemical, and detonation parameters of 3,6-bis(1*H*-1,2,3,4-tetrazol-5-yl-amino)-1,2,4,5-tetrazine (BTATz) along with the consideration of energetic properties of various formulations derived thereof.

Chapter 9 by Philip Pagoria is focused on the achievements in the synthesis of energetic heterocyclic compounds incorporating a nitro-substituted oxadiazole subunit. From the energetic materials science point of view, this research area rose in the twentieth century, but it is still referred to as one of the promising fields in the development of novel high-energy substances for various practical applications such as oxidizers, plasticizers, and solid rocket propellants.

Chapter 10 by Ernst-Christian Koch covers a specific subclass of insensitive explosives incorporating the tetraazapentalene framework, which corresponds to the benzannelated [1,2,3]triazolo[1,2-*a*][1,2,3]triazolium-1-ide heteroaromatic scaffold containing 10 π-electrons. Although such materials require multistep synthetic procedures, their balanced physicochemical and detonation properties make them promising alternatives for common insensitive nitroaromatic-based energetic materials.

Finally, in Chapter 11, Thomas M. Klapötke and coauthors from the University of Münich reviewed advances in the preparation of environmentally benign pyrotechnic formulations based on nitrogen-rich energetic materials. This is quite urgent and a roughly unexplored area of research as such pyrotechnics are important ingredients for various applications, such as marking of aircraft landing positions in emergencies and accidents.

Taking into account the emerging development of chemistry and engineering, there is no doubt that energetic materials will retain their leading position in the framework of materials science. In our opinion, a search of novel powerful high-energy materials is far from being exhausted. We believe that breakthrough achievements in the energetic materials science can be made only by a prospective symbiosis of organic, inorganic, structural, physical, and theoretical chemistry; toxicology; and engineering. A combination of expertise skills in these fields will provide a synergistic effect and will result not only in an estimation of deep insights into structure–property relationships but also will find new industrial applications. A high level of environmental compatibility of newly prepared energetic materials will guarantee the sustainability of chemical processes. Finally, the editors of this book hope that it will stimulate researchers for further work in this field.

Russia
28 July 2022

Michael Gozin
School of Chemistry
Faculty of Exact Sciences
Tel Aviv University
Tel Aviv 6997801
Israel

Center for Nanoscience and Nanotechnology
Tel Aviv University
Tel Aviv 6997801
Israel

Center for Advanced Combustion Science
Tel Aviv University
Tel Aviv 6997801
Israel

Leonid L. Fershtat
N.D. Zelinsky Institute of Organic Chemistry
Moscow

About the Editors

Dr. Michael Gozin is a professor at the School of Chemistry and a founder of the Center for Advanced Combustion Science, Tel Aviv University (Israel). He received his academic degrees from Hebrew University and the Weizmann Institute of Science (Israel). He was a visiting scholar at Stanford University (USA), the Beijing Institute of Technology (China), the Indian Institute of Technology Bombay (Mumbai), and the Dresden University of Technology (Germany). He is serving as the co-editor-in-chief of the *Energetic Materials Frontiers* journal (Elsevier), a member of the Editorial Advisory Board of the *Defense Technology* journal (Elsevier), and a member of the International Scientific Advisory Board of the New Trends in Research of Energetic Materials Conferences (Czechia). He also served as a chairman of the National Committee in Chemistry and Materials of the Israeli National Institute of Standards. His research interests encompass the development of novel energetic materials, catalysts, and green chemistry processes. He published more than 100 articles and reviews, 20 patents, and his *h*-index is 32 (Scopus).

Prof. Leonid L. Fershtat is the head of the Laboratory of Nitrogen Compounds at the N.D. Zelinsky Institute of Organic Chemistry, Russian Academy of Sciences. He graduated from the D.I. Mendeleev University of Chemical Technology in 2012 (faculty at the Higher Chemical College of the Russian Academy of Sciences) and received his PhD degree from the N.D. Zelinsky Institute of Organic Chemistry in 2015 under the supervision of Prof. N.N. Makhova. In 2020, he finished his habilitation studies in the same institute. Dr. Fershtat is serving as an associate editor of the *Energetic Materials Frontiers* journal (Elsevier), a member of the Editorial Advisory Boards of the *Defense Technology* and *FirePhysChem* journals (Elsevier), and a member of an Early Career Advisory Board in *Mendeleev Communications* (Elsevier). He also served as a member of the Organizing Committee of All-Russian Conference on "Chemistry of Nitro Compounds and Related Nitrogen-Oxygen Systems." His scientific interests include the creation of novel methodologies in the synthesis and transformations of polynitrogen and nitrogen–oxygen heterocycles and search for pharmacologically active and high-energy heterocyclic derivatives. He published over 70 articles and reviews and holds three patents and provisional patents.

1

Chemistry of Pentazole

Ming Lu, Pengcheng Wang, Yuangang Xu, and Qiuhan Lin

Nanjing University of Science and Technology, School of Chemistry and Chemical Engineering, Xiaolingwei 200#, Nanjing, 210094, PR China

1.1 Introduction

The pentazolate anion, usually referred to as "*cyclo*-N_5^-," is an all-nitrogen-containing five-membered-ring compound that has recently attracted significant levels of attention; *cyclo*-N_5^- represents an important member of the azole ring family and is an isoelectronic nitrogen analogue of the cyclopentadienide anion [1, 2]. Furthermore, the decomposition of *cyclo*-N_5^- into dinitrogen (N_2) and N_3^- is exothermic by 14.3 kcal mol^{-1}, while the activation energy for this decomposition is 27.2 kcal mol^{-1}. The heat of formation, ΔH_f (298 K, N_5^-), is 59.6 kcal mol^{-1} [3–5], while the vertical ionization energy of *cyclo*-N_5^-, which refers to the formation of neutral N_5 while maintaining the geometry of *cyclo*-N_5^-, has been calculated to be 5.6 eV [5]. These data indicate that *cyclo*-N_5^- is very likely to exist after isolation, potentially becoming the next polynitrogen species in the field of energetic materials. The unique properties of the *cyclo*-N_5^- ring have motivated extensive efforts toward exploring this unknown structure by synthetic chemists [4, 6].

1.2 Substituted Pentazoles

The exploration of *cyclo*-N_5^- has been a difficult and time-consuming process [7]. Experimental progress stagnated at the substituted pentazole stage for a long period of time because the aromaticity of the *cyclo*-N_5 ring is strongly affected and even broken through the attachment of substituents. Consequently, substituted pentazoles are too sensitive for use under most laboratory conditions, and researchers have to investigate them by in-depth computational studies, including the stabilities, as well as the formation and decomposition mechanisms. Here, we classify substituted pentazoles according to their substituted groups.

Nitrogen-Rich Energetic Materials, First Edition. Edited by Michael Gozin and Leonid L. Fershtat.

1.2.1 Arylpentazoles (ArN_5)

For a long time, arylpentazoles (ArN_5), which are usually stored at low temperatures, were the only experimentally known substituted pentazoles. The first attempt to synthesize ArN_5 was carried out in 1903; however, it was only in the 1950s that the existence of ArN_5 was established through kinetic and gas-volumetric methods [8, 9]. The crystal of *p*-dimethylaminophenylpentazole (DMAP-N_5), which owns a room-temperature decomposition point, is the first and only direct evidence for the existence of substituted pentazoles [10].

The stabilities of para and ortho/para-substituted phenylpentazoles (PhN_5) were explored. It was confirmed computationally that ArN_5 is stabilized by the addition of an electron-donating group and destabilized by the addition of an electron-withdrawing group at the para position of the benzene ring. The stabilizing effects of electron-donating groups have been shown to arise through resonance interactions with the electron-deficient pentazole ring, while the decomposition of ArN_5 to N_2 and the corresponding azide was established to be a first-order reaction. The calculated dissociation energies of the N—N bonds in these substituted ArN_5 were determined to be 109–117 kJ mol^{-1}, which are higher than those of PhN_5, but are far lower than the activation energy (362–402 kJ mol^{-1}) for cleavage of the C—N bond that links the benzene and pentazole rings (Figure 1.1) [12].

Derivatives of PhN_5 substituted with alkali metals (M = Li, Na, and K) at the para-, meta-, or ortho-positions of the benzene ring were designed by Gong et al. The metal substituents improved the energy barrier for decomposition of the N_5 ring of PhN_5 by 19.3–65.0 kJ mol^{-1} (K > Na > Li) [12–14]. In particular, a new ionic N—M

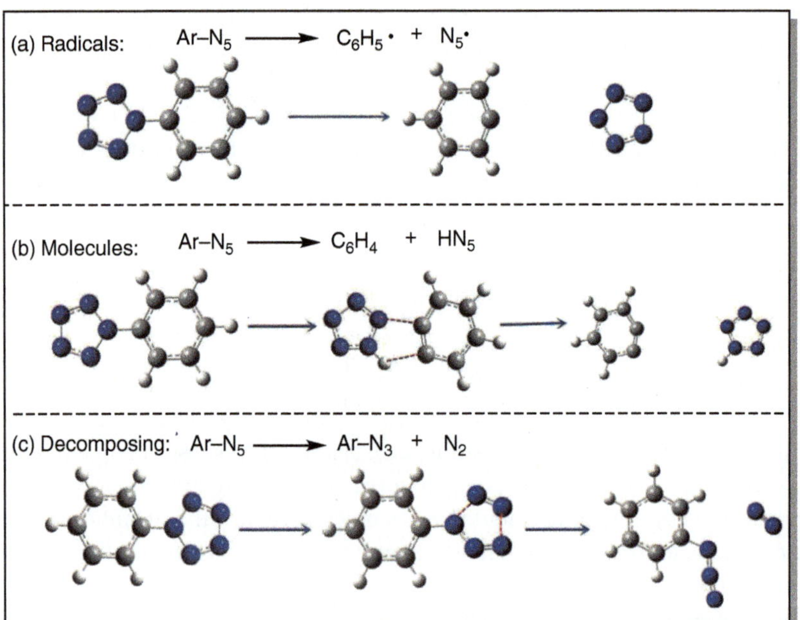

Figure 1.1 (a, b) Two possible pyrolysis paths of the C–N bond of ArN_5; (c) decomposition of the *cyclo*-N_5^- ring of ArN_5. Source: Bondarchuk [11]/with permission of Royal Society of Chemistry.

bond is observed between the metal and N_5 in the ortho-substituted PhN_5, which is not observed for the para- or meta-substituted derivatives. The strong intramolecular interactions in the ortho-substituted PhN_5 generate more negative charge on the pentazole ring, which strengthens dispersion interactions, resulting in higher pentazole-ring stability.

1.2.2 Heterocycle Pentazoles

It appears to be much more difficult to attach heterocyclic rings to pentazole, compared to aryl rings. Tetrazolylpentazole was the second type of pentazole derivative prepared, but it decomposes quickly at −50 °C and can only be confirmed by low-temperature ^{15}N-NMR spectroscopy. Calculations show that the tetrazole substituent has a similar conjugation effect to that of aryl, as is commonly observed for other azoles [15]. Since it is possible to remove an N–H proton from azolylpentazolate (AzN_5), anionic AzN_5 and its derivatives have been investigated. Anionic AzN_5 and its derivatives were found to exhibit lower decomposition energies and higher activation energies for decomposition than the corresponding neutral compounds, due to the higher electron densities of their pentazole rings.

The potential of pyridylpentazole and its pyridazylpentazole, triazinylpentazole, tetrazinylpentazole, and pentazinylpentazole derivatives to replace PhN_5 as a source of the pentazolate anion has been assessed [16]. Replacement of the aryl group of PhN_5 with a six-membered heterocycle weakens both the pentazole ring and the central C—N bond, as evidenced by the lower activation energies required to fragment the pentazole ring (8.4–30.8 kJ mol^{-1}) and the central C—N bond dissociation energies (11.3–21.0 kJ mol^{-1}). However, decreases in the C—N bond stability (2.1–7.8%) caused by the heterocyclic ring and its substituents are less obvious than those of the N_5 ring (12.7–23.7%); therefore, *cyclo*-N_5 substituted with six-membered heterocyclic rings are not likely to be better sources of *cyclo*-N_5^- than PhN_5.

1.2.3 Benzylpentazole (BzN_5)

BzN_5 and its derivatives have a methylene spacer between the pentazole and the benzene ring, which can be substituted by electron-donating ($-NH_2$, −NHMe, $-NMe_2$, and −OH) and electron-withdrawing ($-NO_2$, −CN, $-CF_3$, and $-NF_2$) groups, thereby endowing these molecules with a range of stabilities (Table 1.1). Substituents, especially electron-withdrawing groups, simultaneously lower the stabilities of the pentazole ring and the C—N bond in substituted BzN_5. Compared to ArN_5, computational studies show BzN_5 and its derivatives mostly possess lower activation energies for pentazole-ring decomposition (81.2–101.7 kJ mol^{-1}) and high C—N bond dissociation energy (385.6–452.8 kJ mol^{-1}), which is not suitable to replace ArN_5 as precursors of *cyclo*-N_5^- [13].

1.2.4 Other Substituted Pentazoles

Pentazoles have been substituted with other groups, such as $-CH_3$, −CN, and $-NH_2$, in either neutral (R–N_5) or cationic (R–N_5H^+) forms, and as *N*- and *N,N′*-oxide pentazole-based derivatives [17, 18] (Figure 1.2). Some of these compounds exhibit

Table 1.1 Difference between E_a of BzN_5 derivatives and PhN_5.

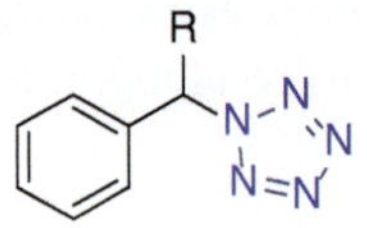

R=–H R=–NH$_2$ R=–NHCH$_3$ R=–N(CH$_3$)$_2$ R=–OH R=–NO$_2$ R=–CN R=–CF$_3$ R=–NF$_2$

	E_a (kJ mol^{-1})	$\Delta E'_a$ (kJ mol^{-1})	$\frac{\Delta E'_a}{E_a}$ (PhN_5) (%)
BzN_5	101.7	13.0	14.7
-NH_2	97.2	8.5	9.6
-NHMe	96.2	7.5	8.5
-NMe_2	100.0	11.3	12.7
-OH	92.3	3.6	4.1
-NO_2	81.2	−7.5	−8.5
-CN	90.0	1.3	1.5
-CF_3	93.5	4.8	5.4
-NF_2	88.6	−0.1	−0.1

very attractive properties based on computational chemistry. For example, neutral methylpentazole is predicted to be as stable as the aromatic pentazole derivative; the pentazolium–nitrate pair provides high amounts of released energy, with a heat of reaction of ~130 kcal mol^{-1}; oxidized or nitrated pentazoles could serve as energetic oxidative materials. There is the strong possibility that these pentazole derivatives will lead to HEDMs; however, these molecules are not available with current experimental techniques; they are also potentially hazardous because of their high formation enthalpies and weak N—N bonds.

1.2.5 Polypentazoles Clusters

The pentazole ring is comparatively stable for a polynitrogen compound since it benefits from conjugation arising from electronic delocalization in the ring [19], and the separation of the σ- and π-electron systems [20]; therefore, the pentazole

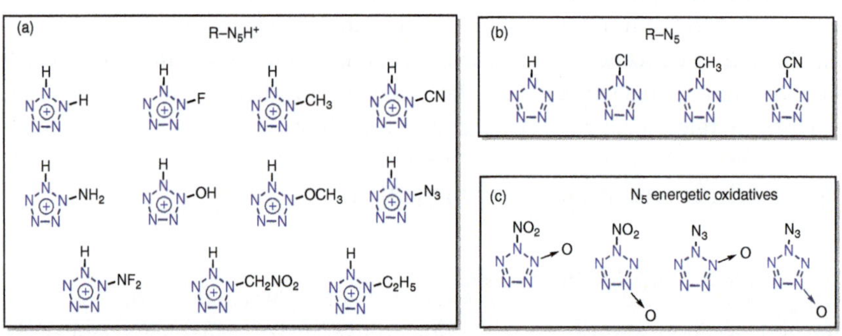

Figure 1.2 Other substituted pentazoles: (a) R−N_5H^+; (b) R−N_5 (R = H, Cl, CH_3, CN); (c) N_5 energetic oxidatives.

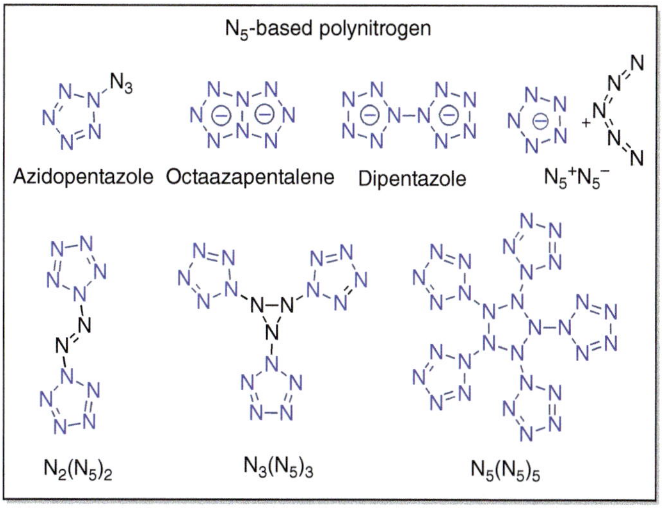

Figure 1.3 Polypentazoles clusters.

ring is a potentially useful building block for producing higher polynitrogen clusters (N_n). In the case of $5 < n \leq 10$, N_8 and N_{10} clusters involving pentazole ring represent minimum energy structures (Figure 1.3) [11, 21, 22]. Among them, $N_5^+N_5^-$, in particular, was previously highly hoped for, but subsequent calculations have shown that it is impossible to exist [5]. Born–Haber cycle-based stability estimates that $N_5^+N_5^-$ will decompose to vibrationally unstable species upon neutralization, by using adiabatic ionization potentials and electron affinities. Other pentazole-ring-containing polynitrogen (N_n) clusters ($n > 10$), which include N_{11}–N_{15}, and N_{18} multiple-ring structures ($N_3(N_5)_3$), N_{20} ($(N_5)_4$), N_{30} ($N_5(N_5)_5$), are all theoretically more stable as clusters than as open-chain structures. The N_{11}–N_{15} structures are composed of one or two pentazole rings connected by chains of three, four, or five nitrogen atoms. These clusters are 5–7 kcal mol^{-1} lower in energy than their corresponding open-chain structures, and their decomposition barriers are as low as 5–10 kcal mol^{-1} under most conditions. The structures containing multiple pentazole rings, namely N_{18}, N_{20}, and N_{30}, are 200–400 kcal mol^{-1} lower in energy than their corresponding cages, and their dissociation energies were determined to be 50–60 kcal mol^{-1}. The improved stabilities of the multiple pentazole-ring-containing structures compared to the single ring structures suggest that N_{60} should exist. N_{60} is analogous to C_{60} and is predicted to have a heat of formation of 6790 kJ mol^{-1} [2].

1.3 Strategies for the Preparation of *cyclo*-N_5^-

There are two routes for preparing pentazolate anion. One way is cleaving the C—N bond in ArN_5 to form *cyclo*-N_5^-, the other way is directly generating through *in situ* [3+2] cyclization between N_3^- and N_2. However, both approaches have great technical bottlenecks; the previous one needs to overcome the problem that the N—N bond is more unstable than the C—N bond and the pentazole ring is easy to destroy during the cleaving process and the second way requires great energy to activate N≡N triple bond so that it can react with N_3^-. So far, many attempts have been reported, and we can classify them into three categories.

1.3.1 Using High-Energy Electrons to Cleave the C—N Bond in ArN_5

Strategies for the preparation of *cyclo*-N_5^- with high-energy electrons have been designed as follows: (i) ArN_5 bearing highly electron-donating substituents in the para-position of the benzene ring should be prepared as starting materials. In this manner, the aromaticity and stability of the *cyclo*-N_5^- ring can be increased by transferring maximum amounts of electron density to the *cyclo*-N_5^- ring, while simultaneously weakening the connective C—N bond. (ii) The C—N bond should be selectively cleaved by high-energy electrons while maintaining intact N—N bonds in the *cyclo*-N_5^- ring. High-energy electrons can be obtained from tandem mass spectrometry (MS–MS) using collision-induced dissociation (CID) at high collision voltages, laser irradiation, and strong UV–vis/Raman spectra.

In 2002, *cyclo*-N_5^- was first detected in mass-spectrometry experiments (Figure 1.4) [7] in which a ArN_5 was dissolved in strongly polar solvents and injected into the mass spectrometer with a syringe pump. The desired *cyclo*-N_5^--ion peak was observed by tandem mass spectrometry (MS–MS), using CID at high collision voltages, and at low collision voltages fragmentation to the azide was seen. This method was further modified by Östmark, who observed the *cyclo*-N_5^- signal directly from a solid sample of ArN_5 by laser irradiation in a vacuum [23].

The two results elucidated a possible reaction pathway for a single electron transfer mechanism (SET) and the radical anion acted as an important intermediate. Hence, researchers attempted to photochemically produce *cyclo*-N_5^- in either solid or liquid phases, using different types of lasers. With UV–vis and Raman spectra providing *in situ* analytical technologies, the important decomposition intermediates and overall quantum yields, as well as the end product distributions, were determined [24–27]. These researchers found that, although the photochemistry of ArN_5 is characterized

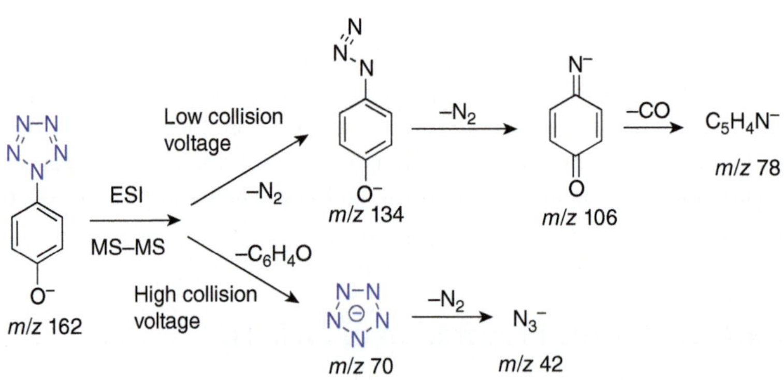

Figure 1.4 Experimental detection of the pentazolate anion. ESI–MS–MS fragmentation observed for the mass-selected ^{15}N singly labeled 4-pentazolylphenolate anion at low and high collision voltages.

by a range of reaction pathways when performed under a variety of conditions, the azide and azophenyl derivatives were the major products in all cases. It appears that the critical factor that drives the reaction is actually the formation of the radical through electron transfer, rather than the photoexcitation step itself.

1.3.2 Cleaving the C—N Bond of ArN_5 Using Redox Chemistry

The redox reaction is the most commonly used method for electron transfer and further leading to chemical cleavage in synthetic chemistry. Ceric ammonium nitrate (CAN) was the first used oxidant and SET reagent, while $Zn(NO_3)_2$ was added to stabilize and capture *cyclo*-N_5^-. ^{15}N-NMR spectroscopy revealed a novel signal at -10.0 ± 2.0 ppm. According to the calculated chemical shifts of $Zn(N_5)Cl$ and $ZnN_5(NO_3)$ (−16 and −17 ppm, respectively), and other possible ions present in the solution, $ZnN_5(NO_3)$ was concluded to have been generated. But repeated experiments attributed the signal at −10.0 ppm to the NO_3^- anion because it could also be observed in the absence of any other nitrogen-containing compound [28]. To confirm whether *cyclo*-N_5^- is generated, the ^{15}N-labeled ArN_5 is prepared and then tested. Although no signal for the N_5^- anion was detected, ^{15}N-labeled N_3^- ions were detected for all samples, confirming that they arose from the degradation of N_5^- in solution [29].

Later, N_5^- was accurately directly detected for the first time in 2016, by alkali metal treatment of ArN_5 at low temperature. Since they readily donate electrons, alkali metals are good reducing agents; consequently, they can convert phenylpentazole into the corresponding radical anion under mild conditions, after which thermal dissociation produces *cyclo*-N_5^-, as evidenced by an MS signal at $m/z = 70$. The assignment of this signal to N_5 is further supported by single and double labeling with ^{15}N, which led to signals at $m/z = 71$ and 72, respectively. A dissociation product with $m/z = 42$, which could be assigned to N_3^-, was observed through further MS analysis of the $m/z = 70$ signal (Figure 1.5) [30, 31].

Another successful example of oxidative cleavage was reported in 2017 [32, 33] in which the N_5^- anion was obtained through cleavage of ArN_5 in a ferrous bisglycinate $Fe(Gly)_2$/*m*-chloroperbenzoic acid (*m*-CPBA) system (Figure 1.6). $Fe(Gly)_2$ and *m*-CPBA form a peroxide intermediate, which functions both as the redox site and a source of electrons, a consequence of the co-existence of Fe and the peroxide group. In contrast to the previous heterogeneous Na/Na_2O system in acetonitrile, $Fe(Gly)_2$/*m*-CPBA is soluble in methanol, resulting in a homogeneous reaction as a consequence. This represents a significant improvement in conditions that realized the cleavage of the *cyclo*-N_5^- anion as well as the isolation of a pentazole as the NaN_5 as a solid product in 9% yield after careful post-treatment and purification. The *cyclo*-N_5^- produced in this manner exhibited acceptable stability under common conditions and decomposed at 116.8 °C, which is superior to that of N_5^+ (above 70 °C) but inferior to that of N_3^- (above 400 °C) [34]. Theoretical study suggested a new proton-transfer pathway for this reaction system involving the radical anion, but the pathway involving direct cleavage of the C—N bond in ArN_5 still dominates kinetically [35–37].

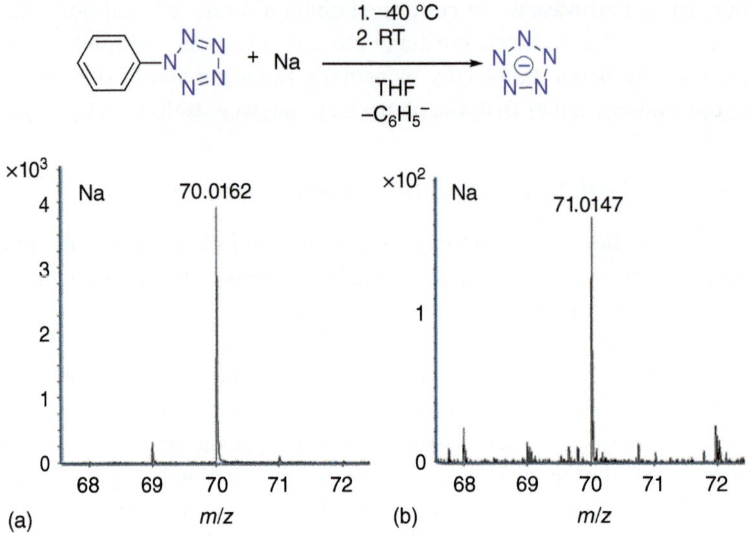

Figure 1.5 The HPLC/MS mass spectrum observed upon heating a solution of the PhN_5 radical anion in THF to room temperature for a short period of time (30–300 seconds); the spectrum was recorded 2.1 minutes after injection into the HPLC column: (a) using the all-^{14}N PhN_5, the spectrum is assigned to *cyclo*-$^{14}N_5^-$, (b) using $^{14}N_4{}^{15}N_1$-phenylpentazole, the spectrum is assigned to *cyclo*-$^{14}N_4{}^{15}N^-$. Source: Butler et al. [9]/with permission of John Wiley & Sons.

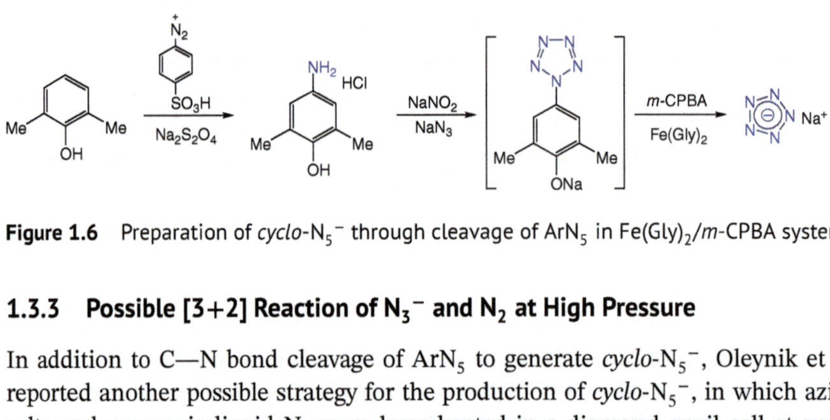

Figure 1.6 Preparation of *cyclo*-N_5^- through cleavage of ArN_5 in $Fe(Gly)_2$/*m*-CPBA system.

1.3.3 Possible [3+2] Reaction of N_3^- and N_2 at High Pressure

In addition to C—N bond cleavage of ArN_5 to generate *cyclo*-N_5^-, Oleynik et al. reported another possible strategy for the production of *cyclo*-N_5^-, in which azide salts and cryogenic liquid N_2 were laser heated in a diamond anvil cell at super high pressure at 60 GPa. The laser provided sufficiently high energy to overcome the [3+2]-cyclization barrier, while the high pressure prevented the decomposition of *cyclo*-N_5^- [38–42]. From then on, many research groups followed up the study of high-pressure preparation pentazole and its congeners [43–46]. The extreme conditions limited direct experimental characterization; consequently, DFT-calculated XRD patterns and Raman spectra provided evidence for the formation of *cyclo*-N_5^-. Unfortunately, *cyclo*-N_5^- salts were not observed when the pressure in the diamond anvil cell was released, which is inconsistent with the results from the organic-synthesis strategy under ambient conditions.

1.4 Complexes of Metal and *cyclo*-N_5^-

The theoretical study of metal and *cyclo*-N_5^- complexes has a long history. Their results indicate that metal ions can stabilize *cyclo*-N_5^- in different forms. The physicochemical properties of the assembled metal complexes were computationally studied. These studies show that *cyclo*-N_5^- metal complexes have high kinetic stability and high energy, which means that the complexes have great potential as HEDMs.

This part will introduce the synthesis of metal-pentazole compounds at low temperature and atmospheric pressure, and the structure and related physicochemical properties of series metal-pentazole compounds.

Since metals, especially transition metals, usually coordinate effectively to nitrogen, metal-pentazolate compounds were further experimentally investigated in order to access the newly synthesized *cyclo*-N_5^- anion. To date, 14 types of metal–N_5 complexes have been prepared (Figure 1.7). Here, we divide these complexes into salts (0D) and metal–inorganic frameworks (MIFs, 1D and 3D) and coordination polymers (CPs) according to their dimensions.

1.4.1 Metal Salts of *cyclo*-N_5^- (0D)

When appropriate metal salts are dissolved into ethanol solutions containing the sodium salt of *cyclo*-N_5^-(**1-1**), $[Mn(H_2O)_4(N_5)_2]\cdot 4H_2O$ (**1-2**), $[Fe(H_2O)_4(N_5)_2]\cdot 4H_2O$ (**1-3**), $[Co(H_2O)_4(N_5)_2]\cdot 4H_2O$ (**1-4**), $[Zn(H_2O)_4(N_5)_2]\cdot 4H_2O$ (**1-5**), or $[Mg(H_2O)_6(N_5)_2]\cdot 4H_2O$ (**1-6**) formed through metathesis reactions between NaN_5 and $MnCl_2$, $FeCl_2$, $Co(NO_3)_2$, $Zn(NO_3)_2$, or $MgCl_2$, respectively [32, 47]; the single-crystal structures of these compounds are shown in Figure 1.8.

Complexes **1-2** to **1-5** crystallize in the orthorhombic *Fmmm* space group, with almost identical unit-cell arrangements (a: 12.144–12.342 Å; b: 17.114–17.444 Å; c: 6.465 6.570 Å; α: 90°; β: 90°; γ: 90°; z: 4). The pentagonal *cyclo*-N_5^- rings bind to the metal atoms (Mn, Fe, Co, or Zn) through η^1-coordination and appeared to be perfectly coplanar. The N—N bonds in the *cyclo*-N_5^- rings of the various complexes are not of equal length; the average N—N bond distances in **1-2** to **1-5** are 1.320, 1.329, 1.314, and 1.328 Å, respectively. Each metal ion (Mn, Fe, Co, or Zn in sp^3d^2 hybridization) contributes six empty orbitals to accommodate the lone pairs of electrons from the *cyclo*-N_5^- ligands, which are coordinated in a quadrilateral bipyramidal environment in which the axial positions are formed by two nitrogen atoms from two *cyclo*-N_5^- rings, and the equatorial planes are formed by four oxygen atoms from four coordinated water molecules. The ball-and-stick packing diagrams of **1-2** to **1-5** reveal that the metal–N_5 fragments are all stacked face to face, with interlayer distances of 3.269, 3.285, 3.233, and 3.268 Å (equal to $c/2$), respectively.

The structure of complex **1-6** contains a magnesium atom that is hydrated by six H_2O molecules and cannot coordinate to *cyclo*-N_5^-; therefore, each N atom in each *cyclo*-N_5^- is hydrogen bonded to H_2O, with the surrounding H_2O molecules providing a hydrophilic and polar environment that helps to stabilize and protect the *cyclo*-N_5^- rings in the crystal.

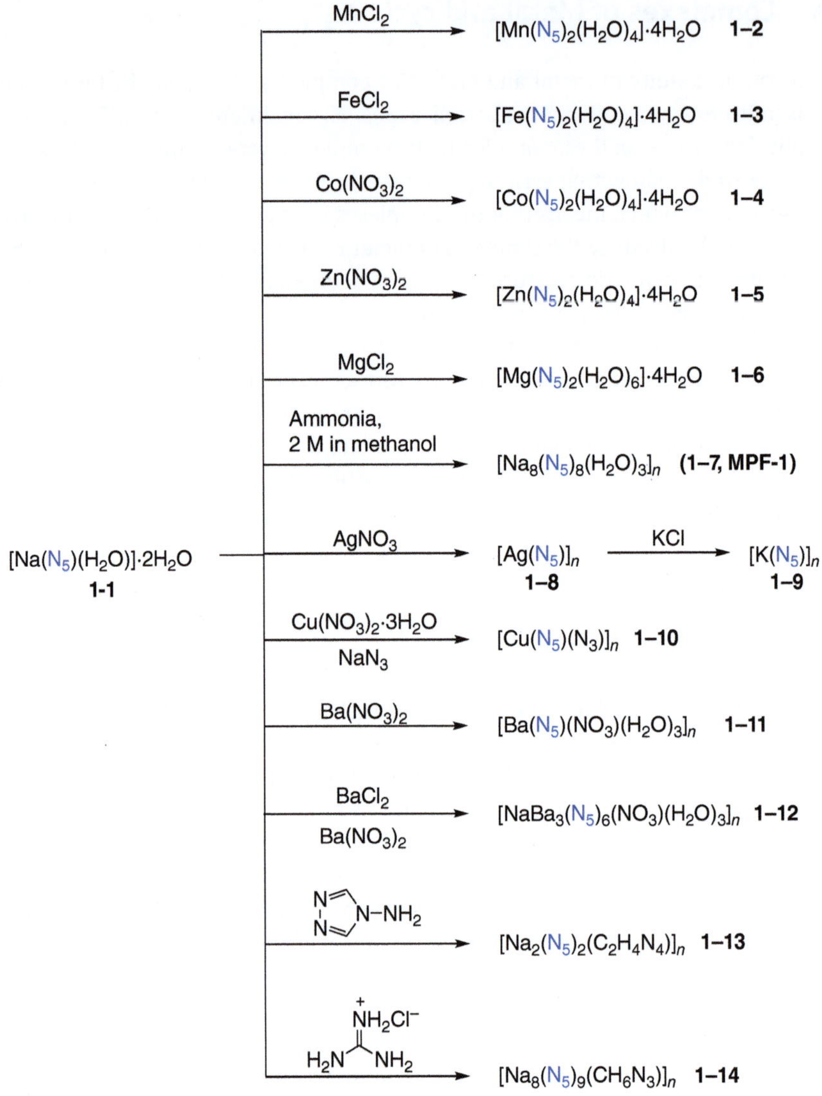

Figure 1.7 Synthesis route of *cyclo*-N_5^--metal complexes.

1.4.2 Metal–Inorganic Frameworks (MIFs) of *cyclo*-N_5^-

Adapting the procedure for C—N bond cleavage provided an air-stable, off-white NaN_5 hydrate, which was recrystallized to produce $[Na(H_2O)(N_5)]{\cdot}2H_2O$ (**1-1**) [32]. Compound **1-1** exhibits a 1D MIF structure with orthorhombic C_{mcm} symmetry, with the pentagonal *cyclo*-N_5^- units adopting perfectly planar arrangements, as well as perfectly coplanar arrangements with adjacent sodium atoms. The *cyclo*-N_5^- anions and sodium ions in **1-1** are sequentially connected to form a saw-tooth-like 1D structure (Figure 1.9a–c).

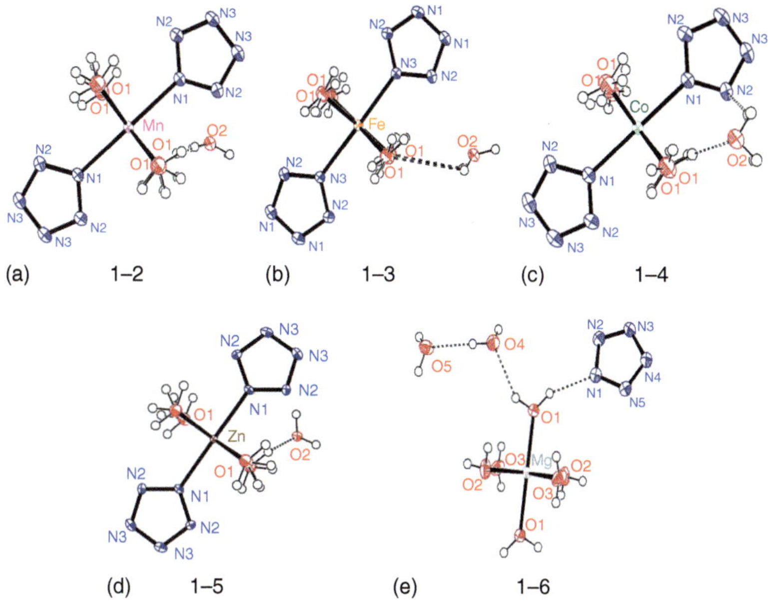

Figure 1.8 Single-crystal X-ray analyses of **1-5**, ellipsoids represent the 50% probability level. Source: Zhang et al. [12, 13]/with permission of Springer Nature. (a–c, e) (d) Xu et al. [47]/with permission of Royal Society of Chemistry.

A zeolite-like MIF with a fascinating 3D structure, namely $[Na_8(N_5)_8(H_2O)_3]_n$ (**1-7**), was successfully constructed from the 1D sodium-N_5^- framework by changing the pH conditions to weakly alkaline during recrystallization (Figure 1.9d–i) [48]. Twenty-four sodium ions, twelve *cyclo*-N_5^- rings, and twelve oxygen atoms form a drum-like nanocage, with one sodium ion in the center of the cage. Only one type of nanocage ($Na_{24}N_{60}O_{12}$) was observed in the single-crystal structure of **1-7**, while Zhang et al. synthesized a similar 3D framework (MPF-1) that exhibited two types of nanocages, namely $Na_{20}N_{60}$ and $Na_{24}N_{60}$ [49]. As shown in Figure 1.9j–l, the small $Na_{20}N_{60}$ cage is composed of 12 pentagons and 28 hexagons; the diameter of the $Na_{20}N_{60}$ nanocage is about 1.2 nm, while the larger $Na_{24}N_{60}$ nanocage contains 12 pentagons and 24 hexagons, and has a diameter of about 1.4 nm. The packing of the two types of nanocages results in a 3D framework with a zeolitic MEP topology (Figure 1.9m,n). When viewed in the *a*, *b*, and *c* directions, MPF-1 exhibits circular channels with 8.4-Å-sized pores. The total accessible volumes of **1-7** and **MPF-1** are 33.5% and 33.3% as calculated. As porous materials, **1-7** and **MPF-1** have calculated N_2-adsorption capacities of 2.59 and 2.5 mmol g^{-1} (1 atm and 273 K).

Aside from Na-based MIFs, *cyclo*-N_5^- and/or N_3^- and NO_3^- have been used as ligands, and self-assembly synthesis has been used to obtain additional 3D MIFs, such as $[Ag(N_5)]_n$ (**1-8**), KN_5 (**1-9**), $[Cu(N_5)(N_3)]_n$ (**1-10**), $[Ba(N_5)(NO_3)(H_2O)_3]_n$ (**1-11**), and $[NaBa_3(N_5)_6(NO_3)(H_2O)_3]_n$ (**1-12**) (Figure 1.10) [50, 51]. Each *cyclo*-N_5^- in **1-11** and **1-12** adopts a tridentate chelating mode to bridge three Ba cations, while *cyclo*-N_5^- adopts a quadridentate chelating mode to bridge four metal cations in **1-8**

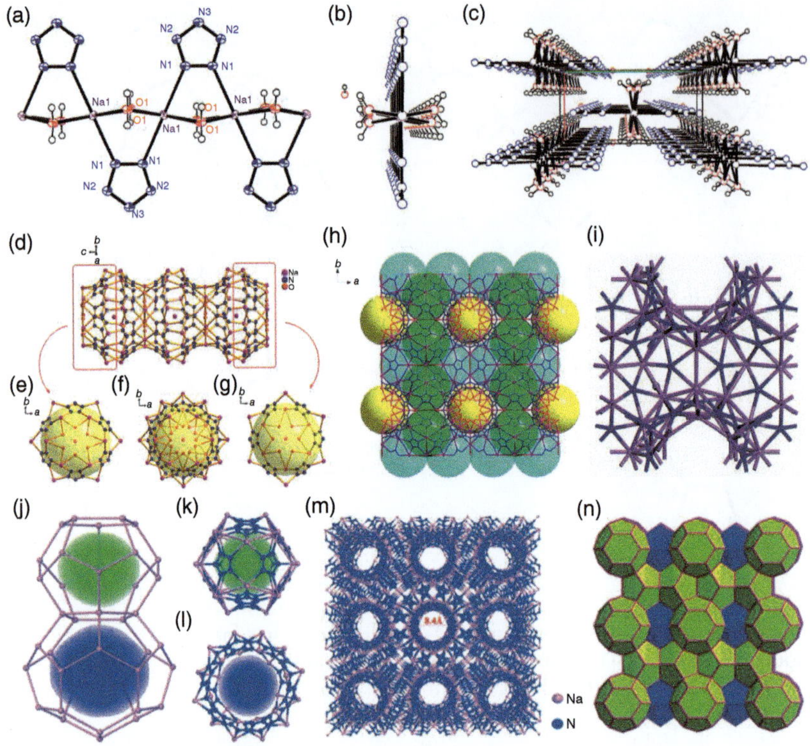

Figure 1.9 The structures of 1D MIF **1-1** (a–c), 3D porous MIFs **1-7** (d–i) and **MPF-1** (j–n). (a, b) Extended molecular view of **1-1** normal to (100) and (001), respectively; (c) Ball-and-stick packing diagrams of **1-1** viewed normal to (001) with hydrogen bonds; (d) Selected three $Na_{24}N_{60}O_{12}$ nanocages in **1-7** viewed along the *a* axis; (e) Half of a cage viewed along the *c** axis; (f) Selected three nanocages viewed along the *c* axis; (g) Half of a cage viewed along the *c* axis; (h) The 3D framework of **1-7**; (i) A schematic representation of the topology for **1-7**, in which the N_5 ligands are represented by blue spheres and the Na ions are represented by purple spheres; (j) The connected mode between $Na_{20}N_{60}$ and $Na_{24}N_{60}$ nanocages in **MPF-1**; (k) The structure of $Na_{20}N_{60}$ nanocage; (l) The structure of $Na_{24}N_{60}$ nanocage; (m) The 3D open framework of **MPF-1**; (n) the zeolitic MEP topology of **MPF-1**. Source: (a–c) Zhang et al. [12]/with permission of Springer Nature; (d–i) Zhang et al. [14]/with permission of Royal Society of Chemistry; (j–n) Zhang et al. [16]/with permission of John Wiley & Sons.

and **1-10**. The most interesting one is **1-9**, in which five N atoms in one *cyclo*-N_5^- coordinate with seven K ions.

1.4.3 Coordination Polymers

The guanidinium cation, $CN_3H_6^+$ (Gu^+), and 3-amino-1,2,4-triazole $C_2H_4N_4$ (ATz) were used instead of Na^+ to prepare nonmetallic salts, but they eventually formed CPs, namely $(NaN_5)_5[(CH_6N_3)N_5](N_5)_3^-$ (**1-13**) and $(NaN_5)_2(C_2H_4N_4)$ (**1-14**) (Figure 1.11). The crystal structures reveal the coexistence of the chelating *cyclo*-N_5^- and the organic component in the solid state. Na^+ and N_5^- are

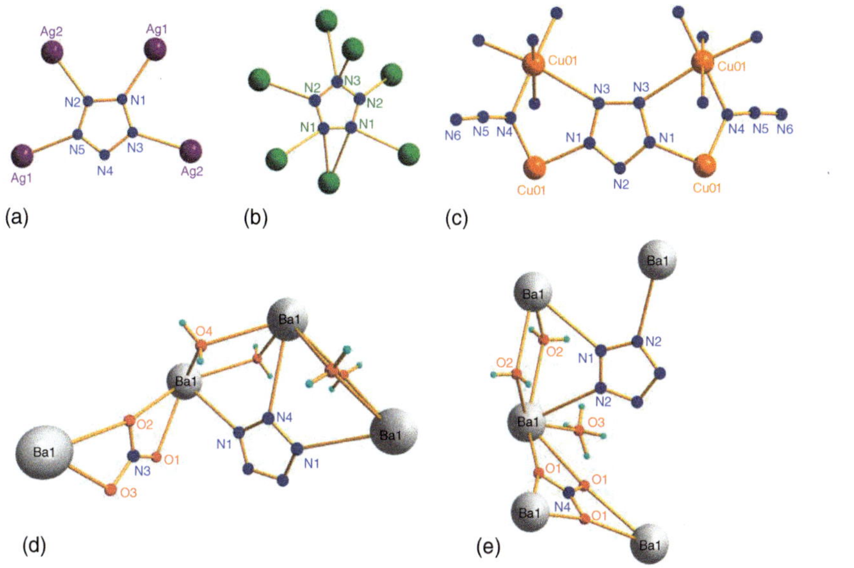

Figure 1.10 Coordination bond of *cyclo*-N_5^- captured from single-crystal X-ray analyses of **1-8** (a), **1-9** (b), **1-10** (c), **1-11** (d), and **1-12** (e). Source: Rahm and Brinck [17]/with permission of John Wiley & Sons.

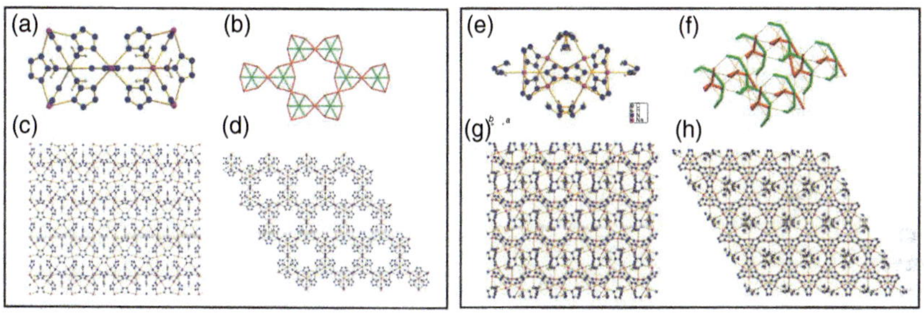

Figure 1.11 Single-crystal X-ray analyses of **1-13** (a–d) and **1-14** (e–h). (a) The coordination mode of Na^+, (b) A topological net of simplified unit cell, (c, d) Unit cell stacking diagram view along *b* and *c* axes (4*4*4) of **1-13**. (e) The coordination mode of Na^+, (f) A topological net of simplified unit cell, (g, h) Unit cell stacking diagram view along *b* and *c* axes (4*4*4) of **1-14**. Source: Zhang et al. [49]/with permission of Springer Nature.

coordinated to form a cage structure that traps the guanidinium cation $[C(NH_2)_3]^+$ in the $(NaN_5)_5[(CH_6N_3)N_5](N_5)_3^-$ CP. A mixed-ligand system is observed in the $(NaN_5)_2(C_2H_4N_4)$ CP; N_5^- and ATz are separately coordinated to Na^+ to form two independent but interwoven nets. Moreover, no crystallized H_2O is formed in these products, which are anhydrous compounds of *cyclo*-N_5^- salts with high heats of formation (>800 kJ mol^{-1}). The nitrogen contents and heats of formation of the two CPs are much higher than those of traditional energetic materials (TNT, RDX, and HMX). However, their detonation properties are not particularly good when compared to those of TNT, RDX, HMX, and CL-20. The poor detonation performance of the two CPs is attributed to their low densities, a consequence of their porous structures. From another perspective, these porous structures could encapsulate other smaller energetic materials in order to improve energetic performance [52].

1.4.4 Structural Characterization

1.4.4.1 MS and NMR Characterization

The *cyclo*-N_5^- can give a signal of $m/z = 70$ in mass spectrometry and no fragments are observed, indicating that the structure has good stability. A concentrated reaction solution containing ^{15}N labeled *cyclo*-N_5^- rings marked using $Na^{15}NO_2$ was dissolved in deuterated methanol (CD_3OD-d_4) for liquid NMR analysis (Figure 1.12). The results showed that the N-NMR peak of *cyclo*-N_5 appeared at the chemical shift of −3.0 ppm. The chemical shift of the peak coincides with that of the theoretically calculated results in the literature (−6.5 to 0 ppm) [53–55]. Residual nitrate was detected at −12.4 ppm, which also agreed with previous results at -10 ± 2 ppm [56, 57].

1.4.4.2 Topology Analysis

The topology of *cyclo*-N_5^- in eight complexes (**1-7** to **1-14**) that own three-dimensional structures was analyzed. Their topology types and legends are shown in Table 1.2 and Figure 1.13. From the result, it can be found that *cyclo*-N_5^- constructs new topological types except complexes **1-8**. For complexes **1-7**, **1-13**, and **1-14**, they have the same metal and all have *cyclo*-N_5^- as the main ligand, but their topological types vary greatly.

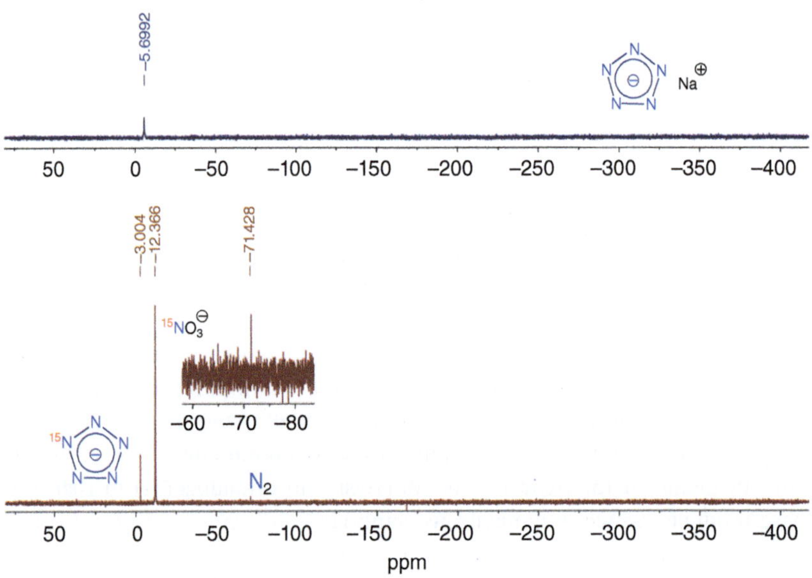

Figure 1.12 NMR spectra of the reaction solution of complex **1-1** and ^{15}N labeled NaN_5.

Table 1.2 Topology types of complexes **1-7** to **1-14**.

Complex	Topological type
1-7	$\{3.4^4.5^2.6^3\}_{12}\{3^2.4^4.5^2.6^2\}_{12}\{3^4.4^8.5^6.6^3\}_{12}\{3^8.4^8.5^8.6^4\}_3\{4^{12}.6^3\}_8$
1-8	PtS, $\{4^2.8^4\}$
1-9	$\{4^{15}.6^6\}\{4^{17}.6^4\}$
1-10	$\{3^2.6^2.7^2\}\{3^4.4^2.6^4.7^5\}$
1-11	$\{3^2.4\}\{3^4.4^3.5^2.6^{10}.7^2\}$
1-12	$\{3^2.4\}_3\{3^4.4^3.5^2.6.8^4.9\}_3\{8^3\}$
1-13	$\{3.4^4.5^2.6^3\}\{3^2.4^8.5^3.6^2\}$
1-14	$\{4^{12}.6^3\}_2\{4^2.6\}_6\{4^3.6^2.8\}_3\{4^6.8^4.10^5\}_3$

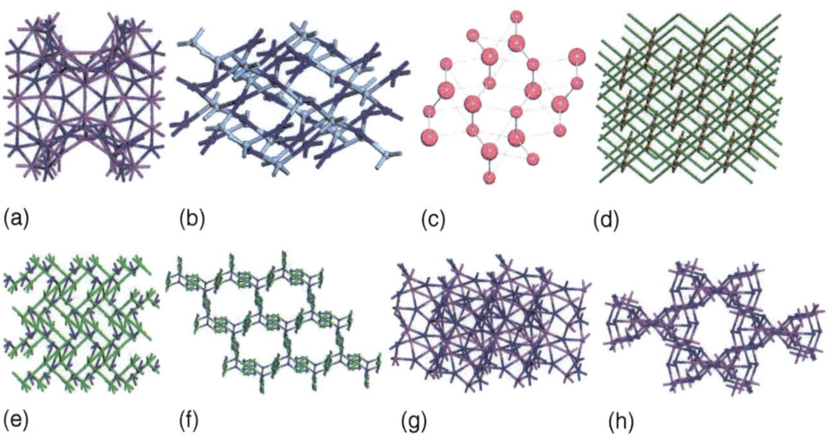

Figure 1.13 Topology types of complexes **1-7** to **1-14**.

1.4.4.3 Raman and IR Analysis

The *cyclo*-N_5^- was predicted to have a D_{5h} symmetry (A1′ + E1′ + 2E2′ + E2″), but indeed showed C_{2v} symmetry (A1 + (B2) + 2(A1 + B2) + (A2 + B1)). It is because that the geometry of *cyclo*-N_5^- in most complexes was affected by coordination mode and the N—N bond length changed. Among them, (A2 + B1) mode derived from E2″ has neither infrared nor Raman activity; (A1 + B2) mode derived from E1′ is infrared activity; the A1 and 2(A1 + B2) mode derived from A1′ and 2E2′ are Raman active (Figure 1.14). So, in theory, the *cyclo*-N_5^- ring with C_{2v} symmetry should have two infrared absorption peaks and five Raman peaks.

The infrared spectra of the 14 kinds of metal complexes measured experimentally are shown in Figure 1.15. The non-three-dimensional complexes (**1-1** to **1-6**) have two absorption peaks: 1244–1258 cm^{-1} (B2) and 1219–1238 cm^{-1} (A1). But the two absorption peaks of *cyclo*-N_5^- in three-dimensional complex (**1-7** to **1-14**) are not clearly split, presented as only one absorption peak, whose wave number was at 1217–1248 cm^{-1}.

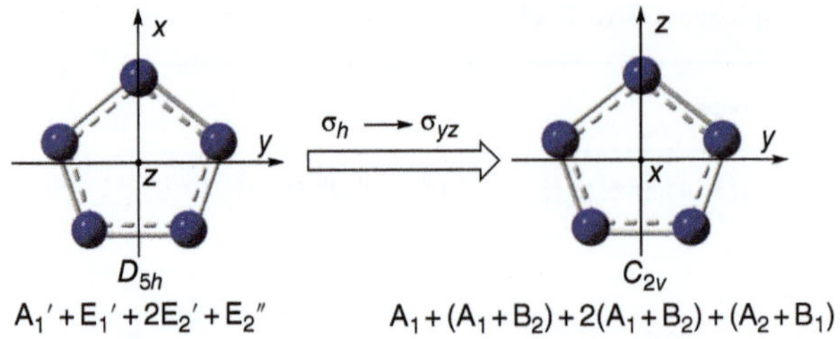

Figure 1.14 Transformation of D_{5h} to C_{2v} symmetry.

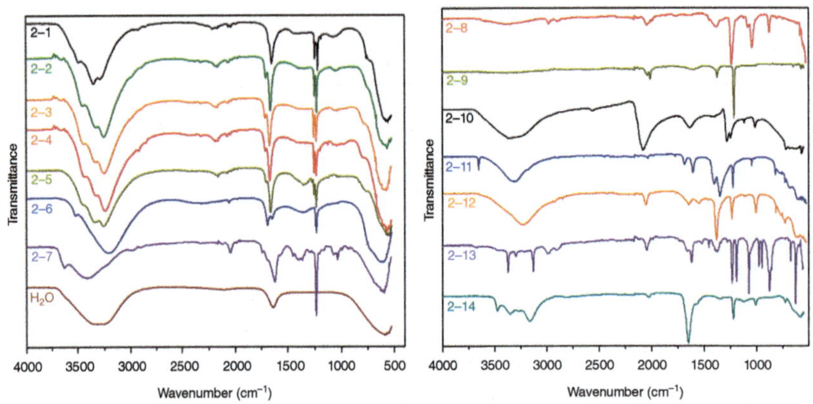

Figure 1.15 Infrared spectra of complexes **1-1** to **1-14**.

The Raman spectra of *cyclo*-N_5^- in these three-dimensional complexes (**1-7** to **1-14**) also could not clearly be defined. Since the three-dimensional complexes are microscopic porous structures, there will be solvents, N_3^-, NO_3^- in the cell gaps and pores, and thus the Raman spectrum is complex. The spectra of non-three-dimensional complexes (**1-1** to **1-6**) can be defined and shown in Figure 1.16 for Raman vibrations analysis. Theoretically, five Raman peaks of *cyclo*-N_5^- were corresponding to A1 and 2(A1 + B2) mode. The experimental results show that all six complexes have a strong peak in the range of 1188–1199 cm^{-1}, which was attributed to the A1 mode, 2(A1 + B2) modes should show four peaks, but three of them with low intensity are observed in Raman spectra, at a range of 1234–1239, 1120–1124, and 1005–1037 cm^{-1}.

1.4.4.4 Stability Analysis

Thermal stability is one of the most important properties of energetic materials. The most common methods to measure thermal stability are differential scanning calorimetry (DSC) and thermogravimetric (TG) analysis. Most of the *cyclo*-N_5^- complexes contain water in their crystals, and these water-containing samples have obvious dehydration and endothermic peaks before decomposition. The complexes **1-1** to **1-5** and **1-12** have only one obvious endothermic process before 150 °C, while for compounds **1-6** and **1-11**, three and two endothermic peaks were observed before

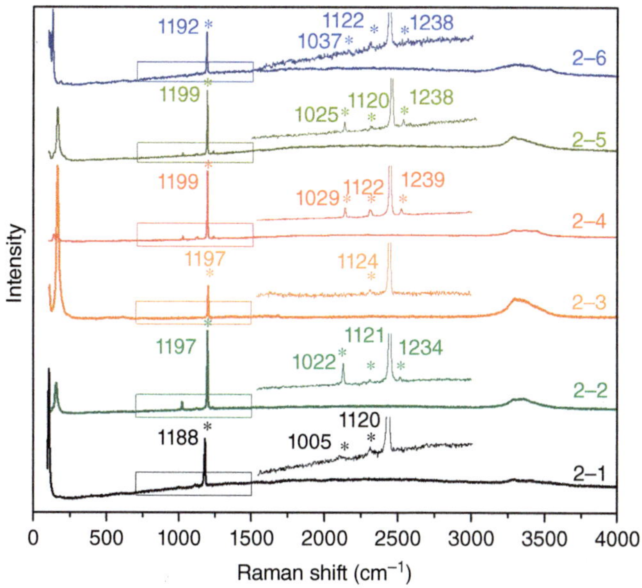

Figure 1.16 Raman spectra of non-three-dimensional complexes of *cyclo*-N_5^-.

the first exothermic peak (Figure 1.17). Except for **1-4** and **1-9**, all samples showed at least two exothermic stages. The first exothermic peak is similar and at a range of 90–130 °C, but the position of the second exothermic peak varies greatly. Among them, the first exothermic peak of the **1-11** and **1-12** complexes is continued to the previous endothermic peak, which means that it is not a simple decomposition process. For all the 14 complexes, the best thermal stability is complex **1-7** with a decomposition temperature up to 129.0 °C and the following two complexes are **1-13** and **1-14**. All these three compounds show three-dimensional structures and have sodium ion as the metal center and *cyclo*-N_5^- as the main ligand. Their thermal behaviors are also very similar; the first decomposition process of **1-7** has the highest heat release (373 J g^{-1}), followed by **2-13**(304 J g^{-1}) and **2-14**(280 J g^{-1}).

Based on a summary of the above studies on the thermal decomposition of *cyclo*-N_5^- complexes, it can be concluded that the thermal decomposition of the *cyclo*-N_5^- ring can be divided into two stages. The first stage is the breaking of two N—N bonds in the *cyclo*-N_5^- ring to decompose into stable N_2 and relatively stable azide ions; the second stage is the further decomposition of azide ions. Therefore, two obvious exothermic peaks can be observed in the DSC diagram of most complexes. In addition, there are two exceptions – complexes **1-4** and **1-9**. The first peak of complex **1-4** includes the exothermic process of water loss and the endothermic process of *cyclo*-N_5^- ring breaking. As thermal effects of water loss are greater than those of *cyclo*-N_5^- ring breaking, the comprehensive performance shows an endothermic peak. For complex **1-9**, after the first step decomposition of *cyclo*-N_5^- around 110–150 °C, the generated KN_3 shows good stability and undergoes only a melting process, which corresponds to an endothermic peak at

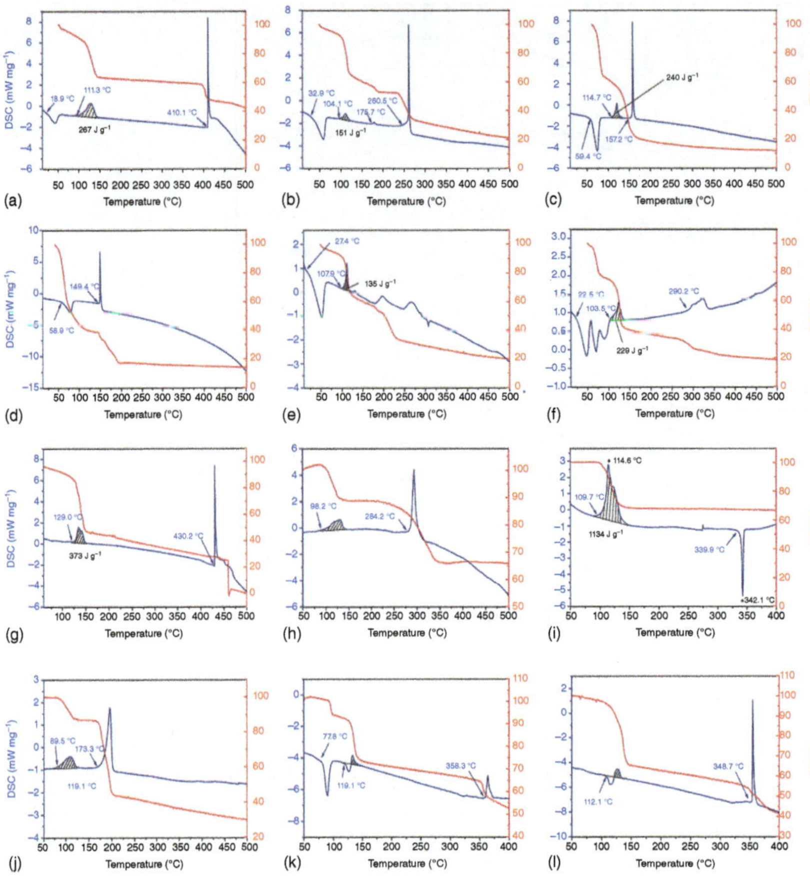

Figure 1.17 DSC and TG of complexes **2-1** to **2-12** (a–l).

342 °C without weight loss in TG. Meanwhile, no other exothermic peak is found even up to 400 °C.

The first decomposition process of these complexes, corresponding to *cyclo*-N_5^- decomposition, is undoubtedly the key step that has been the focus of attention. Further thermal kinetics tests were performed on **1-1**, **1-2**, **1-3**, and **1-6** (Figure 1.18). Based on DSC curves at different heating rates, the thermal decomposition activation energies *cyclo*-N_5^- in **1-1**, **1-2**, **1-3**, and **1-6** were obtained by the Kissinger method as 98.4, 106.6, 108.4, and 118.2 kJ mol^{-1} (Figure 1.19). The average thermal decomposition activation energy of *cyclo*-N_5^- complexes is about 108 kJ mol^{-1}, lower than that of nitroglycerin (209 kJ mol^{-1}), PETN (197 kJ mol^{-1}), picric acid (240 kJ mol^{-1}), etc.

1.4.4.5 Bonding Study

To investigate the relationship between the stability and structure of *cyclo*-N_5^- complex, the bonding condition in its crystal structure was analyzed. The formed coordination bonds and hydrogen bonds of each N atom in *cyclo*-N_5^- ring are shown in Figures 1.20 and 1.21. According to the bonding situation, all the complexes can be

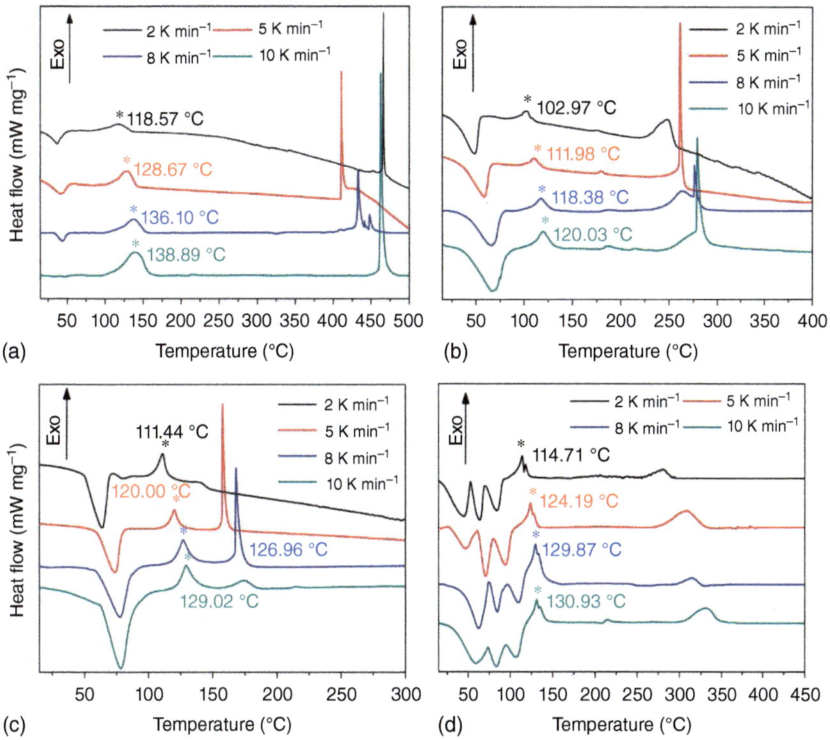

Figure 1.18 DSC curves of complexes **1-1**, **1-2**, **1-3**, and **1-6** (a–d) at different heating rates.

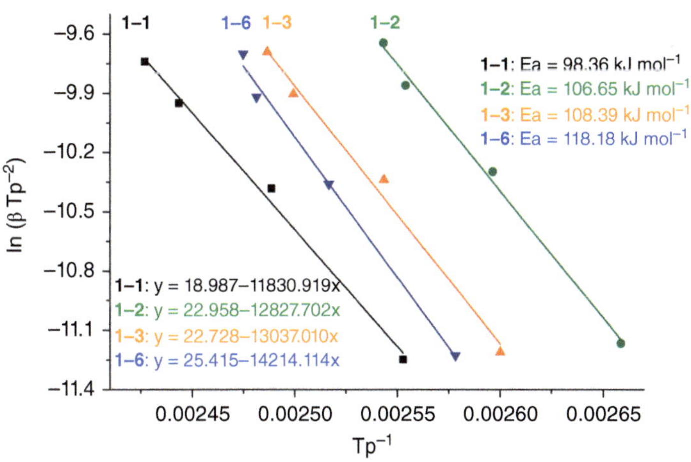

Figure 1.19 Apparent activation energy of first exothermic peak of complexes **1-1**, **1-2**, **1-3**, and **1-6** calculated with Kissinger method.

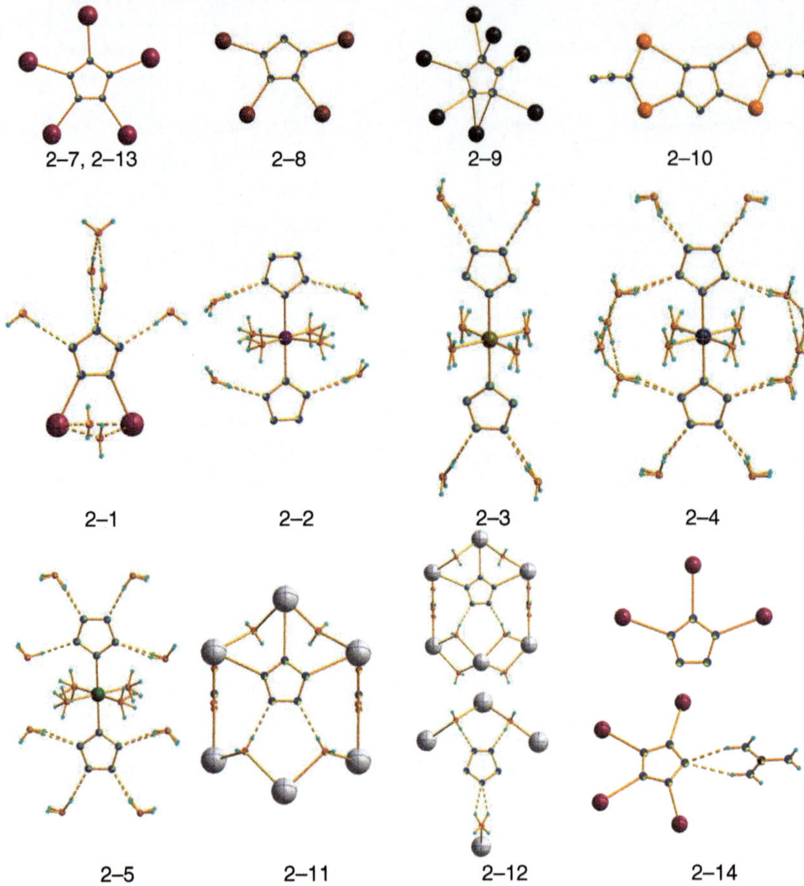

Figure 1.20 Bonding form of *cyclo*-N_5^- in complexes **1-1** to **1-5**, **1-7** to **1-14**.

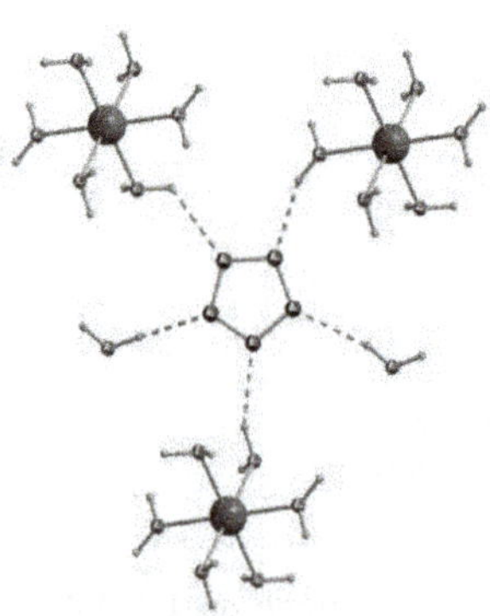

Figure 1.21 Bonding form of *cyclo*-N_5^- in complexes **1-6**.

divided into three classes: (i) coordination bond only, (ii) mixed coordination and hydrogen bond, and (iii) hydrogen bond only.

(1) **Coordination bond.** For complexes **1-7** to **1-10** and **1-13**, there is only a coordination bond found around *cyclo*-N_5^- in their crystals. In complexes **1-7** and **1-13**, the five N atoms in *cyclo*-N_5^- form η^1–σ coordination bonds with five metal ions. In complexes **1-8** and **1-10**, the four N atoms in *cyclo*-N_5^- participate in bonding in η^1–σ form with four metal ions. In complex **1-9**, the five N atoms in *cyclo*-N_5^- form six η^1–σ and two η^2–σ coordination bonds with seven metal ions.
(2) **Mixed coordination and hydrogen bonds.** For hydrous complexes **1-1** to **1-5**, **1-11**, and **1-12**, except for η^1–σ coordination bonds, there are also weak hydrogen bonds interaction between coordination or crystalline water and N atom in *cyclo*-N_5^-. Each *cyclo*-N_5^- prefers to form 1–3 coordination bonds with metal ions and then fills up with 2–4 O—H…N hydrogen bonds with water molecules. Whenever the two kinds of bond combine, *cyclo*-N_5^- ring always maintains the N_5 ring centered axisymmetric structure. Complex **1-12** has two types of *cyclo*-N_5^-, one type has the same bonding condition as that in **1-11**, and the other only has O—H…N hydrogen bonding. Complex **1-14** also contains two types of *cyclo*-N_5^-, one has three N atoms involved in coordination, the other has four N atoms involved in coordination, and one N atom forms N—H…N hydrogen bonds with guanidine cations.
(3) **Hydrogen bonding only.** In complex **1-6**, each N atom in *cyclo*-N_5^- forms a O—H…N hydrogen bonds with water molecules and is *cyclo*-N_5^- immobilized in a hydrogen bond network, which is similar to the second type in **1-12**. For many aqueous complexes, the loss of water will cause the decomposition of *cyclo*-N_5^-, indicating that water plays a very important role in the stability of *cyclo*-N_5^-.

To analyze the stabilizing effect of coordination water on the complex, the molecular orbitals of **1-5** with and without coordination water are calculated by quantum chemistry. It can be found that coordination water was involved in the coordination process by forming new orbitals with metals and pentazoles (Figure 1.22). Complexes **1-2**, **1-3**, and **1-4** were studied by the same method, and the same conclusion can be obtained.

1.4.4.6 Detonation Properties

Solid-state heat of formations (HOFs) is the most important parameter to calculate the detonation properties of energetic salts. For M_pX_q-type energetic complexes, the solid-state HOFs can be calculated according to the Born–Haber energy cycle [58]. For coordination polymers, the solid-state HOFs calculated based on experimental combustion enthalpy are more suitable. According to the combustion enthalpy and experimental HOFs of combustion products, the standard HOFs of coordination polymers can be calculated [59–61].

The detonation parameters such as detonation velocity can be calculated by EXPLO5 software [62] or K–J equation according to the calculated HOFs and the density under 298.15 K. Take **1-8** and **1-9** as examples: the calculated explosive

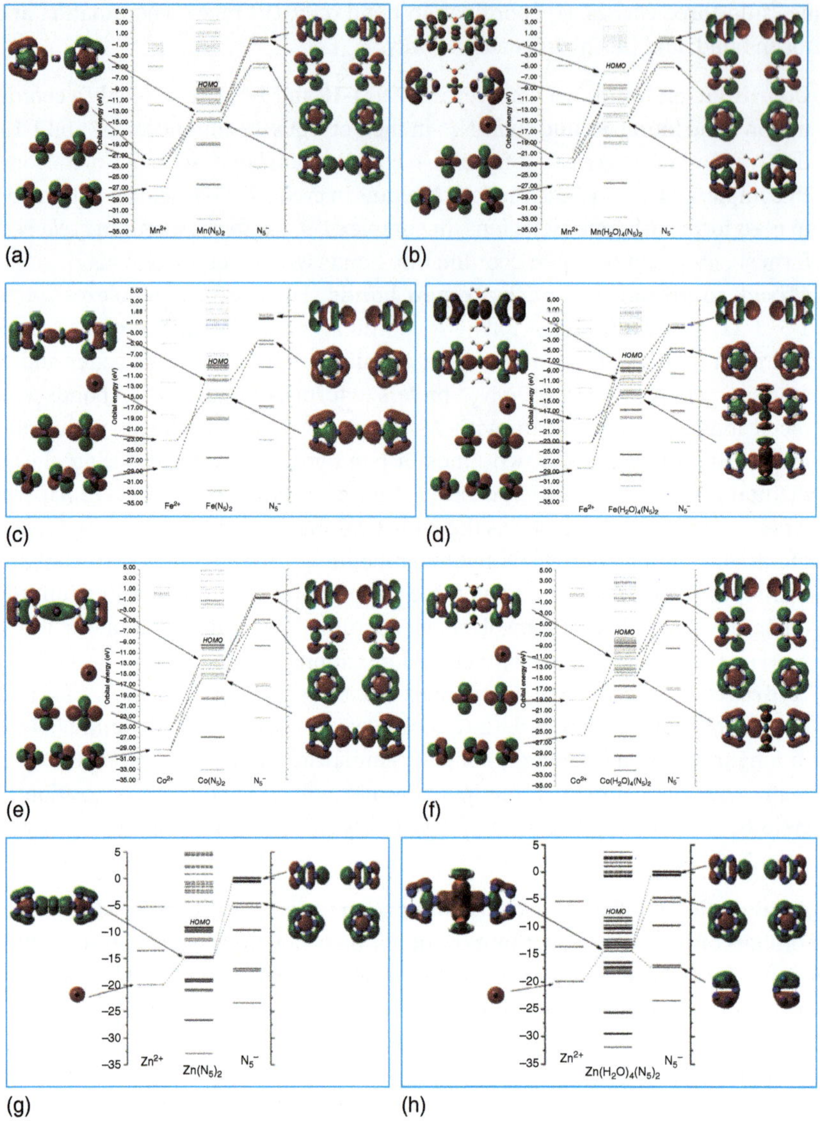

Figure 1.22 The molecular orbitals of **2-2** to **2-5** with and without coordination water.

heat (Q) of **1-8** is 1.110 kcal g^{-1}, the detonation velocity (D) is 7782 m s^{-1}, and the detonation pressure (P) is 34.67 GPa; the Q of **1-9** is 0.75 kcal g^{-1}, D is 6636 m s^{-1}, and P is 19.81 GPa. The detonation performance of **1-8** calculated by EXPLO5 is higher than that calculated by the K–J equation (D = 6977 m s^{-1}, P = 20.9 GPa). Their performances are superior to the reported two-dimensional frame material CHHP and ZnHHP [63], Ag complexes $[(atrz)_{1.5}(NO_3)]_n$ [64], and $[Ag(atz)]_n$ [59] as well as the detonation properties of explosive TNT [65] (Table 1.3).

Table 1.3 Physical and chemical properties of prepared pentazole derivatives.

Compound	Molecule formation	d[a] ($g\ cm^{-3}$)	T_d[b] (°C)	N[c] (%)	OB[d] (%)	ΔH_f[e] ($kJ\ mol^{-1}$)/ ($kJ\ g^{-1}$)	Q[f] ($kcal\ g^{-1}$)	D[g] (m s-1)	P[h] (GPa)	IS[i] (J)	FS[j] (N)	Reference
1-1	$[Na(H_2O)(N_5)]\cdot 2H_2O$	1.44/ 1.471(170 K)	111	47.62	−5.4							[32]
1-2	$[Mn(H_2O)_4(N_5)_2]\cdot 4H_2O$	1.59/ 1.608(205 K)	104	41.30	−4.7							[32]
1-3	$[Fe(H_2O)_4(N_5)_2]\cdot 4H_2O$	1.58/ 1.599(205 K)	115	41.19	−4.7							[32]
1-4	$[Co(H_2O)_4(N_5)_2]\cdot 4H_2O$	1.66/ 1.696(170 K)	59	40.82	−4.7							[32]
1-5	$[Zn(H_2O)_4(N_5)_2]\cdot 4H_2O$	1.65/ 1.669(205 K)	108	40.07	−4.6							[47]
1-6	$[Mg(H_2O)_6(N_5)_2]\cdot 4H_2O$	1.42/ 1.437(205 K)	104	40.66	−4.6							[32]
1-7	$[Na_8(N_5)_8(H_2O)_3]_n$	1.26/ 1.301(100 K)	129	70.19	−8.0	879/ 1.10	0.664	5.011	8.8			[48]
1-8	$[Ag(N_5)]_n$	2.96/ 3.015(173 K)	98	39.37	−4.5	827/ 4.65	4.644	7782	34.7	0.5	1	[42]
1-9	KN_5	1.96/ 2.002(150 K)	110	64.17	−7.3	136.8/ 1.25		6977	20.9	5	40	[51]
1-10	$[Cu(N_5)(N_3)]_n$	2.57/ 2.623(173 K)	90	63.81	−9.1							[50]
1-11	$[Ba(N_5)(NO_3)(H_2O)_3]_n$	2.52/ 2.592(100 K)	130[k]	25.91	+9.9	−914/ −2.83						[50]

(Continued)

Table 1.3 (Continued)

Compound	Molecule formation	d[a] (g cm^{-3})	T_d[b] (°C)	N[c] (%)	OB[d] (%)	ΔH_f[e] (kJ mol^{-1})/(kJ g^{-1})	Q[f] (kcal g^{-1})	D[g] (m s-1)	P[h] (GPa)	IS[i] (J)	FS[j] (N)	Reference
1-12	$[NaBa_3(N_5)_6(NO_3)(H_2O)_3]_n$	2.27/ 2.338(100 K)	121[k]	44.71	−0.8	−6249/ −6.43						[50]
1-13	$[(NaN_5)_5(CH_6N_3)N_5](N_5)_3^-$	1.19/ 1.214(170 K)	118	83.48	−14.9	1015/ 1.26	0.329	4407	6.5			[52]
1-14	$(NaN_5)_2(C_2H_4N_4)$	1.70/ 1.728(173 K)	127	72.59	−41.5	902/ 6.68	1.650	7863	26.4			[52]
MPF-1	—	1.23/ 1.270(100 K)	129	72.15	−2.7							[49]
	FOX-12	1.75	201	46.89	−3.8	−355/ −1.70		8323	26.6	30	350	
	RDX	1.80	205	37.84	0.0	80/ 0.36	1.386	8748	34.9	7	120	
	TNT	1.65	244	18.50	−24.7	−59/ −0.26	0.897	7178	20.5	15	>353	

a) Recalculated from low-temperature X-ray densities $\{d_{298K} = d_T/[1 + 1.5 \times 10^{-4}(298 - T)]\}$. Values after slashes were crystal densities.
b) Onset decomposition temperature (5 °C min^{-1}).
c) Nitrogen content.
d) Oxygen balance. For carbon-containing compound, $OB = 1600(c - a - b/2)/M_W$, M_W = molecular weight.
e) Calculated heat of formation in solid state.
f) Calculated heat of detonation.
g) Calculated detonation velocity.
h) Calculated detonation pressure.
i) Impact sensitivity.
j) Friction sensitivity.
k) The exothermic peak is connected to the previous endothermic peak.

1.5 *cyclo*-N_5^--Based Nonmetallic Ionic Salts

1.5.1 Synthesis

The nitrogen-rich pentazolate salts, *N*-carbamoylguanidinium, and oxalohydrazinium, 4,4′,5,5′-tetraamino-4*H*,4′*H*-[3,3′-bi(1,2,4-triazol)]-1-ium, and 3,9-diamino-6,7-dihydro-5*H*-bis([1,2,4]triazolo)[4,3-*e*:3′,4′-g][1,2,4,5]tetrazepine-2,10-diium ($C_5H_{10}N_{10}^{2+}$) pentazolate [66, 67], can be obtained *via* metathesis reactions using sodium pentazolate $[Na(H_2O)(N_5)]·2H_2O$ [32] and the corresponding chlorides, respectively, can be considered, where the reaction is driven by the low solubility of sodium chloride (Figure 1.23). Among them, 4,4′,5,5′-tetraamino-4*H*,4′*H*-[3,3′-bi(1,2,4-triazol)]-1-ium pentazolate·$3H_2O$ (($C_4H_9N_{10}^+$)(N_5^-)·$3H_2O$, **1-17**) and $(C_5H_9N_{10}^+)_2(Cl^-)(N_5^-)·3.5H_2O$ (**1-18**) can be synthesized in one pot.

The $[Na(H_2O)(N_5)]·2H_2O$ [32] could be transformed into AgN_5 precipitate by reaction with $AgNO_3$ in aqueous solution. Ammonium (**1-20**), hydroxylammonium (**1-21**), hydrazinium (**1-22**), aminoguanidinium (**1-23**), diaminoguanidinium (**1-24**), biguanidinium (**1-25**), 3,4-diamino-1,2,4-triazolium (**1-26**), and 3,6,7-triamino-7*H*-[1,2,4]triazolo[4,3-*b*][1,2,4]triazol-2-ium (**1-27**) pentazolates were synthesized by metathesis reactions of AgN_5 with stoichiometric amounts of the respective chloride salts [51] (Figure 1.24). The metathesis reactions were driven by the precipitation of AgCl, which has a very low solubility in water. So, high yields (>90%) of target pentazolate salts can be obtained by simple filtration followed by the concentration of the filtrate.

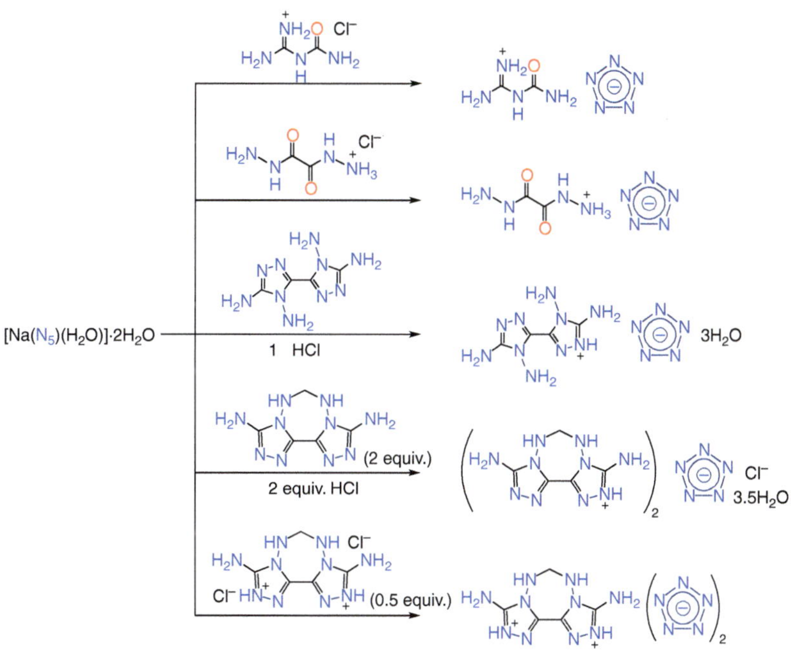

1–15

1–16

1–17

1–18

1–19

Figure 1.23 Synthesis of compounds **1-15** to **1-19**.

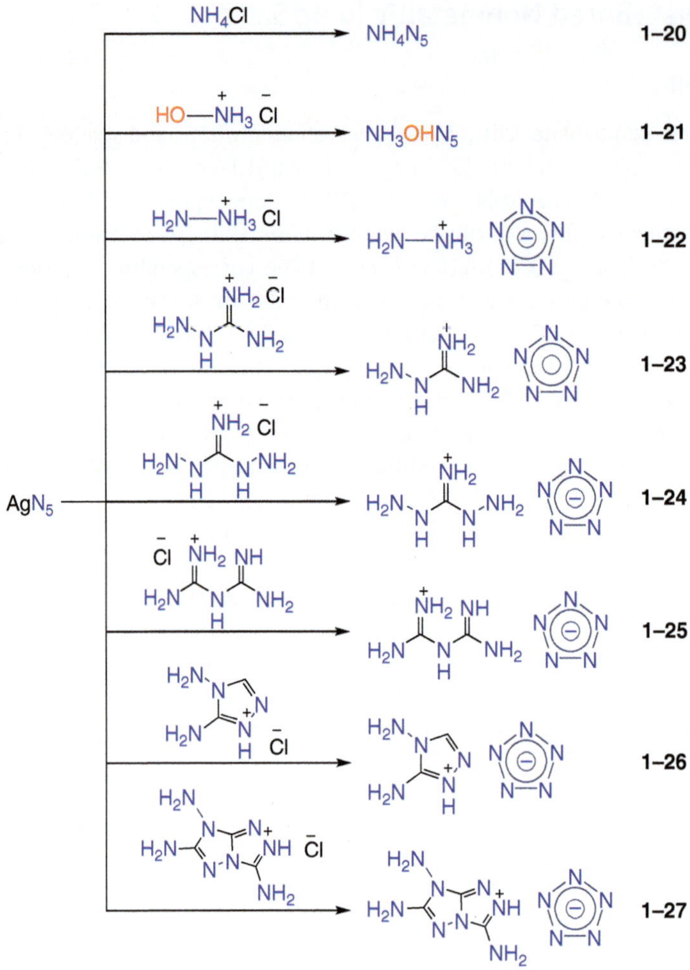

Figure 1.24 Synthesis of compounds **1-20** to **1-27**.

1.5.2 Single-Crystal X-Ray Structures

Salt **1-15** ($C_2H_7N_9O$) crystallizes in the monoclinic $P2_1/c$ space group with four molecules in the unit cell (Figure 1.25). Its calculated density is 1.596 g cm^{-3} at 173 K. As shown in Figure 1.25, the *N*-carbamoylguanidinium cation together with *cyclo*-N_5^- anion is approximately in the plane, which can be seen from the torsion angle C1–C2–N5–N1 of 0.43° and N4–N1–N5–N2 of 0.2°. The C—N bond lengths and the N—C—N bond angles in the *N*-carbamoylguanidinium cation are in the range of 1.313(2)–1.356(2) and 113.20(14)–126.22(16) Å, respectively, which are typical values for the *N*-carbamoylguanidinium cation [68]. The N—N bond lengths in the *cyclo*-N_5^- are in the range of 1.299(2)–1.335(2) Å, with an average N—N bond distance of 1.318 Å, which is similar to that of $[Na(H_2O)(N_5)]{\cdot}2H_2O$ (1.316 Å), $[Mg(H_2O)_6(N_5)_2]{\cdot}4H_2O$ (1.316 Å), and $[Mn(H_2O)_4(N_5)_2]{\cdot}4H_2O$ (1.320 Å). Since the

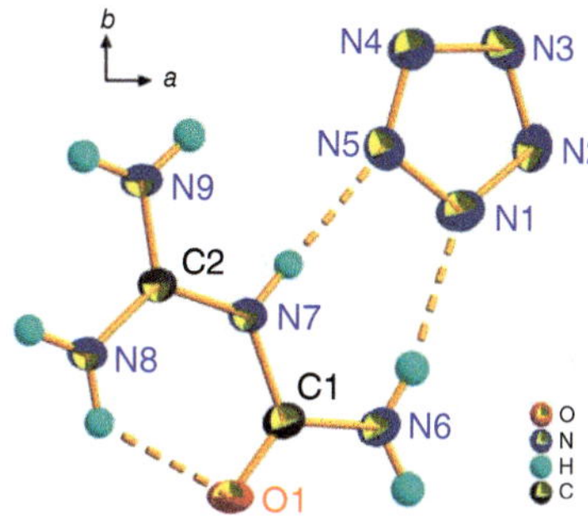

Figure 1.25 Crystal structure of salt **1-15**.

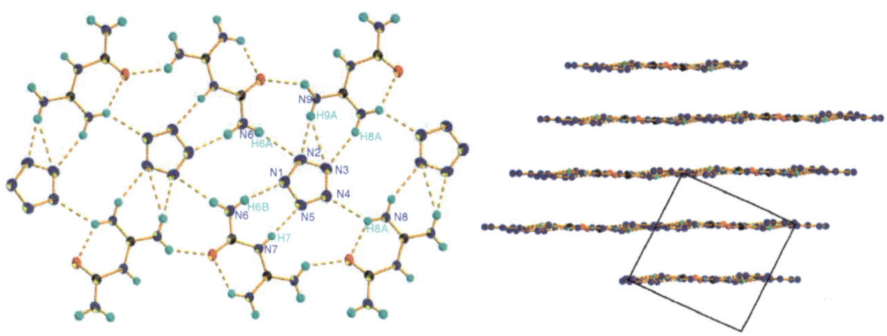

(a) (b)

Figure 1.26 (a) 2D layered intermolecular interactions in the crystal structure of salt **1-15**; (b) 3D structural layer network in salt **1-15**.

amino groups in *N*-carbamoylguanidinium are excellent hydrogen-bonding donors, the discrete *cyclo*-N_5^- rings are linked into a stable 2D layer by the hydrogen-bonding interactions between cations and anions (Figure 1.26). The face-to-face stacking of **1-15** shows that there are no significant interactions bctwccn adjacent layers.

Salt **1-16** ($C_2H_7N_9O_2$) crystallizes in the same space group (monoclinic $P2_1/c$) with **1-15** and eight molecules in the unit cell (Figure 1.27). Its crystal density ($1.681\,g\,cm^{-3}$ at 173 K) is higher than that of **1-15**. Two forms of *cyclo*-N_5^- can be found in **1-16**, with the N—N bond lengths in the range of 1.294(5)–1.339(5)

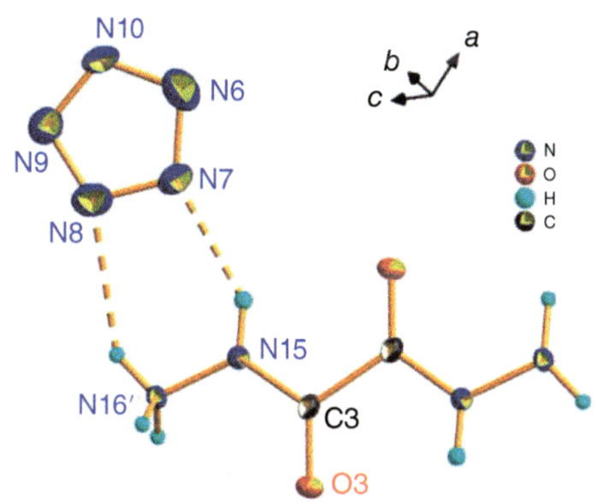

Figure 1.27 Crystal structure of **1-16**.

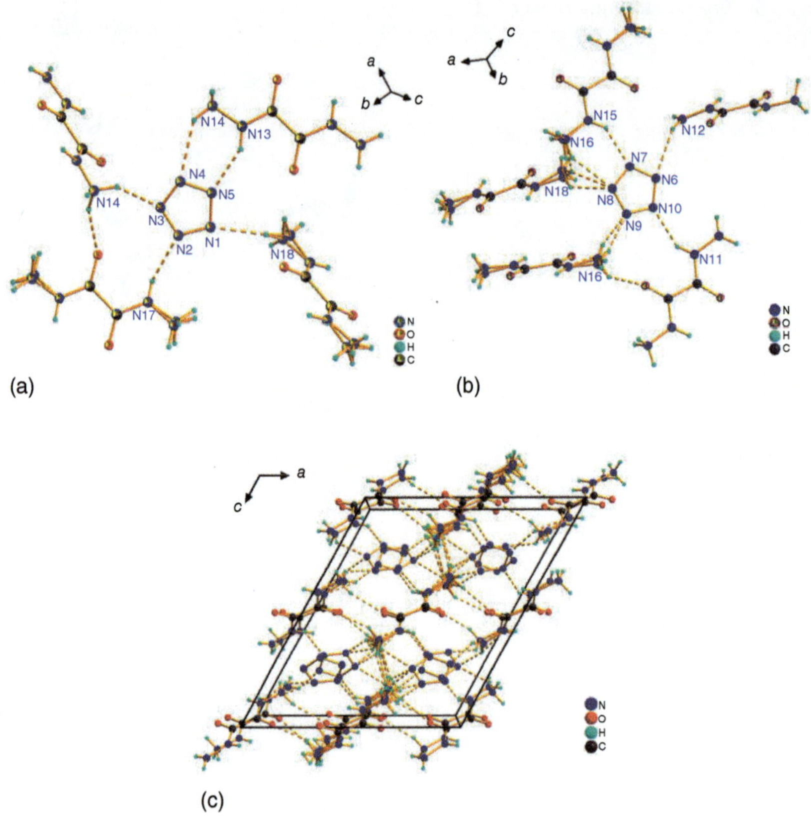

Figure 1.28 (a, b) The hydrogen bonds of *cyclo*-N_5^- in **1-16**; (c) A view of the packing down the *b*-axis of the unit cell of **1-16**. Dashed lines indicate strong hydrogen bonding.

and 1.301(5)–1.333(5) Å and the largest torsion angle N2–N1–N5–N4 of 0.64° and N10–N6–N7–N8 of 1.31°. Every *cyclo*-N_5^- ring participates in at least five hydrogen bonds (Figure 1.28). The lengths of the hydrogen bonds are shorter than the sum of the van der Waals radii ($r_W(N) + r_W(N) = 3.20$ Å); thus, a strong hydrogen-bond network is formed.

Salt **1-17** ($C_4H_{15}N_{15}O_3$) crystallizes in the triclinic space group *P*-1 with a density of 1.520 g cm^{-3} at 173 K (Figure 1.29). Unlike most complexes, the *cyclo*-N_5^- ring in **1-17** lacks perfect planarity, with the largest torsion angle (N13–N11–N12–N15) of 0.64° which is shorter than that of $[Mg(H_2O)_6(N_5)_2]{\cdot}4H_2O$ (0.85°) [32]. We can see from Figure 1.30 that the crystal is layer by layer stacked and the interlayer distance is 3.42 Å. The adjacent layers are connected by two kinds of hydrogen bonds (O2—O3 HBs and O1—N1 HBs).

Salt **1-18** crystallizes from water in the triclinic space group *P*-1 with a density of 1.617 g cm^{-3} at 173 K and four fused $C_5H_9N_{10}^+$ cations, two Cl^-, two *cyclo*-N_5^-, and seven water moieties per unit cell (Figures 1.31 and 1.32). The N—N bond lengths (1.390(3) and 1.396(3) Å) of the unprotonated 1,2,4-triazoles are slightly longer than those of the protonated triazoles (1.382(3) Å). All the amino groups participate in

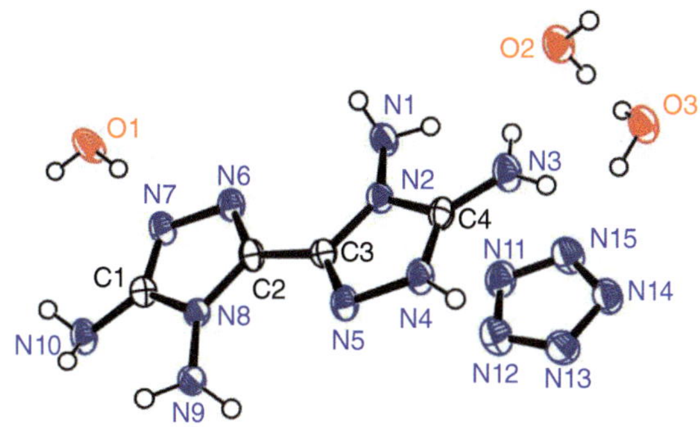

Figure 1.29 Crystal structure of **1-17**.

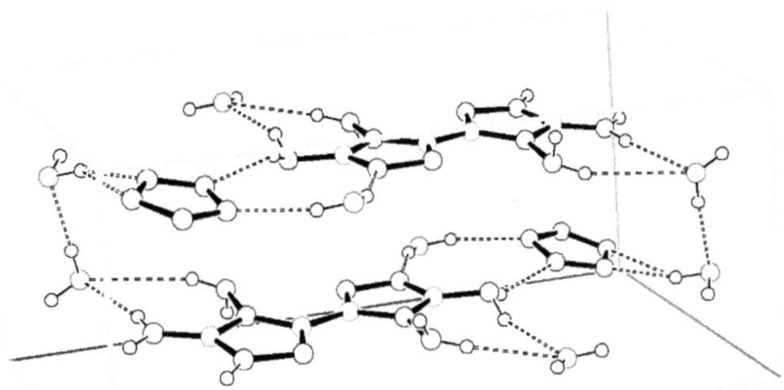

Figure 1.30 The 3D structure of **1-17** constructed by hydrogen-bonding networks.

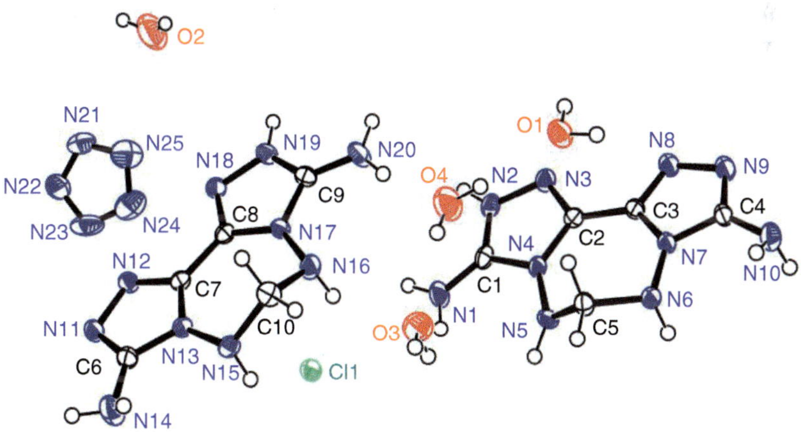

Figure 1.31 Crystal structure of **1-18**.

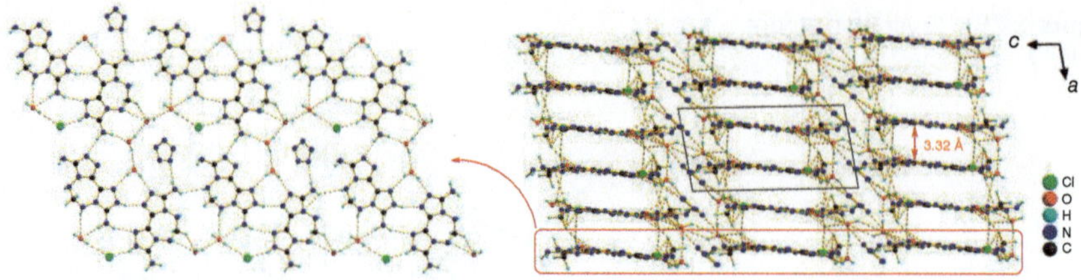

Figure 1.32 The 3D structure of **1-18** constructed by hydrogen-bonding networks.

hydrogen bonds and act as donors, whereas in the two fused cations only N3, N9, N11, N12, and N18 act as hydrogen-bond acceptors. Numerous hydrogen bonds around *cyclo*-N_5^- can be seen from the *b*-axis (*ac* plane) of the packing diagram. The interlayer distance in **1-18** is 3.32 Å, which is slightly shorter than that of **1-17** (3.42 Å).

Salt **1-19** crystallizes in the monoclinic space group *C*2/*c* with a density of 1.660 g cm^{-3} at 173 K and four molecules in the unit cell (Figure 1.33). The cation DABTT^{2+} exhibits a trans structure with an angle of 36.68° between two amino-triazole planes (Figure 1.34). The C3—N5 bond length of C-amino

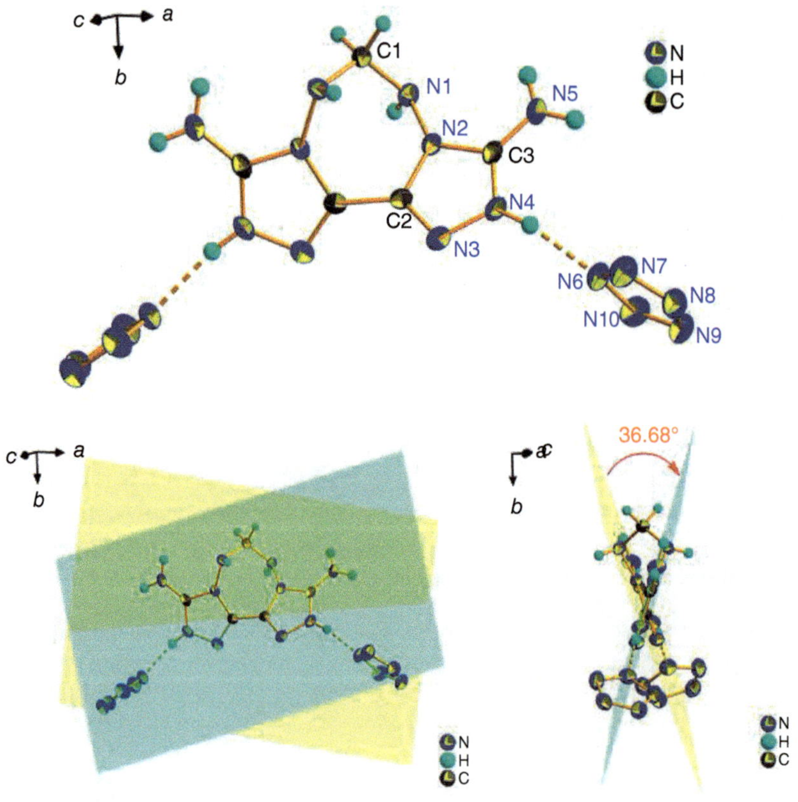

Figure 1.33 Crystal structure of **1-19**.

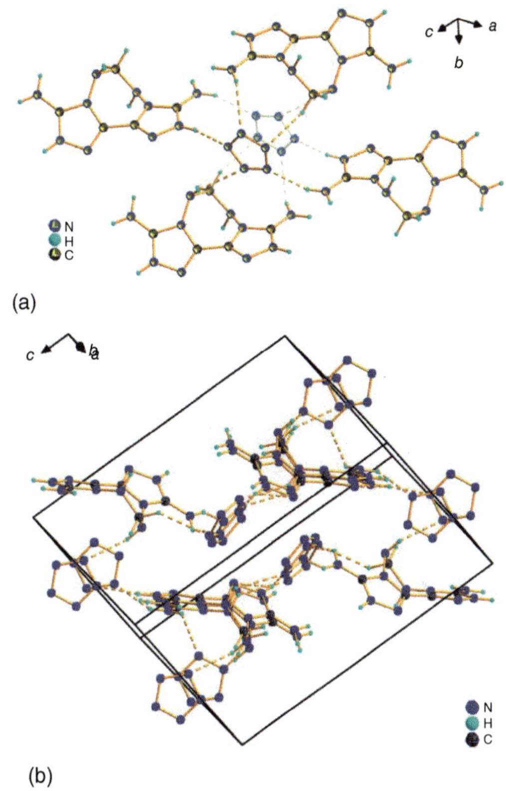

Figure 1.34 (a) The hydrogen bonds of *cyclo*-N_5^- in **1-19**; (b) A view of the packing of the unit cell of **1-19**.

group is 1.313(2) Å, which is slightly shorter than three C—N bonds (C2—N2: 1.379(2) Å; C3—N2: 1.3558(19) Å; C3—N4: 1.334(2) Å) and longer than one C—N bond (C2—N3: 1.291(2) Å) of the triazole moiety. With the bond lengths of 1.291(2)–1.385(2) Å, the distances between the ring atoms of the triazole moiety lie between the length of formal C—N and N—N single and double bonds, which indicates the aromaticity of the ring system [69].

Salt **1-20** crystallizes in the orthorhombic *Pcca* space group with four molecules per unit cell (Figures 1.35 and 1.36). The N—N bond lengths for *cyclo*-N_5^- in **1-20** are 1.309(2), 1.309(2), 1.316(2), 1.316(2), and 1.316(3) Å; the average N—N bond distance (1.3132 Å) is shorter than that of most reported metal-N_5^- compounds (1.314–1.329 Å). The structure of the *cyclo*-N_5^- in **1-20** is almost planar (N3^{ii}-N3-N2-N1 = 0.02°, N2-N2^{ii}-N1-N3^{ii} = 0.01°). All hydrogenated

Figure 1.35 Crystal structure of salt **1-20**.

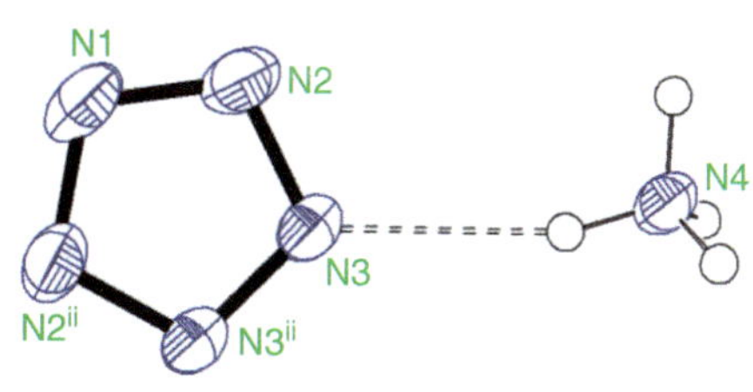

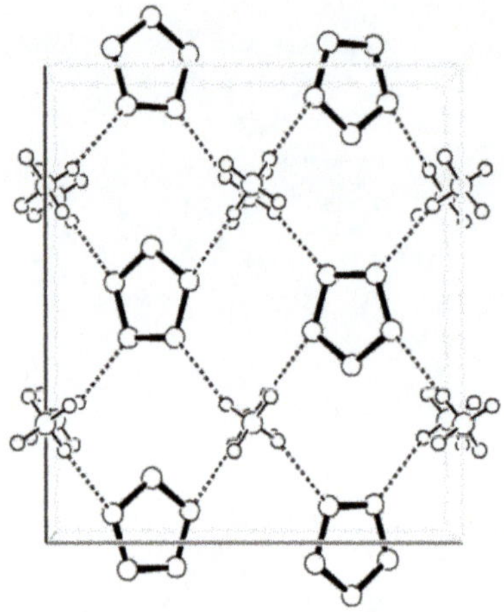

Figure 1.36 Ball-and-stick packing diagram of **1-20** viewed down the *b* axis.

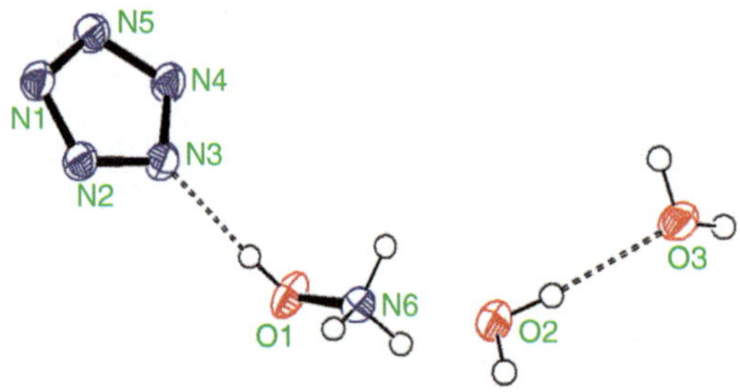

Figure 1.37 Crystal structure of **1-21**·$2H_2O$.

nitrogen atoms (N4) participating in hydrogen bonds act as donors, while only four nitrogen atoms (except for N1) of each *cyclo*-N_5^- act as hydrogen-bond acceptors.

Both **1-21**·$2H_2O$ and **1-22** crystallize in the monoclinic *P*121/*c*1 space group with four molecules per unit cell (Figures 1.37–1.40). The average N—N bond distances in **1-21**·$2H_2O$ and **1-22** are 1.3152 and 1.3227 Å, which are slightly longer than that of **1-20**. Unlike **1-20**, the five nitrogen atoms of *cyclo*-N_5^- in **1-21**·$2H_2O$ and **1-22** are all not completely coplanar, which could be seen from the largest torsion angle of 0.29°

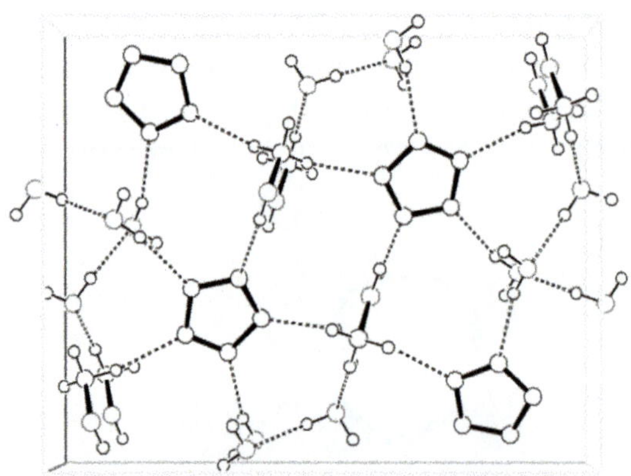

Figure 1.38 Ball-and-stick packing diagram of **1-21**·$2H_2O$.

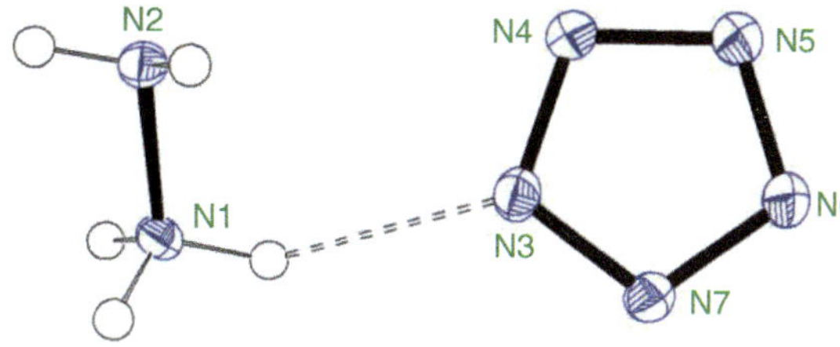

Figure 1.39 Crystal structure of **1-22**.

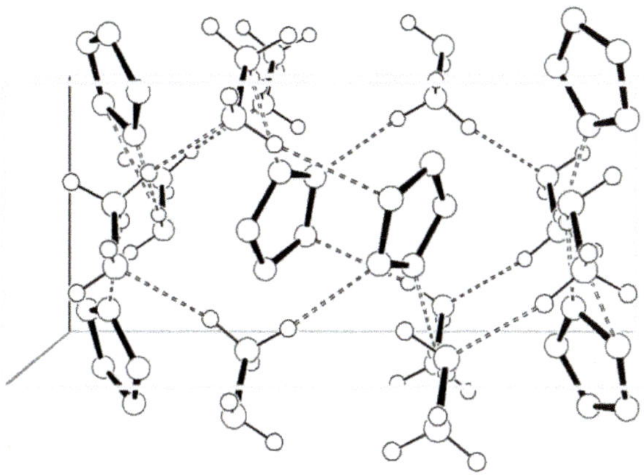

Figure 1.40 Ball-and-stick packing diagram of **1-22**.

(N3–N4–N5–N1) for **1-21**·2H_2O and 0.45° (N5–N7–N6–N3) for **1-22**. All nitrogen atoms of *cyclo*-N_5^- in **1-21**·2H_2O and **1-22** act only as hydrogen-bond acceptors. The N6 and O1 atoms of hydroxylammonium cations act only as donors, while the hydrazinium cations act as not only donors (N1) but also acceptors (N2). Thus, it is not surprising that the *cyclo*-N_5^- anions are controlled by extensive hydrogen-bond networks.

Salt **1-23** crystallizes in the monoclinic *P*121/*n*1 space group with four molecules per unit cell ($Z = 4$) and a calculated density of 1.476 g cm^{-3} at 150 K (Figure 1.41). The aminoguanidinium cations and the corresponding *cyclo*-N_5^- anions are coplanar with torsion angles of 11.777° (N4–N3–N6–N7), 7.480° (N8–N2–N7–N1), and 10.077° (N6–N3–N2–N1). In addition, a large number of hydrogen bonds between aminoguanidinium cations and *cyclo*-N_5^- anions form a 2D layer (Figure 1.42). Salt **1-23** is layer-by-layer stacking and the interlayer distance is 3.18 Å. The adjacent layers are connected by two kinds of hydrogen bonds (N9–H9A…N5 and NH9–H9B…N3).

Salt **1-24** crystallizes in the triclinic *P*-1 space group with two molecules per unit cell and a calculated density of 1.465 g cm^{-3} at 173 K (Figures 1.43 and 1.44). The diaminoguanidinium cation in **1-24** is almost planar with the torsion angles of 3.49° (N10–N8–C1–N9), 14.45° (N10–C1–N7–N6), and 0.05° (N9–N6–N7–N8). Unlike **1-23**, the diaminoguanidinium cations are not coplanar with the *cyclo*-N_5^- rings, they are almost perpendicular to each other with an angle of 79.657°. In addition to hydrogen bonds between *cyclo*-N_5^- and cations, strong intramolecular hydrogen

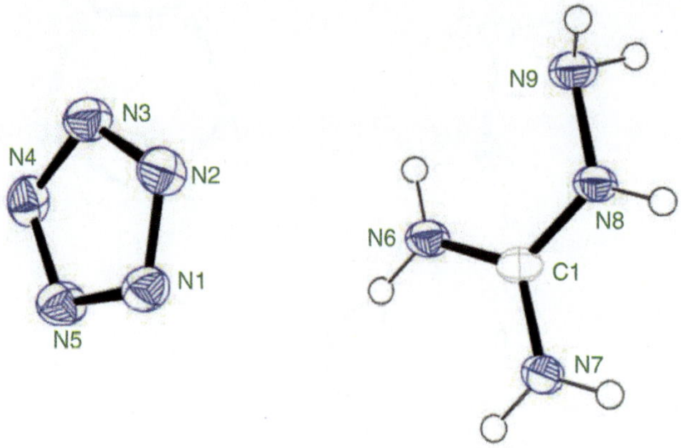

Figure 1.41 Crystal structure of **1-23**.

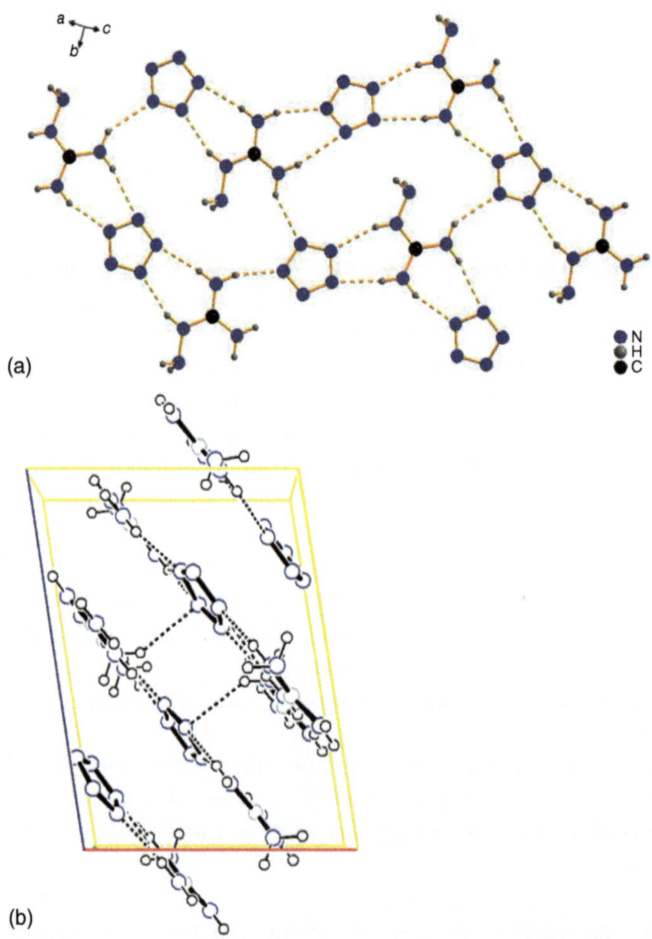

Figure 1.42 (a) Hydrogen bonds in **1-23**; (b) Ball-and-stick packing diagram of **1-23**.

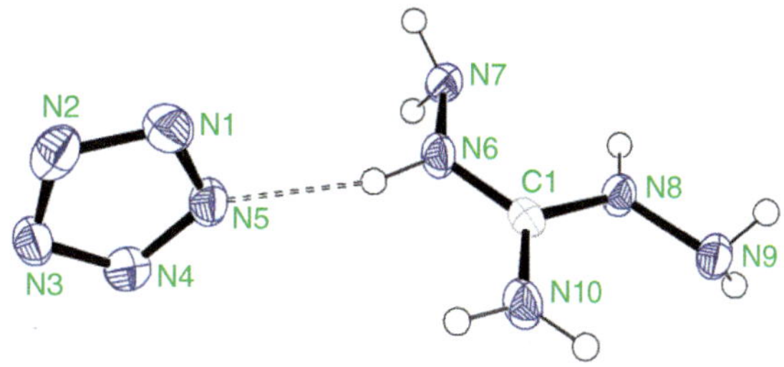

Figure 1.43 Crystal structure of **1-24**.

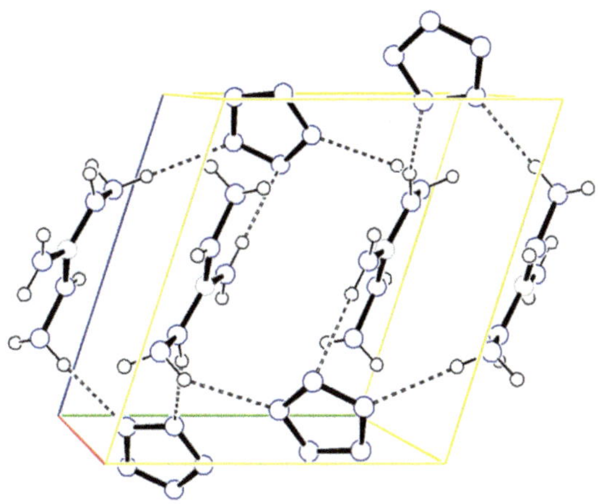

Figure 1.44 Ball-and-stick packing diagram of **1-24**.

bonds (N8—H8...N7 = 2.6856 Å; N10—H10A...N9 = 2.6536 Å) are present in each cation.

Salt **1-25** crystallizes in the monoclinic *P*121/*c*1 space group with four molecules per unit cell ($Z = 4$) (Figures 1.45 and 1.46). It has a higher calculated density of 1.524 g cm^{-3} (150 K) than **1-23** and **1 24**. The crystal structure of biguanidinium cation in **1-25** does not show hydrogen on the bridging nitrogen (N8), which is consistent with the literature [70]. The biguanidinium cation cannot maintain a planar structure due to the steric hindrance of two adjacent amino groups (N7 and N10). The dihedral angle between plane through N6, N7, N8 and plane through

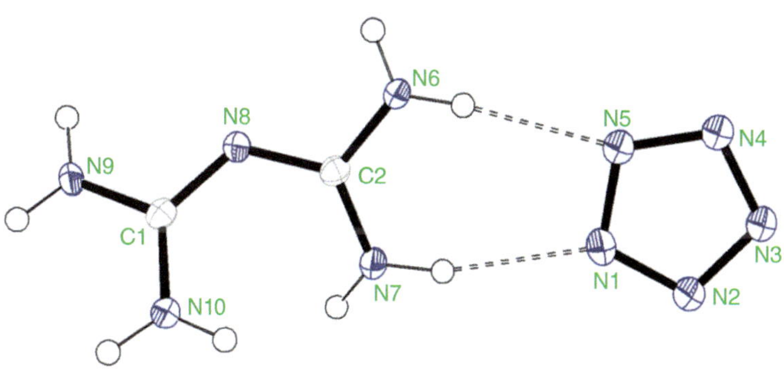

Figure 1.45 Crystal structure of **1-25**.

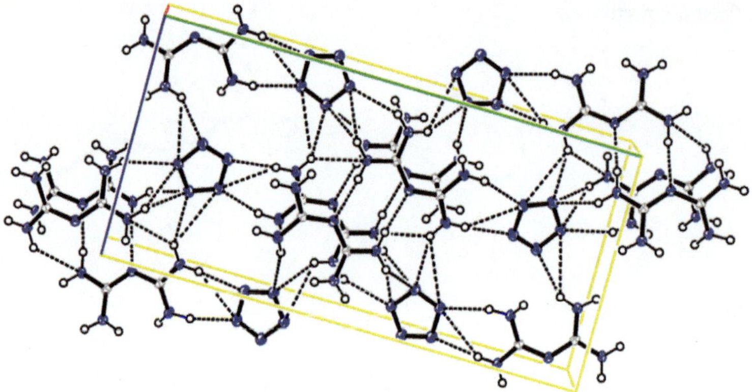

Figure 1.46 Ball-and-stick packing diagram of **1-25**.

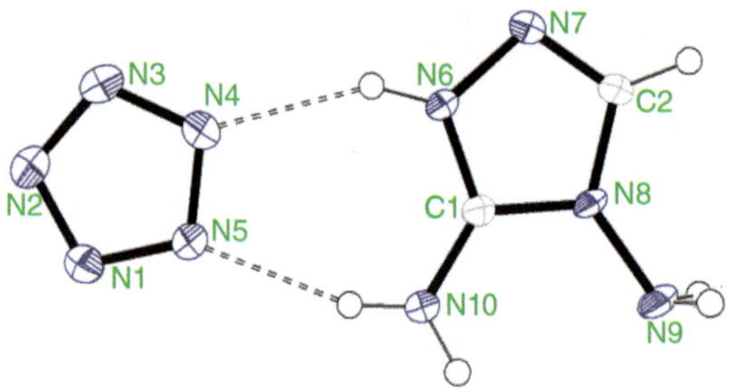

Figure 1.47 Crystal structure of **1-26**.

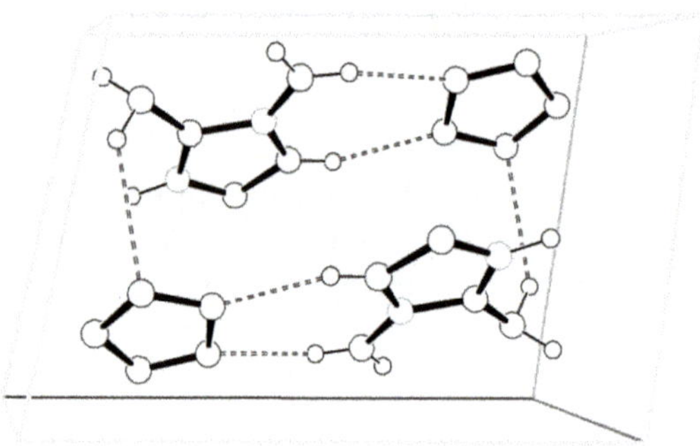

Figure 1.48 Ball-and-stick packing diagram of **1-26**.

N8, N9, N10 is 33.042°. Each biguanidinium cation is substantially coplanar with one of the two *cyclo*-N_5^- rings adjacent thereto which could be seen from the N6–N1–N5–N7 torsion angle of 4.39°.

Salt **1-26** crystallizes in the triclinic *P*-1 space group with two molecules per unit cell and has a calculated density of 1.618 g cm^{-3} (173 K) (Figure 1.47). As shown in Figure 1.48, all carbon and nitrogen atoms and most hydrogen atoms lie in the same plane as supported by the N10—H10A…N5 (2.958(3) Å) and N6—H6…N4 (2.956(3) Å) hydrogen bonds to form a unit. The distance between two adjacent planar units is 3.19 Å.

Salt **1-27** (Figures 1.49 and 1.50) is described by the monoclinic space group *P*121/*c*1 with a density of 1.645 g cm^{-3} at 173 K and four molecules per unit cell.

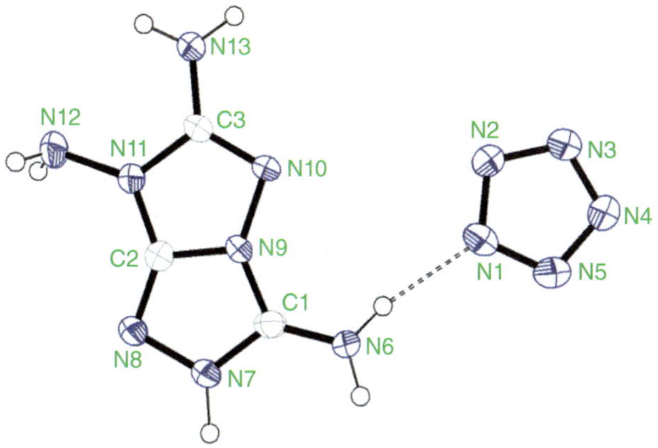

Figure 1.49 Crystal structure of **1-27**.

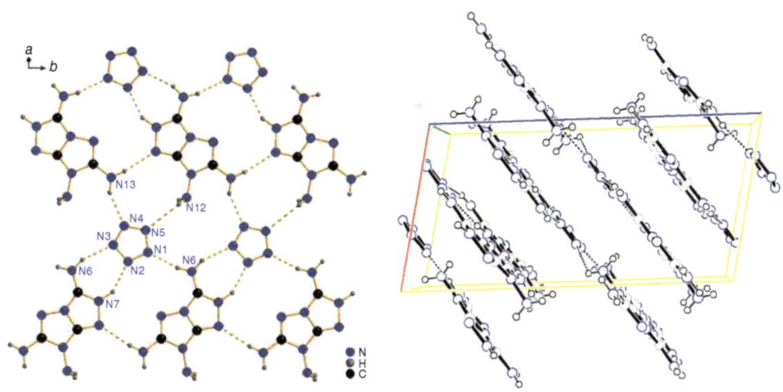

(a) (b)

Figure 1.50 (a) Hydrogen bonds in **1-27**; (b) Ball-and-stick packing diagram of **1-27**.

Every *cyclo*-N_5^- ring participates in five hydrogen bonds to form a 2D layer. We can see from Figure 1.10b that **1-27** is layer-by-layer stacking and the interlayer distance is 3.01 Å which is shorter than that of **1-23** (3.18 Å).

1.5.3 Properties

1.5.3.1 ^{15}N NMR

The ^{15}N NMR spectrum of salt **1-15** was measured in dimethyl sulfoxide (DMSO)-D_6 with respect to CH_3NO_2 as an external standard. Nitrogen signals were assigned based on the literature [68] and Giao NMR calculation with Gaussian 09 [71]. As shown in Figure 1.51, the nitrogen atoms of *cyclo*-N_5^- (N1) resonate at −5.95 ppm, consistent with the $[Na(H_2O)(N_5)]{\cdot}2H_2O$ (−5.7 ppm) [32]. Apart from the signal of the *cyclo*-N_5^- anion, the three kinds of hydrogen-attached nitrogen atoms have higher negative resonances. The N2 nitrogen atom (NH) resonates at the lowest field of all three at −269.19 ppm, whereas the two different kinds of NH_2 groups have very similar shifts at −291.28 ppm (N3) and −296.49 ppm (N4). In

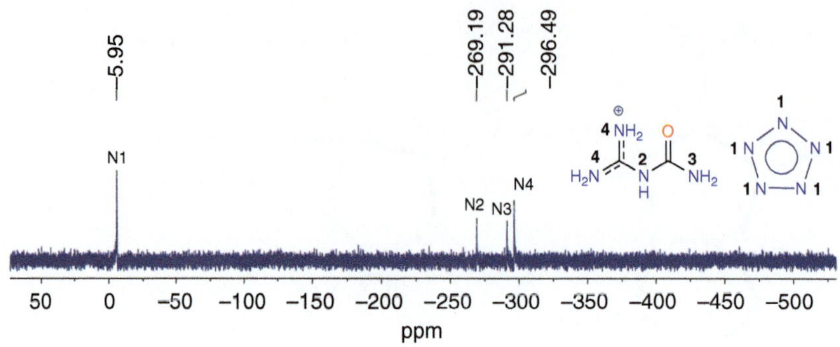

Figure 1.51 ^{15}N NMR spectrum of salt **1-15**.

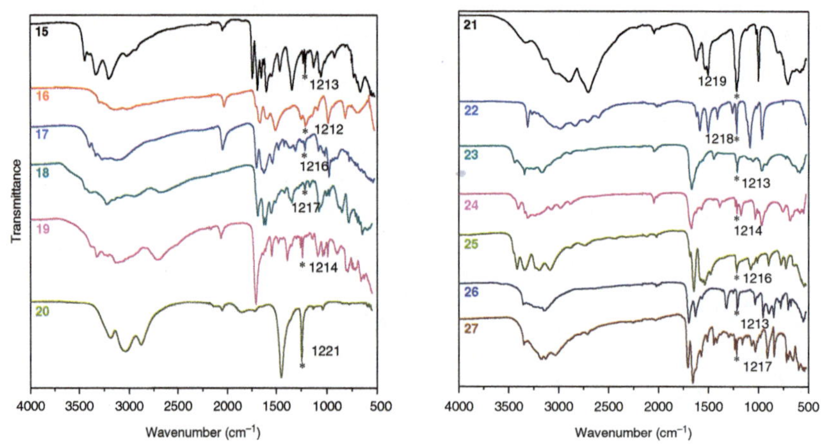

Figure 1.52 IR spectra of **1-15–1-20** to **1-21–1-27**.

contrast to the spectrum of *N*-carbamoylguanidinium nitrate [68], the signals for the cation (N2, N3, and N4) were shifted about 2 ppm to lower fields.

1.5.3.2 IR Spectrum

In the IR spectra (Figure 1.52), the strong *cyclo*-N_5^- bands were found at 1213–1221 cm^{-1}, which are red-shifted by about 20 cm^{-1} than metal salts $[Na(H_2O)(N_5)]\cdot 2H_2O$, $[M(H_2O)_4(N_5)_2]\cdot 4H_2O$ (M = Mn, Fe, Co, and Zn), and $[Mg(H_2O)_6(N_5)_2]\cdot 4H_2O$ [32, 47]. The stretching vibration of the N—H bonds is observed between 3500 and 3000 cm^{-1}. The strongest band observed in the IR spectra of **1-15**, **1-17** to **1-19**, **1-23** to **1-27** is the C=N stretch at 1667, 1668, 1644, 1629, and 1652 cm^{-1}, respectively. In addition, many combined stretching and deformation as well as torsion stretching modes could be observed in the fingerprint region between 1000 and 500 cm^{-1}.

1.5.3.3 Thermal Stability

The thermal stability of *cyclo*-N_5^- based compounds has long been an important concern by chemists. Here, TG and DSC of **1-15** to **1-27** were carried out to

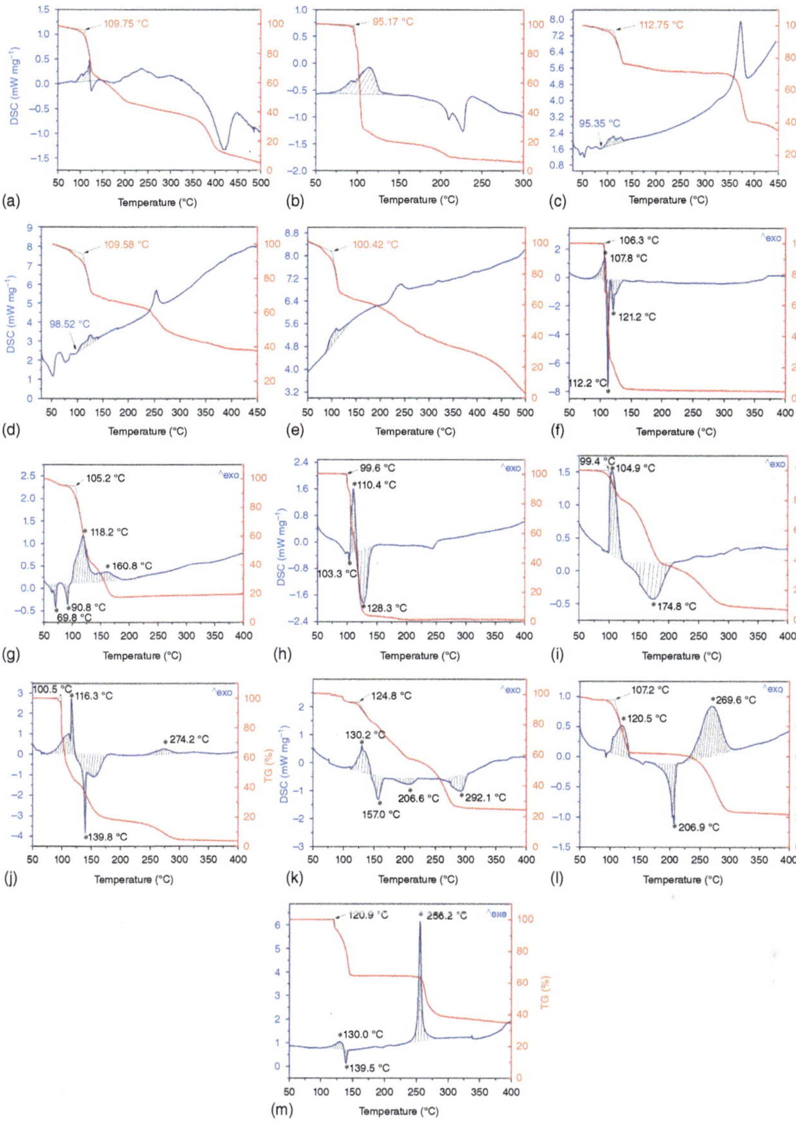

Figure 1.53 DSC and TG of salts **1-15** to **1-27**.

determine their thermal behaviors at a heating rate of 5 °C min^{-1}. As shown in Figure 1.53, the *cyclo*-N_5^- based compounds all decomposed without melting. The *cyclo*-N_5^- rings in the nonmetallic salts are decomposed into N_3^- and N_2 at about 100 °C. For most salts, the exothermic peak of each compound follows an obvious endothermic peak which should be caused by the volatilization of hydrazoic acid (HN_3).

The onset decomposition temperatures (T_d) taken from their TG curves indicate that these salts have good thermal stabilities (T_d > 95.2 °C). In addition to compounds **1-16**, **1-22**, and **1-23**, other new synthesized *cyclo*-N_5^- based salts have decomposition temperatures above 100 °C. Especially, salts **1-25** and **1-27** show excellent thermal stabilities with onset decomposition temperatures of 124.8 and 120.9 °C, respectively. After the decomposition of the *cyclo*-N_5^-, their weight loss processes depend on the type of counter cations: when the cations are simple and carbon-free ammonium, hydroxylammonium, and hydrazinium, the TG curves show significant weight losses in a narrow range from 90 to 170 °C; when the cations are guanidinium, triazolium, and fused cations, the TG curves show obvious weight losses in a relatively wider range from 90 to 300 °C.

1.5.3.4 Sensitivity

For safety purposes, the sensitivities toward impact (IS) and friction (FS) were measured by using standard BAM methods [72]. Salts **1-16** and **1-21** are the most impact and friction-sensitive energetic materials (IS = 6 and 5.5 J, FS = 80 and 50 N) among these *cyclo*-N_5^- compounds. The impact and friction sensitivities of salts **1-15**, **1-19**, **1-20**, and **1-22** are measured to be 7.4–14 and 7.4 J, 120–160 N, respectively, which are comparable with those of RDX and HMX (IS = 7.4 J, FS = 120 N). Salts **1-24** and **1-25** are less impact and friction sensitivities with IS values of 25 and 30 J and FS values of 240 and 300 N, which are comparable with FOX-12 (IS = 30 J, FS = 350 N). Salts **1-23**, **1-26**, and **1-27** were determined to be insensitive to impact and friction (IS > 40 J, FS = 120 N), which highlight their likely application potential as insensitive energetic materials.

With respect to Hirshfeld surfaces (Figure 1.54), salts **1-23**, **1-26**, and **1-27** appear in plate shapes, which indicate insensitivity as it represents planar conjugated molecular structures. And their red dots (predominant intermolecular interactions) are located on the surface edges; in this way, the layers are supported by π–π stacking and thus sustain the maximum external *stimuli*.

1.5.3.5 Density

The densities of these salts fall between 1.312 and 1.650 g cm^{-3} (Table 1.4), less than those of the predicted polynitrogen compounds (2.0–3.9 g cm^{-3}) and TNT (1.65 g cm^{-3}). Fused triazolium pentazolates (**1-19** and **1-27**) have higher densities than other structures. Monocyclic triazolium pentazolate has a lower density than fused triazolium pentazolates. In addition, comparing compounds **1-15** and **1-25**, we found that oxygen-containing cations are beneficial to increase the density of the pentazolate salts.

1.5.3.6 Heat of Formation and Detonation Performance

The solid-phase heats of formation for **1-15** to **1-27** are calculated using the gas phase heat of formation and heat of phase transition (lattice energy) according to Hess's law of constant summation [73] (Born–Haber energy cycle). The heats of formation of *cyclo*-N_5^- anion and most cations involved in the calculations are referenced from the literature [58, 74, 75]. Others were calculated by the Gaussian 09 (Revision A.02) suite of programs [71]. Except for pentazolate salts containing crystal water and Cl^-, all the *cyclo*-N_5^- based energetic salts have relatively high positive heats of formation

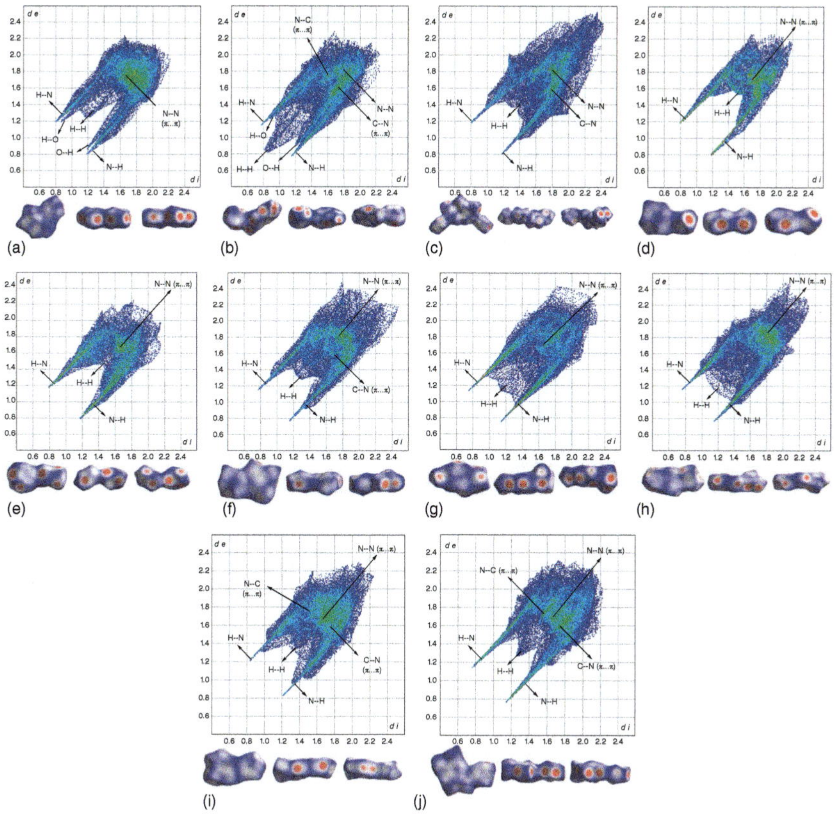

Figure 1.54 Two-dimensional fingerprint plots and Hirshfeld surfaces for **1-15** (a), **1-16** (b), **1-19** (c), **1-20** (d), and **1-22** to **1-27** (e–j).

(ΔH_f) falling between 203.4 (**1-15**) to 1362.0 (**1-25**) kJ mol^{-1} (Table 1.4), which exceed TNT (−67 kJ mol^{-1}), RDX (80 kJ mol^{-1}), and HMX (104.8 kJ mol^{-1}). Among them, the biguanidinium salt **1-25** has the highest heat of formation (1362.0 kJ mol^{-1}) due to the very high heat of formation of biguanidinium cation (1618 kJ mol^{-1}). Therefore, the incorporation of a cation with high heat of formation can still be considered as a useful way to increase the heat of formation of a *cyclo*-N_5^- based energetic ion salt.

The detonation performances of these salts were calculated using the EXPLO5 program (version 6.01) [76] based on the calculated heats of formation and densities at ambient temperature. The calculated detonation velocity (D) and pressure (P) values (Table 1.4) fall in the range of 6920–9257 m s^{-1} and 18.9–33.0 GPa, respectively, which are higher than those of TNT (6.881 km s^{-1}, 19.5 GPa). With respect to D, the values of **1-22** and **1-21** (8796 and 9005 m s^{-1}) exceed FOX-12 and RDX (8323 and 8795 m s^{-1}) and salt **1-25** remarkably exceeds those values at 9257 m s^{-1}, which is higher than HMX (9144 m s^{-1}). In terms of P, the values for **1-21**, **1-22**, and **1-25** (32.7, 30.8, 33.0 GPa) are higher than those of FOX-12 (26.6 GPa) and comparable with those of RDX (34.9 GPa).

Table 1.4 Physicochemical and energetic properties of **1-15** to **1-27**, RDX, HMX, and FOX-12.

Compound	$\rho^{a)}$ (g cm^{-3})	$\Delta H_f^{b)}$ (kJ mol^{-1}/ kJ g^{-1})	$D^{c)}$ (m s^{-1})	$p^{d)}$ (GPa)	IS$^{e)}$ (J)	FS$^{f)}$ (N)	OB$^{g)}$ (%)	$N^{h)}$ (%)
1-15	1.567/ 1.596(173 K)	203.4/ 1.18	6920	18.9	14	160	−41.6	72.81
1-16	1.650/ 1.681(173 K)	388.1/ 2.05	8320	27.1	6	80	−29.6	66.65
1-17	1.492/ 1.520(173 K)	—	—	—	—	—	—	65.40
1-18	1.587/ 1.617(173 K)	—	—	—	—	—	—	59.66
1-19	1.629/ 1.660(173 K)	1341.2/ 3.83	7615	23.6	10	120	−45.7	79.98
1-20	1.486/ 1.519(150 K)	269.1/ 3.06	7757	23.2	8	130	−36.4	95.42
1-21	1.601/ 1.636(153 K)	327.6/ 3.15	9005	32.7	5.5	50	−15.4	80.75
1-22	1.583/ 1.618(150 K)	429.6/ 4.17	8796	30.8	7.4	120	−38.8	95.11
1-23	1.444/ 1.476(150 K)	392.1/ 2.70	7189	19.6	>40	>360	−49.7	86.86
1-24	1.438/ 1.465(173 K)	508.0/ 3.17	7505	21.2	25	240	−50.0	87.46
1-25	1.491/ 1.524(150 K)	1362.0/ 7.91	9257	33.0	35	300	−55.8	81.36
1-26	1.583/ 1.618(150 K)	639.7/ 3.76	7824	24.5	>40	>360	−47.06	82.33
1-27	1.615/ 1.645(173 K)	853.8/ 3.79	7791	24.6	>40	>360	−46.2	80.86
RDX	1.806	80.0/ 0.36	8795	34.9	7.4	120	0	37.84
HMX	1.905	74.8/ 0.25	9144	39.2	7.4	120	0	37.84
FOX-12	1.754	−355.0/ −1.59	8323	26.6	30	350	−3.8	46.89

a) Density, recalculated from low-temperature X-ray densities $\{\rho_{298K} = \rho_T/[1 + 1.5 \times 10^{-4}(298 - T)]\}$. Values after slashes are crystal densities.
b) Heat of formation calculated in solid state.
c) Detonation velocity calculated by EXPLO5 V6.01.
d) Detonation pressure calculated by EXPLO5 V6.01.
e) Impact sensitivity.
f) Friction sensitivity.
g) Oxygen balance (based on CO) for $C_aH_bO_cN_d$, $1600(c - a - b/2)/M_w$; M_w = molecular weight.
h) Nitrogen content.

1.6 Conclusions

In summary, a comprehensive overview of the chemistry and performance of the pentazole-based energetic materials was performed. It should be emphasized that the pentazole chemistry is a rather new area of research and all presented data correspond to the twenty-first century. Long-term computational investigations revealed a high heat of formation of the pentazole ring and demonstrated a huge potential of pentazole derivatives as desirable high-energy compounds. Recent synthetic efforts provided reliable routes to the preparation of various pentazoles including their metal-free and metal-containing derivatives. Multidisciplinary investigations of functional properties of pentazolate-derived ionic compounds showed their application potential both as primary explosives and as components of energetic formulations. Importantly, such compounds have a high level of environmental compatibility, due to the presence of the fully nitrogen heterocyclic core, which upon detonation releases molecular nitrogen as a main decomposition product. There is no doubt that investigations on pentazole-based energetic materials will be of increased attention in near future and should be an excellent platform for the development of novel advanced energetic materials.

References

1 Klapötke, T.M. (2012). *Chemistry of High-Energy Materials*, vol. 2. Berlin: De Gruyter.
2 Christe, K.O. (2007). *Propellants Explos. Pyrotech.* 32: 194–204.
3 Nguyen, M.T., McGinn, M.A., Hegarty, A.F., and Elguéro, J. (1985). *Polyhedron* 4: 1721–1726.
4 Nguyen, M.T. (2003). *Coord. Chem. Rev.* 244: 93–113.
5 Dixon, D.A., Feller, D., Christe, K.O. et al. (2004). *J. Am. Chem. Soc.* 126: 834–843.
6 Klapötke, T.M. and Hammerl, A. (2008). *Comprehensive Heterocyclic Chemistry III*, 739–757. Oxford: Elsevier.
7 Vij, A., Pavlovich, J.G., Wilson, W.W. et al. (2002). *Angew. Chem. Int. Ed.* 41: 3051–3054.
8 Huisgen, R. and Ugi, I. (1956). *Angew. Chem.* 68: 705–706.
9 Butler, R.N., Fox, A., Collier, S., and Burke, L.A. (1998). *J. Chem. Soc., Perkin Trans. 2* 2243–2248.
10 Yang, Y., Li, Y., Zhang, R. et al. (2014). *RSC Adv.* 4: 57629–57634.
11 Bondarchuk, S.V. (2020). *J. Phys. Chem. Lett.* 11: 5544–5548.
12 Zhang, X., Yang, J., Lu, M., and Gong, X. (2014). *RSC Adv.* 4: 56095–56101.
13 Zhang, X., Ma, C., Zhang, Y., and Liu, G. (2018). *Struct. Chem.* 29: 267–274.
14 Zhang, X. and Gong, X. (2016). *J. Mol. Model.* 22: 106.
15 Hammerl, A. and Klapötke, T.M. (2002). *Inorg. Chem.* 41: 906–912.
16 Zhang, X., Yang, J., Lu, M., and Gong, X. (2015). *RSC Adv.* 5: 27699–27705.
17 Rahm, M. and Brinck, T. (2010). *Chem. Eur. J.* 16: 6590–6600.

18 Frison, G., Jacob, G., and Ohanessian, G. (2013). *New J. Chem.* 37: 611–618.
19 Liang, Y.H., Luo, Q., Guo, M., and Li, Q.S. (2012). *Dalton Trans.* 41: 12075–12081.
20 Noyman, M., Zilberg, S., and Haas, Y. (2009). *J. Phys. Chem. A* 113: 7376–7382.
21 Yang, J., Gong, X., and Wang, G. (2015). *Struct. Chem.* 26: 1077–1082.
22 Liu, S.J., Zhao, L., Yao, M.G. et al. (2020). *Adv. Sci.* 7.
23 Östmark, H., Wallin, S., Brinck, T. et al. (2003). *Chem. Phys. Lett.* 379: 539–546.
24 Geiger, U., Elyashiv, A., Fraenkel, R. et al. (2013). *Chem. Phys. Lett.* 556: 127–131.
25 Bazanov, B. and Haas, Y. (2015). *J. Phys. Chem. A* 119: 2661–2671.
26 Geiger, U. and Haas, Y. (2015). *J. Phys. Chem. B* 119: 7338–7348.
27 Geiger, U. and Haas, Y. (2016). *J. Phys. Chem. B* 120: 6208–6214.
28 Schroer, T., Haiges, R., Schneider, S., and Christe, K.O. (2005). *Chem. Commun.* 1607–1609.
29 Butler, R.N., Hanniffy, J.M., Stephens, J.C., and Burke, L.A. (2008). *J. Org. Chem.* 73: 1354–1364.
30 Bazanov, B., Geiger, U., Carmieli, R. et al. (2016). *Angew. Chem. Int. Ed.* 55: 13233–13235.
31 Bazanov, B., Geiger, U., Grinstein, D. et al. (2017). *J. Phys. Chem. A* 121: 6727–6731.
32 Xu, Y., Wang, Q., Shen, C. et al. (2017). *Nature* 549: 78–81.
33 Zhang, C., Sun, C., Hu, B. et al. (2017). *Science* 355: 374–376.
34 Christe, K.O. (2017). *Science* 355: 351–351.
35 Yu, T., Ma, Y., Lai, W. et al. (2018). *R. Soc. Open Sci.* 5: 172269.
36 Shang, F.J., Liu, R.Z., Liu, J.Y. et al. (2020). *J. Phys. Chem. Lett.* 11: 1030–1037.
37 Ren, G.H., Liu, R.Z., Zhou, P.W. et al. (2019). *Chem. Commun.* 55: 2628–2631.
38 Steele, B.A. and Oleynik, I.I. (2016). *Chem. Phys. Lett.* 643: 21–26.
39 Steele, B.A., Stavrou, E., Crowhurst, J.C. et al. (2017). *Chem. Mater.* 29: 735–741.
40 Williams, A.S., Steele, B.A., and Oleynik, I.I. (2017). *J. Chem. Phys.* 147: 234701.
41 Laniel, D., Weck, G., Gaiffe, G. et al. (2018). *J. Phys. Chem. Lett.* 9: 1600–1604.
42 Steele, B.A. and Oleynik, I.I. (2017). *J. Phys. Chem. A* 121: 1808–1813.
43 Li, J.F., Sun, L., Wang, X.L. et al. (2018). *J. Phys. Chem. C* 122: 22339–22344.
44 Xia, K., Zheng, X.X., Yuan, J.A. et al. (2019). *J. Phys. Chem. C* 123: 10205–10211.
45 Zhou, M., Sui, M.H., Shi, X.H. et al. (2020). *J. Phys. Chem. C* 124: 11825–11830.
46 Liu, Z., Li, D., Tian, F.B. et al. (2020). *Inorg. Chem.* 59: 8002–8012.
47 Xu, Y., Wang, P., Lin, Q., and Lu, M. (2017). *Dalton Trans.* 46: 14088–14093.
48 Xu, Y., Wang, P., Lin, Q. et al. (2018). *Dalton Trans.* 47: 1398–1401.
49 Zhang, W., Wang, K., Li, J. et al. (2018). *Angew. Chem. Int. Ed.* 57: 2592–2595.
50 Xu, Y., Lin, Q., Wang, P., and Lu, M. (2018). *Chem. Asian J.* 13: 1669–1673.
51 Xu, Y., Tian, L., Li, D. et al. (2019). *J. Mater. Chem. A* 7: 12468.
52 Wang, P., Xu, Y., Lin, Q., and Lu, M. (2018). *Sci. China Mater.* 62: 122–129.
53 Burke, L.A., Butler, R.N., and Stephens, J.C. (2001). *J. Chem. Soc., Perkin Trans. 2* 9: 1679–1684.
54 Tsipis, A.C. and Chaviara, A.T. (2004). *Inorg. Chem.* 43: 1273–1286.

55 Perera, S.A., Gregušová, A., and Bartlett, R.J. (2009). *J. Phys. Chem. A* 113: 3197–3201.

56 Bayar, I., Khedhiri, L., Soudani, S. et al. (2018). *J. Mol. Struct.* 1161: 185–193.

57 Crawford, M.J. and Mayer, P. (2005). *Inorg. Chem.* 44: 8481–8485.

58 Gao, H., Ye, C., Piekarski, C.M., and Shreeve, J.M. (2007). *J. Phys. Chem. C* 111: 10718–10731.

59 Qu, X., Zhang, S., Wang, B. et al. (2016). *Dalton Trans.* 45: 6968–6973.

60 Zhang, Y., Zhang, S., Sun, L. et al. (2017). *Chem. Commun.* 53: 3034–3037.

61 Wang, Y., Zhang, J., Su, H. et al. (2014). *J. Phys. Chem. A* 118: 4575–4581.

62 Sućeska, M. (2013). EXPLO5 6.01[Z]. Brodarski Institute, Croatia.

63 Bushuyev, O.S., Peterson, G.R., Brown, P. et al. (2013). *Chem. Eur. J.* 19: 1706–1711.

64 Li, S., Wang, Y., Qi, C. et al. (2013). *Angew. Chem. Int. Ed.* 52: 14031–14035.

65 Zhang, S., Yang, Q., Liu, X. et al. (2016). *Coord. Chem. Rev.* 307: 292–312.

66 Xu, Y., Lin, Q., Wang, P., and Lu, M. (2018). *Chem. Asian J.* 13: 924–928.

67 Xu, Y., Tian, L., Wang, P. et al. (2019). *Cryst. Growth Des.* 19: 1853–1859.

68 Klapötke, T.M. and Sabaté, C.M. (2010). *Z. Anorg. Allg. Chem.* 636: 163–175.

69 Dippold, A. and Klapötke, T.M. (2012). *Chem. Eur. J.* 18: 16742–16753.

70 Bharatam, P.V., Patel, D.S., and Iqbal, P. (2005). *J. Med. Chem.* 48: 7615–7622.

71 Frisch, M.J., Trucks, G.W., Schlegel, H.B. et al. (2009). *Gaussian 09, Revision A.02*. Wallingford, CT: Gaussian, Inc.

72 (a) Tests were conducted according to the UN Recommendations on the Transport of Dangerous Goods(2009). *Manual of Tests and Criteria*, 5e. New York: United Nations Publication. (b) 13.4.2 Test 3 (a) (ii) BAM Fallhammer, pp. 75–82; (c) 13.5.1 Test 3 (b) (i): BAM friction apparatus, pp. 104–107.

73 Jenkins, H.D.B., Roobottom, H.K., Passmore, J., and Glasser, L. (1999). *Inorg. Chem.* 38: 3609–3620.

74 Zhang, Y., Huang, Y., Parrish, D.A., and Shreeve, J.M. (2011). *J. Mater. Chem.* 21: 6891–6897.

75 Chand, D., Parrish, D.A., and Shreeve, J.M. (2013). *J. Mater. Chem. A* 1: 15383–15389.

76 Sućeska, M. (2013). EXPLO5 6.01. Brodarski Institute, Zagreb, Croatia.

2

Aromatic Fused-Ring-Based Energetic Compounds

Kangcai Wang and Qinghua Zhang

Institute of Chemical Material, China Academy of Engineering Physics, Mianyang 621000, PR China

2.1 Introduction

Energetic materials are extremely important and widely used in national defense and aerospace technologies as large amounts of chemical energy stored in these materials are rapidly released by the application of an appropriate external stimulation [1–3]. According to different applications, these materials can be categorized into three groups: explosives, propellants, and pyrotechnics [4–6]. Generally, the use of these materials necessitates typical requirements including a high-energy output, good safety, and a feasible process. Currently, with the occurrence of several major safety accidents caused by the accidental explosion of energetic materials, increasing attention has been focused on their safety [7, 8].

Search for novel energetic materials with "high energy" and "good security" has attracted considerable attention worldwide as the currently used energetic materials, e.g. RDX, HMX, and CL-20, are extremely sensitive to external stimulation and easily cause accidental explosions, leading to serious accidents and consequences [9–13]. The balance of their energy and sensitivity is the key to the design and preparation of high-energy insensitive materials. Thus far, TATB is the only example of an insensitive energetic compound. Gaining in-depth insights into its structure–energy–sensitivity correlations can aid in the development of high-energy insensitive compounds. As a promising high-energy insensitive compound, 2,4,6-triamino-5-nitropyrimidine-1,3-dioxide (ICM-102) exhibits molecular and crystal structures similar to those of TATB [14] (Figure 2.1). The energetic molecules in the crystal of ICM-102 exhibit a "planar" structure, which also can be found in some low-sensitive explosives (TATB, LLM-105, and FOX-7) [15–17], indicating that a "planar" structure can be a crystal structural feature of high-energy insensitive compounds.

In the past decades, numerous research institutions, such as Livermore National Laboratory (United States), University of Idaho (United States), University of Munich (Germany), Beijing University of Technology (China), Nanjing University

Nitrogen-Rich Energetic Materials, First Edition. Edited by Michael Gozin and Leonid L. Fershtat.

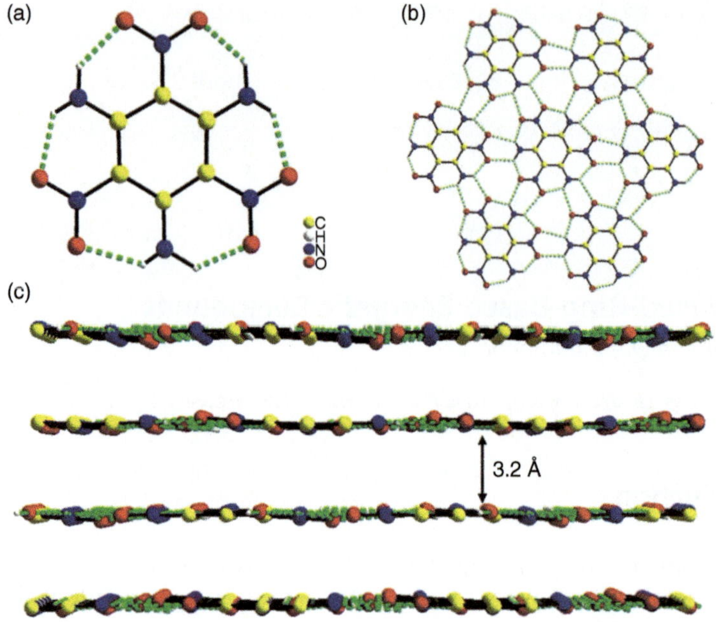

Figure 2.1 (a) Molecular structure of TATB. (b) Layer structure in the TATB crystal. (c) Crystal structure of TATB.

of Technology (China), Xi'an Institute of Modern Chemistry (China), and Institute of Chemical Materials of Chinese Academy of Sciences (China), have devoted increasing attention toward the development of high-energy insensitive materials, and considerable key advances have been made toward the design, synthesis, and performance of energetic materials. The construction of a planar molecular structure, with the regular alternating arrangement of nitro and amino groups around the edge of the molecular skeleton, and the formation of hydrogen bonds among the energetic molecules, as well as the construction of a "face-to-face" supramolecular stacking structure, are confirmed to be effective strategies for developing high-energy insensitive compounds.

Compared to monocyclic backbones, fused-ring aromatic backbones exhibit a planar conjugated structure, which is beneficial for constructing a "planar" molecular structure, attaining high heats of formation, which guarantees high energy, and an increased number of modifiable chemical sites, which are beneficial in adjusting the molecular structure to construct a hydrogen-bonding network and form "face-to-face" crystal structures [18]. With the report of these backbones for energetic materials, they have attracted immense attention, which are expected to demonstrate immense potential for the development of high-energy insensitive materials.

In this chapter, energetic compounds will be categorized into three groups according to the size of fused-ring backbones, the chemistry of these energetic compounds will be partly introduced, and the structure–energy–sensitivity correlations will be discussed.

2.2 Fused-Ring Aromatic Energetic Compounds

2.2.1 [5,5]-Bicyclic-Backbone-Based Energetic Compounds

Among fused-ring backbones, [5,5] bicycles are the smallest skeletons that are typically used to synthesize energetic compounds. Among the reported [5,5]-bicyclic energetic compounds, 3,6-dinitropyrazolo[4,3-*c*]pyrazole (DNPP) is a well-known example with a low sensitivity (impact sensitivity, IS: 15 J, friction sensitivity, FS: 160 N) and excellent thermal stability (T_d: 336 °C) [19]. DNPP also exhibits a promising density and energetic performance besides low sensitivity. The calculated detonation velocity is 8250 $m\,s^{-1}$; this value is 1369 $m\,s^{-1}$ greater than that of the traditional explosive TNT. Two symmetrical chemical sites (N–H) are present in the molecular structure of DNPP, which are sufficiently active to be modified by energetic groups. To further improve the energetic properties of DNPP, in recent years, a series of new energetic compounds have been prepared by energy functionalization of the N–H chemical sites. Scheme 2.1 shows the corresponding synthesis routes of those energetic compounds [19, 20].

Several energetic compounds have been obtained by the introduction of energy functionalities into the DNPP skeleton (Scheme 2.1). With this strategy, the energetic performance of those derivatives is significantly improved. Compounds **2** and **3** exhibit a considerably high energetic performance, and their detonation velocities are greater than those of high-energetic explosives of RDX and HMX and even comparable to that of CL-20 (Table 2.1). However, their mechanical sensitivities (impact sensitivity and friction sensitivity, respectively) sharply decreased, adversely hindering their application potential in weapon systems. As a high-energy insensitive derivative of DNPP, compound **6** has been reported in 2020. Its detonation velocity is 8721 $m\,s^{-1}$, and its IS (15 J) is comparable to that of TNT (IS = 15 J). On the other hand, **6** exhibits excellent thermal stability with a decomposition temperature as 281 °C.

Sensitivity of energetic compounds is affected by molecular structure, crystal structure, and particle size, etc. It is difficult to obtain in-depth insights into the structure–sensitivity relationship. Fortunately, the similar molecular structures and different performances of DNTP and its derivatives provide an opportunity to examine this relationship. Hence, the molecular structure and performance relationship of the energetic derivatives of DNTP are extensively investigated. In-depth examination of the molecular structures of DNTP and its energetic derivatives revealed that polycyclic energetic compounds exhibit low sensitivity if molecules exhibit strong aromaticity. Furthermore, the bond strengths of $C{-}NO_2$, $C{-}NHNO_2$, and $C{-}N_3$ also considerably affect the sensitivity; sensitivity decreases with the decrease in the bond length.

As another fused-ring aromatic backbone with three amino groups, 3,6,7-triamino-7*H*-[1,2,4]triazolo[4,3-*b*][1,2,4]triazole (TATOT) is a well-known energetic compound due to its simple preparation, high nitrogen content, and high energy and safety performance. TATOT can be synthesized in two steps (Figure 2.2a) [24].

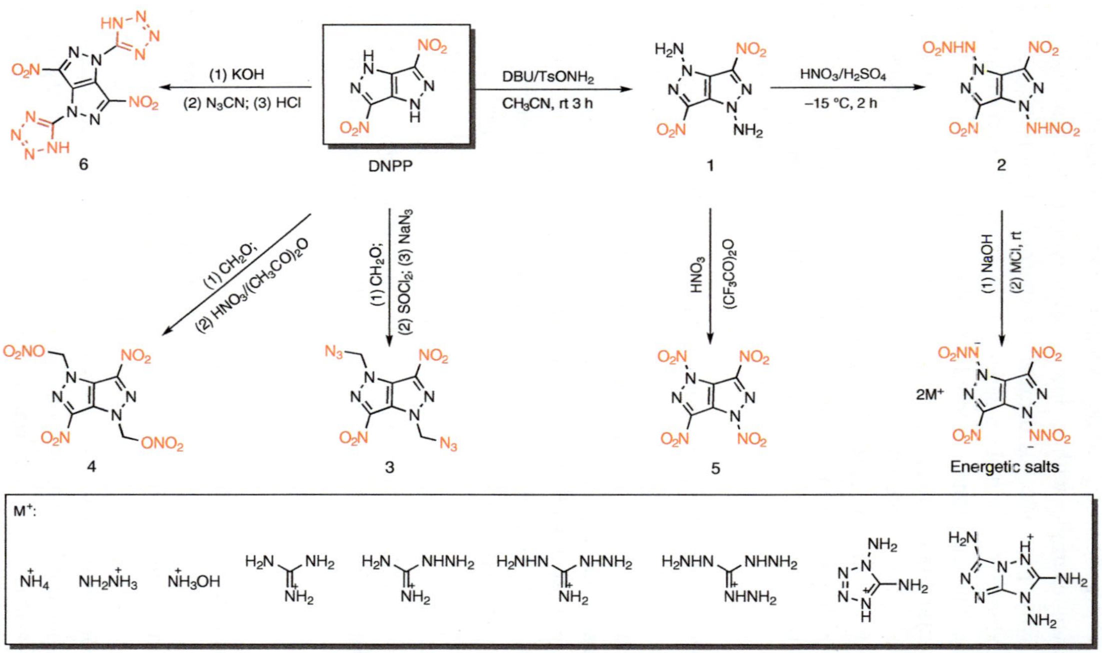

Scheme 2.1 Synthesis routes of energetic derivatives of DNPP.

Table 2.1 Physical and detonation properties of DNPP, its energetic derivatives, and some explosives [19, 20].

Compound	T_{dec} a) (°C)	ρ b) (g cm^{-3})	$\Delta_f H$ c) (kJ mol^{-1})	v_D d) (m s^{-1})	P e) (GPa)	IS f) (J)	FS g) (N)
DNPP	336	1.85	322.6	8250	27.4	15	160
1	230	1.84	467.0	8864	33.9	7	120
2	128	1.93	595.2	9507	41.8	2	20
3	196	1.71	946.3	8292	27.6	11	120
4	206	1.82	133.7	8674	33.1	10	120
5	145	1.95	550.9	9460	40.9	3	20
6	281	1.79	1111.5	8721	30.9	15	192
RDX [21]	204	1.80	70.3	8795	34.9	7	120
HMX [22]	287	1.91	74.8	9059	39.2	7	120
ε-CL-20 [23]	215	2.04	365.4	9455	46.7	4	48

a) Decomposition temperature.
b) Density.
c) Heat of formation.
d) Detonation velocity.
e) Detonation pressure.
f) Impact sensitivity.
g) Friction sensitivity.
Source: Adapted from Refs. [19, 20].

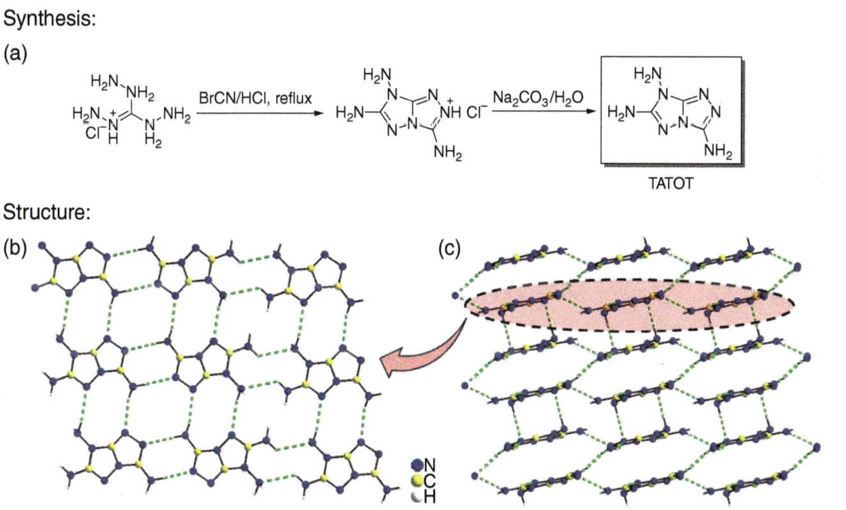

Figure 2.2 (a) Synthesis route of TATOT. (b) Molecular layer structure in the TATOT crystal. (c) 3D supramolecular structure of TATOT. Green dotted lines represent hydrogen bonds.

TATOT exhibits a high nitrogen content of up to 72.70% and a density of $1.73\,g\,cm^{-3}$. The calculated detonation velocity and detonation pressure of TATOT are $8580\,m\,s^{-1}$ and 25.9 GPa, respectively. In addition, TATOT exhibits excellent sensitive performance. Its IS and FS are 40 J and 360 N, respectively. The high energetic performance can be explained by its high nitrogen content and high heat of formation ($\Delta_f H^c = 446.7\,kJ\,mol^{-1}$), while the insensitivity property is possibly related to its special molecular and crystal structures. Each TATOT molecule interacts with four nearby molecules through double hydrogen bonds to form a planar layer structure (Figure 2.2b). If each TATOT molecule is considered as a four connect node, the layer structure exhibits a sql topology. The layer structures further stack to form a 3D supramolecular structure of TATOT, affording an intensive hydrogen-bonding net among the molecules in the crystal (Figure 2.2c). Such a planar layer structure as well as the presence of intensive hydrogen bonds significantly affect its sensitivity.

To improve the energy of TATOT, one of the effective strategies involves the use of protonated TATOT as counter ions for incorporation with high energetic anions to form energetic salts. By this strategy, Prof. T.M. Klapötke and Prof. J.M. Shreeve have synthesized numerous high-performance energetic salts [24, 25]. Among those energetic salts, several exhibit high energy and high safety. The high performance is related to the increase in the oxygen content as well as the presence of intensive hydrogen-bonding interactions among the cations and anions in the crystals. For example, compared with that of TATOT, the oxygen balance in the energetic salt of $[TATOT^+][NO_3^-]$ increased from −93.43% to −47.89% with the incorporation of $[TATOT^+]$ and $[NO_3^-]$. On the other hand, the unique crystal structure of $[TATOT^+][NO_3^-]$ is beneficial for the improvement of its energy and safety. Each protonated TATOT molecule connects with two nearly protonated TATOT molecules and four nitrates, and each nitrate interacts with four adjacent protonated TATOT molecules through hydrogen bonds (Figure 2.3a). A planar layer structure is formed through these interactions. In this layer structure, all non-hydrogen atoms occupy a unique plane (Figure 2.3b). The parallel layers stack in the ABAB mode to form a 3D

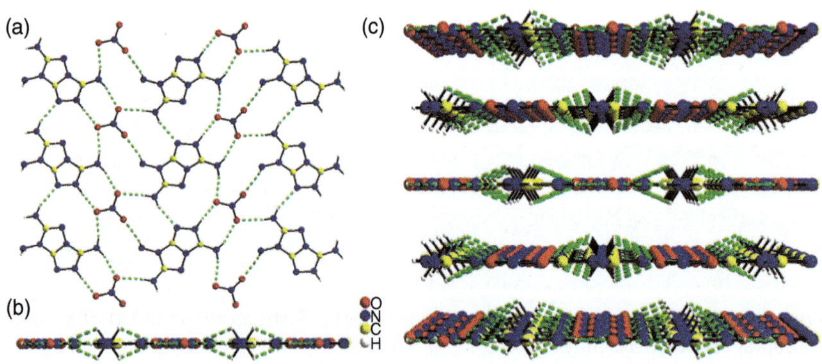

Figure 2.3 (a) Layer structure of $[TATOT^+][NO_3^-]$. (b) Side view of the layer structure. (c) 3D supramolecular structure of $[TATOT^+][NO_3^-]$. Green dotted lines represent hydrogen bonds.

supramolecular structure of [TATOT$^+$][NO_3^-]. The distance between two adjacent layers is ~3.1 Å, which is slightly shorter than that of the well-known safety explosive TATB (3.2 Å). The planar layer structures significantly affect the stability of [TATOT$^+$][NO_3^-], especially to resist external mechanical stimulations.

2.2.2 [5,6]-Bicyclic Energetic Compounds

Compared to [5,5]-bicyclic structures, [5,6]-bicyclic backbones exhibit a six-membered ring, which can provide an increased number of modifiable chemical sites. Therefore, based on [5,6]-bicyclic backbones, an increased number of aesthetic structures can be designed, and their performance should be relatively easy to adjust. Currently, more than 20 covalent [5,6]-bicyclic energetic compounds have been designed and prepared. Among those novel energetic compounds, different structures with interesting performance were obtained.

As a high-energy insensitive compound, 4-amino-3,7,8-trinitropyrazolo-[5,1-*c*][1,2,4]triazine (PTX) is constructed on the basis of the [5,6]-bicyclic skeleton. PTX was first synthesized by Prof. I.L. Dalinger et al. in 2010; its performance and application potential were later evaluated by Prof. D.E. Chavez and coworkers [26, 27]. PTX can be prepared by the route shown in Scheme 2.2, with a total yield of ~30%. The density of PTX is up to 1.946 g cm^{-3}, which is greater than those of most of the high energetic compounds, e.g. HMX and RDX, etc. and it is the highest value known for fused-ring-based aromatic energetic compounds. The high density of PTX could be related to the planar molecular configuration, high nitrogen content, presence of extensive intra- and intermolecular hydrogen bonds, and strong π–π interactions between two adjacent parallel PTX molecules in its crystal.

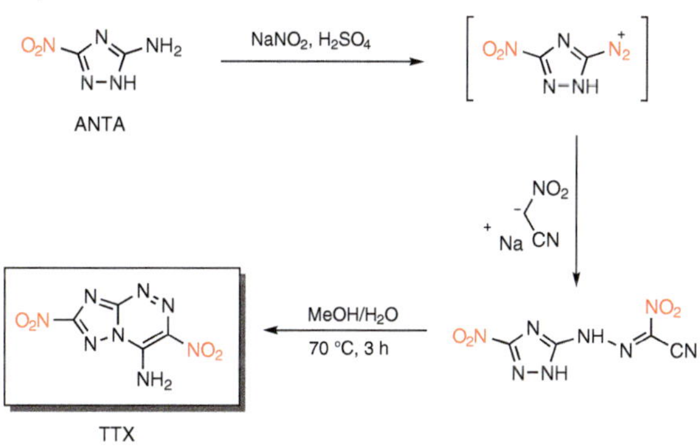

Scheme 2.2 Synthesis route of TTX.

PTX exhibits a planar skeleton structure, and all atoms, except two oxygen atoms, originating from the same nitro groups, are presumably held in the plane of the fused-ring skeleton (Figure 2.4a,b). The planar molecular configuration considerably contributes to its high density. In the crystal, each PTX molecule interacts with three adjacent molecules through hydrogen bonds (Figure 2.4c). Furthermore, besides hydrogen-bonding interactions, strong π–π interactions also exist among

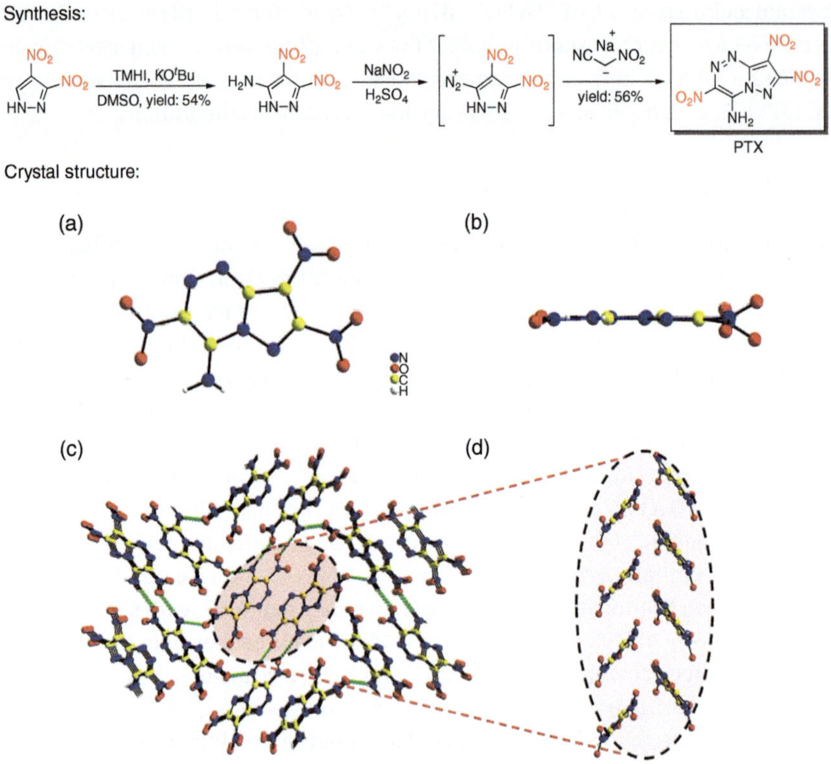

Figure 2.4 Synthesis route and crystal structure of PTX: (a) molecular structure; (b) side view of the molecule; (c) 3D supramolecular structure; (d) arrangement of nearby molecules. Green dotted lines represent hydrogen bonds.

the nearby molecules. A "T"-type structure is constructed from two adjacent molecules through π–π interactions (Figure 2.4d). The presence of hydrogen bonds and π–π interactions promote the construction of the PTX crystal structure. These strong interactions render an ultrahigh density of 2.006 $g\,cm^{-1}$ at 100 K for PTX. Notably, the detonation velocity and pressure of PTX are 8898 $m\,s^{-1}$ and 36.04 GPa, respectively, which are closely related to its high density. Furthermore, PTX exhibits excellent mechanical sensitivity (IS: 58.4 cm, 50% load Bruceton up/down method; FS: 324 N); these values are considerably less than those of HMX (IS: 24.9 cm, 50% load Bruceton up/down method; FS: 150 N). The high performance of PTX can satisfy a majority of requirements for high-energy insensitive materials, but its complex synthesis route adversely limits its practical application potential.

In 2016, 4-amino-3,7-dinitro-[1,2,4]triazolo[5,1-*c*][1,2,4]triazine (TTX) was prepared by Prof. J.M. Shreeve and coworkers [28]. TTX can be synthesized in two steps by using 5-amino-3-nitro-1*H*-1,2,4-triazole (ANTA) as the starting material (Scheme 2.2), with a reported total yield of ~65 wt% (calculated on the basis of ANTA). In fact, the synthesis of TTX is difficult as described in the literature due to the complex synthesis of ANTA and sodium cyano(nitro)methanide. On the other hand, TTX exhibits high thermal stability, and its decomposition temperature is up to 272 °C, which is greater than that of RDX (205 °C) and comparable to that of HMX (280 °C). The detonation velocity and pressure of TTX are 8580 m s^{-1} and 31.2 GPa, respectively, which are similar to those of RDX (8795 m s^{-1} and P = 34.9 GPa). Notably, TTX is insensitive to mechanical stimulation of impact and friction (IS: >60 J, FS: >360 N). To further enhance the energetic performance of TTX, N-oxidation of TTX with hypofluorous acid in acetonitrile is conducted. By this reaction, a higher energetic compound, i.e. 4-amine-3,7-dinitro-[1,2,4]triazolo[5,1-*c*][1,2,4]triazine-1-oxide, is prepared [29]. The detonation velocity and pressure of this new compound are 8970 m s^{-1} and 35.4 GPa, respectively. However, after the N-oxidation of TTX, mechanical sensitivities (IS and FS) are significantly decreased, as well as its thermal stability. The decomposition temperature of this N-oxidation product is only 138 °C, significantly limiting its application potential.

As a notable [5,6]-bicyclic skeleton for high-energetic insensitive compounds, 3,6-diamino-1,2,4-triazolo[4,3-*b*][1,2,4,5]tetrazine (TTA) can be prepared by the route shown in Scheme 2.3 [30]. This compound exhibits high thermal stability, and its decomposition temperature is 345 °C. The nitrogen content of this compound is up to 73.66 wt%. Based on this backbone, numerous high-energetic insensitive compounds have been synthesized by Prof. J.M. Shreeve and coworkers. Several energetic derivatives of 3,6-diamino-1,2,4-triazolo[4,3-*b*][1,2,4,5]tetrazine have been synthesized by N-oxidation, N-nitrification, and N-diazotization (Scheme 2.3). The detonation velocities of these derivatives are greater than 8800 m s^{-1}, and the IS and FS are greater than 25 J and 240 N, respectively. Table 2.2 lists the detailed performance data of these compounds [31, 32].

All derivatives exhibit a high energetic performance (Table 2.2). The detonation velocities of all derivatives are greater than 8540 m s^{-1}, even reaching the suprahigh velocity of 9712 m s^{-1}; this value is greater than that of HMX (9144 m s^{-1}) and even greater than that of CL-20 (9706 m s^{-1}). On the other hand, most of these derivatives exhibit low sensitivity. The IS and FS of all compounds, except **8**, **9b**, and **12**, are greater than 25 J and 240 N, respectively. The high energy and low-sensitivity performance of these derivatives could be reasonably explained by their molecular and crystal structures.

A series of intermolecular hydrogen bonds of N—H···O and N—O···H are formed in the crystal structure of compound **7** among the neighboring molecules along different directions (Figure 2.5a). Notably, some nearby molecules do not exhibit hydrogen-bonding interactions (Figure 2.5b). Among those molecules with no hydrogen-bonding interactions, intensive π–π interactions are present. Hirshfeld surface analysis is beneficial to examine supramolecular interactions, which is calculated by using the Crystal Explorer program [33]. In the Hirshfeld surface image, red and blue dots represent the high and low close contact populations, respectively. The Hirshfeld surface of compound **7** is almost shaped as plates, which is in good agreement with its fused-ring backbone configuration (Figure 2.5c–e). Most of the red dots in the Hirshfeld surface of this compound are located on side faces of the plate, while the blue parts are located on the front face. The red dots correspond to intermolecular hydrogen-bonding interactions, and the blue parts

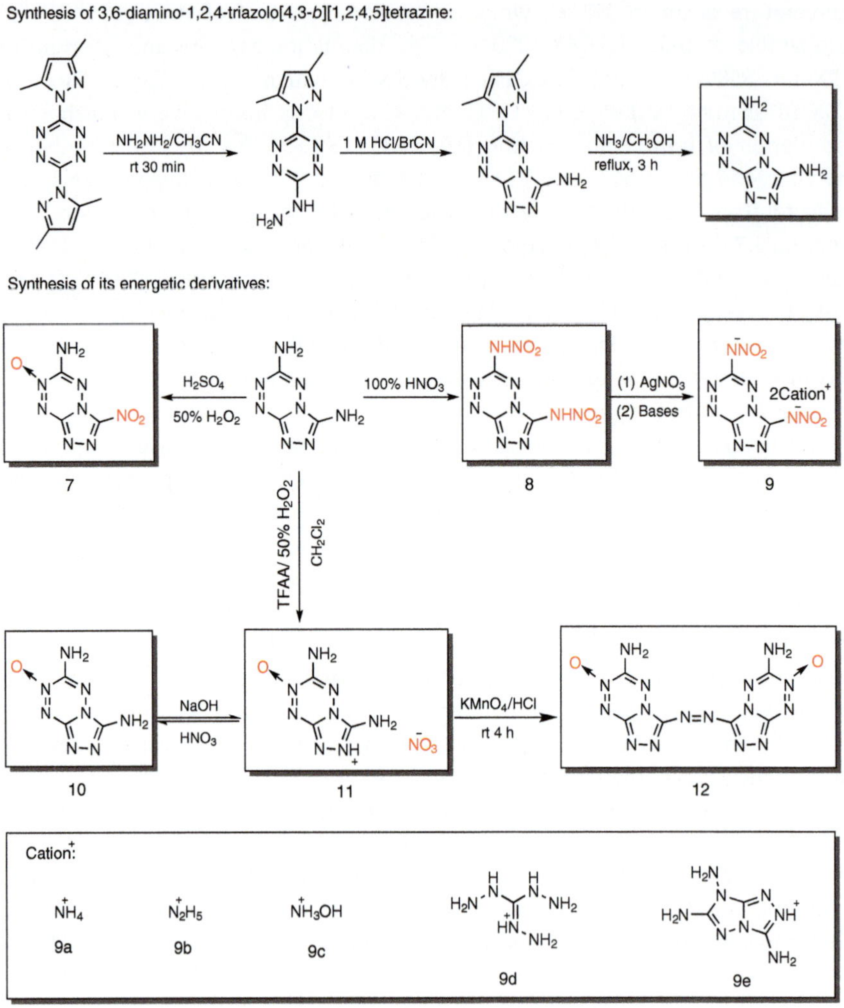

Scheme 2.3 Synthesis route of 3,6-diamino-1,2,4-triazolo[4,3-*b*][1,2,4,5]tetrazine and its energetic derivatives.

correspond to O–C, C–C, and C–N interactions, indicative of the presence of π–π stacking. The high energy and insensitivity of compound **7** are closely related to its planar molecular configuration, intense hydrogen bonding, and π–π interactions in the crystal structure.

A planar layer structure for compound **9c** is constructed from the cations and anions of compound **9c** and water molecules (Figure 2.6a). Each energetic anion interacts with five hydroxylammonium cations and two water molecules, while each water molecule interacts with two energetic anions through hydrogen-bonding interactions. Two types of hydroxylammonium cations are present in the layer structure, which interact with two and three energetic anions, respectively. The presence of intensive hydrogen bonds in the layer structure is one of the main

Table 2.2 Physical properties of the energetic derivatives of 3,6-diamino-1,2,4-triazolo [4,3-*b*][1,2,4,5]tetrazine.

Compound	T_{dec} [a)] (°C)	ρ [b)] (g cm^{-3})	$\Delta_f H$ [c)] (kJ mol^{-1})	v_D [d)] (m s^{-1})	P [e)] (GPa)	IS [f)] (J)	FS [g)] (N)
7	220	1.86	744	9384	39.1	25	240
8	138	1.91	740.9	9301	38.3	3	>5
9a	213	1.84	543.5	9301	35.4	32	>360
9b	154	1.82	871.8	9625	37.9	13	>240
9c	154	1.92	638.9	9712	42.9	25	>360
9d	150	1.70	1301.6	9061	30.8	30	>360
9e	244	1.76	1780.9	8792	29.8	>40	>360
10	196	1.82	492	9008	34.7	>40	>360
11	191	1.81	502.3	8808	30.1	>40	>360
12	210	1.82	1062	8542	28.8	22	40

a) Decomposition temperature.
b) Density.
c) Heat of formation.
d) Detonation velocity.
e) Detonation pressure.
f) Impact sensitivity.
g) Friction sensitivity.

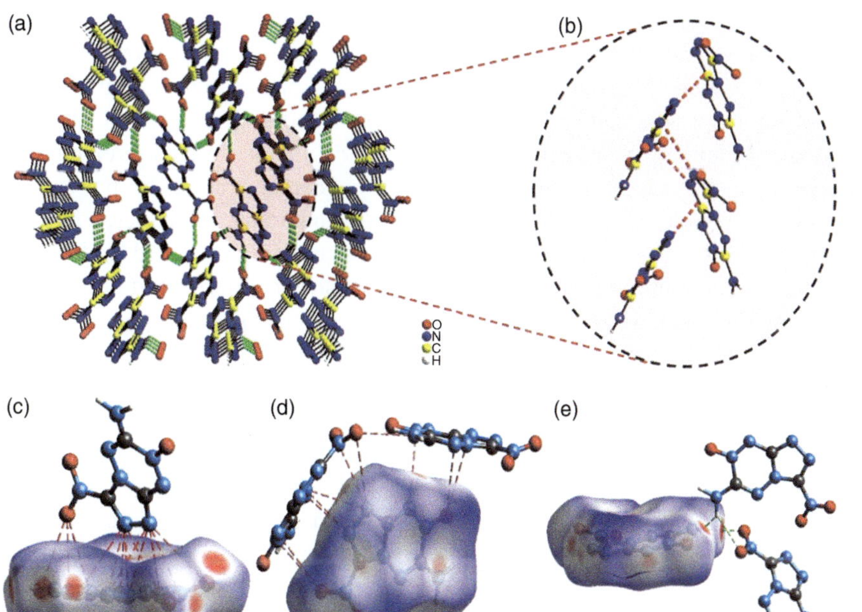

Figure 2.5 (a) 3D supramolecular structure of compound **7**. (b) Presence of $\pi-\pi$ interactions among the nearby molecules in the crystal of compound **7**. (c)–(e) Hirshfeld surface of compound **7** and interactions among molecules. Green and red dotted lines represent hydrogen bonds and $\pi-\pi$ interactions, respectively.

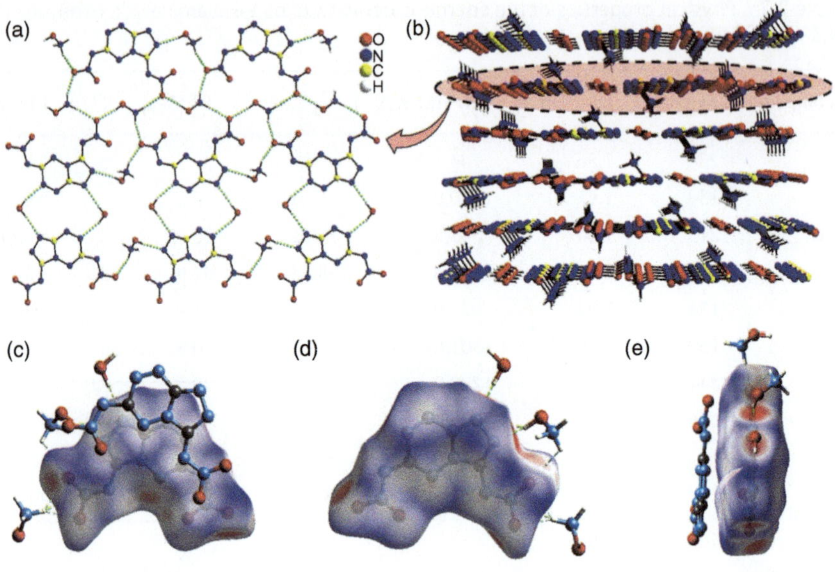

Figure 2.6 (a) Layer structure in **9c**. (b) 3D supramolecular structure of **9c**. (c)–(e) Hirshfeld surface of **9c** and interactions among the molecules. Green dotted lines represent hydrogen-bonding interactions.

reasons for the high stability and low insensitivity of compound **9c**. Moreover, the planar layer structures are further stacked to form a 3D supramolecular structure of **9c** in the "face-to-face" stacking mode (Figure 2.6b). The "face-to-face" stacking mode is beneficial for improving the resistance against external stimulations. The Hirshfeld surface of compound **9c** does not exhibit a planar shape (Figure 2.6c–e). A red dot is not observed on the planar face, demonstrating the absence of strong interactions among the adjacent parallel energetic anions. Intensive red dots are observed on the edges of the Hirshfeld surface, demonstrating the presence of strong interactions among the cations, anions, and water molecules. The high energy and insensitivity of compound **9c** can be reasonably explained by its planar molecular configuration, the presence of an intensive hydrogen-bond net, and the "face-to-face" stacking mode in the crystal structure of compound **9c**.

6-Nitro-7-azido-pyrazol[3,4-*d*][1,2,3]triazine-2-oxide (ICM-103) is designed and synthesized by a facile two-step reaction using commercially available reagents (Scheme 2.4), which exhibits immense application potential as a green primary explosive with IS and FS of 4 J and 60 N, respectively [34]. Table 2.2 lists its physical

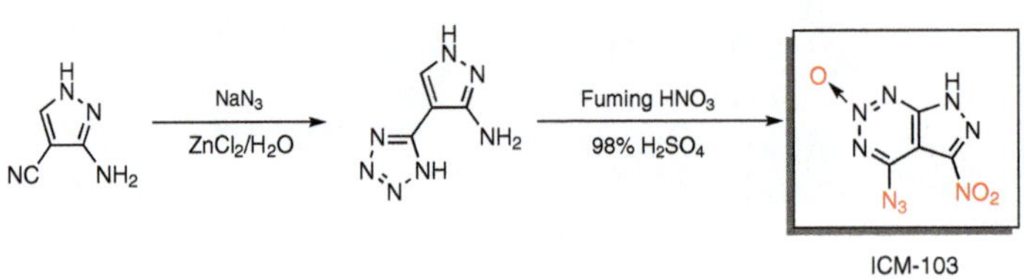

Scheme 2.4 Synthesis route of 6-nitro-7-azido-pyrazol[3,4-*d*][1,2,3]triazine-2-oxide.

and detonation properties, as well as the currently used primary explosives of LA and DDNPA.

Compared to the currently used organic primary explosive of DDNP, ICM-103 exhibits a higher density (Table 2.3). The higher density corresponds to a higher-energy output. The detonation velocity and pressure of ICM-103 are 9111 m s^{-1} and 35.1 GPa, respectively, which are considerably greater than those of LA (5920 m s^{-1} and 33.8 GPa) and DDNP (6900 m s^{-1} and 24.2 GPa). Moreover, ICM-103 exhibits environmental resistance. The currently used organic primary explosive of DDNP is well known to exhibit photosensitivity. It rapidly decomposes under solar

Table 2.3 Properties of ICM-103, LA, and DDNPAs.

Items	LA[a]	DDNP[b]	ICM-103[c]
Formula	PbN_6	$C_6H_2N_4O_5$	$C_4HN_9O_3$
M[d] (g mol^{-1})	291.3	210.1	223.1
N[e] (%)	28.9	26.7	56.5
Ω_{CO}[f] (%)	−11.0	−15.23	−10.75
ρ[g] (g cm^{-3})	4.80	1.72	1.86
$\Delta_f H_m$[h] (kJ mol^{-1})	450	321	744
D[i] (m s^{-1})	5920	6900	9111
p[j] (GPa)	33.8	24.2	35.1
IS[k] (J)	2.5–4	1	4
FS[l] (N)	0.1–1	24.7	60
EDS[m] (mJ)	<5	1.8	130
FlameS[n] (cm)	<8	17	>60
T_{dec}[o] (°C)	315.0	157.0	160.3
PD[p] (MPa)	<118	<60	>100
MPC/AC[q] (mg)	10/20–30	70/230–280	60/90–100

a) LA: lead azide.
b) DDNP: 2-diazo-4,6-dinitrophenol.
c) ICM-103: 6-nitro-7-azido-pyrazol[3,4-*d*][1,2,3]triazine-2-oxide.
d) Formula weight.
e) Nitrogen content.
f) Oxygen balance based on CO.
g) Crystal density.
h) Heat of formation.
i) Calculated detonation velocities.
j) Calculated detonation pressure.
k) Impact sensitivity.
l) Friction sensitivity.
m) Electrostatic discharge sensitivity.
n) Flame sensitivity.
o) Decomposition temperature.
p) Pressure durability, below which primary explosives are dead-pressed.
q) Minimum primary charge (MPC) and actual charge (AC).

irradiation, thereby significantly increasing its storage and transport costs. Notably, ICM-103 is extremely stable to sunlight. After exposing it under sunlight for several days, the H^1 NMR spectrum of ICM-103 did not change, demonstrating its strong environmental resistance. Furthermore, after keeping it under humid conditions for 24 hours, the hygroscopicity of ICM-103 is only 0.18 wt%. Under the same conditions, the hygroscopicity of DNNP is 0.36 wt%, demonstrating that the hygroscopicity of ICM-103 is better than that of DNNP. Safety is one of the most crucial parameters of primary explosives. ICM-103 exhibits considerably suitable sensitivity toward mechanical and electrostatic stimuli (Table 2.3). The IS, FS, and electrostatic discharge sensitivity of ICM-103 are 4 J, 60 N, and 130 mJ, respectively, which are less than those of LA and DDNP; this result reflects its higher safety. The excellent safety properties of ICM-103 can be reasonably explained by its molecular and crystal structures.

ICM-103 exhibits a fused [5,6]-heterocyclic skeleton, the energetic groups of which hang on the edges of the skeleton (Figure 2.7a,b). All atoms of ICM-103 are almost arranged in a plane to form a nearly flat molecular geometry. Two nearby molecules are connected with each other through double hydrogen bonds to form a dimer structure (Figure 2.7d), followed by the stacking of these dimers to construct a 3D supramolecular structure of ICM-103 (Figure 2.7c). Hydrogen-bonding interaction is not observed among the adjacent ICM-103 dimers, with the distance between azide and nitro energetic groups being ~3.2 Å. These structure characteristics should partly account for the high sensitivities of ICM-103.

The nitro and azide groups in the molecular structure of ICM-103 exhibit a partial negative charge and positive charge, respectively, while the N-oxygen and

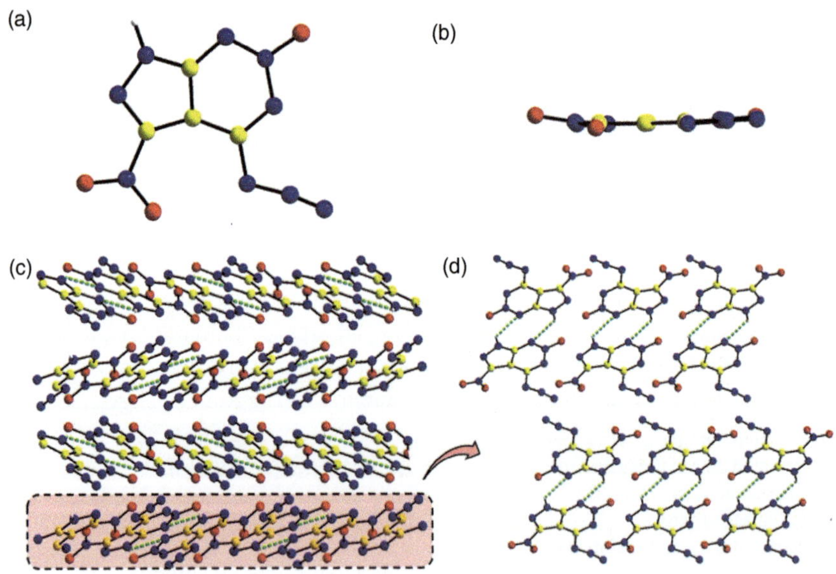

Figure 2.7 (a) Molecular structure of ICM-103. (b) Side view of the molecular structure. (c) 3D supramolecular structure of ICM-103. (d) Molecular stacking mode of ICM-103 in the ICM-103 crystal structure. Green dotted lines represent hydrogen bonds.

imino groups are potential hydrogen-bond acceptors and donors, respectively. If the positions of nitro and imino groups are exchanged with each other, a planar fused-ring skeleton with a "cross-shaped" configuration will be constructed. This designed molecular configuration is similar to that of TATB. On the other hand, together with the "cross-shaped" configuration and intermolecular interactions between N-oxygen and imino groups from the nearby molecules, a layer stacking crystal structure should be constructed.

Based on the above analysis, an isomer of ICM-103, 4-nitro-7-azido-pyrazol-[3,4-*d*]-1,2,3-triazine-2-oxide (NAPTO), is designed and prepared [35]. Scheme 2.5 shows the synthesis route of NAPTO. As expected, NAPTO exhibits a high detonation performance and excellent sensitivity. The detonation velocity and pressure data values of this novel explosive are 9120 m s^{-1} and 35.1 GPa, respectively. The corresponding IS and FS values are 18 J and 325 N. These performance parameters are better than those of most of the high-energy explosives, e.g. RDX and HMX. NAPTO also exhibits a high density (1.852 g cm^{-3}) and thermal stability (T_d: 203 °C) besides its high energy and safety performance. The excellent property of this novel energetic compound is closely related to its special molecular and crystal structures.

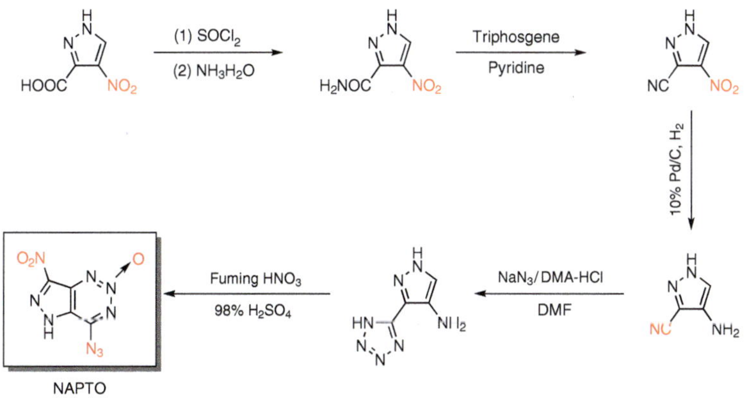

Scheme 2.5 Synthesis route of 4-nitro-7-azido-pyrazol-[3,4-*d*]-1,2,3-triazine-2-oxide.

The NAPTO molecule exhibits a "cross-shaped" configuration as nitro and azide groups as well as N-oxygen and imino groups are hanging on the side of the fused-ring skeleton in opposite positions (Figure 2.8a,b). This "cross-shaped" configuration is beneficial to improve the stability of the molecule. All atoms of NATOP are present in a planar layer. The planar molecules interact with each other to form a 1D infinite linear structure through hydrogen bonds (Figure 2.8d). Then, the linear structures are stacked to construct a 2D supramolecular planar layer structure. The 2D planar structures are further stacked to form a 3D supramolecular structure of NAPTO (Figure 2.8c). This planar stacking mode can absorb a certain amount of mechanical energy and hence slightly prevent decomposition. In other words, this stacking mode is beneficial to enhance the safety of energetic materials, and it can be observed in the classical safety explosive of TATB.

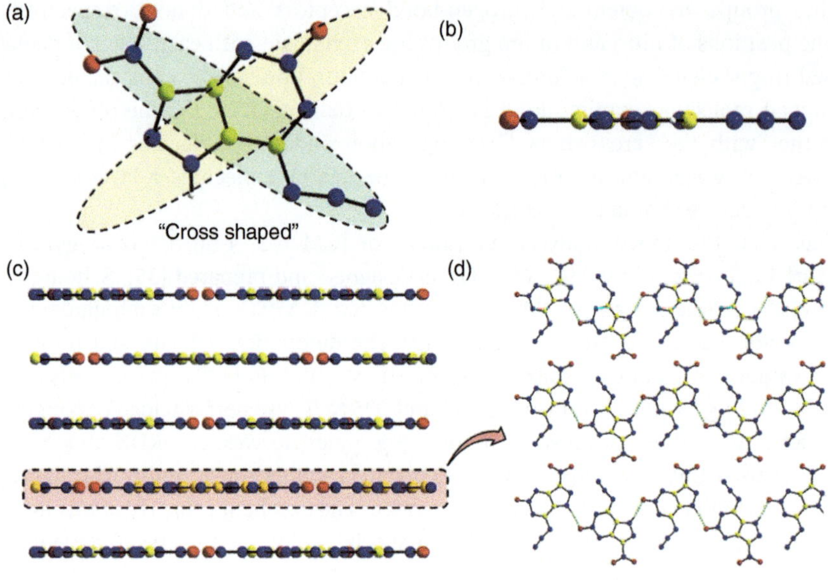

Figure 2.8 (a) "Cross-shaped" molecular structure of NAPTO. (b) Side view of the molecular structure. (c) 3D supramolecular structure of NAPTO. (d) Molecular stacking mode of NAPTO molecules in the crystal structure. Green dotted lines represent hydrogen bonds.

As another high-energetic insensitive compound, 6,8-diamino-7-nitrotetrazolo [1,5-*b*]pyridazine (TPAN) is constructed on the basis of the [5,6]-bicyclic skeleton [36]. TPAN prepared by Prof. Qinghua Zhang and coworkers in three steps was obtained with a high total yield of ~75% (Scheme 2.6).

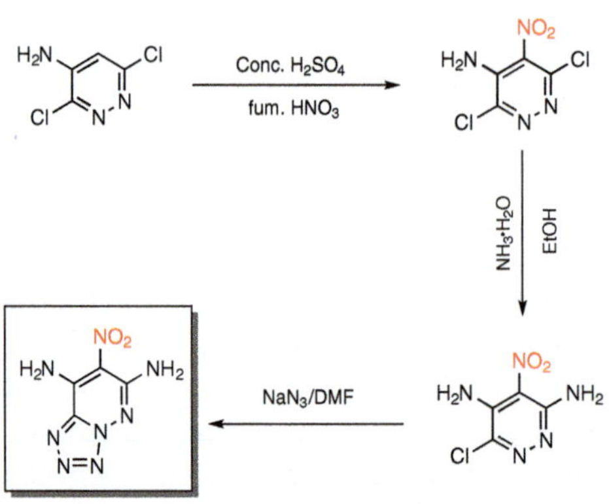

Scheme 2.6 Synthesis route of 6,8-diamino-7-nitrotetrazolo[1,5-*b*]pyridazine (TPAN).

TPAN exhibits a high density of 1.87 g cm^{-3}, which is greater than that of RDX (1.81 g cm^{-3}) and comparable to that of HMX (1.91 g cm^{-3}). The calculated detonation velocity and detonation pressure of TPAN are 8899 m s^{-1} and 30.3 GPa, respectively, which are greater than those of RDX. Furthermore, TPAN exhibits

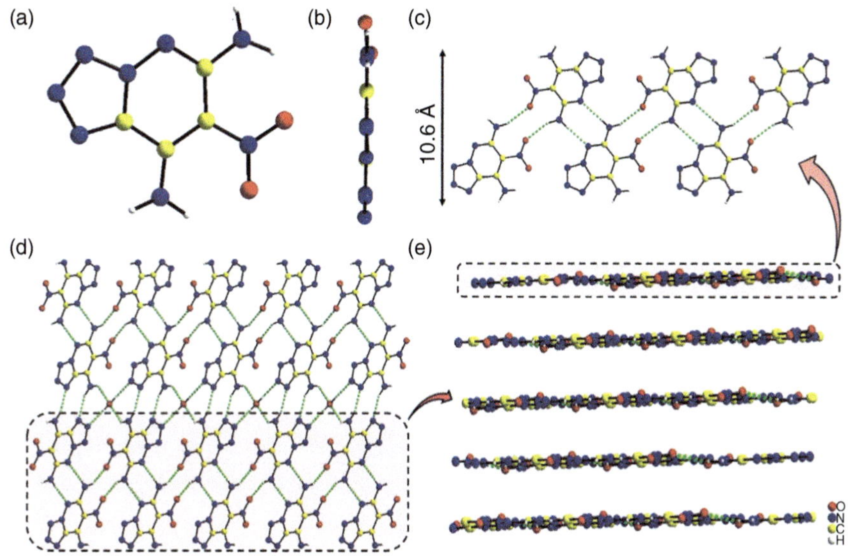

Figure 2.9 (a) Molecular structure of TPAN. (b) Side view of the molecular structure. (c) Infinite linear structure of TPAN. (d) 3D supramolecular structure of TPAN. (e) "Short-range order" face-to-face stacking mode of TPAN contained in the crystal. Green dotted lines represent hydrogen bonds.

excellent sensitive performance. The IS and FS are >40 J and >360 N, respectively. The high energetic performance should be related to its high heat of formation ($\Delta_f H^c = 398\ \text{kJ mol}^{-1}$) and high density, while the insensitivity capacity should be closely related to its special molecular and crystal structures.

TPAN exhibits a planar configuration with amino and nitro groups alternately hanging on the edge of the six-membered-ring side (Figure 2.9a,b). Notably, in the crystal structure, each energetic molecule interacts with two nearby molecules to form a 1D infinite linear structure through double hydrogen bonding (Figure 2.9c). Viewed along the [001] direction, the band-shaped structures further stack to construct a thick layer structure with a height of ~10.6 Å in the "face-to-face" stacking mode (Figure 2.9e). Then, the thick layer structures stack to build a 3D supramolecular structure of TPAN in the ABAB mode, and crystal water molecules are embedded into the space of two nearby thick layer structures (Figure 2.9d). The planar molecular structure, alternate appearance of amino and nitro groups along the edge of the fused N-heterocyclic skeleton, and "face-to-face" stacked thick layer structure are crucial factors for the low sensitivity of TPAN.

2.2.3 Other Multicyclic Energetic Compounds

Besides the [5,5] and [5,6] bicyclic backbones, multicyclic skeletons, such as [6,6], [5,6,5], and [5,7,5] skeletons, and some others are used to design and prepare energetic compounds. Among these polycyclic backbones, the most well-known energetic compound is [1,2,3,4]tetrazino[5,6-*e*][1,2,3,4]tetrazine-1,3,6,8-tetraoxide (TTTO). Scheme 2.8 shows the synthesis route of TTTO.

TTTO exhibits a [6,6]-bicyclic backbone, which belongs to the so-called family of "compounds with alternating charges." TTTO was designed in the last century, and it attracted immense attention in the scientific community due to its high energy and esthetic butterfly-like structure [37, 38]. The calculated heat of formation, density, detonation velocity, and detonation pressure of TTTO are 826 kJ mol^{-1}, 1.98 g cm^{-3}, 9710 m s^{-1}, and 43.2 GPa, respectively. In 2016, TTTO was first prepared through a series of 10-step reactions by Prof. Churakov and coworkers (Scheme 2.7). In their study, TTTO was obtained as a yellow powder with a decomposition temperature of ~183 °C. The preparation of TTTO is one of the important advances in energetic materials. However, the complex synthesis method and extremely low overall yield adversely limit its practical use.

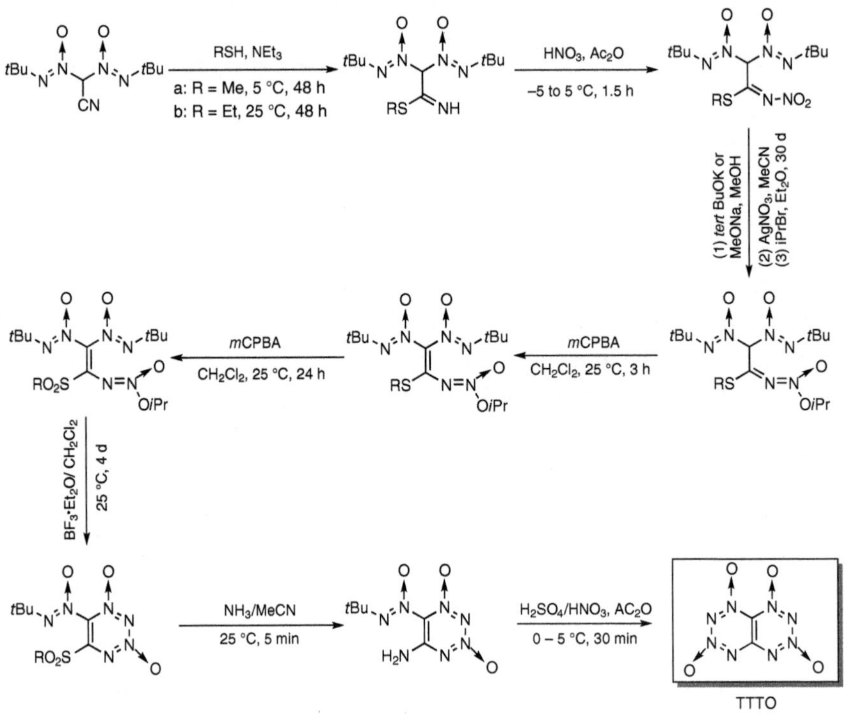

Scheme 2.7 Synthesis route of [1,2,3,4]tetrazino[5,6-*e*][1,2,3,4]tetrazine-1,3,6,8-tetraoxide (TTTO).

2,3,5,6-Tetranitro-4*H*,9*H*-dipyrazolo[1,5-*a*:5′,1′-*d*][1,3,5]triazine (PTAN) exhibits a [5,6,5]-tricyclic backbone, which can be prepared by four-step reactions shown in Scheme 2.8 by using LLM-116 as the starting materials [12]. PTAN exhibits a high density of 1.90 g cm^{-3}, which is greater than those of conventional high energetic compounds, e.g. RDX and HMX. Owing to its high density, the detonation velocity and pressure of PTAN are 8792 m s^{-1} and 34.4 GPa, respectively. On the other hand, PTAN exhibits a low sensitive performance. The IS and FS are 15 J and 240 N, respectively. Owing to its excellent density, high energy, and

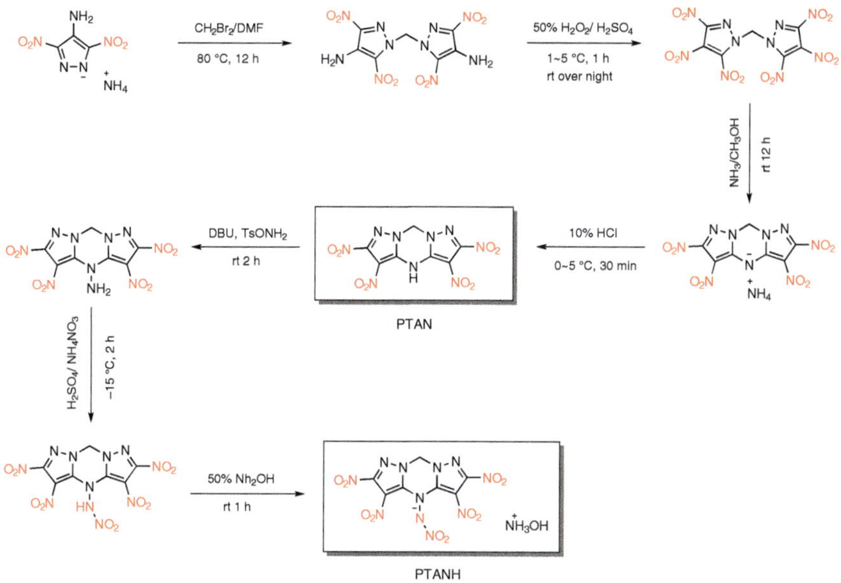

Scheme 2.8 Synthesis route of PTAN and PTANH.

low sensitivity, PTAN demonstrates immense potential as high-energy-density materials.

The fused tricyclic backbone of PTAN is constructed by the incorporation of the 1,3,5-triazinane ring with a polynitro pyrazole structure (Figure 2.10a). Except two oxygen atoms which originate from one nitro group, all nonhydrogen atoms of PTAN are present in the same plane. The nearly planar molecular structure of PTAN is beneficial for improving its thermal stability and sensitivity. The nearby PTAN molecules interact with each other through double hydrogen bonds to form a "zigzag" chain, followed by stacking to construct a layer structure (Figure 2.10d). The layer structures are further stacked to form a 3D supramolecular structure of PTAN. This stacking mode observed in the PTAN crystal is key to its high-energy-density performance.

DFP is a well-known [5,6,5] tricyclic energetic backbone with a high planar structure. DFP can be synthesized by four-step reactions by using 3,4-diaminofurazan and oxalic acid as the starting materials (Scheme 2.9). Based on DFP, several high-energetic insensitive materials have been prepared [39, 40]. As one of the high-energetic derivatives of DFP, 4,8-diaminodifurazano[3,4-*b*,*e*]pyrazine (DADFP) is prepared by the N-amination of DFP. The detonation velocity and pressure of DADFP are 9524 $m\,s^{-1}$ and 43.26 GPa, respectively. 4,8-Di(1*H*-tetrazol-5-yl)-difurazano[3,4-*b*,*e*]pyrazine is another high-energetic derivative of DFP, which is prepared according to the route shown in Scheme 2.9, with an acceptable overall yield of ~55% based on DFP. Its crystal density is 1.84 $g\,cm^{-3}$. The detonation velocity and pressure of this derivative are 8884 $m\,s^{-1}$ and 31.4 GPa, respectively. On the other hand, this derivative exhibits immense safety as the IS and FS are 20 J and 360 N, respectively. Several energetic salts have been designed and prepared on

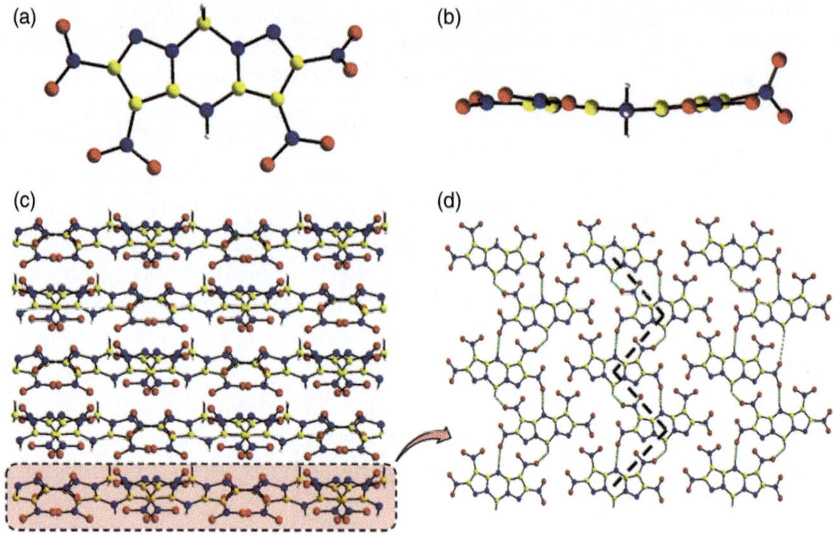

Figure 2.10 (a) Molecular structure of PTAN. (b) Side view of the molecular structure. (c) 3D supramolecular structure of PTAN. (d) Layer structure in the PTAN crystal. Green dotted lines represent hydrogen bonds.

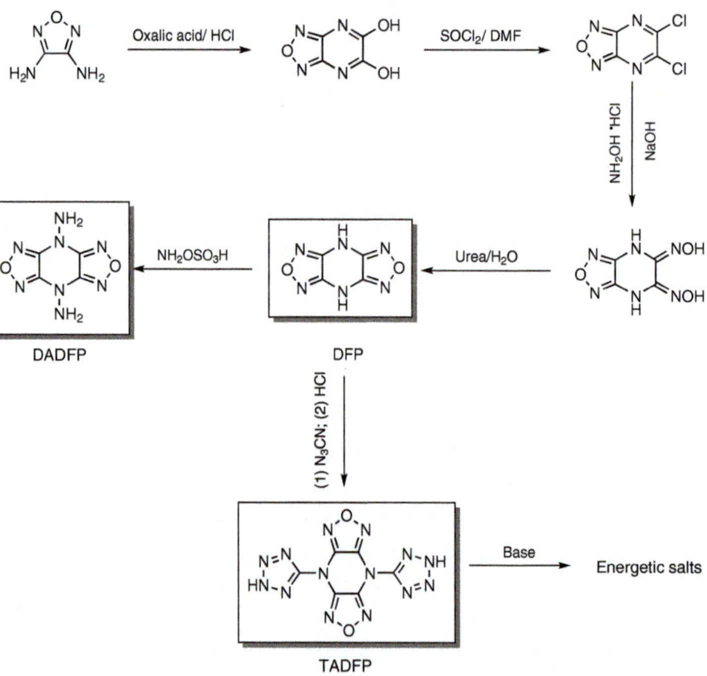

Scheme 2.9 Synthesis route of DFP and its energetic derivatives.

the basis of 4,8-di(1*H*-tetrazol-5-yl)-difurazano[3,4-*b*,*e*]pyrazine; most of those salts exhibit high energetic performance and high safety. Notably, most of the known DFP energetic derivatives exhibit excellent performance, which is closely related to their planar molecular structures.

As a well-known hydrogen-free explosive, benzotrifuroxan (BTF) exhibits a poly-ring heterocyclic skeleton comprising one benzene and three furoxan rings. BTF can be synthesized by a two-step reaction using 1,3,5-trichlorobenzene as the starting material (Figure 2.10). BTF exhibits a high density and melting point of 1.90 g cm^{-3} and 197 °C, respectively. Moreover, the energetic performance of BTF is comparable to that of HMX. Unfortunately, BTF exhibits a considerably high mechanical sensitivity, which is closely related to its molecular and crystal structures.

With characteristics of a fused-ring backbone, BTF exhibits a planar molecular structure (Figure 2.11a,b). Notably, hydrogen is absent in the molecular structure of BTF. Therefore, no hydrogen-bonding interactions exist among the BTF molecules, only weaker interactions (compared to hydrogen-bonding interactions) can exist

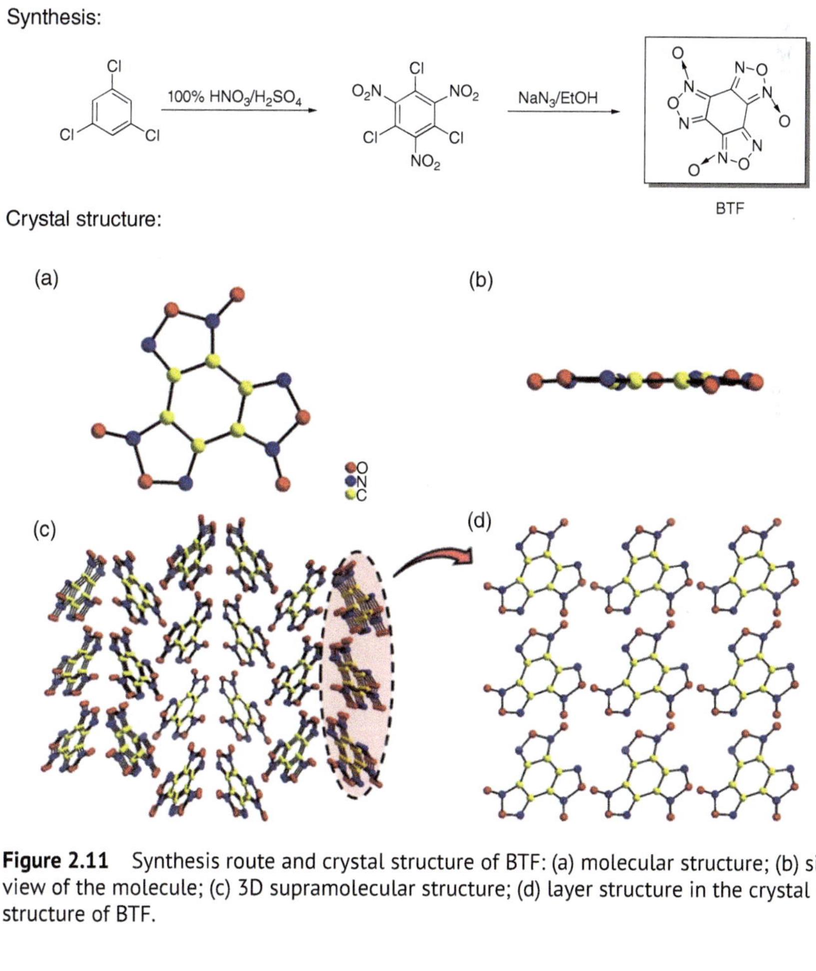

Figure 2.11 Synthesis route and crystal structure of BTF: (a) molecular structure; (b) side view of the molecule; (c) 3D supramolecular structure; (d) layer structure in the crystal structure of BTF.

among the molecules in the BTF crystal (Figure 2.11c and d). All molecular and crystal structure characteristics without strong intermolecular interactions make it difficult for BTF to exhibit resistance to an external stimulus. The high sensitivity of BTF limits its application potential, and its high energy and thermal stability also attract considerable attention. To improve the sensitivity, several cocrystals toward BTF have been designed and prepared. The sensitivity of BTF is well improved by the cocrystal engineering strategy [41, 42]. In the structures of cocrystals, BTF molecules interact with, and are separated by, the low-sensitive components. Under the influence of external stimuli on these cocrystals, BTF molecules cannot interact with each other, leading to improved sensitivity.

2.3 Conclusions

Fused-ring aromatic energetic compounds exhibit immense potential for the preparation of high-energy insensitive materials and attract considerable attention. In this chapter, the chemistry of fused-ring aromatic energetic compounds, including the synthesis, structure, performance, and structure–energy–sensitivity correlations, is introduced and discussed. Energetic materials exhibit strong practical applications in national defense and aerospace technologies. Therefore, the design and preparation of energetic compounds in terms of performance and cost need to be considered.

Recent studies have investigated high-performance parameters of high density, high energy, high thermal stability, and low sensitivity of fused-ring-based energetic compounds, which can provide basic knowledge and data for the design and preparation of novel energetic compounds. Oxygen balance is the basic composition index for correlating energy and sensitivity, which is also directly related to the stability of the molecule. Generally, a high oxygen balance indicates a low molecular stability and high sensitivity. Meanwhile, with similar oxygen balance, energetic compounds exhibit low sensitivities if intramolecular hydrogen bonds are present. Therefore, to design an energetic compound with an expected high performance, oxygen balance and intramolecular hydrogen bonds are two key parameters. In addition, the presence of a hydrogen-bond net, π–π interactions among molecules, and a planar layer structure, and a "face-to-face" stacking mode is also beneficial to improve sensitivity.

The unique planar conjugated structural characteristic of fused-ring aromatic skeletons makes it comparatively easy to design energetic compounds with a suitable oxygen balance as well as intra- and intermolecular hydrogen bonds. However, it is considerably difficult to construct fused-ring aromatic skeletons as their synthesis requires several step reactions. The complex synthesis methods not only increase the cost but also decrease the total yield of the target compounds. Therefore, a significant challenge of developing high-performance fused-ring-based energetic compounds is to develop a backbone, which can fulfill requirements for target molecules as well as be a simple preparation method. Only if there are less than three reaction steps, the target compound can have a higher yield and lower cost, which will immensely benefit their practice application.

In fact, the development of high-performance energetic materials exhibits several challenges. Fused-ring backbones provide a new approach and chance to the development of novel energetic materials. In recent years, considerable efforts have been devoted to this research area, and an increasing number of fused-ring-based energetic compounds have been designed and prepared. It is hypothesized that in-depth insights into the chemistry of fused-ring aromatic energetic compounds will accelerate the development of high-performance energetic materials, and fused-ring aromatic energetic compounds will become one of the main research areas in energetic materials.

References

1 O'Sullivan, O.T. and Zdilla, M.J. (2020). Properties and promise of catenated nitrogen systems as high-energy-density materials. *Chem. Rev.* 120: 5682–5744.

2 Wu, J.T., Xu, J., Li, W., and Li, H.B. (2020). Coplanar fused heterocycle-based energetic materials. *Propellants Explos. Pyrotech.* 45: 536–545.

3 Yin, P., Zhang, Q., and Shreeve, J.M. (2016). Dancing with energetic nitrogen atoms: versatile N-functionalization strategies for N-heterocyclic frameworks in high energy density materials. *Acc. Chem. Res.* 49: 4–16.

4 Gouzman, I., Grossman, E., Verker, R. et al. (2019). Advances in polyimide-based materials for space applications. *Adv. Mater.* 31: 1807738.

5 Gayathri, S. and Reshmi, S. (2017). Nitrato functionalized polymers for high energy propellants and explosives: recent advance. *Polym. Adv. Technol.* 28: 1539–1550.

6 Steinhauser, G. and Klapoter, T.M. (2008). "Green" pyrotechnics: a chemists' challenge. *Angew. Chem. Int. Ed.* 47: 3330–3347.

7 Marlair, G. and Kordek, M.A. (2005). Safety and security issues relating to low capacity storage of AN-based fertilizers. *J. Hazard. Mater.* 123: 13–28.

8 Oran, E.S., Chamberlain, G., and Pekalski, A. (2020). Mechanisms and occurrence of detonations in vapor cloud explosions. *Prog. Energy Combust. Sci.* 77: 100804.

9 Wang, Y., Song, S., Huang, C. et al. (2019). Hunting for advanced high-energy-density materials with well-balanced energy and safety through an energetic host-guest inclusion strategy. *J. Mater. Chem. A* 7: 19248–19257.

10 Guo, T., Wang, Z., Tang, W. et al. (2018). A good balance between the energy density and sensitivity from assembly of bis(dinitromethyl) and bis(fluorodinitromethyl) with a single furazan ring. *J. Anal. Appl. Pyrolysis* 134: 218–230.

11 Li, G. and Zhang, C. (2020). Review of the molecular and crystal correlations on sensitivities of energetic materials. *J. Hazard Mater.* 398: 122910.

12 Yin, P., Zhang, J., Imler, G.H. et al. (2017). Polynitro-functionalized dipyrazolo-1,3,5-triazinanes: energetic polycyclization toward high density and excellent molecular stability. *Angew. Chem. Int. Ed.* 56: 8834–8838.

13 Liu, J., Liu, L., and Liu, X. (2019). Development of high-energy-density materials. *Sci. China Technol. Sci.* 63: 195–213.

14 Wang, Y., Liu, Y., Song, S. et al. (2018). Accelerating the discovery of insensitive high-energy-density materials by a materials genome approach. *Nat. Commun.* 9: 2444.

15 Kolb, J.R. and Rizzo, H.F. (1979). Growth of 1,3,5-triamino-2,4,6-trinitrobenzene (TATB) I. Anisotropic thermal expansion. *Propellants Explos. Pyrotech.* 4: 10–16.

16 Averkiev, B.B., Antipin, M.Y., Yudin, I.L., and Sheremetev, A.B. (2002). X-ray structural study of three derivatives of dinitropyrazine. *J. Mol. Struct.* 606: 139–146.

17 Bemm, U. and Ostmark, H. (1998). 1,1-Diamino-2,2-dinitroethylene: a novel energetic material with infinite layers in two dimensions. *Acta Crystallogr., Sect. C: Cryst. Struct. Commun.* 54: 1997–1999.

18 Gao, H., Zhang, Q., and Shreeve, J.M. (2020). Fused heterocycle-based energetic materials (2012–2019). *J. Mater. Chem. A* 8: 4193–4216.

19 Yin, P., Zhang, J., Mitchell, L.A. et al. (2016). 3,6-Dinitropyrazolo[4,3-*c*]pyrazole-based multipurpose energetic materials through versatile N-functionalization strategies. *Angew. Chem. Int. Ed.* 55: 12895–12897.

20 Xia, H., Zhang, W., Jin, Y. et al. (2019). Synthesis of thermally stable and insensitive energetic materials by incorporating the tetrazole functionality into fused ring 3,6-dinitropyrazolo-[4,3-*c*]pyrazole framework. *ACS Appl. Mater. Interfaces* 11: 45914–45921.

21 Ma, J., Tang, J., Yang, H. et al. (2019). Polynitro-functionalized triazolylfurazanate triaminoguanidine: novel green primary explosive with insensitive nature. *ACS Appl. Mater. Interfaces* 11: 26053–26059.

22 Fu, W., Zhao, B., Zhang, M. et al. (2017). 3,4-Dinitro-1-(1*H*-tetrazol-5-yl)-1*H*-pyrazol-5-amine (HANTP) and its salts: primary and secondary explosives. *J. Mater. Chem. A* 5: 5044–5054.

23 Zhang, W., Zhang, J., Deng, M. et al. (2017). A promising high-energy-density material. *Nat. Commun.* 8: 181.

24 Yin, P., Zhang, J., Parrish, D.A., and Shreeve, J.M. (2015). Energetic fused triazoles – a promising C–N fused heterocyclic cation. *J. Mater. Chem. A* 3: 8606–8612.

25 Klapötke, T.M., Schmid, P.C., Schnell, S., and Stierstorfer, J. (2015). 3,6,7-Triamino-[1,2,4]triazolo[4,3-*b*][1,2,4]triazole: a non-toxic, high-performance energetic building block with excellent stability. *Chem. Eur. J.* 21: 9219–9228.

26 Dalinger, I.L., Vatsadse, I.A., Shkineva, T.K. et al. (2010). Synthesis and transformations of 5-amino-3,4-dinitropyrazole. *Russ. Chem. Bull.* 59: 1631–1638.

27 Schulze, M.C., Scottb, B.L., and Chavez, D.E. (2015). A high density pyrazolo-triazine explosive (PTX). *J. Mater. Chem. A* 3: 17963–17965.

28 Kumar, D., Imler, G.H., Parrish, D.A., and Shreeve, J.M. (2017). A highly stable and insensitive fused, triazolo-triazine explosive (TTX). *Chem. Eur. J.* 23: 1743–1747.

29 Piercey, D.G., Chavez, D.E., Scott, B.L. et al. (2016). An energetic triazolo-1,2,4-triazine and its N-oxide. *Angew. Chem. Int. Ed.* 55: 15315–15318.

30 Chavez, D.E. and Hiskey, M.A. (1998). Synthesis of the bi-heterocyclic parent ring system 1,2,4-triazolo[4,3-*b*][1,2,4,5]tetrazine and some 3,6-disubstituted derivatives. *J. Heterocyclic Chem.* 35: 1329–1332.

31 Hu, L., Yin, P., Zhao, G. et al. (2018). Conjugated energetic salts based on fused rings: insensitive and highly dense materials. *J. Am. Chem. Soc.* 140: 15001–15007.

32 Hu, L., Yin, P., Imler, G.H. et al. (2019). Fused rings with N-oxide and $-NH_2$: good combination for high density and low sensitivity energetic materials. *Chem. Commun.* 55: 8979–8982.

33 Spackman, M.A. and Jayatilaka, D. (2009). Hirshfeld surface analysis. *CrystEngComm* 11: 19–32.

34 Deng, M., Feng, Y., Zhang, W. et al. (2019). A green metal-free fused-ring initiating substance. *Nat. Commun.* 10: 1339.

35 Feng, Y., Deng, M., Song, S. et al. (2020). Construction of an unusual two-dimensional layered structure for fused-ring energetic materials with high energy and good stability. *Engineering* https://doi.org/10.1016/j.eng.2020.01.013.

36 Chen, S., Liu, Y., Feng, Y. et al. (2020). 5,6-Fused bicyclic tetrazolo-pyridazine energetic materials. *Chem. Commun.* 56: 1493–1496.

37 Rezchikova, K.I., Churakov, A.M., Shlyapochnikov, V.A., and Tartakovsky, V.A. (1999). A quantum-chemical study of 1,2,3,4,5,6,7,8-octaazanaphthalene and its N-oxides. *Russ. Chem. Bull.* 48: 870–872.

38 Klenov, M.S., Guskov, A.A., Anikin, O.V. et al. (2016). Synthesis of tetrazino-tetrazine 1,3,6,8-tetraoxide (TTTO). *Angew. Chem. Int. Ed.* 55: 11472–11475.

39 Li, W., Tian, J., Qi, X. et al. (2018). Synthesis of 4,8-dinitraminodifurazano[3,4-*b,e*]pyrazine derived nitrogen-rich salts as potential energetic materials. *ChemistrySelect* 3: 849–854.

40 Li, W., Wang, Y., Qi, X. et al. (2018). Design and synthesis of polycyclic DFP-based low-sensitivity energetic materials with excellent thermal stability. *Chin. J. Energetic Mater.* 26: 901–909.

41 Yang, Z., Li, H., Zhou, X. et al. (2012). Characterization and properties of a novel energetic-energetic cocrystal explosive composed of HNIW and BTF. *Cryst. Growth Des.* 12: 5155–5158.

42 Zhang, H., Guo, C., Wang, X. et al. (2013). Five energetic cocrystals of BTF by intermolecular hydrogen bond and π-stacking interactions. *Cryst. Growth Des.* 13: 679–687.

3

Advances in Computations of Nitrogen-Rich Materials

Lei Zhang[1,2] *and Chuang Yao*[3]

[1] *CAEP Software Center for High Performance Numerical Simulation, Beijing 100088, PR China*
[2] *Institute of Applied Physics and Computational Mathematics, Beijing 100088, PR China*
[3] *Yangtze Normal University, Key Laboratory of Extraordinary Bond Engineering and Advance Materials Technology (EBEAM) of Chongqing, School of Materials Science and Engineering, Chongqing 408100, PR China*

3.1 Why Computation and What Role It Plays?

Computation is the use of digital technology to solve the mathematical equations defining a particular theory or model, which is enormously attractive to be a companion to experiments [1]. In the field of high-energy density materials (HEDMs), the experimental research is extraordinarily hazardous and expensive, and the lifecycle of the characterization, manufacturing, testing, and inspection of HEDMs is very long [2]. In contrast, the computational approach is relatively safe, economical, and time saving. The synergy between computational and experimental approaches is very promising to speed up the discovery, synthesis, and optimization of advanced HEDMs. First, computations could be used to better understand or interpret the ambiguity of existing experimental results. Second, the computations could be employed in a simultaneous fashion to optimize an experiment's design and progress. Last, the computation could be used to predict properties that might be especially difficult, dangerous, or costly to measure experimentally, such as detonation velocity, detonation pressure, and shock sensitivity of a novel HEDM, prior to these compounds scaled-up synthesis [3].

Many HEDMs have large-scale crystal systems, with each primitive cell containing tens or hundreds of atoms. Performing quantum mechanical calculations for these HEDMs has been known to be time and resource consuming. To make the quantum mechanical calculations more efficient, chemists widely simplified the periodically repeated components of a solid-state crystal to be an isolated free cluster. Recently, with the development of algorithms for solving Schrödinger equation specialized for HEDMs, the quantum mechanical computations have become more reliable to reflect the solid-state physics, rather than just properties of a single-molecule [4–6]. In particular, the development of advanced parallel software infrastructure

Nitrogen-Rich Energetic Materials, First Edition. Edited by Michael Gozin and Leonid L. Fershtat.

and the revolution of programming mode-based platform effectively synchronized the upgrade of the produced software and the development of modern supercomputer hardware [7–9]. The resulting supercomputing capabilities allowed high-throughput quantum mechanical calculations of HEDMs, and the calculated data of physicochemical properties, such as crystal structures, interspecies energetics, kinetic dynamics, mechanical properties, detonation performance, and sensitivity metrics of HEDMs, are accumulating faster than ever. Over the past decade, machine learning (ML) algorithms have emerged, rapidly developed, and extensively used in materials research [10–15], pushing the computation of HEDMs to an even more critical role. Compared to the traditional physics-based or knowledge-driven approach, ML algorithms are data driven. They have an essential trait of efficiently handling and deciphering complex datasets, providing accurate predictions, and new knowledge. Therefore, the computational approach is very promising to promote the development of crystal engineering of HEDMs to a new era of theory-guided rational design of these materials [16].

3.2 Why Nitrogen-Rich HEDMs and How They Work?

Traditional nitrogen-rich HEDMs mainly refer to the CHNO composed HEDMs, with nitrogen content higher than 50% [17]. The feature of nitrogen-rich HEDMs is that many nitrogen atoms are bound to each other *via* single or double bonds [18]. The bond energy for a N—N single bond is 38 kcal mol^{-1} and for a N=N double bond is 100 kcal mol^{-1}, in the unreacted HEDMs, while the bond energy for a N≡N triple bond in the N_2 gaseous molecule in the reaction products [19] is 223 kcal mol^{-1}. Obviously, one-third of the bond energy of N≡N triple bond is higher than the bond energy of N—N single bond, and it is also higher than half of the bond energy of N=N double bond. The exceptionally high bond energy (or high stability) of the N≡N triple bond is responsible for the fact that nitrogen-rich HEDMs generate a large amount of heat when exothermically decomposing into molecular dinitrogen [20].

Besides the characteristic of satisfactory high-energy density, a considerable amount of nitrogen-rich HEDMs also show low shock sensitivity, improved burning rate, and other advanced advantages [18]. Therefore, nitrogen-rich HEDMs are very promising candidates to address the long-standing performance-sensitivity contradiction issue of HEDMs [18, 21–25].

In the early stage, the preparation of the first-generation high-nitrogen HEDMs was focused on increasing the number of nitrogen atoms in heterocycles (Figure 3.1). Therefore, these compounds were often lack of oxygen and had relatively low oxygen balance (OB). Compounds with low OB combust incompletely, producing toxic gases and CO as their products, and therefore have relatively low explosive performance. On the other hand, excess oxygen in molecules leads to the generation of O_2 product that takes away a great deal of energy during the explosion, also resulting in a reduced explosive performance of HEDMs. Therefore, the development of the second-generation nitrogen-rich HEDMs was accompanied by optimization in their

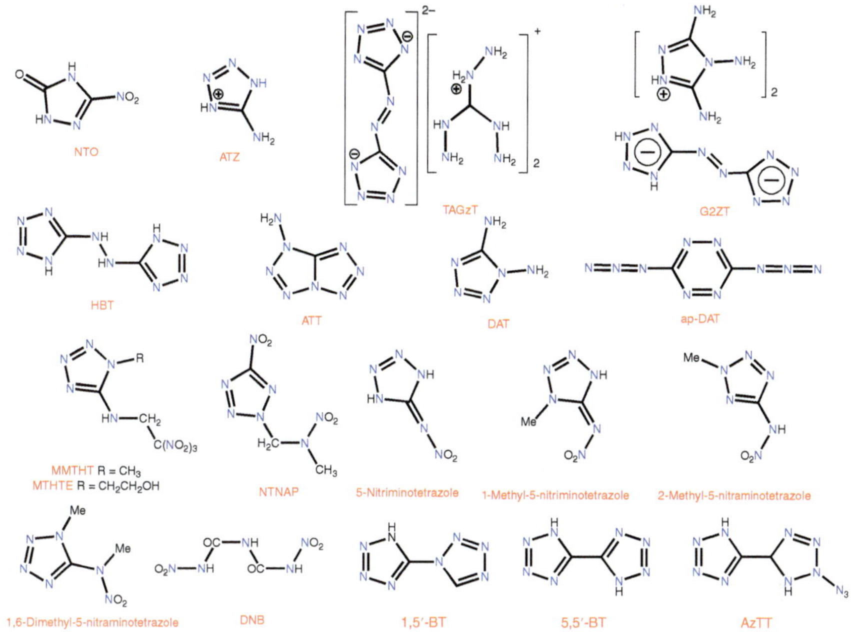

Figure 3.1 Nitrogen-rich HEDMs: first generation.

OB. This strategy has substantially improved the comprehensive properties of the second-generation nitrogen-rich HEDMs, as we will detail in this chapter.

Over the past decade, HEDMs composed of all nitrogen atoms were receiving increased attention. Although the nitrogen content of some of these materials derivatives could be lower than 50%, due to the introduction of metal atoms in their structures, these HEDMs have a prominent characteristics of a high nitrogen molar ratio and can be classified as the third-generation of nitrogen-rich HEDMs. The pursuit of crystal structures composed of all nitrogen atoms is an ongoing frontier subject for computational and experimental research. In this chapter, we will mainly focus on the review of advances in computations related to all three generations of nitrogen-rich HEDMs.

3.3 Advances in Computation of First-Generation Nitrogen-Rich HEDMs

One of the most typical nitrogen-rich HEDMs is 3-nitro-1,2,4-triazole-5-one (NTO). It was synthesized in 1905 and proved to be of high-energy density in 1987, and later found its application in gas generators for automobile inflatable airbags. NTO has been demonstrated to have good explosive properties (comparable to RDX) and low sensitivity (lower than HMX). Nowadays, NTO has become a very important ingredient in the polymer-bonded explosive (PBX) formulations for insensitive munitions (IMs) [17, 26, 27]. For example, the pressable formulation P15636, based on NTO and HMX, is known to be successful and is characterized against the extremely insensitive substance (EIS) tests. Recently, a new IM composition, P16945, made up of NTO and RDX, was developed and qualified. The comprehensive evaluation of the IM behavior is still in progress, and this composition may meet the STANAG 4439 standard [28].

Nevertheless, NTO has strong acidity, which brings serious erosion problems to the metallic shells of guns and munitions. To address the erosion problem of NTO, novel cocrystals like NTO/5,6,7,8-tetrahydro-tetrazolo[1,5-*b*][1,2,4]-triazine (TZTN) and NTO/HMX [29, 30] have been synthesized and characterized.

The density functional theory (DFT) method was employed to study the interatomic interaction–atomistic structure–macroscopic property interrelationships of NTO and its cocrystals, as shown in Figure 3.2 [31]. The interatomic and intermolecular interactions were studied by calculating the interatomic close contact distances, Mulliken bond orders, Hirshfeld surfaces, and intermolecular binding energies. The atomistic structure and chemical compositions were characterized by packing coefficient (PC), density, and OB. The calculated macroscopic properties included mechanical properties, thermodynamic properties, detonation properties, and impact sensitivity. The calculations results showed that the co-crystallization of NTO and TZTN molecules enhances the intermolecular binding force, which thermodynamically drives the in-crystal mixture of the two components. However, co-crystallization decreases the molecular packing density, reduces the density, and deteriorates the OB, leading to a reduction in the detonation performance.

The mechanical sensitivity is linked to the molecular vibration's energy distributions, thereby interpreting the rank of the impact sensitivity of NTO < NTO/TZTN < TZTN. Later on, *ab initio* molecular dynamics (AIMD) simulations of shock wave effect on HEDM crystals, each composed of up to 1000 atoms, were realized [32]. Through the AIMD simulations, a theoretical indicator, $t_{\text{initiation}}$, the delay of decomposition reaction under shock, was proposed to characterize the shock sensitivity of HEDM crystals, which was satisfactorily consistent with experimental measured $h_{50\%}$. The AIMD simulations suggested that the coupling of heat and pressure drives the initiation of the shock reaction, wherein the vibrational spectra, the specific heat capacity, and the strength of the trigger bonds were the determinants of the shock sensitivity. Abundant intermolecular hydrogen bonds were found to effectively buffer the system from heating, thereby delaying the decomposition reaction and reducing the shock sensitivity of the HEDMs [32].

The atomistic structure–macroscopic property interrelationships study helps to figure out optimal ranges of the key parameters, providing experimenters a theoretical guide for the rational design of advanced HEDMs with a good balance of energy density and safety. The DFT study of charge distribution and interatomic interactions of HEDMs, on the other hand, help theorists to capture the physics and reveal the underlying mechanism. The entire computational circle, as shown in Figure 3.2, serves as a satisfactory complementary to experiments, overall accelerating the development of novel advanced HEDMs.

The decomposition mechanism of various triazole compounds was explored by computational studies. In 1998, Politzer [35] investigated the decomposition of

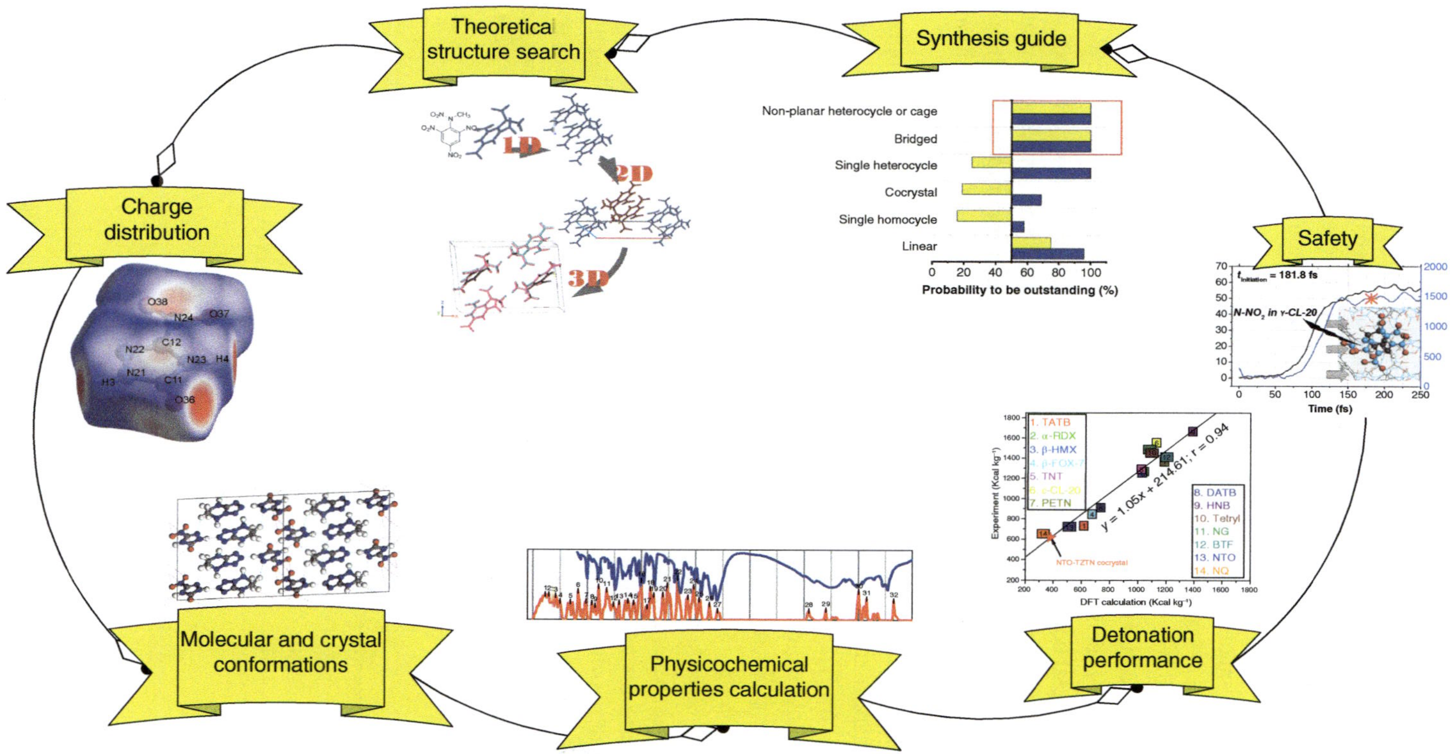

Figure 3.2 DFT study of interaction-structure–property interrelationship of NTO and other HEDMs. Source: Adapted from Refs. [31–34].

4-nitro-1,2,3-triazole using a DFT method. The authors found that activation energy of 52 kcal mol^{-1} was sufficient to initiate this compound decomposition process. It was much less than that typically needed energy to rupture the trigger linkage C–NO_2 of many HEDMs, and being in correlation with an experimental observation that 4-nitro-1,2,3-triazole is highly impact sensitive. The decomposition starts with the opening of the rings, followed by the formation of N_2. da Silva employed the G3B3 composite theoretical method to examine the thermodynamics and kinetics of five-membered nitrogen-containing heterocycles such as 1,2,3-triazole, 1,2,4-triazole, 1,2,5-triazole, 1,3,4,-triazole, 1*H*-tetrazole, and 2*H*-tetrazole [36]. He found that the retro [3+2] cycloaddition is an important initial reaction step in these compounds decomposition, forming into a combination of the three-membered diene products, including hydrazoic acid NH=N=H, nitrilimine NH=N=CH, carbodiimide, NH=C=NH, and ketenimine NH=C=CH_2, as well as two-membered dienophile products, including acetylene, hydrogen cyanide, and N_2 (Figure 3.3). Calculated kinetic parameters and experimentally observed data indicated that the dissociation reactions are essential in the decomposition of tetrazole and 1,2,3-triazole in the temperature range of 300–3000 K.

Kowhakul performed a systematic and detailed study on the thermal decomposition mechanism and stability criteria of 1*H*-1,2,4-triazole derivatives, using a DFT method, at the B3LYP accuracy level, for the ground state calculations of geometry optimizations and frequency calculations. He used the unrestricted B3LYP/6-31+G* accuracy level for the excited singlet state calculations for the optimized geometries and energies [37]. Using the lowest energy change of the decomposition pathway model of intramolecular proton transfer, combined with bond cleavage, the onset temperatures T_{DSC} of 1*H*-1,2,4-triazole and its derivatives were determined very close to the experimentally measured T_{DSC}. This work advances the understanding of the thermal hazard correlations to computationally screen functional types and groups of azoles [37].

Korolev performed a systematic investigation of the thermal decomposition pathways of nitro-1,2,4-triazoles, using DFT computational method at the B3LYP/6-31G*

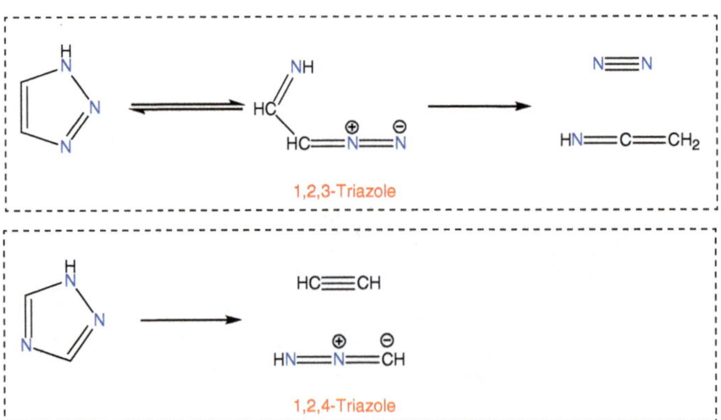

Figure 3.3 Retro-[3+2]-cycloaddition mechanism for decomposition of 1,2,3-triazole and 1,2,4-triazole. Source: da Silva et al. [36]/with permission of American Chemical Society.

accuracy level [38]. It was found that more thermolysis products like H_2O, CO, NO_2, cyanamide, cyanuric acid, and melamine may be formed, in addition to the experimentally detected species including N_2, N_2O, NO, CO_2, HCN, HNCO, 1,2,4-triazole, 3(5)-nitroso-1,2,4-triazole, and 1,2,4-triazolone. The energy gap width was found to correlate with the thermostability of the nitro-1,2,4-triazoles: a smaller energy gap of the compound corresponds to a lower thermostability, and *vice versa*.

Zhou carried out a thorough study on the initial decomposition mechanism of triazoles and their substituted derivatives by using high-accuracy level methods. More specifically, the MP2 method was used for geometry optimization and DLPNO-CCSD(T) method for energy calculation [39]. It was found that the proton transfer is the primary dissociation pathway for 1,2,4-triazole, and it is consistent with the conclusion derived by Kowhakul [37]. Ring-opening path, which is the initial decomposition step of 1*H*-1,2,3-triazole, has ~7 kcal mol^{-1} lower energy barrier than the proton transfer in 1*H*-1,2,4-triazole, thereby interpreting the experimental observation that 1*H*-1,2,4-triazole is more stable than 1,2,3-triazole (Figure 3.4). The presence of nitro group was found to enrich the decomposition pattern with C–NO_2 dissociation, nitro-nitrite isomerization,

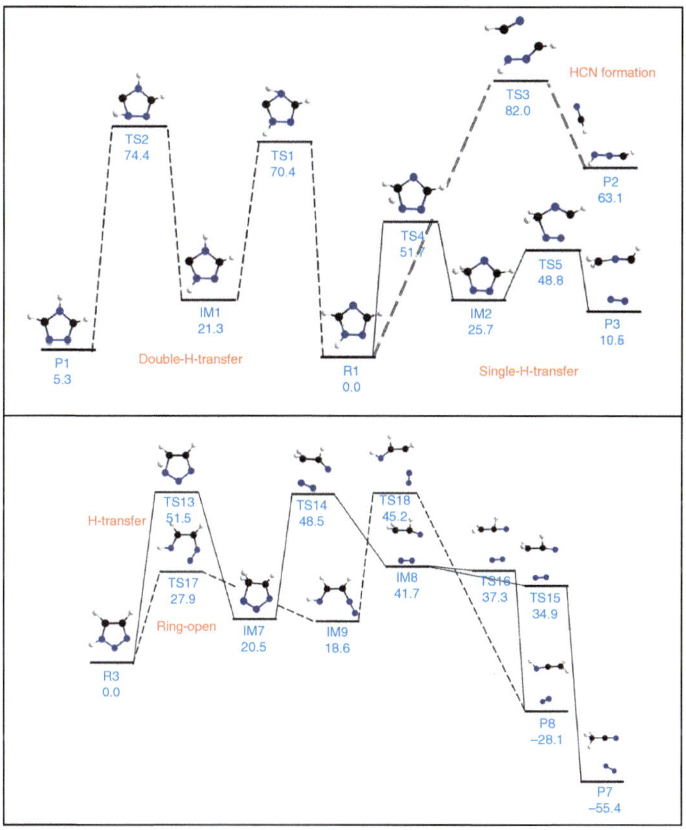

Figure 3.4 Initial decomposition pathways of 1*H*-1,2,4-triazole and 1*H*-1,2,3-triazole. Source: Lu et al. [39]/with permission of American Chemical Society.

and OH and HONO channels, hindering the proton transfer, but promoting the ring-opening process [39].

Recently, it was found that tetrazole rings can well balance the molecule's stability and explosive performance. These findings were extensively utilized in using tetrazole building blocks in the design of novel nitrogen-rich HEDMs. For example, 5-aminotetrazolium (ATZ) is quite insensitive to shock [40], whereas triaminoguanidinium azotetrazolate (TAGzT) can increase the heat generation and burning rates of propellants over a wide pressure range [18]. 5,5′-Hydrazinebistetrazole (HBT) and bis(3,4,5-triamino-1,2,4-triazolium)-5,5′-azotetrazolate (G_2ZT) contain up to ~80 wt% of nitrogen and exhibit relatively high heat of combustion ($-2396\,cal\,g^{-1}$ for HBT, and $-2775\,cal\,g^{-1}$ for G_2ZT) [41].

For 1-amino-tetrazolo-[4,5-*b*]tetrazole (ATT) [42] and 1,5-diaminotetrazole (DAT) [43], DFT method was extensively employed to better understand their molecular properties, such as the optimized geometry, NMR data, electrostatic potential, electron density topology, vibration analysis (thermo-chemistry and IR spectroscopy), natural bond orbital (NBO), lowest unoccupied molecular orbital (LUMO), and highest occupied molecular orbital (HOMO). For other recently synthesized nitrogen-rich HEDMs, like 3,6-diazido-1,2,4,5-tetrazine (*ap*-DAT) [44], ammonium 5-azidotetrazolate ($[NH_4]^+[CN_7]^-$), hydrazinium 5-azidotetrazolate ($[N_2H_5]^+[CN_7]^-$), guanidinium 5-azidotetrazolate ($[CH_6N_3]^+[CN_7]^-\cdot H_2O$) and 1-aminoguanidinium 5-azidotetrazolate ($[CH_7N_4]^+[CN_7]^-$) [45], the crystal properties were investigated by the DFT method, wherein the periodic boundary condition was taken into account.

Kiselev performed a theoretical study on the thermal decomposition of tetrazoles using the W1 high-level procedure and proposed that a new mechanism indicating mutual interconversion of the tetrazole tautomers is of vital importance [46]. The mutual interconversion of the 1*H*-tetrazole, 2*H*-tetrazole, and an N-heterocyclic carbene 14H was driven by the concerted double H-transfer in hydrogen-bonded complexes, which has a very low energy barrier of ~18–28 kcal mol^{-1}. The computational Arrhenius parameters were found in a good agreement with experimental Arrhenius parameters, and the values of E_a and $\log A$ of the Arrhenius parameters demonstrated that the decomposition of carbene isomer 14H dominates the tetrazole decomposition. The investigators also found that if tautomerism was neglected, the rate of tetrazole decomposition would be underestimated (Figure 3.5).

In recent years, energetic molecules constructed by two linked azoles including furazan, 1,2,4-triazole, and tetrazole were found advantageous due to the vast possibilities for tailoring their energetic behavior and performance. Typical examples are 5,5′-bistetrazole (5,5′-BT), 1,5′-bistetrazole (1,5′-BT), and 5-(5-azido-(1 or 4)*H*-1,2,4-triazol-3-yl)tetrazole (AzTT) (Figure 3.1). Bernstein carried out the unimolecular decomposition mechanism study of the three tetrazoles initiated by electronic excitation, by using both experimental observations and the CASSCF computational method [47]. Conical intersections, providing non-adiabatic radiationless internal conversion between electronic states on the femtosecond time scale and placing the undissociated molecule on a new part of the ground state potential surface with all its excitation energy in the ground state vibrations, are available for bond breaking.

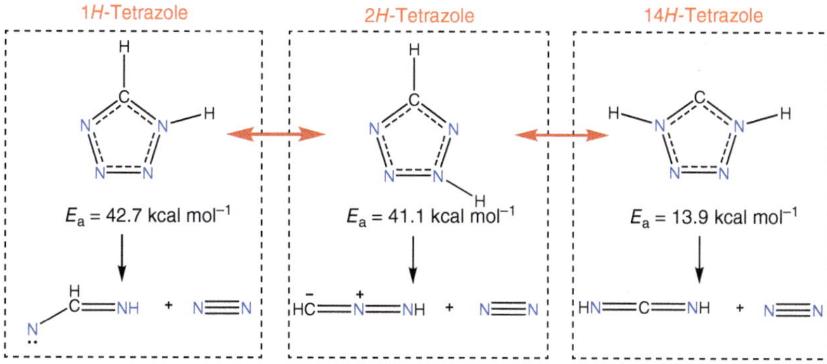

Figure 3.5 Tautomerism and thermal decomposition mechanism of tetrazole by high-level *ab initio* study. Source: Adapted from Kiselev et al. [46].

These are the key points for the theoretically derived mechanisms. It was revealed that the N_2 molecule is an initial decomposition product from these three materials. In 1,5′-BT and 5,5′-BT, tetrazole rings open at various positions on the S_1 excited state and N_2 product is created afterward on the ground state. In AzTT, N_2 is formed by N—N bond rupture of the azide group on the ground state, which has a much lower energy barrier than tetrazole ring opening.

3.4 Advances in Computation of Second-Generation Nitrogen-Rich HEDMs

Realizing the importance of OB in the performance of HEDMs, the development of nitrogen-rich HEDMs entered the second stage. The development of the second-generation of nitrogen-rich HEDMs was focused on the simultaneous improvement of nitrogen content increase and OB optimization. For example, recently developed *N*-trinitro-ethyl derivatives of nitrogen-containing compounds (MMTHT and MTHTE), nitrated aminotetrazoles, such as 5-nitriminotetrazole, 1-methyl-5-nitriminotetrazole, 2-methyl-5-nitraminotetrazole, and 1,6-dimethyl-5-nitraminotetrazole, as well as neutral nitramine compounds, including dinitrobiuret (DNB), contain a larger amount of oxygen in comparison to the first-generation nitrogen-rich HEDMs. The second generation of nitrogen-rich energetic materials presents advantages in their gaseous products, high heat of formation, high propulsive/explosive power, high specific impulse, and high flame temperatures [48].

In particular, Klapötke synthesized 5,5′-bistetrazole-1,1′-diolate (TKX-50) in 2012 [49]. TKX-50 combines desirable high nitrogen content with satisfactory OB and presents other attractive properties [49]. The detonation performance of TKX-50 (D = 9.70 km s^{-1}, $p_{\text{C-J}}$ = 42.4 GPa) was calculated to be close to CL-20, and its mechanical and thermal sensitivity was found to be lower than currently used explosives, such as RDX, HMX, and 2,4,6,8,10,12-hexanitro-2,4,6,8,10,12-hexaazatetracyclo-[5.5.0.0^{3,11}.0^{5,9}]-dodecane (CL-20) [49]. TKX-50 also exhibited low toxicity, making it one of the most promising salts for possible replacement of more toxic RDX for practical applications.

Driven by the application prospects of TKX-50, extensive theoretical research on its properties under extreme conditions has been carried out [50–52]. For example, the electronic, structural, and mechanical properties of TKX-50 over a broad pressure range (up to 100 GPa) were examined using the DFT method [52]. The calculated lattice constants, axial compressibility, and equation of state of TKX-50 satisfactorily reproduce the available experimental results. The uniaxial compressibility of the TKX-50 single crystal was calculated to be $b > c > a$ and its stiffness constants were calculated to be C_{33} (61.9 GPa) > C_{11} (58.1 GPa) > C_{22} (36.9 GPa). At the ambient conditions, the calculated Hirshfeld surfaces and their 2D fingerprints indicate that N…H close contacts contribute 42% of the intermolecular interactions of TKX-50, where the bandgap was calculated to be 3.40 eV. As the pressure increases, the intermolecular contact distances become shorter, the intermolecular repulsion increases, and the hydrogen bond ratio decreases. The bandgap decreases with the pressure increase, reaching 3.10 eV at 50 GPa and 2.75 eV at 100 GPa. The decrease of the bandgap implies that the electron transfer from HOMO to LUMO becomes easier, corresponding to the increase of sensitivity when the compound is compressed [52].

The anisotropic impact sensitivity and shock-induced plasticity of TKX-50 single crystals were investigated by empirical molecular dynamics (MD) simulations, using a half-million atoms model [50]. From the large-scale MD simulations, the TKX-50 single-crystal shows anisotropic impact sensitivity, wherein [010] is the most sensitive direction, and [100] is the least sensitive one. The plastic deformations in shock compression along the [100] direction primarily arise from the (001)/[210] and (010)/[001] slip systems. For the [010] shock, the primary slip systems are (100)/[021] and (001)/[210]. For [001], no obvious slip system is observed [50].

The atomistic reaction mechanisms of initial thermal decomposition of TKX-50 crystals have been investigated using the MD method, where the forces were obtained from the density functional-based tight-binding calculations (DFTB). As illustrated in Figure 3.6, the chemistry of TKX-50 is initiated by proton transfer from the cation to the dianion. Continuous heating of TKX-50 crystals leads to dissociation of the protonated or deprotonated bistetrazole to release N_2 and N_2O gases. DFT calculation indicated that transferring one proton to the bistetrazole dianion reduces 18% of the reaction barriers for generating both N_2 and N_2O gases [53]. Subsequently, Zhang et al. found that increasing pressure has an effect of preventing the decomposition of TKX-50, in contrast to the reported accelerating effect of high pressure on the decomposition of common HEDMs [54]. DFT results revealed that increasing pressure impedes the H-transfer *via* the enhanced long-range electrostatic repulsion of $H^{+\delta}$…$H^{+\delta}$ of neighboring NH_3OH^+, with blue shifts of the intermolecular H-bonds. The suppression of the intermolecular proton transfer thereby leads to the delay of decomposition of TKX-50 under high pressure [54].

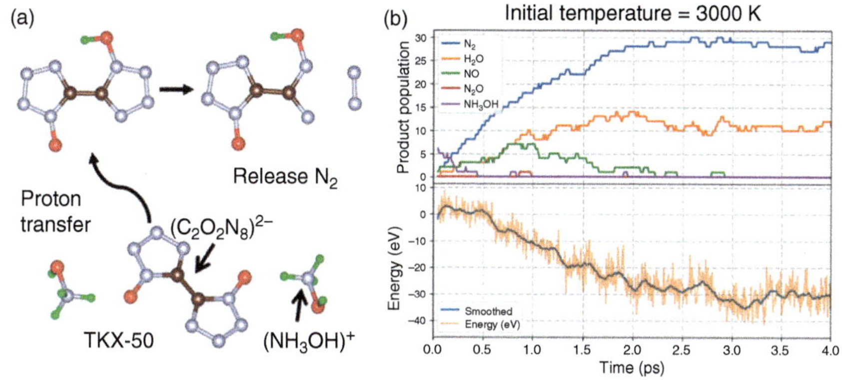

Figure 3.6 Thermal decomposition mechanism of TKX-50 crystals revealed by atomistic simulations. Source: Adapted from Yu et al. [51] and An et al. [53].

Later on, AIMD method was employed to investigate the entire decomposition process of TKX-50 under different pressures and temperatures [51]. The computational results suggested three main processes of N_2 formation: two processes are derived from the break of the tetrazole ring, and the third process is related to the ammonium–triazole interspecies interaction. The former processes can be accelerated by elevated temperatures, whereas they are insensitive to pressure variation. The latter process, which significantly depends on the diffusion of the intermediate clusters, can be accelerated by increasing the temperature and lowering the pressure. The synergy of the three processes makes the reaction rate of TKX-50 increase with the increase of the temperature, whereas it decreases with the increase of pressure [51].

Besides TKX-50, diammonium salt of 5,5′-bistetrazole-1,1′-diolate (ABTOX) is another advanced tetrazole compound that is thermally insensitive, non-toxic, and relatively simple to prepare [55]. It is known as a side product in the one-pot five-step synthesis of TKX-50, starting from glyoxime [55], and it is also an important intermediate of the decomposition of TKX-50 [56, 57]. Thermal analysis (DSC and TGA) confirmed that ABTOX is more thermally stable than TKX-50. The underlying mechanism of this behavior was interpreted with the aid of DFT calculations, which showed that the calculated value of bandgap of ABTOX is 0.013 au higher than that of TKX-50, corresponding to a higher thermostability, according to the frontier molecular orbital theory [55].

Zhang et al. designed novel nitrogen-rich HEDMs, using tetrazole derivatives as components and simultaneously optimizing OB. Using DAT as building blocks, they proposed three novel polynitro-substituted tetrazolo-triazine-based compounds: TNTzTr, DNOTzTr, and HNBTzTr (Figure 3.7) [58]. The density, enthalpy of formation, and explosive performance of these three compounds were estimated with the aid of DFT method. TNTzTr presents a positive enthalpy of formation (1386.00 $kJ\ mol^{-1}$), a high density (2.00 $g\ cm^{-3}$), and remarkable detonation performance ($D = 9.82\ km\ s^{-1}$, $p_{C\text{-}J} = 45.45$ GPa). The calculated performance of HNBTzTr was found to be even higher, with the corresponding values to be 1625.31 $kJ\ mol^{-1}$,

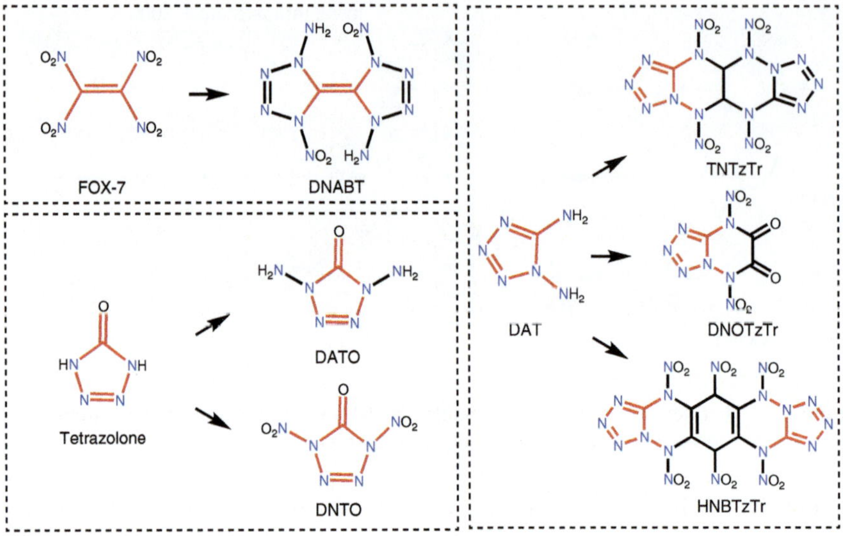

Figure 3.7 Theoretical design of nitrogen-rich HEDMs. Source: Adapted from He et al. [58].

2.05 g cm^{-3}, $D = 9.94$ km s^{-1}, and $p_{C\text{-}J} = 47.30$ GPa. This work showed that the design of compounds containing tetrazole building blocks and having OB close to zero is an effective way to improve the detonation performance of HEDMs. Similarly, 1,1-diamino-2,2-dinitroethylene (FOX-7) [59] and tetrazolone [60] that have a satisfactory balance between low sensitivity and high detonation performance were used as building blocks for the design and preparation of nitrogen-rich HEDMs, such as DNABT [61], DATO, and DNTO [62]. DNABT has OB = −12.30% and has a predicted detonation velocity of $D = 9.53$ km s^{-1}. DATO and DNTO have positive enthalpies of formation (+395.79 and +342.77 kJ mol^{-1}), better detonation performance ($D = 8.80$ km s^{-1}, $p_{C\text{-}J} = 33.69$ GPa; $D = 8.89$ km s^{-1}, $p_{C\text{-}J} = 34.98$ GPa) than RDX, and similar sensitivity to NTO [62]. These theoretical studies are necessary for developing and verifying innovative strategies for the design of advanced nitrogen-rich HEDMs, which are inevitable pioneering exploration for practical experimental synthesis.

3.5 Advances in Computation of Third-Generation Nitrogen-Rich HEDMs: Polynitrogen Materials

Polynitrogen material (N_n) is a special HEDM with a nitrogen content of 100%, with its reaction products consisting solely of N_2. The energy density of N_n is expected to have large positive formation enthalpy, large propulsive or explosive power, very high specific impulse, very high reaction temperature, smoke-free combustion, minimal signature off a rocket engine, and low erosion of gun barrels [17]. Therefore, N_n is an extremely appealing HEDM that is green, nonpolluting, and ultrahigh energetic.

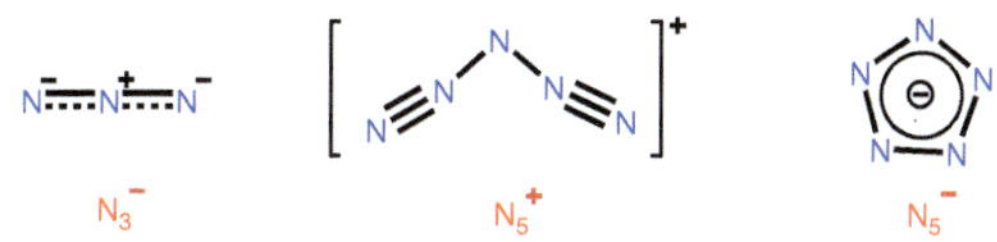

Figure 3.8 Successfully synthesized polynitrogen HEDMs stable at ambient conditions.

For more than one century, scientists have been devoting to a search for stable N_n isomers, possibly stable at various conditions, with n ranging from 3 to 60. To date, three species of polynitrogen energetic materials have been successfully isolated and stabilized by experiments at ambient conditions, which are azide anion N_3^-, pentanitrogen cation N_5^+, and pentazole anion *cyclo*-N_5^- (Figure 3.8) [63–65]. However, these synthesized polynitrogen isomers cannot stay stable alone. They need periodic intermolecular interactions from non-nitrogen counterparts to construct a crystal structure stable at ambient conditions. Such crystal structures include $Pb(N_3)_2$, NaN_3, $N_5^+AsF_6^-$, $N_5^+SbF_6^-$, $(N_5)_6(H_3O)_3(NH_4)_4Cl$, NH_3OHN_5, and $[Ag(N_5)]_n$, etc. [20, 64–67]. Therefore, the nitrogen content in these compounds is usually low, especially in those metal- and metalloid-containing compounds. For example, the nitrogen content in $Pb(N_3)_2$, $N_5^+AsF_6^-$, and $[Ag(N_5)]_n$ is 28.9%, 27.0%, and 39.4%, respectively. Nonetheless, the mole ratio of nitrogen atoms in the metal-azide and metal-pentazolate compounds is generally higher than 70%. We review the recent progress of the computation of these polynitrogen-related compounds, with an emphasis on the development of the recently synthesized pentazole anion *cyclo*-N_5^-.

3.5.1 Nitrogen-Rich HEDMs Composed of N_3^+

In 1980, Curtius explored azide chemistry utilizing aqueous solutions of HN_3, which marked the first detection of a stable N_3^- species [63]. At the beginning, Curtius considered the N_3^- structure to be cyclic. However, Douglas found that HN_3 presents a 270 nm absorption in flash photolysis, which identifies the transition as $^2\Sigma_u^+ \leftarrow ^2\Pi_g$. The ground $^2\Pi_g$ has a linear geometry, with a terminal-to-terminal nitrogen length of 2.363 Å, consisting of two equal N—N bonds of 1.1818 Å. Later, FTIR work confirmed that the ground state of the N_3 radical has a linear symmetry with a bond length of 1.1815 Å [68]. Then, it is now broadly accepted that the lowest energy conformer of the azide anion N_3^- is linear.

Azide salts, such as $Pb(N_3)_2$, AgN_3, NaN_3, KN_3, RbN_3, and CsN_3, can decompose with the release of nitrogen gas and found wide application as energetic materials. For example, the first pyrotechnic mixture used in airbag gas generators was based on NaN_3, which was later replaced due to its toxicity [69]. Heavy metal salts like $Pb(N_3)_2$, AgN_3, and $Ba(N_3)_2$ are widely used in detonators. They are primary high explosives detonable when heated or shaken, decomposing into corresponding metals and nitrogen gas. $Pb(N_3)_2$ has a detonation rate of $D = 5.18\ \text{km s}^{-1}$ and is susceptible to shock and friction.

$[NH_4]N_3$, with a nitrogen content of up to 93.3%, is one of the most remarkable nitrogen-rich HEDMs. Over the past decades, considerable theoretical efforts have been made to investigate the structure of $[NH_4]N_3$ and its application in

HEDMs [70–74]. NH_4N_3 exhibits strong hydrogen bonding features under compression and undergoes a phase transition to $(NH)_4$ at 36 GPa [72, 75]. Although the nitrogen content of $[NH_4]N_3$ is very high, some of the nitrogens turned into NH_3 reaction products, instead of nitrogen gas. Therefore, the detonation performance of $[NH_4]N_3$ is not very high, with $D = 6.45\,km\,s^{-1}$ and $p_{C\text{-}J} = 15.16\,GPa$ [76]. Owing to its high vapor pressure, $[NH_4]N_3$ has not yet found much practical application, but it is expected to be used as a gas generator [77].

3.5.2 Nitrogen-Rich HEDMs Composed of N_5^+

Due to the appealing application perspectives of N_5^+ as a HEDM, extensive theoretical work was carried out for a better understanding of the electronic structure, stabilization mechanism, and reaction chemistry of N_5^+ [78–80].

In 1991, Pyykkö and coworkers performed HF and MP2 calculations, with a 6-31G* basis, on the geometries and frequencies of a series of ABCBA-type compounds, wherein N_5^+ was included. The authors predicted N_5^+ to have a planar *C2v*-symmetric structure with a B–C–B angle of 110.78 [81]. Nguyen and Ha carried out *ab initio* MO and DFT calculations, at the CCSD(T) and B3LYP level, to study the structure, energies, and properties of N_5^+. N_5^+ cation was found to have a minimal triplet–singlet separation, in spite of its symmetrical open structure and a closed-shell singlet electronic ground state. The energy difference between the electronic ground state of N_5^+ and the separated $[N_3^+(\text{triplet}) + N_2]$ fragments was calculated to be $-44\,kJ\,mol^{-1}$, and the difference from the separated $[N_3^+(\text{singlet}) + N_2]$ fragments was $-181\,kJ\,mol^{-1}$. The heat of formation for singlet N_5^+ was calculated to be $351.3 \pm 3.6\,kcal\,mol^{-1}$, and the ionization energy was $7.4 \pm 0.2\,eV$ [78]. Gagliardi et al. studied the dissociation of N_5^+ to N_2 and N_3^+ at the B3LYP and CASPT2 levels. The barrier was calculated to be zero to the formation of N_5^+ from N_2 and N_3^+ on the singlet surface. N_5^+ was estimated to be a weakly bound ion, with the lowest electronic singlet state being $36\,kcal\,mol^{-1}$ [79].

In 1999, Christe et al. carried out high-level accuracy quantum mechanical calculations on geometries and vibrational properties of N_5^+ [64]. The structure of N_5^+ is found to be an open symmetrical V-shaped chain with short terminal nitrogen bonds and long inner nitrogen bonds, which confirmed the results of Pyykkö and coworkers [81]. The calculation results verified the vibrational stability of N_5^+, and the IR, Raman, and NMR spectra were calculated to aid in the identification and characterization. In the same work, they eventually synthesized N_5^+ for the first time in the form of $N_5^+AsF_6^-$ salt, which was characterized by low-temperature Raman, ^{14}N and ^{15}N NMR spectrometries, and X-ray crystallography [64].

Since then, efforts were made to combine the N_5^+ with non-energetic counterions, such as NO, NO_2, Br_2, NO^+, NO_2^+, Br_2^+, Cl_2, O_2, and Xe, and energetic counterions, such as N_3^-, NO_3^-, ClO_4^-, $B(N_3)_4^-$, and $P(N_3)_6^-$. Till 2007, 12 N_5^+ salts have been isolated and characterized one after another [82]. Among them, the $N_5^+SbF_6^-$ salt was found to be thermostable up to 70 °C. Highly energetic compounds of $[N_5][P(N_3)_6]$ and $[N_5][B(N_3)_4]$ were successfully prepared, regretfully, being too sensitive to be used practically [82].

3.5.3 Nitrogen-Rich HEDMs Composed of *cyclo*-N_5^-

3.5.3.1 Synthesis and Related Mechanism at Ambient and High Pressures

The synthesis of *cyclo*-N_5^- started in 1903 [83] and experienced failure [84], misunderstanding [85], and dispute [86] in the following decades. In 2002, *cyclo*-N_5^- was first detected with electrospray ionization mass spectrometry but existed only fleetingly in a gas phase [87]. From 2003 to 2009, *cyclo*-N_5^- was proved to be formed in solution, using the ^{15}N NMR, yet with a short lifespan of just a few seconds [88, 89]. da Silva extrapolated the Marcus equation relationship and theoretically predicted *cyclo*-N_5^- to decompose to dinitrogen and NHNN in a retro-[3+2]-cycloaddition fashion, with an activation enthalpy of only 19.5 kcal mol^{-1} and a half-life of only 14 seconds, at 298 K [36]. Such a low barrier suggests that it is unlikely that *cyclo*-N_5^- would exist as a stable compound at ambient conditions. In 2016, *cyclo*-N_5^- was demonstrated to be stably present in THF solution, at temperatures below −40 °C [90]. In 2017, a room temperature-stable pentazolate $(N_5)_6(H_3O)_3(NH_4)_4Cl$ crystal was eventually isolated from a mild acidic solution, by introducing *m*-chloroperbenzoic acid (*m*-CPBA) and ferrous bis-glycinate $[Fe(Gly)_2]$ [65]. Since then, the experimental and computational research on the topic of *cyclo*-N_5^- was greatly enriched. The experimental research mainly focused on the design and synthesis of various compounds composed of *cyclo*-N_5^-, such as metal pentazolate hydrates, and nanocages, as well as the self-assembled energetic 3D metal–organic framework [91–95].

However, due to the insufficient understanding of the stabilization mechanism of *cyclo*-N_5^-, a methodology for scaled-up synthesis is still unavailable. The low synthetic yields of $(N_5)_6(H_3O)_3(NH_4)_4Cl$ salt have promoted theorists to explore the formation mechanisms of *cyclo*-N_5^- in solutions [96]. Liu and coworkers carried out thorough time-dependent density functional theory (TDDFT) calculations, and UV–vis absorption spectra experiment validation and revealed that the formation of *cyclo*-N_5^- is resulted from the selective C—N bond cleavage of arylpentazole with ferrous bisglycinate and *m*-CPBA [96]. The direct C—N bond activation by *m*-CPBA was computed to have a high reaction barrier and is therefore kinetically inaccessible. Instead, an alternate reaction of $[Fe(Gly)_2] \rightarrow [Fe(Gly)_2(H_2O)] \xrightarrow{m\text{-CPBA}} [Fe(O)\text{–}(Gly)_2]$ is much more favorable. The intermediate $[Fe(O)\text{–}(Gly)_2]$ was proved to be a suitable oxidizer for the C—N bond activation of arylpentazole at low temperature, allowing the *cyclo*-N_5^- formation. The authors stated that further improvement of the oxidation ability of Fe(IV)–oxo complex may be achieved by coordinating ferrous complex with binding ligands other than glycinate, and such oxidation ability improvement is expected to facilitate an improvement in the yield of *cyclo*-N_5^-.

Besides the direct synthesis of pentazole compounds at ambient pressure, synthesizing a high-pressure stable pentazole compound and quenching it back to ambient conditions is an alternate route to stabilize *cyclo*-N_5^-. Theoretical structure searching methods, in a combination of DFT calculations, have been proved to be useful in searching for stable crystal structures [97–101]. The characteristic of the structure searching method is to traverse as many as possible combinations of stoichiometry, symmetry, orientation, spatial packing, and assumed pressure. By calculating the total energies of the structures under all considered pressures and plotting their energy landscape, the structures with the lowest energies are screened out and considered as candidates that might be stable under assumed conditions. Then, these structures are usually sent for phonon calculations and MD simulations to further check their stability. If no imaginary frequency presents in the phonon dispersion curves, and the candidate compound can keep its structure under applied pressures and temperatures for some time, the candidate structure is theoretically considered to be stable.

Alkali metals are of highly electropositive characteristics, promoting the electron transfer from the alkali metal elements to *cyclo*-N_5^-, thus stabilizing the complex. In addition, elements like Be and Li are light, facilitating the high weight ratio of nitrogen in these materials and their high-energy density [102]. Theorists carried out extensive computations to search for stable alkali metal pentazolate salts under a wide range of pressures. Using the DFT method and unbiased structure searching techniques, a LiN_5 crystal was predicted to be stable at pressures above 9.9 GPa and remains mechanically stable under ambient conditions [97]. The stabilization of LiN_5 crystal is derived from the strong anion–cation interactions. The behavior of LiN_5 is expected to be highly exothermic when this material is decomposing ($2.72\ kJ\ g^{-1}$), making it a promising candidate for HEDM. Using the DFT method and variable-composition evolutionary techniques, Shen and coworkers predicted the geometries of various LiN_5 compounds and their corresponding stability at various pressures [98]. The $P2_1/c$ structure of LiN_5 is stable at 15 GPa and transforms into the $C2/c$ phase at the pressure of 65 GPa. Another LiN_5 structure with a $C2/c$ symmetry is found to remain stable at least up to 100 GPa. In the $P2_1/c$ structure, the N–N distance of the *cyclo*-N_5^- ring is slightly different from each other, while in the $C2/c$ structure, the *cyclo*-N_5^- ring presents a nearly isosceles pentagon.

The above theoretical work was soon proved to be used as a guide for further synthetic experiments. In 2018, Laniel et al. achieved the synthesis of solid LiN_5 by compressing and laser-heating lithium embedded in N_2 at around 45 GPa, with the recovery of LiN_5 under ambient conditions [103]. The Raman spectroscopy of vibrational modes and the mass spectroscopy results confirmed the presence of pentazole anion in the synthesized compound. The X-ray diffraction measurements were in agreement with the theoretical predictions of the LiN_5 structures, and the transparency of the laser-heated sample fitted the predicted large bandgap conductivity nature of the LiN_5 compound, which is intrinsic to its ionic interactions.

3.5.3.2 Pentazole Compounds with Non-nitrogen Stabilizer

So far, none of the synthesized *cyclo*-N_5^- compounds has reported detonation parameters measured by experiments, mainly due to the insufficient amount of the sample required by a detonation test. Therefore, computations were needed to evaluate the detonation performance of the synthesized compounds [6, 67, 94, 95, 104–107]. In the synthesized metal pentazolates, such as $[Na(H_2O)(N_5)]\cdot 2H_2O$, $[Mg(H_2O)_6(N_5)_2]\cdot 4H_2O$, $[M(H_2O)_4(N_5)_2]\cdot 4H_2O$ (M = Mn, Fe, Co, and Zn), metal–organic framework $[Na_8(N_5)_8(H_2O)_3]_n$, and metal–inorganic framework $[Ag(NH_3)_2]^+[Ag_3(N_5)_4]^-$, the non-energetic moieties like H_2O, NH_4^+ and H_3O^+ were present. These non-energetic moieties constructed a hydrogen-bonding framework in these crystal structures to stabilize them. However, the presence of these non-energetic moieties makes the detonation performance of the pentazolate compounds non-competitive. Later on, *cyclo*-pentazolate compounds without non-energetic components such as $[Cu(N_5)(N_3)]_n$ and $[Ag(N_5)]_n$ were synthesized. Unfortunately, the detonation performance of the $[Ag(N_5)]_n$ was theoretically predicted to be lower than of the HMX, with a detonation pressure of 34.67 GPa and a detonation velocity of 7.78 km s^{-1} [95]. The quantum mechanical calculations predicted that the $[N_2H_5][N_5]$ has the best detonation performance among the *cyclo*-pentazolate salts, with $p_{C\text{-}J} = 37.00$ GPa and $D = 10.40$ km s^{-1} [107], falling within the range of traditional CHNO energetic materials.

To improve the performance of the pentazole compounds, theoreticians began searching for pentazolate salts, devoid of non-energetic components. In 2020, Yao and Zhang designed four metal pentazolate salts, which were NaN_5, $Mg(N_5)_2$, $Co(N_5)_2$, and $Mn(N_5)_2$ (Figure 3.9) [34]. These salts were found to have slightly lower stability but were found to be significantly more powerful than the corresponding hydrates. In particular, $Mn(N_5)_2$ was calculated to have a better detonation performance than CL-20 and outstanding crystal stability. The detonation parameters of the designed $Mn(N_5)_2$ crystal were predicted to be $D = 9.2$ km s^{-1} and $p_{C\text{-}J} = 49.3$ GPa. Its crystal stability was evaluated by electron density topological analysis, lattice energy, vibrational spectra, and AIMD simulations. The designed $Mn(N_5)_2$ crystal was predicted to have approximately double the lattice energy of NaCl. The theoreticians suggested that synthetic chemists should design pentazole crystals with a large *cyclo*-N_5^- coordination number and an atomic crystal type, which were believed to facilitate comprehensive properties of pentazole compounds with high stability and explosive power.

The initiation mechanisms of the designed $Mn(N_5)_2$ crystal, under various shock waves, were further studied using a combination method of the AIMD simulation

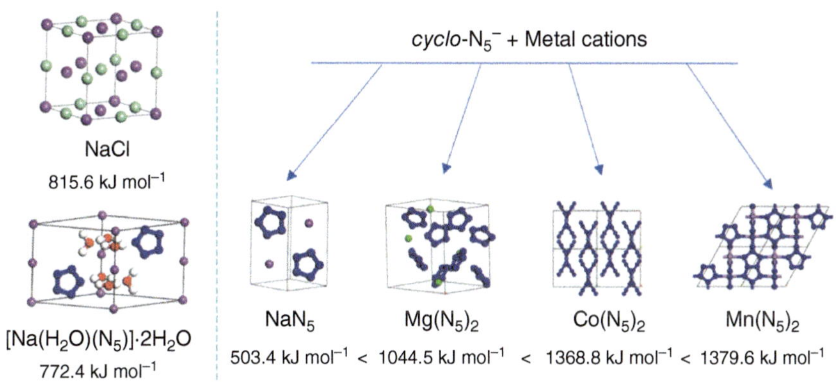

Figure 3.9 Lattice energies of the designed NaN_5, $Mg(N_5)_2$, $Co(N_5)_2$, and $Mn(N_5)_2$, with the NaCl and $[Na(H_2O)(N_5)]\cdot 2H_2O$ crystals for comparison. Source: Yao et al. [34]/with permission of American Chemical Society.

and the multiscale shock technique (MSST) [33]. It was found that chemical reactions should not occur within 5000 fs when shock speed $v_{shock} < 10\,km\,s^{-1}$, and there is only a slight compression of the volume (<10%). When $v_{shock} = 10\,km\,s^{-1}$, N–N starts to rupture uniformly, within the space of the perfect crystal at 512.8 fs. When v_{shock} continues to increase to 11 and $12\,km\,s^{-1}$, the reaction initiation occurs earlier, and the reaction process accelerates further. The effect of microdefects on the shock response, chemical decomposition, and damage evolution of $Mn(N_5)_2$ crystal was investigated by using a model with 3% vacancy defects. The vacancy defect was found to greatly advance the N—N bond rupture time and makes the N–N rupture to take place in the vicinity of the vacancy. The vacancy defect was proved to be one of the early nucleation structures of hot spots. The stress in the vicinity of the vacancy promotes the cascade decomposition of the surrounding pentazolate anions, causing the damage to grow and propagate, which in turn ignites the energetic materials (Figure 3.10). Therefore, diminishing the vacancy defects in pentazole compounds is expected to improve the safety of these compounds.

3.5.3.3 Argument on the Protonation of *cyclo*-N_5^- in $(N_5)_6(H_3O)_3(NH_4)_4Cl$

The successful synthesis of $(N_5)_6(H_3O)_3(NH_4)_4Cl$ salt has initiated intense arguments regarding the protonation state of *cyclo*-N_5^- in various environments and conditions, including isolated clusters, in various solutions, crystal structures, temperatures, and pressures [108, 109]. For example, Huang and coworkers performed computations at the B3LYP/6-311++G(d,p) level on the electronic structures and energies of the $HN_5...H_2O$ cluster and the $N_5^-...H_3O^+$ cluster, modeled in the vacuum. Their results indicated that neutral HN_5 is favored over the N_5^- anion, so the authors questioned the correctness of the reported $(N_5)_6(H_3O)_3(NH_4)_4Cl$ crystal structure [104]. Zhang et al. carried out crystal-level DFT and AIMD calculations for $(N_5)_6(H_3O)_3(NH_4)_4Cl$ compound, and also performed the same calculations as Huang for comparison [6, 108, 109]. The results suggested completely opposite

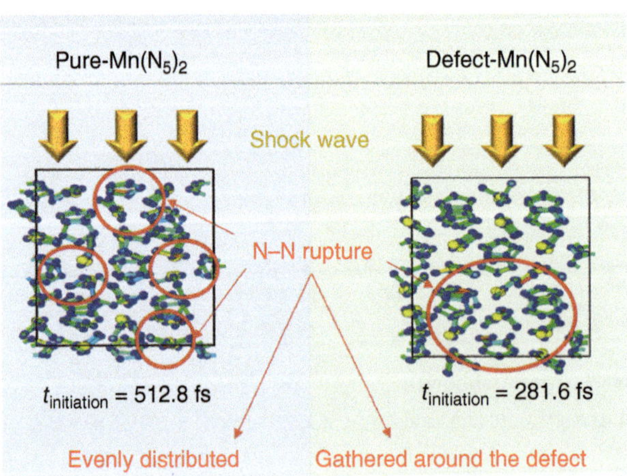

Figure 3.10 Effect of the microdefect vacancy presence in the pentazolate salt crystal. Source: Yao et al. [33]/with permission of Chinese Journal of Energetic Materials.

transfer directions of *cyclo*-N_5^- in the $(N_5)_6(H_3O)_3(NH_4)_4Cl$ crystal and in the isolated N_5^-...H_3O^+ cluster. The proton transfers to H_2O in the crystal, in contrast, it transfers to N_5^- anion in the isolated cluster. Both proton transfers proceed freely, with each energy barrier being zero. In addition, the electronic structure and energetics calculations of the $(N_5)_6(H_3O)_3(NH_4)_4Cl$ crystal also suggest the stability of $(N_5)_6(H_3O)_3(NH_4)_4Cl$ crystal, thereby supporting the original publication by Hu and Lu [65].

Inspired by the above-mentioned debate, more theoretical studies were carried out, searching for the reactivity of *cyclo*-N_5^- to acid radicals. Huang employed DFT and AIMD methods to explore the proton transfer mechanism in $(N_5)_6(H_3O)_3(NH_4)_4Cl$ and proposed that temperature should be the key factor in the activation of the proton transfer from H_3O^+ to *cyclo*-N_5^- [110]. At a temperature below 123 K, the proton transfer from H_3O^+ to *cyclo*-N_5^- is not activated, so the originally reported $(N_5)_6(H_3O)_3(NH_4)_4Cl$ crystal structure should be stable. However, when the temperature increases to room temperature or above, the proton transfer may happen at a higher probability, and HN_5 molecule is speculated to form in the crystal structure of $(N_5)_6(H_3O)_3(NH_4)_4Cl$.

3.5.3.4 Stabilization Mechanism of Dual-Aromatic *cyclo*-N_5^-

Besides the above augment on the protonation of *cyclo*-N_5^- under various conditions, the reason why *cyclo*-N_5^- could be isolated only from acidic solutions at ambient pressure is unclear. Therefore, Zhang and coworkers studied the stabilization mechanism of *cyclo*-N_5^- by calculating the electron density, molecular orbitals, structure, magnetic shielding, energetics, and dynamics of the *cyclo*-N_5^- anion in various acidic surroundings. These calculations lead to the discovery of the abnormal dual aromaticity (π and σ) of *cyclo*-N_5^-, which accounts for its instability in nature, but enhances stability in acidic solutions (Figure 3.11) [109].

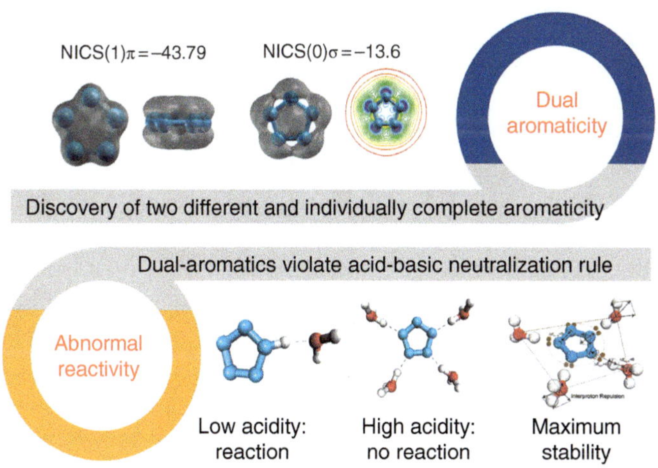

Figure 3.11 Discovery of dual-aromaticity of *cyclo*-N_5^- and its abnormal reactivity violating acid-basic neutralization rule. Source: Adapted from Refs. [108, 109, 111–113].

The "dual" aromaticity of *cyclo*-N_5^- has two meanings: (i) two types of electrons and (ii) two sets of aromatic systems. In addition to the usual π-aromaticity, which is similar to the prototypical benzene, five lone pairs are delocalized in the equatorial plane of *cyclo*-N_5^-, forming additional σ-aromaticity. This discovery is supported by nine types of aromaticity criteria, i.e. harmonic oscillator model of aromaticity (HOMA), Bird index (BI), nucleus-independent chemical shifts (NICS), isochemical shielding surface (ICSS), electron localization function (ELF), Shannon aromaticity (SA), and the curvature of electron density perpendicular to the ring plane at the ring critical point (RCP ρ curvature). The NBO analysis and the quantum theory of atoms in molecule (QTAIM) method were used to probe the contributions to the *cyclo*-N_5^- anion's aromaticity solely in terms of π or lone-pair electrons [109].

The dual aromaticity in *cyclo*-N_5^- is very different from the double aromaticity of metal atoms containing compounds. The dual aromaticity of, for example, Al_4^{2-} dianion means that four σ electrons and two π electrons together form a single aromatic system, due to the electron deficiency of the dianion [114–116]. In contrast, the dual aromaticity of the *cyclo*-N_5^- anion is derived from two separate aromatic systems, with independently delocalized π-electrons and σ(LP)-electrons, like those present in 3,5-dehydrophenyl cation [117], saturated inorganic rings [118], and probably in pnictogen five-membered rings like P_5^- and As_5^- anions [119, 120]. The dual aromaticity discovered in *cyclo*-N_5^- is expected to be a universal nature of the five-membered pnictogen ring systems and five-membered C–N heterocycles. However, it is not present in six-membered C–N heterocycles [109, 111, 119, 120].

The competition between the dual aromaticity of *cyclo*-N_5^- and the proton affinity of the nitrogen lone pairs drives the *cyclo*-N_5^- to show distinct reactivity under different surroundings. If the surrounding protons of *cyclo*-N_5^- are less than three, the system will sacrifice its dual aromaticity to allow protonation. If the surrounding protons are more than three, the nitrogen lone pairs will give up their proton affinity to keep the dual aromaticity intact. Then, the *cyclo*-N_5^- turns to be unsusceptible to that electrophilic attack but will violate the acid-basic neutralization rule, and the *cyclo*-N_5^- will remain to be unprotonated. Comparison between the calculated and experimental vibrational spectra has confirmed the unprotonated state of the *cyclo*-N_5^- anion in the crystal structure of the pentazolate salt $(N_5)_6(H_3O)_3(NH_4)_4Cl$, where the structures of metal-containing pentazolate hydrate crystals $[Na(H_2O)(N_5)]\cdot 2H_2O$, $[Co(H_2O)_4(N_5)_2]\cdot 4H_2O$ and $[Mg(H_2O)_6(N_5)_2]\cdot 4H_2O$ verify the above-described theory.

Only in a sufficiently acidic solution, the *cyclo*-N_5^- can have all its five N—N bonds strengthened and gain extra stability through the formation of the hydrogen-bonded complex from surrounding electrophiles. Otherwise, the *cyclo*-N_5^- cannot be isolated. To realize high-yield synthesis of the *cyclo*-N_5^- compounds, the proton concentration (H_3O^+ or NH_4^+) should be controlled and kept three times higher (optimally four times higher) than that of *cyclo*-N_5^- in the solution.

In polar organic solution of THF or neutral organic solution of benzene, the hydrogen bonding interactions with the *cyclo*-N_5^- were calculated to be very weak. There is no such reverse of reactivity of *cyclo*-N_5^- with the increase of the solvents. Consequently, even if the surrounding THF or benzene molecules are more than three, they have little influence on the strengthening of the structure of *cyclo*-N_5^-. Above reactivity and mechanism found for *cyclo*-N_5^- were also applicable to tetrazole anion $N_4C_1H_1^-$, and 1,2,4-triazole anion $N_3C_2H_2^-$ [112].

Further investigation revealed that it is the type and number of the adjacent counter ions of the *cyclo*-N_5^- anion, but not the crystal packing compression, that fundamentally accounts for its protonation state reverse in the vacuum and in the crystal [108]. The net charge $-e$ is found to be crucial to the formation of the dual aromaticity of pnictogen five-membered ring systems [108]. Once ionized to be neutral, more lone-pair electrons will transfer from the second-HOMO to the HOMO. Such electron redistribution, at the energy levels, will significantly increase the total energy of the system, making the neutral *cyclo*-N_5 lose its stability, and its molecular structure will quickly break down into $N_2 + N_3$ fragments [108].

3.5.4 All-Nitrogen Cluster N_4

N_4 is an up-and-coming HEDM candidate, which is not only of great interest as an explosive but also as a rocket propellant. Its specific impulse was predicted to be 408 seconds, and the combustion chamber temperature was predicted to be 7500 K [17]. In 1996, Korkin and coworkers carried out *ab initio* calculations, including coupled-cluster theory for N_4, a 28-electrons tetraatomic molecule. The authors predicted that tetraazatetra-hedrane might be metastable and long lived. The dissociation of N_4 into two N_2 molecules is very exothermic, releasing energy of about 800 kJ mol^{-1}. Therefore, N_4 is a promising candidate for HEDMs that are environmentally benign [121]. Experimentally, neutral N_4 was detected in 2002, but its lifetime was found to be extremely short, at the level of few μs [122]. Nevertheless, the reported stabilization of the tetratomic molecules of phosphorus and arsenic, which are other elements in the pnictogen group (including nitrogen), lead us to expect possible stabilization of N_4. In 2018, FeN_4 was synthesized through a direct reaction between Fe and N_2 in a laser-heated diamond anvil cell at 106 GPa [123]. In the product, $[N_4^{2-}]_n$ presents a polymeric nitrogen chain structure, which is stabilized with the aid of iron ions. However, FeN_4 was found to lose its stability once the high pressure was removed.

3.5.5 All-Nitrogen Cluster N_6

The N_6 molecule has been intensively investigated theoretically. In 1981, Ha et al. carried out DFT and CI calculations for N_6 structure having D_{6h} symmetry. Their results confirmed that free N_6 is not stable. However, it could be formed in a matrix, at low temperature, during a photochemical reaction, although for a very short lifetime [124].

In 1992, Glukhovtsev and Schleyer performed high-accuracy level quantum mechanical calculations on the structures, bonding, and energies of N_6 isomers. The homodesmotic and hyper-homodesmotic reaction energies were calculated to be −17.6 and −10.4 kcal mol^{-1}, respectively. These results further confirmed that the isomer with D_{6h} symmetry is destabilized. In addition, N_6 was found to prefer a nonplanar twist-boat structure with D_2 symmetry, instead of a planar structure resembling benzine. Nevertheless, the isomer with C_2 symmetry was predicted to have even higher stability by 26.0 kcal mol^{-1} than the one with D_2 symmetry [125].

Very recently, Zhang and coworkers carried out DFT calculations for the molecular orbital, electronic structure, and aromaticity of the N_6 isomer with D_{6h} symmetry. It was found that after geometry optimization, N_6 isomer presented slight distortion from the planar structure (by 0.18 Å). The lowest σ molecular orbital spreads out over each molecule into a flower shape in the equatorial plane, similarly to *cyclo*-N_5^-. Unlike other six-membered C–N heterocyclic molecules, N_6 presents a second σ-aromaticity in addition to the proto-typical π-aromaticity. It is the compatible coexistence of the inter-lone-pair repulsion and inter-lone-pair attraction, within the σ-aromatic system that makes the N_6 unstable [109, 111].

3.5.6 All-Nitrogen Crystals Composed of N_8

N_8 isomer was calculated to have a specific impulse (I_{sp}) of 500 seconds, which is 92 seconds higher than the I_{sp} of N_4 isomer. It was reported that an increase of 20 seconds in the specific impulse can allow a doubling in a payload. Therefore synthesis of this isomer would be of extraordinary value [17].

In 2001, Fau investigated possible products of the end-on addition of N_3^- to N_5^+ and searched for transition states to assess their stability. Quantum mechanical calculations were performed for the structure optimization and energetics evaluations at the B3LYP/aug-cc-pVDZ and MBPT(2)/aug-cc-pVDZ levels of theory. Isolation of covalently bonded N_8 from N_3^- to N_5^+ was found to be difficult because the most likely product had a decomposition barrier of only 18 kcal mol^{-1}. Because one of the approach pathways has great potential for mutual neutralization and subsequent fragmentation, N_8 may not be formed at all. In the same year, Gagliardi carried out quantum mechanical calculations on the possibility of forming N_8 from *cyclo*-N_5^- and N_3^+. However, no stable N_8 form was found [79]. In 2013, Hirshberg reported a theoretical prediction of a molecular crystalline structure consisting of linear N_8 molecules [126]. It was not obtained by compressing N_2 molecules, but was spontaneously generated during the optimization of a tetramer of N_4. The N_8 monomer was found to retain the same structure and bonding pattern as those in the gas phase. The crystal was stabilized by weak van der Waals and electrostatic forces, presenting high cohesive energy of 41.1 kJ mol^{-1}, almost five times stronger than that of the solid N_2. The N_8 molecular crystal was found to be metastable even at ambient pressure. Under pressures lower than 20 GPa, the N_8 molecular crystal was found to be more stable than other reported polymeric forms of nitrogen. The N_8 solid was calculated to be highly energetic upon decomposition into nitrogen molecules, with an energy release of 260 kcal mol^{-1}. The predicted N_8 molecular crystal is a promising candidate as HEDM, but presents a challenge to its synthesis (Figure 3.12).

Besides the above-described linear molecular structure, quantum mechanical calculations suggested that azido-pentazole is possibly the N_8 isomer with global minimum energy. Azidopentazole also has a significant energy barrier with respect to ring-closing deformation and thereby is a possible synthetic target. The most

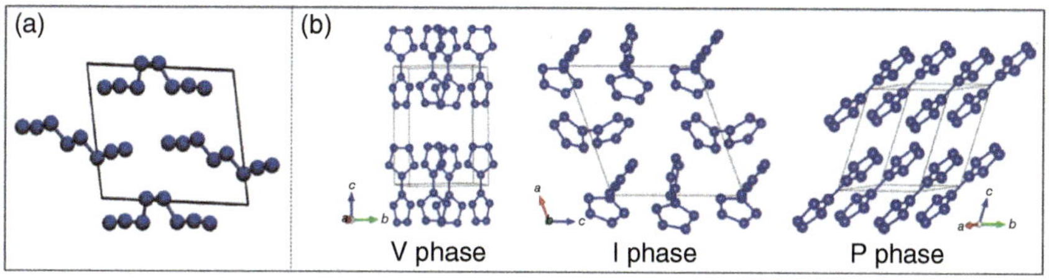

Figure 3.12 (a) Unit cell of the predicted N_8 molecular crystal. Source: Reproduced from Hirshberg et al. [126]/Macmillan Publishers Limited. (b) Energetically favorable structure of V phase, I phase, and P phase of predicted N_{10} molecular crystal. Source: Liu et al. [127]/ John Wiley & Sons/Public Domain CC BY.

promising route was suggested to be the reaction of phenyl-pentazole with a covalent azide compound [17].

3.5.7 All-Nitrogen Crystals Composed of N_{10}

To search for all-nitrogen compounds, with exceedingly high-energy density, theoretical calculations have been carried out to find stable molecular and crystal structures consisting of only nitrogen element [79, 127, 128].

On the molecular level, Manaa carried out high-level accuracy electronic structure calculations for two isomers of bicyclic N_{10} in 2000 [128]. The bridging N—N bond in both N_{10} was strong, yet, it was flexible to allow a facile rotation of one ring with respect to the other. The planar bicyclic structure (D_{2h} symmetry) was a saddle point for N_{10}, and a corresponding orthogonal structure (D_{2d} symmetry) was the equilibrium structure with lower energy by 3–6 kcal mol^{-1} [128]. The N_{10} structure was predicted to be remarkably stable with respect to N–N homolysis, within thc polynitrogen ring, and was postulated to serve as a building block for a potential super-high-energy content N_{60} cluster. In 2001, Gagliardi et al. carried out calculations for N_{10}, using multiconfigurational self-consistent-field wave functions and second-order perturbation theory [79]. N_{10} was predicted to be a stable compound in the form of ion-pair $N_5^+ + N_5^-$ [128]. But, its energy was reported to be by 24 kcal mol^{-1} less stable than the bicyclic form of N_{10} described by Manaa [79, 128].

On the crystal level, Liu et al. carried out search for an all-nitrogen crystal by a combination of DFT calculations. In 2020, they predicted a stable molecular crystal consisting only of bispentazole N_{10} [127]. In each of the N_{10} molecules, two N_5 polynitrogen rings were connected to each other, making the polynitrogens aromatic and stable, without introducing any metal or other guest stabilizer. The crystal structures and the molecular conformations could be changed by the application of an external pressure. Phonon calculations and AIMD simulations have been utilized to confirm the stability of these molecular crystals at room temperature and even at higher temperatures. These predicted molecular crystals were found to be recoverable upon the release of the pressure, indicating that these structures have the possibility of their synthesis. The decomposition processes have been found to start from N–N rupture, within the N_5 polynitrogen ring, releasing a large amount of energy. This theoretical work showed the feasibility of experimental synthesis of all-nitrogen HEDMs (Figure 3.12) [127].

3.5.8 All-Nitrogen HEDMs with Cubic Gauche Structure

Cubic gauche structure (cg-N) refers to a polymeric nitrogen structure with all-cubic gauche dihedral angles, wherein each atom of nitrogen is connected to three neighbors with three single covalent bonds. In 1992, Mailhiot et al. carried out extensive DFT calculations of the equation of state of polymeric nitrogen structures and diatomic nitrogen structures. The diatomic phase of nitrogen ε-N_2 was predicted to keep metastable up to 180 GPa and room temperature. Under 50 ± 15 GPa, the ε-N_2 structure was transformed into the cg-N structure. The authors performed total energy calculation along the selected high-symmetry transformation paths, between the cg-N structure and the decomposition product N_2. All cases produced significant barriers that would inhibit the proceeding of the decomposition. Therefore, the cg-N structure is likely to exit under atmospheric pressure [129].

In 1993, the metastability of the cg-N structure was further studied by Barbee, using linear response theory to calculate the phonon dispersion, indicating the cg-N structure to be mechanically stable at pressures below 360 GPa [130]. In 2003, Alemany et al. performed AIMD simulations and structural optimizations of nitrogen at pressures up to 500 GPa. The stable crystalline phase at pressures between 27 and 205 GPa has been shown to be the cg-N structure [131].

Soon after, Mattson et al. reported extensive DFT calculations and AIMD simulations that predicted five new polynitrogen structures and the transitions of these phases under different conditions [132]. These five structures included α phase, ε phase, cg-N, *Cmcm* chain structure, and a hexagonal molecular form based on N_2 and N_6 units. Among them, α phase with a cubic *Pa*3 structure was predicted to be stable at low pressure; ε phase with a hexagonal structure was predicted to be stable at higher pressure. *Cmcm* chain structure was shown to be energetically competitive with the cg-N structure, and the N_2 and N_6 composed hexagonal molecular phase was not energetically favored structure. However, it could be possibly observed due to large barriers to transitions into lower energy structures. Large energy barriers between structures were considered to be responsible for the hysteresis and metastability of covalent nitrogen solids [132].

Wang et al. performed DFT calculations to explore the stability of different proposed polymeric nitrogen. It was found that at a pressure below 47 GPa, the molecular phase ε-N_2 is the lowest-enthalpy structure in solid nitrogen. With the increase of the pressure, the molecular phase ε-N_2 transfers to cg-N phase at 47 GPa, to *Pba*2 at 170 GPa, and then to *P*212121 at 307 GPa [133].

The cg-N polymeric nitrogen was eventually synthesized in 2004. The cg-N structure was synthesized directly from molecular nitrogen at temperatures above 2000 K and pressures above 110 GPa, using a laser-heated diamond cell, and it was found to be metastable down to 42 GPa at room temperature [134]. All-nitrogen atoms were found to be connected *via* single covalent bonds, forming a cg-N structure that

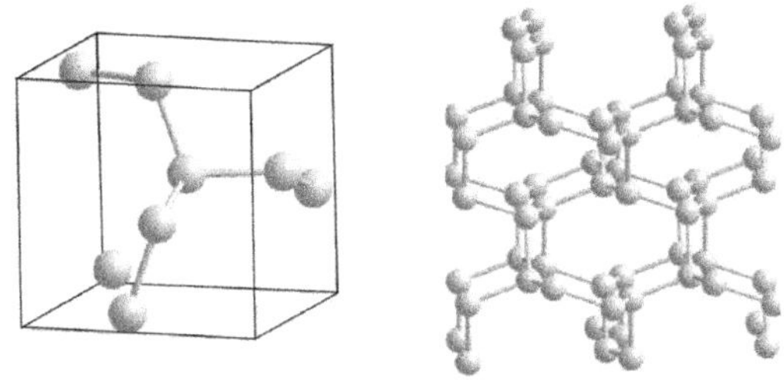

Figure 3.13 cg-N structure: primitive cell and extended structure. Source: Eremets et al. [134]/with permission of Springer Nature.

perfectly agrees with the theoretical prediction. The unique structure of cg-N leads to two important properties. It has been predicted to have extraordinarily high bulk modulus ≥300 GPa and is expected to have more than five times the energy capacity of contemporarily used energetic materials (Figure 3.13) [134].

3.6 Final Remarks

Although there has been tremendous progress in the computation of nitrogen-rich HEDMs, several problems are still needed to be addressed.

The first problem is related to the solid effect, although known to be very important to the understanding of physicochemical properties of nitrogen-rich HEDMs, it was not considered in most of the currently reported research. The quantum mechanical calculations mainly focus on the structure and energy properties of individual molecules, rather than those of crystals, owing to the lack of a reliable approach to handle the periodic structures of crystals and the weak in-crystal intermolecular interactions.

The second problem is related to the calculation results of the detonation parameters for nitrogen-rich HEDMs that are often overestimated in comparison to the real value. The underlying reasons are various and could be very complicated. One of the reasons is that a density value is used regardless of the molecular packing of the assessed compounds. Another reason is the inclusion of rich nitrogen elements in the HEDMs, which makes the empirical Kamlet-Jacobs formula no longer valid to generate reliable detonation velocities and detonation pressures. For novel nitrogen-rich HEDMs, the introduction of metal or elements other than C, H, N, and O makes an accurate description of the equation of states of the detonation products impossible, using most current calculation methods.

Over the past decade, machine learning algorithms have emerged, rapidly developed, and extensively used in materials research. Compared to the traditional physics-based or knowledge-driven approach, machine learning algorithms are data driven and have an important characteristic trait of easily handling very complex datasets, providing accurate predictions, and helping in deciphering of new knowledge. Therefore, machine learning is an extremely promising technique to handle the detonation-stability contradiction conundrum and accelerate the progress in the design of advanced HEDMs. Regretfully, this advanced method has not been extensively utilized yet in the field of HEDMs.

Acknowledgement

This work is supported by the National Natural Science Foundation of China (Grant No. 12072045 and 12222204).

References

1 Cramer, C.J. (2004). *Essentials of Computational Chemistry: Theories and Models*, 2e. Chichester, England: Wiley.
2 Wejsa, J., and U.S. Army Armament Research, Development, & Engineering Center (ARDEC) Energetics. (2014). *Joint Armaments Conference*, Phoenix, Arizona, USA (12–15 May).
3 Rice, B., Byrd, E., and Mattson, W.D. (2007). Computational aspects of nitrogen-rich HEDMs. In: *High Energy Density Materials*, vol. 125, 153–194. Berlin, Heidelberg: Springer.
4 Zhang, L., Jiang, S.-L., Yu, Y. et al. (2016). Phase transition in octahydro-1,3,5,7-tetranitro-1,3,5,7-tetrazocine (HMX) under static compression: an application of the first-principles method specialized for CHNO solid explosives. *J. Phys. Chem. B* 120 (44): 11510–11522.
5 Zhang, L., Jiang, S.-L., Yu, Y., and Chen, J. (2019). Revealing solid properties of high-energy-density molecular cocrystals from the cooperation of hydrogen bonding and molecular polarizability. *Sci. Rep.* 9 (1): 1257.
6 Jiang, C., Zhang, L., Sun, C. et al. (2018). Response to comment on "Synthesis and characterization of the pentazolate anion *cyclo*-N_5^- in $(N_5)_6(H_3O)_3(NH_4)_4Cl$". *Science* 359 (6381): eaas8953.
7 Mo, Z., Zhang, A., Cao, X. et al. (2010). JASMIN: a parallel software infrastructure for scientific computing. *Front. Comput. Sci. China* 4 (4): 480.
8 Mo, Z., Zhang, A., Liu, Q., and Cao, X. (2015). Research on the components and practices for domain-specific parallel programming models for numerical simulation. *Sci. Sin. Inf.* 45 (3): 385–397.
9 Mo, Z., Zhang, A., Liu, Q., and Cao, X. (2016). Parallel algorithm and parallel programming: from specialty to generality as well as software reuse. *Sci. Sin. Inf.* 46 (10): 1392–1410.
10 Zahrt, A.F., Henle, J.J., Rose, B.T. et al. (2019). Prediction of higher-selectivity catalysts by computer-driven workflow and machine learning. *Science* 363 (6424): eaau5631.
11 Sanchez-Lengeling, B. and Aspuru-Guzik, A. (2018). Inverse molecular design using machine learning: generative models for matter engineering. *Science* 361 (6400): 360.
12 Lu, S., Zhou, Q., Ouyang, Y. et al. (2018). Accelerated discovery of stable lead-free hybrid organic–inorganic perovskites via machine learning. *Nat. Commun.* 9 (1): 3405.
13 Zhong, M., Tran, K., Min, Y. et al. (2020). Accelerated discovery of CO_2 electrocatalysts using active machine learning. *Nature* 581 (7807): 178–183.

14 Granda, J.M., Donina, L., Dragone, V. et al. (2018). Controlling an organic synthesis robot with machine learning to search for new reactivity. *Nature* 559 (7714): 377–381.

15 Butler, K.T., Davies, D.W., Cartwright, H. et al. (2018). Machine learning for molecular and materials science. *Nature* 559 (7715): 547–555.

16 Li, C.-Y., Li, H., Zong, H.-H. et al. (2020). Strategies for achieving balance between detonation performance and crystal stability of high energy density materials. *iScience* 23 (3): 100944.

17 Klapötke, T.M. (2015). *Chemistry of High-Energy Materials*, 3e. De Gruyter.

18 Shukla, M.K., Boddu, V.M., Steevens, J.A. et al. (2017). *Energetic Materials: From Cradle to Grave*, vol. 25. Cham, Switzerland: Springer International Publishing.

19 Shukla, M.K., Boddu, V.M., Steevens, J.A. et al. (2017). *Challenges and Advances in Computational Chemistry and Physics*. Cham, Switzerland: Springer Nature.

20 Politzer, P. and Murray, J.S. (2003). Energetic materials, Part 1. Decomposition, crystal, and molecular properties. In: Amsterdam: Elsevier.

21 Sabin, J.R. (2014). *Advances in Quantum Chemistry*. Waltham, USA: Academic Press.

22 Zhang, C., Xue, X., Cao, Y. et al. (2014). Toward low-sensitive and high-energetic co-crystal II: structural, electronic and energetic features of CL-20 polymorphs and the observed CL-20-based energetic-energetic co-crystals. *CrystEngComm* 16 (26): 5905–5916.

23 Yu, Q., Yang, H., Imler, G. et al. (2020). Derivatives of 3,6-bis(3-aminofurazan-4-ylamino)-1,2,4,5-tetrazine: excellent energetic properties with lower sensitivities. *ACS Appl. Mater. Interfaces* 12 (28): 31522–31531.

24 Tang, Y., Huang, W., Imler, G. et al. (2020). Enforced planar FOX-7-like molecules: a strategy for thermally stable and insensitive π-conjugated energetic materials. *J. Am. Chem. Soc.* 142 (15): 7153–7160.

25 Jiao, F., Xiong, Y., Li, H., and Zhang, C. (2018). Alleviating the energy & safety contradiction to construct new low sensitive and high energetic materials through crystal engineering. *CrystEngComm* 20 (13): 1757–1768.

26 Lee, K.-Y., Chapman, L.B., and Cobura, M.D. (1987). 3-Nitro-1,2,4-triazol-5-one, a less sensitive explosive. *J. Energetic Mater.* 5 (1): 27–33.

27 Trzciński, W.A. and Szymańczyk, L. (2005). Detonation properties of low-sensitivity NTO-based explosives. *J. Energetic Mater.* 23 (3): 151–168.

28 Struyve, D. and Songy, C. (2015). Qualification of a new IM composition and first evaluation in 90 mm HESH-T ammunition. *2015 Insensitive Munitions & Energetic Materials Technology Symposium*, Roma, Italy (18–21 May 2015).

29 Wu, J.-T., Zhang, J.-G., Li, T. et al. (2015). A novel cocrystal explosive NTO/TZTN with good comprehensive properties. *RSC Adv.* 5 (36): 28354.

30 Li, J.C., Jiao, Q.J., Gong, Y.G. et al. (2018). Explosive performance of HMX/NTO co-crystal. *IOP Conf. Ser.: Mater. Sci. Eng.* 292: 012032.

31 Zhang, L., Wu, J.-Z., Jiang, S.-L. et al. (2016). From intermolecular interactions to structures and properties of a novel cocrystal explosive: a first-principles study. *Phys. Chem. Chem. Phys.* 18 (38): 26960–26969.

32 Zhang, L., Yu, Y., and Xiang, M.-Z. (2019). A study of the shock sensitivity of energetic single crystals by large-scale ab initio molecular dynamics simulations. *Nanomaterials* 9 (9): 1251.

33 Yao, C., Yang, Y., Yu, Y. et al. (2020). Effect of vacancy defect on shock reaction and damage evolution of pentazolate salt $Mn(N_5)_2$. *Chin. J. Energetic Mater.* 28 (10): 1003–1009.

34 Yao, C., Yang, Y., Yu, Y. et al. (2020). How stable and powerful can metal *cyclo*-pentazolate salts be? An answer through theoretical crystal design. *Cryst. Growth Des.* 20 (7): 4794–4801.

35 Politzer, P., Grice, M.E., and Seminario, J.M. (1997). Density functional analysis of a decomposition of 4-nitro-1,2,3-triazole through the evolution of N_2. *Int. J. Quantum Chem.* 61 (3): 389–392.

36 da Silva, G. and Bozzelli, J.W. (2008). Retro-[3+2]-cycloaddition reactions in the decomposition of five-membered nitrogen-containing heterocycles. *J. Org. Chem.* 73 (4): 1343–1353.

37 Kowhakul, W., Inoue, D., Nakagawa, Y. et al. (2017). Thermal decomposition mechanisms of 1*H*-1,2,4-triazole derivatives: a theoretical study. *J. Loss Prev. Process Ind.* 50: 37–54.

38 Korolev, V.L., Petukhova, T.V., Pivina, T.S. et al. (2006). Thermal decomposition mechanisms of nitro-1,2,4-triazoles: a theoretical study. *Russ. Chem. Bull.* 55 (8): 1388–1410.

39 Lu, M., Zhou, P., Yang, Y. et al. (2020). Thermochemistry and initial decomposition pathways of triazole energetic materials. *J. Phys. Chem. A* 124 (15): 2951–2960.

40 Warner, K.F. and Granholm, R.H. (2011). Synthesis of insensitive 5,5′-azotetrazolate salts. *J. Energetic Mater.* 29 (1): 1–6.

41 Klapötke, T.M. and Sabaté, C.M. (2008). Bistetrazoles: nitrogen-rich, high-performing, insensitive energetic compounds. *Chem. Mater.* 20 (11): 3629–3637.

42 He, P., Zhang, J.-G., Wang, K. et al. (2014). Calculations predict a novel desired compound containing eight catenated nitrogen atoms: 1-amino-tetrazolo-[4,5-*b*]tetrazole. *RSC Adv.* 4 (48): 25302–25309.

43 He, P., Zhang, J.-G., Feng, L.-N. et al. (2014). Theoretical study on the tautomerization of 1,5-diaminotetrazole (DAT). *J. Mol. Model.* 20 (10): 2457.

44 He, P., Zhang, J.-G., Wang, K. et al. (2015). Theoretical study on the azido-cyclization of 3,6-diazido-1,2,4,5-tetrazine (DAT) with the dimethylsulfoxide (DMSO) solvent. *Comput. Theor. Chem.* 1068: 142–148.

45 Wang, K., Zhang, J., Shang, J., and Zhang, T. (2013). Theoretical study about the 5-azido-1*H*-tetrazole and its ion salts. *J. Mol. Model.* 19 (6): 2383–2389.

46 Kiselev, V.G., Cheblakov, P.B., and Gritsan, N.P. (2011). Tautomerism and thermal decomposition of tetrazole: high-level ab initio study. *J. Phys. Chem. A* 115 (9): 1743–1753.

47 Yuan, B., Yu, Z., and Bernstein, E.R. (2015). Initial mechanisms for the decomposition of electronically excited energetic materials: 1,5′-BT, 5,5′-BT, and AzTT. *J. Chem. Phys.* 142 (12): 124315.

48 Klapötke, T.M. (2007). New nitrogen-rich high explosives. In: *High Energy Density Materials* (ed. T.M. Klapötke), 85–121. Berlin, Heidelberg: Springer Berlin Heidelberg.

49 Fischer, N., Fischer, D., Klapötke, T.M. et al. (2012). Pushing the limits of energetic materials – the synthesis and characterization of dihydroxylammonium 5,5′-bistetrazole-1,1′-diolate. *J. Mater. Chem.* 22 (38): 20418–20422.

50 An, Q., Cheng, T., Goddard, W.A., and Zybin, S.V. (2015). Anisotropic impact sensitivity and shock induced plasticity of TKX-50 (dihydroxylammonium 5,5′-bis(tetrazole)-,1′-diolate) single crystals: from large-scale molecular dynamics simulations. *J. Phys. Chem. C* 119 (4): 2196–2207.

51 Yu, Y., Zhang, L., Jiang, S.-L. et al. (2018). Molecular simulation on the nitrogen generation in thermal decomposition of TKX-50. *Chin. J. Energetic Mater.* 26 (1): 75–79.

52 Zong, H.-H., Zhang, W.-B., Li, H.-R., and Zhang, L. (2018). Structural, mechanical and electronic properties of dihydroxylammonium 5,5′-bistetrazole-1,1′-diolate (TKX-50) under high pressures: a first-principles study. *Chin. J. Energetic Mater.* 26: 46–52.

53 An, Q., Liu, W.-G., Goddard, W.A. et al. (2014). Initial steps of thermal decomposition of dihydroxylammonium 5,5′-bistetrazole-1,1′-diolate crystals from quantum mechanics. *J. Phys. Chem. C* 118 (46): 27175–27181.

54 Lu, Z., Zeng, Q., Xue, X. et al. (2017). Does increasing pressure always accelerate the condensed material decay initiated through bimolecular reactions? A case of the thermal decomposition of TKX-50 at high pressures. *Phys. Chem. Chem. Phys.* 19: 23309–23317.

55 Talawar, M.B., Nandagopal, S., Singh, S. et al. (2018). TGA, DSC and DFT studies of TKX-50, ABTOX and their key precursors. *ChemistrySelect* 3 (43): 12175–12182.

56 Muravyev, N.V., Monogarov, K.A., Asachenko, A.F. et al. (2017). Pursuing reliable thermal analysis techniques for energetic materials: decomposition kinetics and thermal stability of dihydroxylammonium 5,5′-bistetrazole-1,1′-diolate (TKX-50). *Phys. Chem. Chem. Phys.* 19 (1): 436–449.

57 Wang, J., Chen, S., Jin, S. et al. (2020). Thermal behavior, compatibility study and safety assessment of diammonium 5,5′-bistetrazole-1,1′-diolate (ABTOX). *J. Therm. Anal. Calorim.* 139 (3): 1771–1777.

58 He, P., Zhang, J.-G., Wang, K. et al. (2015). Combination multinitrogen with good oxygen balance: molecule and synthesis design of polynitro-substituted tetrazolotriazine-based energetic compounds. *J. Org. Chem.* 80 (11): 5643–5651.

59 Latypov, N.V., Bergman, J., Langlet, A. et al. (1998). Synthesis and reactions of 1,1-diamino-2,2-dinitroethylene. *Tetrahedron* 54 (38): 11525–11536.

60 Fischer, D., Klapötke, T.M., and Stierstorfer, J. (2012). Salts of tetrazolone – synthesis and properties of insensitive energetic materials. *Propellants Explos. Pyrotech.* 37 (2): 156–166.

61 He, P., Zhang, J.-G., Wang, K. et al. (2015). Extensive theoretical studies on two new members of the FOX-7 family: 5-(dinitromethylene)-1,4-

dinitramino-tetrazole and 1,1′-dinitro-4,4′-diamino-5,5′-bitetrazole as energetic compounds. *Phys. Chem. Chem. Phys.* 17 (8): 5840–5848.

62 He, P., Zhang, J.-G., Wang, K. et al. (2016). Computational studies on two novel energetic nitrogen-rich compounds based on tetrazolone. *J. Phys. Org. Chem.* 29 (1): 29–34.

63 Curtius, T. (1890). Ueber Stickstoffwasserstoffsäure (Azoimid) N_3H. *Ber. Dtsch. Chem. Ges.* 23 (2): 3023–3033.

64 Christe, K.O., Wilson, W.W., Sheehy, J.A., and Boatz, J.A. (1999). N_5^+: a novel homoleptic polynitrogen ion as a high energy density material. *Angew. Chem. Int. Ed.* 38 (13/14): 2004–2009.

65 Zhang, C., Sun, C., Hu, B. et al. (2017). Synthesis and characterization of the pentazolate anion *cyclo*-N_5^- in $(N_5)_6(H_3O)_3(NH_4)_4Cl$. *Science* 355 (6323): 374–376.

66 Christe, K.O., Dixon, D.A., Vasiliu, M. et al. (2019). How energetic are *cyclo*-pentazolates? *Propellants Explos. Pyrotech.* 44 (3): 263–266.

67 Sun, C., Zhang, C., Jiang, C. et al. (2018). Synthesis of AgN_5 and its extended 3D energetic framework. *Nat. Commun.* 9 (1): 1269.

68 Brazier, C., Bernath, P., Burkholder, J.B., and Howard, C.J. (1988). Fourier transform spectroscopy of the $\nu 3$ band of the N_3 radical. *J. Chem. Phys.* 89 (4): 1762–1767.

69 Meyer, R., Köhler, J., and Homburg, A. (2007). *Explosives*. Weinheim: Wiley-VCH & Co. KGaA.

70 Yedukondalu, N., Vaitheeswaran, G., Anees, P., and Valsakumar, M. (2015). Phase stability and lattice dynamics of ammonium azide under hydrostatic compression. *Phys. Chem. Chem. Phys.* 17 (43): 29210–29225.

71 Yu, H., Duan, D., Tian, F. et al. (2015). Polymerization of nitrogen in ammonium azide at high pressures. *J. Phys. Chem. C* 119 (45): 25268–25272.

72 Hu, A. and Zhang, F. (2010). A hydronitrogen solid: high pressure ab initio evolutionary structure searches. *J. Phys. Condens. Matter* 23 (2): 022203.

73 Frevel, L.K. (1936). The crystal structure of ammonium azide, NH_4N_3. *Z. Kristallogr. – Cryst. Mater.* 94 (1–6): 197–211.

74 Medvedev, S., Palasyuk, T., Trojan, I. et al. (2012). Pressure-tuned vibrational resonance coupling of intramolecular fundamentals in ammonium azide (NH_4N_3). *Vib. Spectrosc.* 58: 188–192.

75 Wu, X., Cui, H., Zhang, J. et al. (2013). High pressure synchrotron X-ray diffraction and Raman scattering studies of ammonium azide. *Appl. Phys. Lett.* 102 (12): 121902.

76 Yedukondalu, N., Ghule, V.D., and Vaitheeswaran, G. (2012). Computational study of structural, electronic, and optical properties of crystalline NH_4N_3. *J. Phys. Chem. C* 116 (32): 16910–16917.

77 Yakovleva, G., Kurbangalina, R.K., and Stesik, L. (1977). Detonation properties of ammonium azide. *Combust. Explos. Shock Waves* 13 (3): 405–407.

78 Nguyen, M.T. and Ha, T.-K. (2000). Theoretical study of the pentanitrogen cation N_5^+. *Chem. Phys. Lett.* 317 (1): 135–141.

79 Gagliardi, L., Orlandi, G., Evangelisti, S., and Roos, B.O. (2001). A theoretical study of the nitrogen clusters formed from the ions N_3^-, N_5^+, and N_5^-. *J. Chem. Phys.* 114 (24): 10733–10737.

80 Dixon, D.A., Feller, D., Christe, K.O. et al. (2004). Enthalpies of formation of gas-phase N_3, N_3^-, N_5^+, and N_5^- from ab initio molecular orbital theory, stability predictions for $N_5^+N_3^-$ and $N_5^+N_5^-$, and experimental evidence for the instability of $N_5^+N_3^-$. *J. Am. Chem. Soc.* 126 (3): 834–843.

81 Pyykkö, P. and Runeberg, N. (1991). Ab initio studies of bonding trends: part 9. The dicyanamide-carbon suboxide-dicyanoether-cyanogen azide isoelectronic series A=B=C=D=E[1]. *J. Mol. Struct. THEOCHEM* 234: 279–290.

82 Christe, K.O. (2007). Recent advances in the chemistry of N_5^+, N_5^- and high-oxygen compounds. *Propellants Explos. Pyrotech.* 32 (3): 194–204.

83 Hantzsch, A. (1903). Ueber Diazoniumazide, Ar.N5. *Ber. Dtsch. Chem. Ges.* 36 (2): 2056–2058.

84 Dimroth, O. and de Montmollin, G. (1910). Zur Kenntnis der Diazohydrazide. *Ber. Dtsch. Chem. Ges.* 43 (3): 2904–2915.

85 Lifschitz, J. (1915). Synthese der Pentazol-Verbindungen. I. *Ber. Dtsch. Chem. Ges.* 48 (1): 410–420.

86 Curtius, T., Darapsky, A., and Müller, E. (1915). Die sogenannten Pentazol-Verbindungen von J. Lifschitz. *Ber. Dtsch. Chem. Ges.* 48 (2): 1614–1634.

87 Vij, A., Pavlovich, J.G., Wilson, W.W. et al. (2002). Experimental detection of the pentaazacyclopentadienide (pentazolate) anion, *cyclo*-N_5^-. *Angew. Chem. Int. Ed.* 41 (16): 3051–3054.

88 Butler, R.N., Stephens, J.C., and Burke, L.A. (2003). First generation of pentazole (HN_5, pentazolic acid), the final azole, and a zinc pentazolate salt in solution: a new N-dearylation of 1-(*p*-methoxyphenyl) pyrazoles, a 2-(*p*-methoxyphenyl) tetrazole and application of the methodology to 1-(*p*-methoxyphenyl) pentazole. *Chem. Commun.* 0 (8): 1016–1017.

89 Perera, S.A., Gregusova, A., and Bartlett, R.J. (2009). First calculations of 15N–15N J values and new calculations of chemical shifts for high nitrogen systems: a comment on the long search for HN_5 and its pentazole anion. *J. Phys. Chem. A* 113 (13): 3197–3201.

90 Bazanov, B., Geiger, U., Carmieli, R. et al. (2016). Detection of *cyclo*-N_5^- in THF solution. *Angew. Chem. Int. Ed.* 55 (42): 13233–13235.

91 Wenquan, Z., Kangcai, W., Juecheng, L. et al. (2018). Stabilization of the pentazolate anion in a zeolitic architecture with $Na_{20}N_{60}$ and $Na_{24}N_{60}$ nanocages. *Angew. Chem. Int. Ed.* 130 (10): 2622–2625.

92 Zhang, C., Yang, C., Hu, B. et al. (2017). A symmetric $Co(N_5)_2(H_2O)_4{\cdot}4H_2O$ high-nitrogen compound formed by cobalt(II) cation trapping of a *cyclo*-N_5^- anion. *Angew. Chem. Int. Ed.* 56 (16): 4512–4514.

93 Xu, Y., Wang, Q., Shen, C. et al. (2017). A series of energetic metal pentazolate hydrates. *Nature* 549 (7670): 78–81.

94 Xu, Y., Wang, P., Lin, Q. et al. (2018). Self-assembled energetic 3D metal–organic framework $[Na_8(N_5)_8(H_2O)_3]_n$ based on *cyclo*-N_5^-. *Dalton Trans.* 47 (5): 1398–1401.

95 Xu, Y., Lin, Q., Wang, P., and Lu, M. (2018). Syntheses, crystal structures and properties of a series of 3D metal–inorganic frameworks containing pentazolate anion. *Chem. Asian J.* 13 (13): 1669–1673.

96 Shang, F., Liu, R., Liu, J. et al. (2020). Unraveling the mechanism of *cyclo*-N_5^- production through selective C—N bond cleavage of arylpentazole with ferrous bisglycinate and *m*-chloroperbenzoic acid: a theoretical perspective. *J. Phys. Chem. Lett.* 11 (3): 1030–1037.

97 Peng, F., Yao, Y., Liu, H., and Ma, Y. (2015). Crystalline LiN_5 predicted from first-principles as a possible high-energy material. *J. Phys. Chem. Lett.* 6 (12): 2363–2366.

98 Shen, Y., Oganov, A.R., Qian, G. et al. (2015). Novel lithium-nitrogen compounds at ambient and high pressures. *Sci. Rep.* 5 (1): 14204.

99 Xia, K., Zheng, X., Yuan, J. et al. (2019). Pressure-stabilized high-energy-density alkaline-earth-metal pentazolate salts. *J. Phys. Chem. C* 123 (16): 10205–10211.

100 Yu, S., Huang, B., Zeng, Q. et al. (2017). Emergence of novel polynitrogen molecule-like species, covalent chains, and layers in magnesium–nitrogen Mg_xN_y phases under high pressure. *J. Phys. Chem. C* 121 (21): 11037–11046.

101 Wei, S., Li, D., Liu, Z. et al. (2017). Alkaline-earth metal (Mg) polynitrides at high pressure as possible high-energy materials. *Phys. Chem. Chem. Phys.* 19 (13): 9246–9252.

102 Xu, Y., Ding, L., Yang, F. et al. (2022). LiN_5: a novel pentazolate salt with high nitrogen content. *Chem. Eng. J.* 429: 132399.

103 Laniel, D., Weck, G., Gaiffe, G. et al. (2018). High-pressure synthesized lithium pentazolate compound metastable under ambient conditions. *J. Phys. Chem. Lett.* 9 (7): 1600–1604.

104 Huang, R.-Y. and Xu, H. (2018). Comment on "Synthesis and characterization of the pentazolate anion *cyclo*-N_5^- in $(N_5)_6(H_3O)_3(NH_4)_4Cl$". *Science* 359 (6381): eaao3672.

105 Luo, J.-H., Chen, L.-Y., Nguyen, D.N. et al. (2018). Dual functions of water in stabilizing metal-pentazolate hydrates $[M(N_5)_2(H_2O)_4]{\cdot}4H_2O$ (M = Mn, Fe, Co, and Zn) high-energy-density materials. *J. Phys. Chem. C* 122 (37): 21192–21201.

106 Wang, P., Xu, Y., Lin, Q., and Lu, M. (2018). Recent advances in the syntheses and properties of polynitrogen pentazolate anion *cyclo*-N_5^- and its derivatives. *Chem. Soc. Rev.* 47: 7522–7538.

107 Yang, C., Zhang, C., Zheng, Z. et al. (2018). Synthesis and characterization of *cyclo*-pentazolate salts of NH_4^+, NH_3OH^+, $N_2H_5^+$, $C(NH_2)_3^+$, and $N(CH_3)_4^+$. *J. Am. Chem. Soc.* 140 (48): 16488–16494.

108 Zhang, L., Yao, C., Yu, Y. et al. (2019). Mechanism and functionality of pnictogen dual aromaticity in pentazolate crystals. *ChemPhysChem* 20 (19): 1–7.

109 Zhang, L., Yao, C., Yu, Y. et al. (2019). Stabilization of the dual-aromatic *cyclo*-N_5^- anion by acidic entrapment. *J. Phys. Chem. Lett.* 10 (10): 2378–2385.

110 Huang, H., Zhong, J., Ma, L. et al. (2019). Reconciling the debate on the existence of pentazole HN_5 in the pentazolate salt of $(N_5)_6(H_3O)_3(NH_4)_4Cl$. *J. Am. Chem. Soc.* 141 (7): 2984–2989.

111 Zong, H.-H., Yao, C., Sun, C.Q. et al. (2020). Structure and stability of aromatic nitrogen heterocycles used in the field of energetic materials. *Molecules* 25 (14): 3232.

112 Li, C., Huang, Y., Sun, C.Q., and Zhang, L. (2021). Acidic stabilization of the dual-aromatic heterocyclic anions. *Catalysts* 11 (7): 766.

113 Gebhardt, J. and Johnston, M.R. (2021). From aryl pentazoles to pentazolate salts—the continued development of *cyclo*-N_5 chemistry. In: *Reference Module in Chemistry, Molecular Sciences and Chemical Engineering*, 4e, vol. 6 (ed. D.S. Black, J. Cossy and C.V. Stevens), 514–527. Elsevier.

114 Santos, J.C., Tiznado, W., Contreras, R., and Fuentealba, P. (2004). Sigma–pi separation of the electron localization function and aromaticity. *J. Chem. Phys.* 120 (4): 1670–1673.

115 Li, X., Kuznetsov, A.E., Zhang, H.-F. et al. (2001). Observation of all-metal aromatic molecules. *Science* 291 (5505): 859–861.

116 Boldyrev, A.I. and Wang, L.-S. (2005). All-metal aromaticity and antiaromaticity. *Chem. Rev.* 105 (10): 3716–3757.

117 Chandrasekhar, J., Jemmis, E.D., and von Ragué Schleyer, P. (1979). Double aromaticity: aromaticity in orthogonal planes. The 3,5-dehydrophenyl cation. *Tetrahedron Lett.* 20 (39): 3707–3710.

118 Li, Z.-H., Moran, D., Fan, K.-N., and Schleyer, P.v.R. (2005). σ-Aromaticity and σ-antiaromaticity in saturated inorganic rings. *J. Phys. Chem. A* 109 (16): 3711–3716.

119 De Proft, F., Fowler, P.W., Havenith, R.W.A. et al. (2004). Ring currents as probes of the aromaticity of inorganic monocycles : P_5^-, As_5^-, S_2N_2, $S_3N_3^-$, $S_4N_3^+$, $S_4N_4^{2+}$, $S_5N_5^+$, S_4^{2+} and Se_4^{2+}. *Chem. Eur. J.* 10 (4): 940–950.

120 Liu, Z.-Z., Tian, W.-Q., Feng, J.-K. et al. (2005). Theoretical study on structures and aromaticities of P_5^- anion, $[Ti(\eta^5\text{-}P_5)]^-$ and sandwich complex $[Ti(\eta^5\text{-}P_5)_2]_2$. *J. Phys. Chem. A* 109 (25): 5645–5655.

121 Korkin, A.A., Balkova, A., Bartlett, R.J. et al. (1996). The 28-electron tetraatomic molecules: N_4, CN_2O, BFN_2, C_2O_2, B_2F_2, CBFO, C_2FN, and BNO_2. Challenges for computational and experimental chemistry. *J. Phys. Chem.* 100 (14): 5702–5714.

122 Cacace, F., de Petris, G., and Troiani, A. (2002). Experimental detection of tetranitrogen. *Science* 295 (5554): 480–481.

123 Bykov, M., Bykova, E., Aprilis, G. et al. (2018). Fe–N system at high pressure reveals a compound featuring polymeric nitrogen chains. *Nat. Commun.* 9 (1): 2756.

124 Ha, T.-K., Cimiraglia, R., and Nguyen, M.T. (1981). Can hexazine (N_6) be stable? *Chem. Phys. Lett.* 83 (2): 317–319.

125 Glukhovtsev, M.N. and von Ragué Schleyer, P. (1992). Structures, bonding and energies of N_6 isomers. *Chem. Phys. Lett.* 198 (6): 547–554.

126 Hirshberg, B., Gerber, R.B., and Krylov, A.I. (2014). Calculations predict a stable molecular crystal of N_8. *Nat. Chem.* 6 (1): 52–56.

127 Liu, S., Zhao, L., Yao, M. et al. (2020). Novel all-nitrogen molecular crystals of aromatic N_{10}. *Adv. Sci.* 7: 1902320.

128 Manaa, M.R. (2000). Toward new energy-rich molecular systems: from N_{10} to N_{60}. *Chem. Phys. Lett.* 331 (2): 262–268.

129 Mailhiot, C., Yang, L.H., and McMahan, A.K. (1992). Polymeric nitrogen. *Phys. Rev. B* 46 (22): 14419–14435.

130 Barbee, T.W. (1993). Metastability of atomic phases of nitrogen. *Phys. Rev. B* 48 (13): 9327–9330.

131 Alemany, M.M.G. and José Luís, M. (2003). Density-functional study of nonmolecular phases of nitrogen: metastable phase at low pressure. *Phys. Rev. B* 68 (2).

132 Mattson, W.D., Sanchez-Portal, D., Chiesa, S., and Martin, R.M. (2004). Prediction of new phases of nitrogen at high pressure from first-principles simulations. *Phys. Rev. Lett.* 93 (12): 125501.

133 Wang, X., Tian, F., Wang, L. et al. (2010). Structural stability of polymeric nitrogen: a first-principles investigation. *J. Chem. Phys.* 132 (2): 024502.

134 Eremets, M.I., Gavriliuk, A.G., Trojan, I.A. et al. (2004). Single-bonded cubic form of nitrogen. *Nat. Mater.* 3: 558.

4

Laser Ignition of Energetic Transition Metal Complexes

Maximilian Wurzenberger[1], Daniel Shem-Tov[2], and Jörg Stierstorfer[1]

[1] *Ludwig Maximilian University of Munich (LMU), Department of Chemistry, Butenandtstr. 5-13, München 81377, Germany*
[2] *Tel Aviv University, School of Chemistry, Faculty of Exact Sciences, Tel Aviv 69978, Israel*

4.1 Introduction

Since the invention of a practical laser by Maiman in 1960 [1], it has become an omnipresent and essential tool and component in countless areas of our daily life, ranging from simple light pointers, readers of optical storage media, to numerous biomedical [2], telecommunication [3], and defense applications [4]. Soon after the introduction of laser systems, they were successfully utilized for the initiation of energetic materials by Brish and Yang [5, 6]. Since the latter report, many more investigations regarding the laser ignition of energetic materials were carried out, studying the nature and the initiation mechanisms of the laser ignition phenomena, and using a variety of experimental setups [7–9]. Although the use of electrically initiated fuses (electro-detonators) for detonation of explosives is by far more prevalent than any other methodologies, the main disadvantage of this technology is related to the sensitivity of electro-detonators to static electricity, stray currents, and electromagnetic interferences. In contrast to the traditional initiation methods, such as heat, spark, impact, and friction, the laser initiation and ignition of energetic coordination compounds (ECCs) is probably one of the most advantageous methods of all, since the isolation of the energetic material from electric impulses may prevent an unintended initiation in most cases. Further benefits over the classical initiation techniques are the very short function times (<1 ms) and the possible utilization of less sensitive ignition charges, which significantly improves the handling and safety of these energetic substances [10, 11]. Although the concept of ECCs was first postulated in 1996 by Sinditskii and Serushkin [12], the investigation of energetic transition metal complexes for their initiation abilities, using laser diodes, started in 1988 by Kunz and Salas with evaluation of pentaamine-(5-cyano-1*H*-tetrazole)Co(III) perchlorate (**CP**) [13]. Due to the use of cyanogen as a starting material for the preparation of **CP**, the latter compound was replaced by tetraamine-*cis*-bis(5-nitro-1*H*-tetrazole)Co(III) perchlorate (**BNCP**) (Figure 4.1) [14].

Nitrogen-Rich Energetic Materials, First Edition. Edited by Michael Gozin and Leonid L. Fershtat.

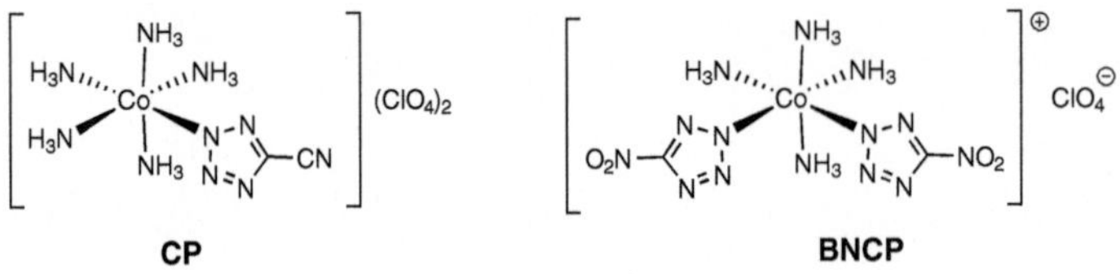

Figure 4.1 Structures of **CP** and **BNCP**.

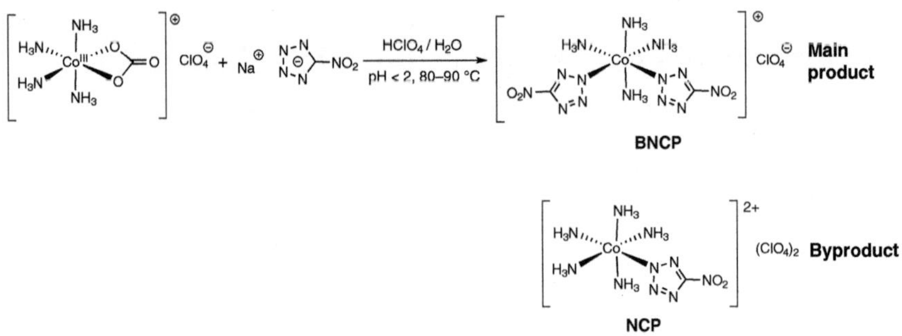

Figure 4.2 Synthesis of **BNCP**.

However, the synthesis of 5-nitrotetrazole and the target Co complex **BNCP** was found to be a challenging process. 5-Nitrotetrazole is usually prepared by the diazotization of 5-aminotetrazole, leading to the formation of the 5-diazotetrazole intermediate, which is a highly sensitive compound that tends to cause micro-detonations when carried out in batch process conditions. To circumvent this severe safety problem, use of a micro-flow reactor was preferred. **BNCP** was then synthesized by the reaction of a tetraaminecarbonato-Co(III) perchlorate complex, though the latter preparation protocol may result in formation of pentaamine-(5-nitro-1*H*-tetrazole)Co(III) perchlorate (**NCP**) as a contaminating byproduct (Figure 4.2) [15].

Despite the very fast reaction time (initiation delay time below 30 μs) and low-energy threshold (below 130 μJ) required for laser initiation, complemented with high thermostability (decomposition temperature above 230 °C), **BNCP** exhibits relatively high sensitivity toward friction (below 5 N), which is in the range of typical primary explosives, thus thwarting its broader use in optical detonators [16], especially, for the use in so-called in-line-detonators, which can be integrated into active protection systems, where materials with lower sensitivity are essential. **NCP** was also extensively investigated in recent years [17, 18]. In comparison to the **BNCP**, **NCP** has a single 5-nitro-1*H*-tetrazole and five amine ligands, which increases the stability of the latter explosive toward mechanical *stimuli* to the values comparable to the PETN [19]. Unfortunately, the biggest drawback of **NCP** was found to be its severe toxicity, emanating from its Co content, which was indicated by the authors as a general problem of all Co-containing ECCs [20, 21]. Subsequently, a search for safer laser-ignitable compounds was shifted toward metal complexes containing ligands with higher content of carbon [22].

Notably, in addition to a possible desensitization of laser-ignitable ECCs by blending them into polymeric matrices, many investigations were conducted regarding the influence of various optical (light absorbing) additives on the initiation energy threshold, particularly by different carbon-based materials, such as graphene [23] and diamond [24].

In general, the concept of the ECCs is based on three different building blocks: a cation/metal, a ligand, and an anion. The variation of the individual building blocks allows the tailoring of the complexes' physical and chemical properties, to meet specific requirements. Therefore, it is possible to control a compound's energetic nature, as well as to make adjustments to its optical characteristics [25]. The cation building block is commonly containing transition metals having the 3d configuration, including Mn, Fe, Co, Ni, Cu, and Zn. There are also more exotic energetic complexes that were reported, based on Ag [26], Cd [27], Au [28, 29], and Hg [30]. In ECCs, the metal center serves as a catalyst and matrix responsible for the cohesion between all components. The ligand component mainly functions as a fuel and frequently comes from the family of nitrogen-rich heterocycles, such as triazoles, tetrazoles, triazines, and tetrazines [31]. Many of these ligands possess a number of potential coordination sites and have a highly positive enthalpies of formation, due to their large number of N—N single and N=N double bonds, and the associated driving force for the formation of molecular nitrogen N_2. The anion building block can be used as a fine-tuning component for the energetic framework. While the anions assortment is commonly limited to chlorate and perchlorate, as was postulated by Sinditskii and Serushkin [12], in recent years it was shown that other anions, such as nitrate and azide, could be also used [26, 32, 33]. In some cases, the nitrogen-rich components can function as both the ligand and the anion [34], which narrows the possible tuning opportunities. The most important benefits of ECCs in laser ignition devices are these compounds relatively low-energy threshold for an energetic response, which in most cases is drastically lower in comparison to other high explosives [5, 6]. In general, primary explosives possess quite low laser initiation thresholds. For example, $Pb(N_3)_2$ (lead azide, LA) requires only 11 mJ cm^{-2} for its initiation, using near-infrared (NIR) laser [35]. Furthermore, the sensitivity of explosives formulations toward laser irradiation was found to be in good correlation with their sensitivity toward mechanical *stimuli*. Different formulations of PETN explosive can have a range of required exposure-activation energy (1.4–21 J cm^{-2}), which depends on various parameters, such as formulation density, light source, and addition of light-absorbing additives [36–38]. In addition to integration of **BNCP** into optical initiation devices, many other tetrazole-based ECCs were evaluated for the same purposes [39]. Zhilin and coworkers observed successful laser-induced detonation of several Co(III) complexes, containing 1*H*-tetrazol-5-amine, and 1*H*-tetrazole-1,5-diamine ligands, while attempts to initiate by laser structurally similar Co(III) complexes from the same $[Co(NH_3)_4(\text{tetrazole ligand})_n][ClO_4]_m$ family, containing 5-methyl-1*H*-tetrazole, *N*-(5*H*-tetrazol-5-ylidene)-nitramide, and 5-methyl-1*H*-tetrazol-1-amine ligands, failed, and complex of 1*H*-tetrazole produced only a combustion [39]. Gerasimov et al. reported synthesis and evaluation of Hg(II) perchlorate polymeric complex (**HMPT**) in which 5-hydrazineyl-1*H*-tetrazole was

Figure 4.3 Structures of **HTMP** and **Cu-HATr**.

used as a ligand (Figure 4.3) [40, 41]. These researchers found that **HMPT** complex dispersed in amount of 10 wt% in optically transparent matrix (**VS2**; comprised 1-methyl-5-vinyl-1*H*-tetrazole and methacrylic acid co-polymer) was ignitable with solid state Nd:YAG laser (operating at 1064 nm) in both a single pulse and in a continuously lasing regimes.

The ignition delay time of the **VS2** formulation in the single pulse regime (a single pulse length τ of 10 ns) did not exceed 400 ns, where in the continuous lasing regime (for 30 μs), the ignition delay varied from 1 to 15 μs, becoming longer when the pulse energy density decreased by 1 order of magnitude. In addition, laser initiation of the **VS2** formulation was also studied by using a blue diode laser of 1 W, operating at wavelength of 445 nm. Despite these impressive results, a high toxicity of Hg metal used in **VS2** composition, as described in the CLRTAP protocol, prevented further development of the **HMPT** [42].

To optimize the sensitivity of laser-ignitable explosives to an acceptable range, different strategies were applied by scientists around the world. In general, it can be stated that optimal properties of new laser-ignitable ECCs should achieve similar requirements from a booster explosive (impact sensitivity larger than 4 J, and friction sensitivity larger than 80 N). However, depending on the specific application, certain deviations from these values are conceivable [43]. Ilyushin investigated several azole-based ECCs [44, 45]. In addition to the **HTMP**, which contained 5-hydrazineyl-1*H*-tetrazole ligand, transition metal complexes containing structurally related ligand with decreased nitrogen content – 3-hydrazino-4-amino-1,2,4-triazole (**HATr**), were investigated [46]. The most promising ECC was found to be the Cu(II) perchlorate complex **Cu-HATr** (Figure 4.3). This compound showed the highest sensitivity toward irradiation by laser operated in a single pulse mode, with an initiation energy threshold as low as 10 mJ cm^{-2} [45]. **Cu-HATr** was found to perform better than other **HATr** complexes in this study, including perchlorate ECCs of Cd^{2+}, Ni^{2+}, and Co^{2+}. Yet, the sensitivity toward mechanical *stimuli* of **Cu-HATP** was found to be comparable with conventional primary explosives, which could be only reduced upon mixing/formulating of the **Cu-HATP** with stabilizing polymers [20, 47]. It should be mentioned that most water-soluble Co salts are classified as toxic, pushing current investigations toward "greener" metals, such as Fe and Cu.

In recent years, researchers at the Los Alamos National Laboratory focused their attention on ECCs containing various tetrazine-based ligands [25]. Chavez and coworkers synthesized a series of such compounds, based on Ni, Co, Cu, and Fe, and investigated a correlation between the metal–ligand charge transfer in these compounds and their laser initiation properties. These studies were also complemented

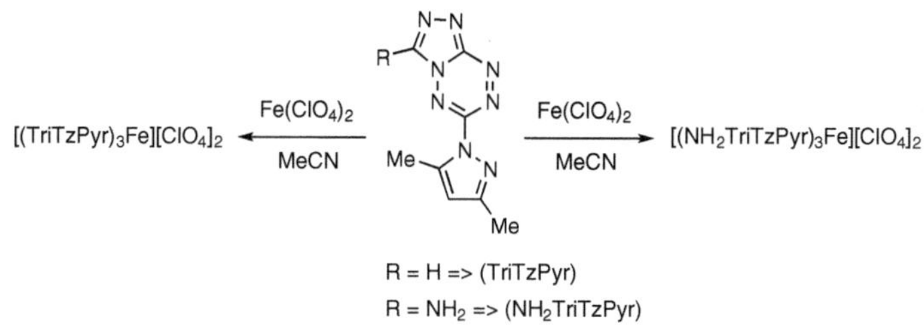

Figure 4.4 Synthesis of $[(TriTzPyr)_3Fe][ClO_4]_2$ and $[(NH_2TriTzPyr)_3Fe][ClO_4]_2$ complexes.

by extensive time-dependent density functional theory calculations [48, 49]. Although these researchers found out that it was impossible to initiate their nitrate complexes with NIR light [24, 50], they successfully proved that both the laser initiation threshold (4.5 J cm^{-2}), as well as the mechanical sensitivity could be reduced upon using Fe(II) perchlorate complexes [25]. The most promising materials consist of 6-pyrazolyl-1,2,4-triazolo[4,3-*b*][1,2,4,5]-tetrazine (TriTzPyr) and 3-amino-6-pyrazolyl-1,2,4-triazolo[4,3-*b*]-[1,2,4,5]-tetrazine (NH_2TriTzPyr), as the ligands, yielding the complexes ($[(TriTzPyr)_3Fe][ClO_4]_2$ and $[(NH_2TriTzPyr)_3Fe][ClO_4]_2$) (Figure 4.4). Upon evaluation of 3-amino-6-(4-nitro-pyrazolyl)-1,2,4-triazolo-[4,3-*b*]-[1,2,4,5]-tetrazine it was possible to show that the threshold is not only related to the absorption and explosive properties of the compound, but also from additional factors, such as the density [51]. Furthermore, this type of complexes could be also successfully applied in optical fuses to detonate PETN [52].

While laser diodes (GaAlAs) with a wavelength of 820 nm were used in the first laser initiation experiments of ECCs [13], most other researchers in around the globe utilized NIR lasers sources (Nd:YAG) operating in the range of 1064 nm [29, 50]. It should be mentioned that there is a high level of diversity with respect of the experimental setups used by various research groups. Evaluation of laser ignition performance of metal organic frameworks (MOFs) containing Cd metal center and bistetrazole-1,1-diolate ligand was conducted using setup operating at 1064 nm [53]. Another very interesting compound was found to be Ag perchlorate complex of 3-amino-1*H*-1,2,4-triazole-5-carbohydrazide, which possesses a high decomposition temperature (230 °C) and has mechanical sensitivity comparable to PETN. This silver ECC was successfully initiated very low energies of 200 mJ, using a femtosecond laser operating at 800 nm [54]. Nevertheless, ignition delay time of 73 ms of the latter compound could be somewhat slow for its potential use in practical applications, which may require ignition delay times <1 ms. The relatively high ignition delay time of the silver compound can be explained with its colorless appearance, resulting in a significant lower absorption than in the case of strongly colored complexes. Whereas most of the reported experiments were conducted using a single-pulse irradiation, for the evaluation of the silver ECC a continuous exposure, with a frequency of 1 kHz, was used. It could be concluded that the initiation was achieved by a heating of the material. In most cases, the absorption of colorless compounds in a visible range inhibits their initiation in single-pulsed laser experiments, unless colorants or dyes are added.

Additional reports describing laser initiation of ECCs used system equipped with a laser diode operating at 980 nm [55]. In these experiments Ni^{2+}, Co^{2+}, and Cu^{2+} carbohydrazide complexes, with nitrate and perchlorate anions, were examined and their ignition delay times were further improved by doping these ECCs with graphene oxide additive. The doping with graphene oxide led to an increased stability, where the ignition delay times were drastically shortened. In general, in contrast to their nitrate analogs, the perchlorate containing ECCs possess short initiation delay times in the range of several hundred microseconds and sensitivity to mechanical impact below the sensitivity of PETN [56].

One of the important issues in the research and development of laser-ignitable ECC's is the use of a suitable laser ignition system, in terms of the evaluated wavelengths, laser power, pulse width (time), sample shape and consistency (such as powder, pressed pellets, crystals, and formulation) and other parameters, allowing more straightforward comparison between results of different investigators. Whether the objective of the reported research is the design and synthesis of laser-ignitable primer cap replacements, study of laser ignition mechanisms, the ignition of pyrotechnic compositions or solid propellants; the unification and standardization in the design of the laser initiation setup is required, with respect to the specifications of the target application. In 1972, Yang and Menichelli described the construction of a laser initiation system that was designed to evaluate the possibility to detonate secondary charges with a laser-induced shock wave (Figure 4.5) [5, 6]. In order to achieve this goal, a continuous-wave laser system, coupled to a camera, was constructed. In this system, a Korad K-1Q ruby laser system, including a KD*P Pockel-cell assembly with a maximum Q-switched energy of 4.5 J in 25 ns, was

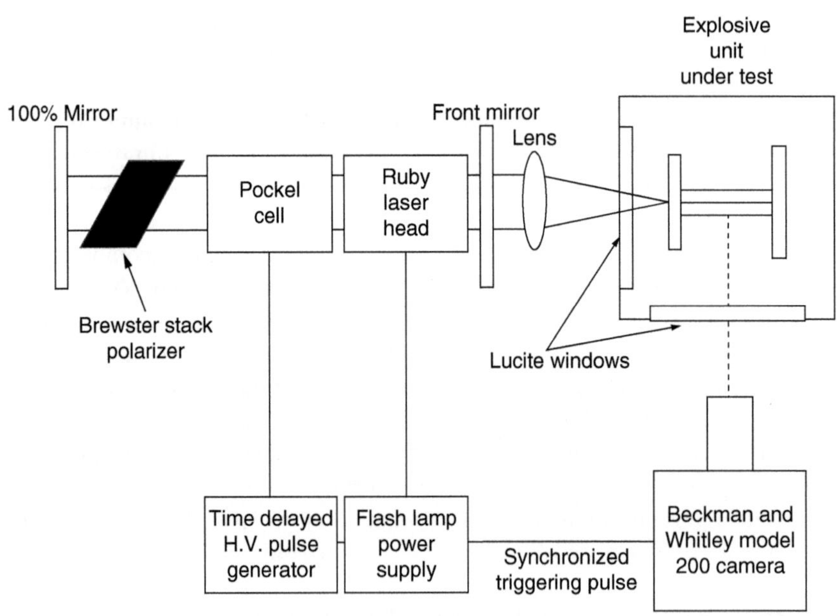

Figure 4.5 Schematic representation of Yang and Menichelli laser initiation system.

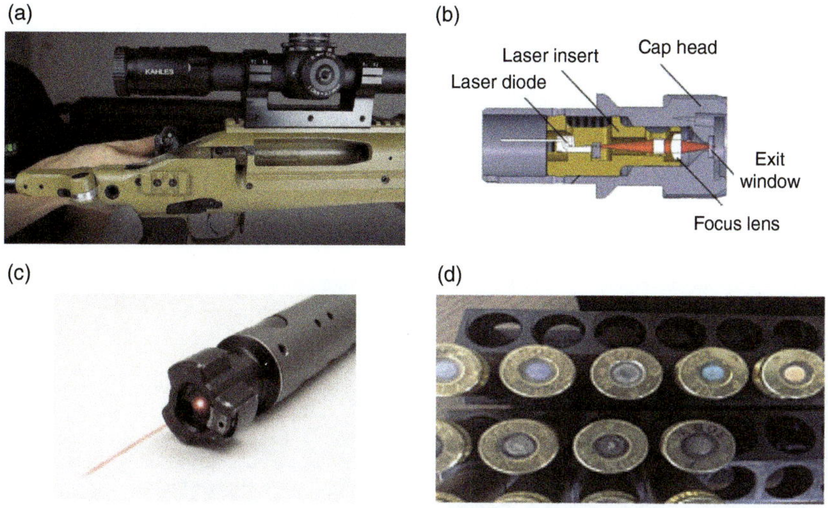

Figure 4.6 (a): VOERE sniper rifle with a laser bolt; (b): laser bolt schematics; (c): image of laser bolt; (d): 7.62 mm bullets containing ECCs in transparent primer caps that are suitable for laser bolt initiation.

used. The radiation was focused by a double convex lens, with a focal length of 30 cm, and directed into a steel chamber through a 2.54 cm thick Lucite window to the explosive unit under the study.

In a search for the replacement of traditional primary explosives used in primer caps, Klapötke studied different ECCs with respect of their toxicity, ignition, and sensitivity properties toward various *stimuli*. More specifically, these studies included hot-needle ignition tests, small-scale reactivity tests (SSRTs), dropping hammer tests, electrostatic discharge (ESD) and friction sensitivity tests, as well as evaluation of the laser ignition properties [34]. For the latter purposes, a tailor-made computer-controlled, fiber-optics-directed laser ignition system was designed [57]. In this system, transparent primer cups were used for the encasing of the evaluated laser sensitive materials. Based on the result of this research, VOERE company successfully demonstrated that optical transparent cups could be used in sniper rifles, in which a mechanic trigger system was replaced with a laser diode (Figure 4.6) (http://www.voere.com). Such light-induced shooting system was intended to prevent the influence of undesired human-generated movements and vibrations during the target-aiming process, to achieve a higher precision long-range shooting accuracy.

Based on requirements of the VOERE sniper rifle bullets, ECC samples of about 15 mg were filled into transparent plastic caps, pressed with a pressure force of 1 kN, and sealed by a transparent UV-cured adhesive (Figure 4.7). The laser initiation experiments were carried out using a 45 W InGaAs laser diode, operating in the single-pulsed mode. The diode was connected to an optical fiber with a core diameter of 400 μm and a cladding diameter of 480 μm. The optical fiber was coupled *via* a SMA type connecter, directly to the laser and to a collimator. This collimator was

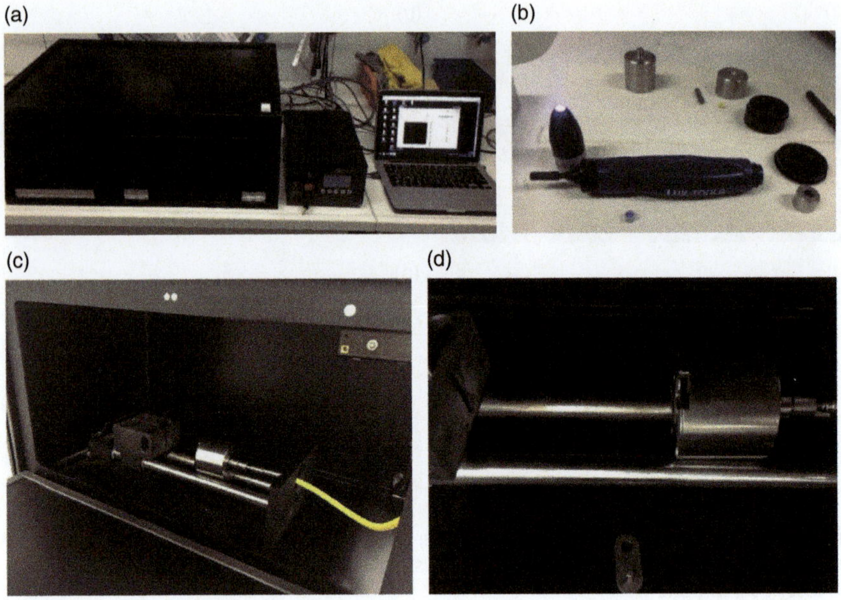

Figure 4.7 (a): Laser ignition system; (b): pressing tools, transparent plastic cap, UV-cured adhesive, and a UV light source; (c): inside view of the safe cabinet of the laser ignition system; (d): a plastic cap sample holder and sapphire protection glass holder.

linked to an optical lens, which was positioned in its focal distance ($f = 29.9$ mm) to the sample. The lens was protected from the explosive by a sapphire glass. The confined samples were irradiated at a wavelength of 915 nm, a pulse length of 15 ms, a varying current of 7–10 A, and a voltage of 4 V, which combined results in an approximate energy output of 26–60 mJ. A series of ECCs, with sensitivity toward mechanical *stimuli* in the range of PETN, were tested in the described laser initiation setup. Although some of these ECCs showed slightly higher sensitivity than PETN, they still were found safe to handle. The investigated ECCs showed different responses to the laser irradiation, depending on the used coordinated metal and counter anion. The researchers also tested whether different current values (7–10 A) could make any significant change in the energetic response of the complexes. However, no significant changes were observed for the examined ECCs.

In the described setup it could only be shown that for less sensitive compounds showing decomposition at 7 A, a deflagration behavior at 10 A was observed. However, different complexes exhibit different laser initiation thresholds following a similar trend as sensitivity of these compounds toward mechanical *stimuli*. Figure 4.8 presents laser initiation of $[Cu(1\text{-}MTZ)_6](ClO_4)_2$ complex (structure presented in Figure 4.29), which undergoes deflagration to detonation transition.

A utilization of the computer-controlled power-tunable diode-pumped solid-state laser system, operating at λ of 915 nm with a maximum power of 8.0 W was recently reported (Figure 4.9) [34]. This system is capable of providing rectangular pulses, with time duration (τ) down to 100 μs and was operated in single- and multiple-pulse

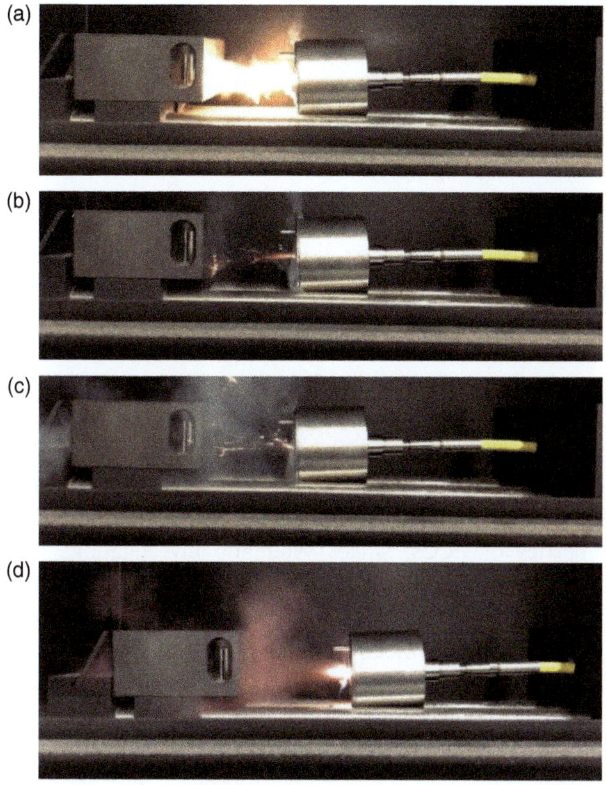

Figure 4.8 (a, b, c): Sequence of selected frames from high speed video clip which presents laser-initiated deflagration of $[Cu(1\text{-}MTZ)_6](ClO_4)_2$ complex; (d): a frame describing a moment of detonation of $[Cu(1\text{-}MTZ)_6](ClO_4)_2$ complex.

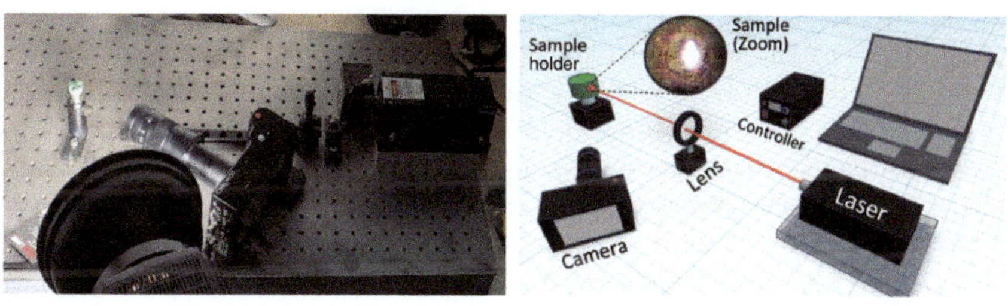

Figure 4.9 Laser ignition system.

modes with controlled intervals between them. The system is also equipped with a focal length of 250 mm, allowing the focus of the laser beam on an examined sample. Furthermore, this system setup contains an amplifier allowing fast settling times at high repetition rates. The system is coupled with a computer-controlled fast camera operating at about 5000 frames per second (fps) and an open sample holder that allows the visual analysis of the ignition process.

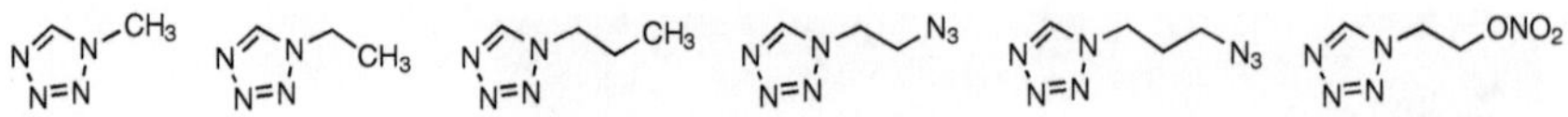

1-MTZ 1-ETZ 1-PTZ 1-AET 1-APT 1-NET

Figure 4.10 Structures of 1-substituted 5*H*-tetrazoles.

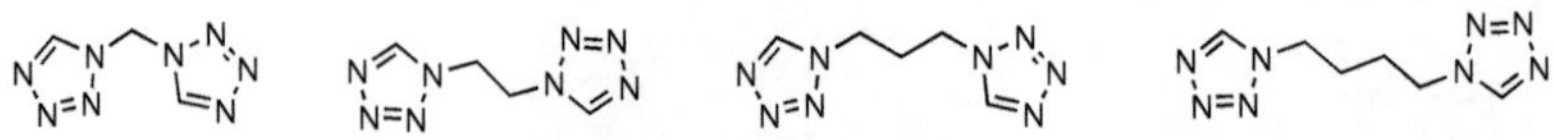

1,1-dtm 1,1-dte 1,1-dtp 1,1-dtb

Figure 4.11 Alkyl-bridged 5*H*-tetrazoles.

4.2 Synthesis of Energetic Coordination Compounds

Hundreds of ECCs were described in literature so far. Only a few of them were investigated for their use in laser ignition systems. It is still not possible to predict the energetic properties of ECCs, since systematic investigations and comparison of the single building blocks were missing. Many of these materials were tested for their laser ignitability, using the setup shown in Figure 4.7. ECCs mostly follow a general formula $(\text{metal})_x(\text{ligand})_y(\text{anion})_z$. The metal is usually taken from the late 3d row (Mn to Zn), since these metals are not too expensive and show a relatively low oxophilicity. However, neither Mn nor Zn complexes could be successfully ignited, likely because of their low colored nature, and as a result of their half- and fully occupied d-orbitals. Furthermore, presence of Zn often causes hygroscopicity to its containing ECCs, and also promotes coordination of water in corresponding crystals, which significantly decreases the energetic properties of these compounds. As ligands, 1-substituted non-acidic 5*H*-tetrazoles ligands were selected in order to show a general trend. 1-Substituted tetrazoles were favored over 2-substituted derivatives, since they typically show better energetic properties, crystallization behavior, lower toxicity, and better synthetic accessibility [58]. Various mono-tetrazole ligands were evaluated, including 1-methyl-5*H*-tetrazole (**1-MTZ**), 1-ethyl-5*H*-tetrazole (**1-ETZ**), 1-propyl-5*H*-tetrazole (**1-PTZ**), 1-azido-ethyl-5*H*-tetrazole (**1-AET**), 1-azidopropyl-5*H*-tetrazole (**1-APT**) and 1-nitratoethyl-5*H*-tetrazole (**1-NET**) (Figure 4.10), as well as alkyl-bridged ditetrazole ligands, such as di(1*H*-tetrazol-1-yl)methane (**1,1-dtm**), 1,3-di(1*H*-tetrazol-1-yl)ethane (**1,1-dte**), 1,3-di(1*H*-tetrazol-1-yl)propane (**1,1-dtp**), and 1,4-di(1*H*-tetrazol-1-yl)butane (**1,1-dtb**) (Figure 4.11). Perchlorate and chlorate counter anions were used in all these complexes since the perchlorate anion is one of the most commonly utilized in light-ignitable ECCs.

4.3 Synthesis of Energetic Tetrazole Ligands

Economic and sustainable synthetic routes toward the construction of energetic materials are desired. The nitrogen-rich heterocycle **1-MTZ** can be synthesized in

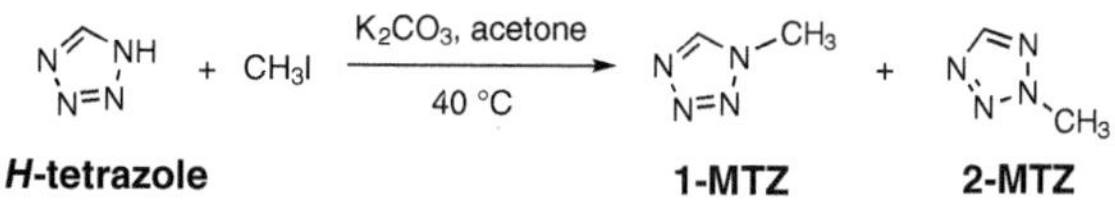

Figure 4.12 Synthesis of **1-MTZ** and **2-MTZ**.

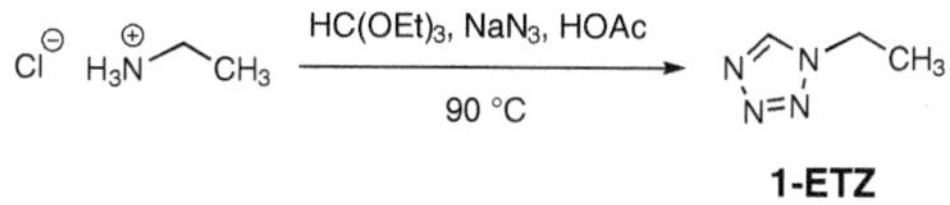

Figure 4.13 Synthesis of **1-ETZ**.

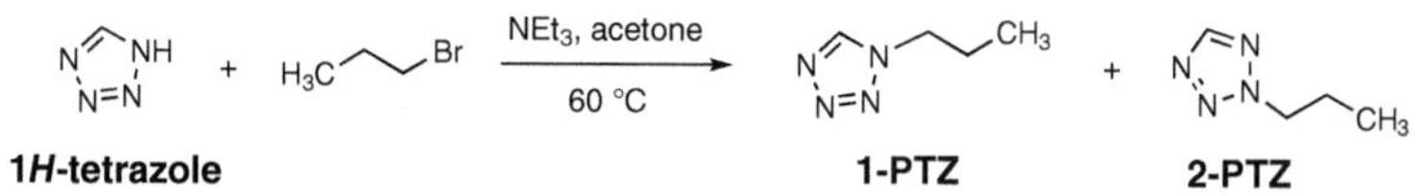

Figure 4.14 Synthesis of **1-PTZ**.

various ways. One route is based on the cycloaddition reaction of sodium azide with triethyl orthoformate and methylamine in acetic acid under reflux [59]. Another method includes the nucleophilic methylation of 1*H*-tetrazole with methyl iodide in acetone at 40 °C under reflux, using potassium carbonate as a base (Figure 4.12) [60]. The methylation leads to an isomeric mixture of **1-MTZ** and 2-methyl-5*H*-tetrazole (**2-MTZ**). The desired product was isolated from this mixture either by flash column chromatography, or by distillation.

1-Ethyl-5*H*-tetrazole (**1-ETZ**) was obtained *via* [3+1+1] ring closure reaction, reacting triethyl orthoformate with NaN_3 in glacial acetic acid at 90 °C (Figure 4.13). After filtration and extraction using ethyl acetate, the product **1-ETZ** was obtained in 75% yield.

Another heterocyclic ligand, 1-propyl-5*H*-tetrazole (**1-PTZ**) was synthesized by substitution of 1,5*H*-tetrazole with 1-bromopropane (Figure 4.14) [61]. The deprotonation of tetrazole with triethylamine, followed by reaction with 1-bromoalkane led to the formation of an isomeric mixture of **1-PTZ** and **2-PTZ** in 70% overall yield. Isomerically-pure **1-PTZ** was isolated in 31% yield, by column chromatography, using ethyl acetate as an eluent.

Over three decades ago, the synthesis of 1-(2-azidoethyl)-1*H*-tetrazole (**1-AET**) was published for the first time by Gaponik (Figure 4.15) [62]. Its coordination to metals was investigated only in the case of $CuCl_2$ and the corresponding crystal structure was elucidated [63]. For synthesizing the **1-AET** ligand, a two-step reaction, starting from a readily available feedstock was performed (Figure 4.15) [64]. The first target molecule was 2-azidoethylamine hydrochloride, which is less volatile than the neutral compound [43, 46]. 2-Chloroethylamine hydrochloride was reacted with NaN_3 in water to form an intermediate in quantitative yield. The crude salt was further reacted *via* a [3+1+1] cyclization to obtain the corresponding substituted alkyltetrazole **1-AET,** by adding an excess of triethyl orthoformate and NaN_3, in glacial acetic acid at elevated temperature. The physical liquid state (at room temperature) of **1-AET** was explained by the liquefying effect of the azidoethyl functional groups [65]. For the synthesis of 1-(2-azidoethyl)-1*H*-tetrazole (**1-APT**),

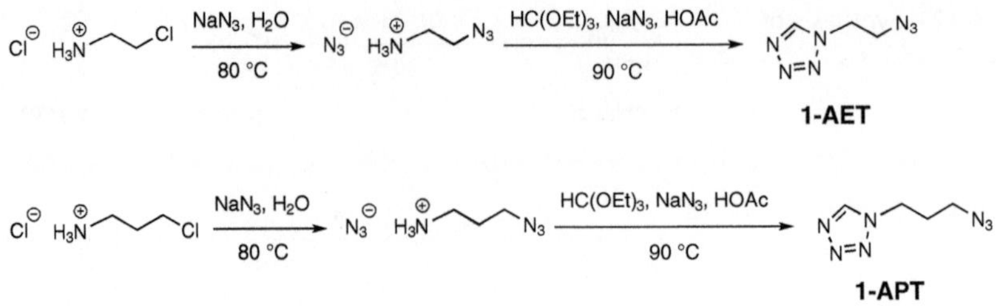

Figure 4.15 Syntheses of **1-AET** and **1-APT**.

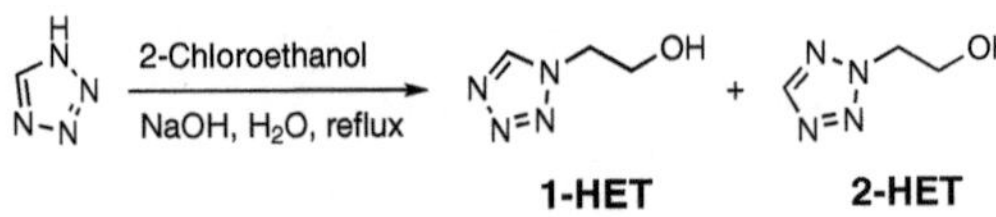

Figure 4.16 Synthesis of **1-HET** and **2-HET**.

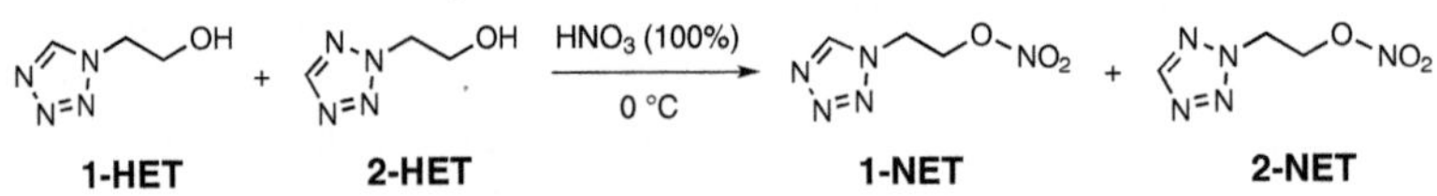

Figure 4.17 Synthesis of **1-NET** and **2-NET**.

a simple two-step procedure was performed similar to that of **1-AET** (Figure 4.15) [63]. The starting commercially available 3-chloropropylamine hydrochloride can be transformed to the corresponding azide with an excess of NaN_3 in almost quantitative yields. A [3+1+1] cyclization of triethyl orthoformate with NaN_3 in acetic acid led to the target **1-APT** molecule in 60% overall yield (Figure 4.15). **1-APT** can also be synthesized in a one-pot reaction, avoiding the isolation and purification of the intermediate.

To obtain the 2-(1*H*-tetrazol-1-yl)ethyl nitrate (**1-NET**), a classic alkylation was applied. Initially, a reaction of 2-chloroethanol with tetrazole using aqueous NaOH as a base, resulted in formation of a mixture of the two isomers 2-(1*H*-tetrazol-1-yl)ethan-1-ol (**1-HET**) and 2-(2*H*-tetrazol-1-yl)ethan-1-ol (**2-HET**) in a 1 : 1 ratio (in overall yield of 88%) (Figure 4.16). No separation of the products was carried out.

The nitration of the isomeric mixture of **1-HET** and **2-HET** was carried out by adding these alcohols to fuming HNO_3. After quenching the nitration reaction with ice and subsequent organic extraction, column chromatography was used to separate the ligands 2-(1*H*-tetrazol-1-yl)ethyl nitrate (**1-NET**; in 45% yield) and 2-(2*H*-tetrazol-2-yl)-ethyl nitrate (**2-NET**; in 43% yield) (Figure 4.17).

Structure of 1-azidopropyl-5*H*-tetrazole (**1-APT**) was elucidated by analysis of infrared (IR) spectra, as well as by ^{1}H and ^{13}C NMR spectroscopies. Two-dimensional $\{^{1}H\text{-}^{15}N\}$ HMBC NMR was also found to be very useful technique for the analysis of alkylated 1*H*-tetrazoles. $\{^{1}H\text{-}^{15}N\}$ HMBC NMR was found to be much faster and more sensitive in comparison to the regular ^{15}N NMR (Figure 4.18). On the other

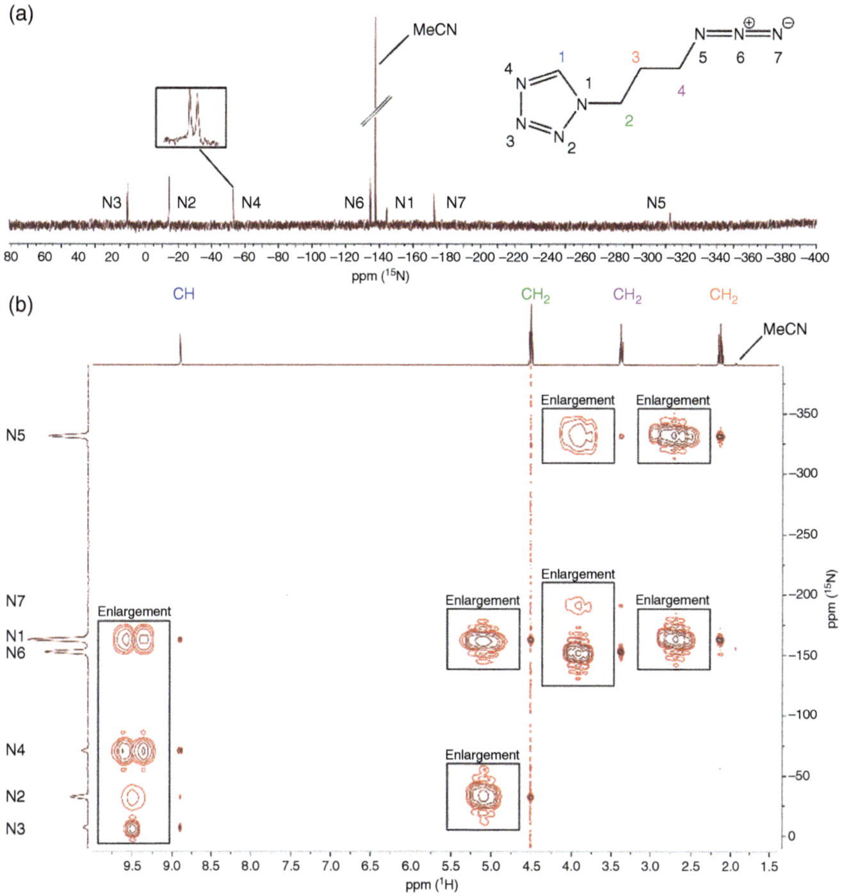

Figure 4.18 (a): Proton-coupled ^{15}N NMR spectra of **1-APT**; (b): 2D {^{1}H,^{15}N}-HMBC NMR spectra of **1-APT**. ^{15}N NMR (CD_3CN, 25 °C): δ 10.7 (N3, s), −14.2 (N2, s), −52.8 (N4, d, J_{N-H} = 11.9 Hz), −134.6 (N6, s), −144.4 (N1, s), −172.6 (N7, s), −312.7 (N5, s) ppm.

hand, in {^{1}H-^{15}N} HMBC NMR, the terminal nitrogen of the azide group was almost invisible, due to its long distance to the closest hydrogen atom (Figure 4.18b). In the case of **1-APT**, researchers mentioned that it was important to conduct both types of ^{15}N NMR measurements.

Although the mono-tetrazole ligands showed excellent complexation properties and their complexes exhibited very good crystallization behavior, this type of ligands also has certain drawbacks. First, when moistened or dissolved in water, these ligands compete with water molecules in metal complexation, which was clearly observed by UV–vis spectroscopy. Sensitive energetic materials are often moistened for handling and transportation, which could yield less energetic aqua complexes. Second, in few cases, the mono-tetrazoles (in particular, volatile ones) underwent cleavage at higher temperatures, as was detected by DSC and TG analyses. Therefore, a development of various synthetic approaches toward the preparation of non-soluble and highly thermostable complexes was conducted. One

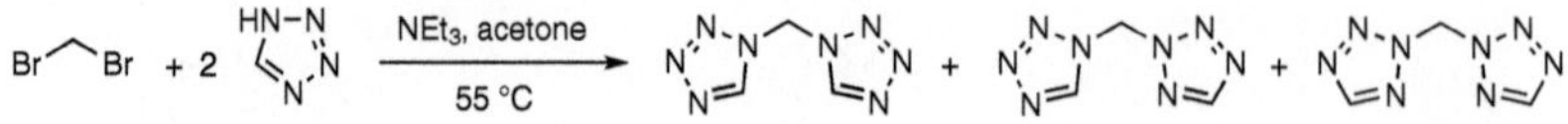

1,1-dtm 1,2-dtm 2,2-dtm

Figure 4.19 Synthesis of ditetrazolylmethane isomers utilizing dibromomethane and 1,5*H*-tetrazole.

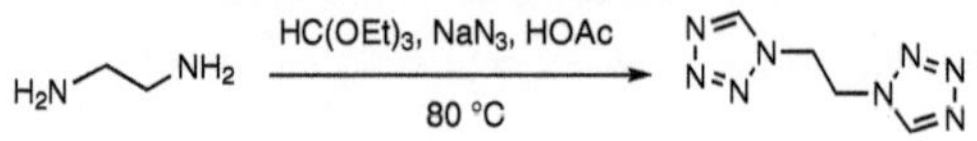

1,1-dte

Figure 4.20 Synthesis of the **1,1-dte**.

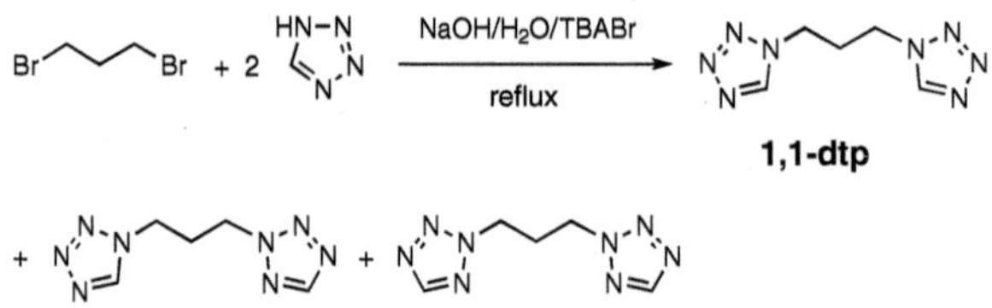

1,1-dtp

1,2-dtp 2,2-dtm

Figure 4.21 Synthesis of the three ditetrazolylpropane isomers.

such approach was focused on the evaluation of alkyl-bridged tetrazole ligands. While ditetrazoles with even numbers of carbon atoms in their bridges were synthesized by the [3+1+1]-cycloaddition, using the suitable diamines, NaN_3 and triethyl orthoformate; ditetrazoles with the odd numbers of carbon atoms in their bridges could only be synthesized by a direct alkylation. In contrast to the mono-tetrazoles, bridged tetrazoles, in all isomeric substitution patterns (1,1′; 1,2′ and 2,2′), were found to be suitable for ECCs preparation. In Figure 4.19, the synthesis of the three ditetrazolylmethane isomers is shown. Due to their lower carbon content, these ligands formed highly energetic, yet sensitive ECCs.

Another good ligand for preparation of photosensitive complexes was found to be 1,2-di(1*H*-tetrazol-1-yl)-ethane (**1,1-dte**) (Figure 4.20). The structures of a few transition metal complexes of Fe, Co, Ni, and Cu, with **1,1-dte** ligand, containing perchlorate or azide anions, were described as polynuclear coordination compounds [27]. Although various coordination compounds based on 1-(2-(2*H*-tetrazol-2-yl)ethyl)-1*H*-tetrazole (**1,2-dte**) ligand have shown promising results during their laser initiation experiments, these ECCs were found to be quite sensitive for practical applications [30].

In order to further increase the stability of the **dte** complexes, ditetrazolyl-propane (**dtp**) ligands, with higher carbon content, were used for the synthesis of new complexes. The main focus was invested into the preparation of the 1,3-di(1*H*-tetrazol-1-yl)propane (**1,1-dtp**), which has been proven to be an extraordinary ligand for the preparation of new energetic materials. The highest yield of 71% toward the three ditetrazolyl-propane isomers was described (Figure 4.21) [66].

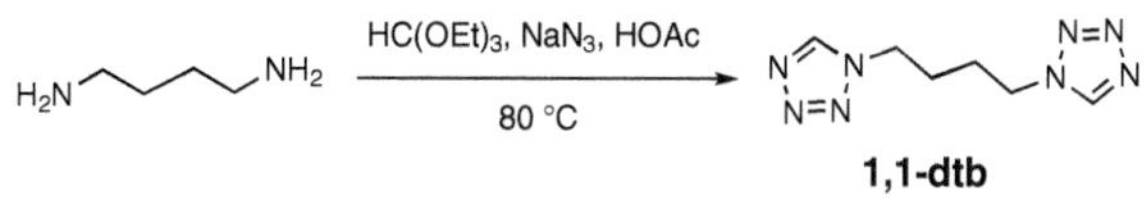

Figure 4.22 Synthesis of **1,1-dtb**.

Another related ligand – 1,1-ditetrazolylbutane (**1,1-dtb**) was synthesized by reacting 1,4-diaminobutane with NaN_3 and triethyl orthoformate in glacial acetic acid (Figure 4.22).

4.4 Synthesis Energetic Coordination Complexes

The synthesis of complexes **11–42** was typically achieved by reacting an aqueous solution of hydrated metal perchlorate with 6 equiv of a ligand (in case of the mono-tetrazoles), or 3 equiv of a ligand (in case of the ditetrazoles) at elevated temperature. The resulting solutions were allowed to cool down to room temperature and left for crystallization (Figures 4.23–4.28).

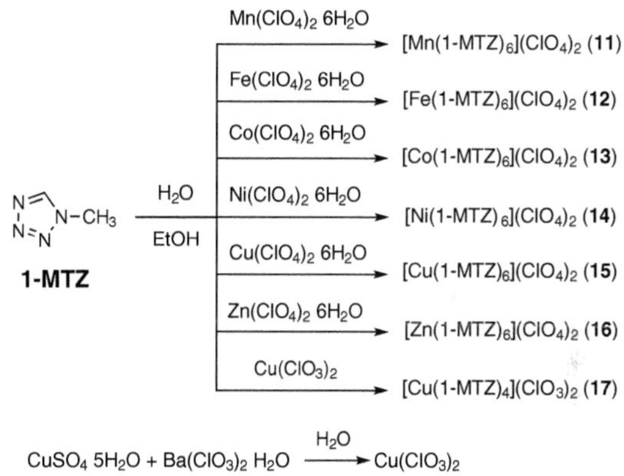

Figure 4.23 Synthesis of ECCs **11–17** based on **1-MTZ** ligand.

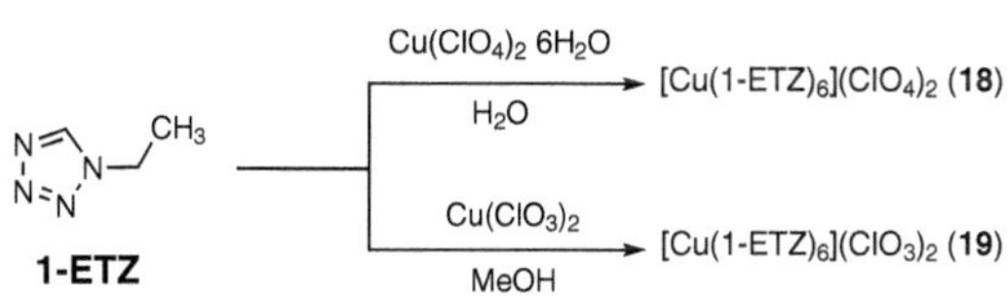

Figure 4.24 Synthesis of copper ECCs **20** and **22** based on **1-ETZ** ligand.

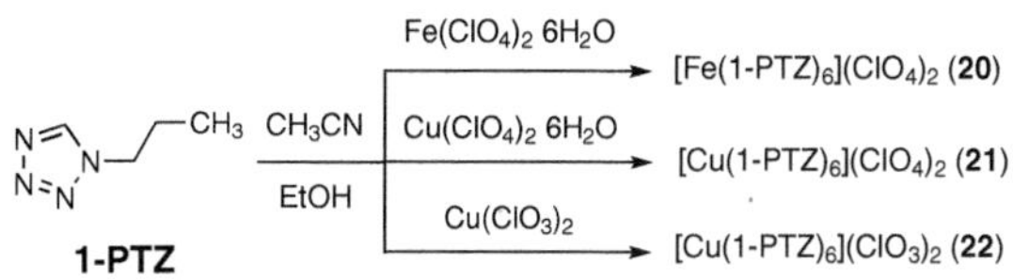

Figure 4.25 Synthesis of ECCs **20–22** based on **1-PTZ** ligand.

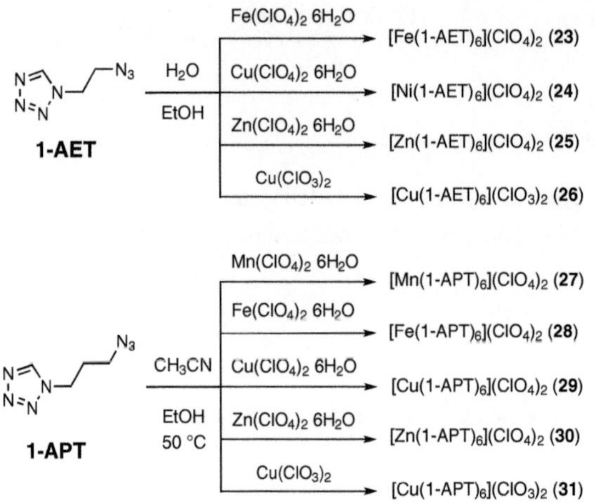

Figure 4.26 Synthesis of ECCs **23–26** based on **1-AET** ligand, and ECCs **27–31** based on **1-APT** ligand.

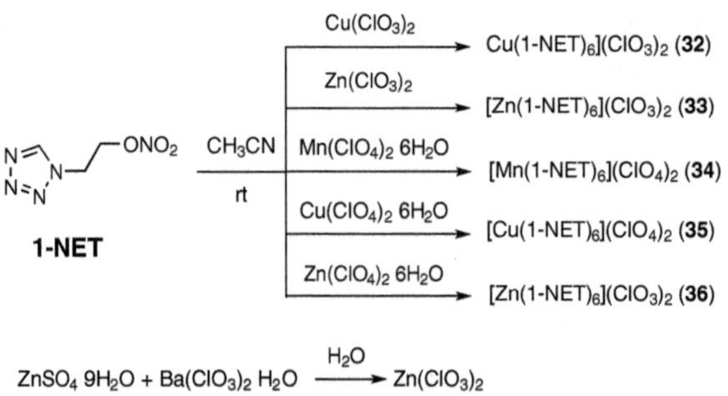

Figure 4.27 Synthesis of ECCs **32–36** based on **1-NET** ligand.

4.5 Examples of Molecular Structures

Many of ECCs were investigated by X-ray crystallography. For example, $[Cu(1\text{-}MTZ)_6](ClO_4)_2$ (**15**) is a typical Cu(II) perchlorate complex containing a mono-dentate **1-MTZ** ligand (Figure 4.29) [67]. Distorted octahedral coordination spheres are following the d^9 Jahn–Teller distortion and were observed in most cases of this type of compounds.

In contrast to copper ECCs, the Fe(II) complexes mostly show a regular octahedral coordination sphere, which can be seen on the structure of $[Fe(1\text{-}AET)_6](ClO_4)_2$ (**23**) (Figure 4.30), in which **1-AET** was utilized as a mono-dentate ligand [64].

When symmetric ditetrazolyl-alkanes, such as **1,1-dtp** were used as ligands, crystal structures with trigonal or hexagonal space groups were often observed (Figure 4.31) [68].

4.6 Energetic Properties of Ligands and Corresponding Energetic Coordination Compounds

Table 4.1 summarizes critical temperatures (endo and exothermic events) obtained by DTA measurements, the impact and friction sensitivities, ESD sensitivities, and response to hot needle and laser irradiation.

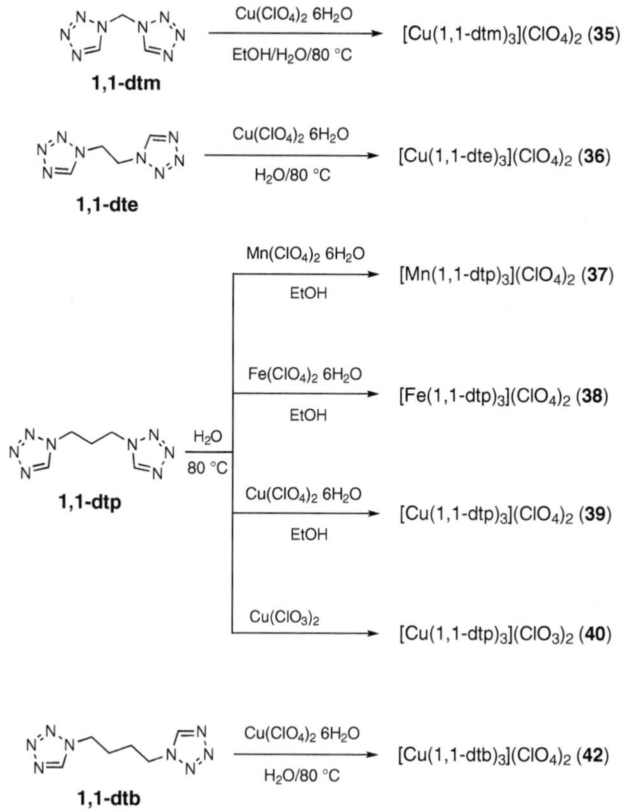

Figure 4.28 Synthesis of ECCs based on 1,1-ditetrazolylalkanes.

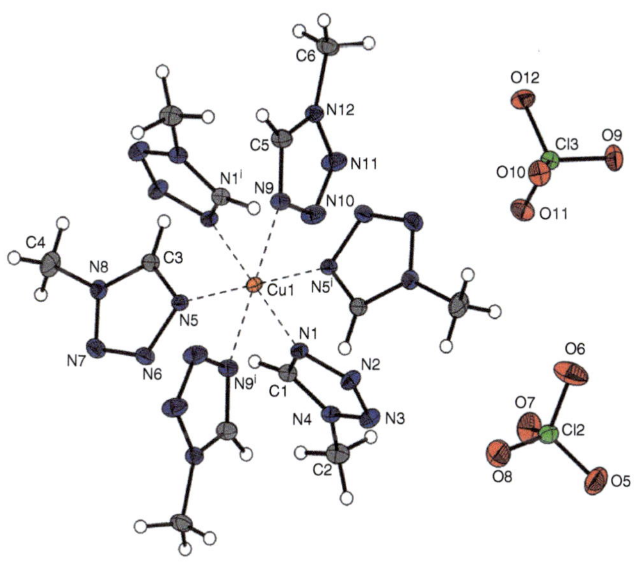

Figure 4.29 Structure of $[Cu(1\text{-}MTZ)_6](ClO_4)_2$ (**15**). Selected bond lengths (Å): Cu1–N1 2.035(2), Cu1–N5 2.014(2), Cu1–N9 2.401(2), Cl2–O5 1.437(2); selected bond angles (°): N1–Cu1–N5 89.83(2), N1–Cu1–N9 91.04(2), N5–Cu1–N9 89.54(2), O5–Cl2–O6 109.32(2). Symmetry codes: (i) $-x$, $2-y$, $-z$; (ii) $1-x$, $-y$, $1-z$; (iii) $1-x$, $1-y$, $1-z$; (iv) $-x$, $1-y$, $-z$.

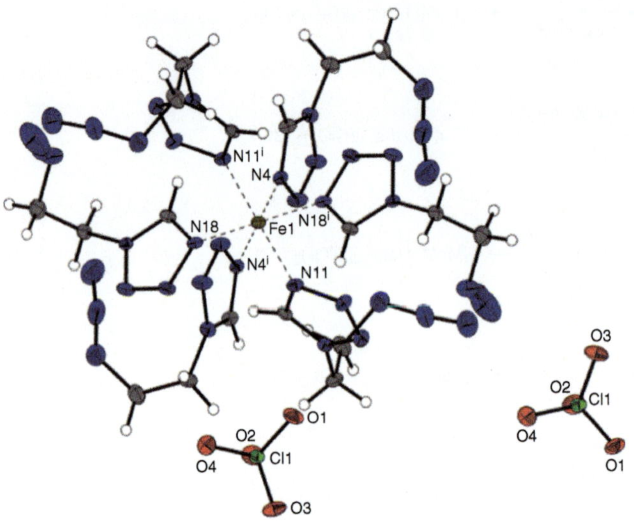

Figure 4.30 Structure of $[Fe(1\text{-}AET)_6](ClO_4)_2$ (**23**). Selected bond lengths (Å): Fe1–N4 1.990(4), Fe1–N11 1.996(4), Fe–N18 1.977(5); selected bond angles (°): N11–Fe1–N11i 180.00, N4–Fe1–N11 89.17(16), N18–Fe1–N18i 180.00, N4–Fe1–N18i 89.96(16), N4–Fe1–N4i 180.00, N11–Fe1–N18 89.41(18). Symmetry codes: −*x*, −*y*, −*z*.

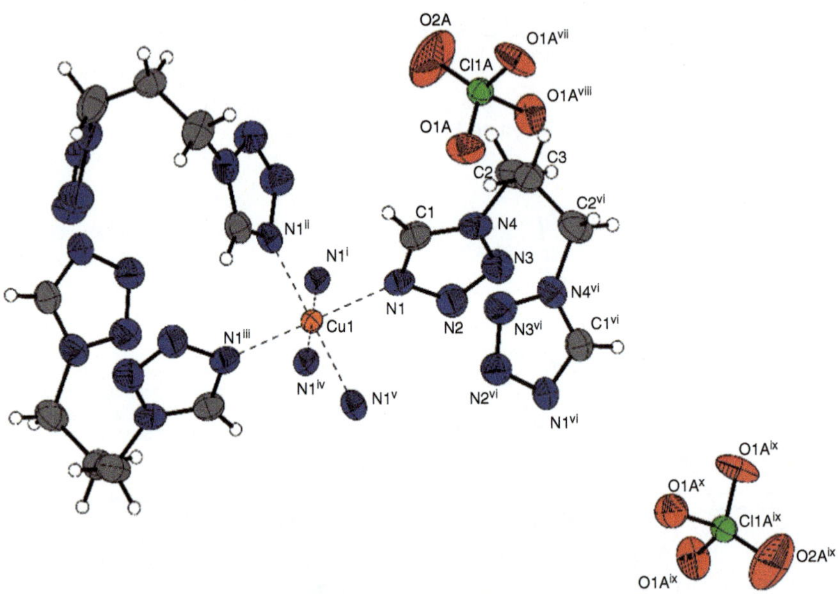

Figure 4.31 Structure of $[Cu(1,1\text{-}dtp)_3](ClO_4)_2$. Selected bond length (Å): Cu–N1 2.131(2); selected bond angles (°): N1–Cu1–N1i 89.33(9), N1–Cu1–N1iii 180.00, and N1–Cu1–N1iv 90.68(9). Symmetry codes: (i) $y, -x+y, 1-z$; (ii) $x-y, x, 1-z$; (iii) $-x, -y, 1-z$; (iv) $-y, x-y, z$; (v) $-x+y, -x, z$; (vi) $-x, -x+y, 1.5-z$; (vii) $-x+y, 1-x, z$; (viii) $1-y, 1+x-y, z$; (ix) $1+x-y, 1-y, 1.5-z$; (x) $y, x, 1.5-z$; (xi) $1-x, -x+y, 1.5-z$.

Table 4.1 Properties of various abovementioned ligands and complexes.

Compound	T_{endo}.(°C)[a]	$T_{exo.}$ (°C)[b]	IS[c] (J)	FS[d] (N)	ESD[e] (mJ)	Flame test[f]	Hot needle[g]	Laser ignition[h]
1-MTZ	39	206	>40	>360	1500	Def.		—
1-ETZ		208	>40	>360	1500	Combustion		—
1-PTZ		206	>40	>360	1500	Dec.		—
1-AET		193	9	>360		Def.		—
1-APT		195	>40	>360		Def.		—
1-NET	26	168	10	>360		Def.		—
1,1-dtm			8	>360	480	Def.		—
1,1-dte			>40	>360	1350	Combustion		—
1,1-dtp·H_2O	68 (H_2O), 126	205	>40	>360	1350	Combustion		—
1,1-dtb			>40	>360	1350	Combustion		—
$[Mn(1\text{-}MTZ)_6](ClO_4)_2$ (**11**)		223	7	120	80	Def.	Def.	—
$[Fe(1\text{-}MTZ)_6](ClO_4)_2$ (**12**)		213	1	60	100	Def.	Def.	Def.
$[Co(1\text{-}MTZ)_6](ClO_4)_2$ (**13**)		253	3	60	500	Def.	Def.	—
$[Ni(1\text{-}MTZ)_6](ClO_4)_2$ (**14**)		271	5	60	100	Def.	Def.	Def.
$[Cu(1\text{-}MTZ)_6](ClO_4)_2$ (**15**)		211	2.5	54	80	Def.	Def.	Def.
$[Zn(1\text{-}MTZ)_6](ClO_4)_2$ (**16**)		221	10	120	400	Def.	Dec.	—
$[Cu(ClO_3)_2(1\text{-}MTZ)_4]$ (**17**)		159	3	7	100	Def.	Def.	Det.
$[Cu(1\text{-}ETZ)_6](ClO_4)_2$ (**18**)	172	210	10	120	960	Def.	Def.	Dec.
$[Cu(1\text{-}ETZ)_6](ClO_3)_2$ (**19**)	111	158	7	60	608	Def.	Def.	Dec.
$[Fe(1\text{-}PTZ)_6](ClO_4)_2$ (**20**)		202	25	192	1220	Dec.	Dec.	—

(continued)

Table 4.1 (Continued)

Compound	T_{endo}.(°C)[a)]	$T_{exo.}$ (°C)[b)]	IS[c)] (J)	FS[d)] (N)	ESD[e)] (mJ)	Flame test[f)]	Hot needle[g)]	Laser ignition[h)]
$[Cu(1\text{-}PTZ)_6](ClO_4)_2$ (**21**)	174	209	30	240	>1500	Dec.	Dec.	Dec.
$[Cu(1\text{-}PTZ)_6](ClO_3)_2$ (**22**)	111	152	>40	168	960	Dec.	Dec.	Dec.
$[Fe(1\text{-}AET)_6](ClO_4)_2$ (**23**)		151	3	3.75	650	Det.	Det.	Det.
$[Cu(1\text{-}AET)_6](ClO_4)_2$ (**24**)	135	158	<1	15	368	Def.	Det.	Det.
$[Zn(1\text{-}AET)_6](ClO_4)_2$ (**25**)	159	196	15	40	368	Def.	Def.	—
$[Cu(1\text{-}AET)_6](ClO_3)_2$ (**26**)	77	146	2.5	4	226	Def.	Det.	Det.
$[Mn(1\text{-}APT)_6](ClO_4)_2$ (**27**)	122	208	4.5	60	368	Def.	Dec.	—
$[Fe(1\text{-}APT)_6](ClO_4)_2$ (**28**)		146	2.5	28	317	Def.	Def.	Def.
$[Cu(1\text{-}APT)_6](ClO_4)_2$ (**29**)	115	165	2.5	32	368	Def.	Def.	Det.
$[Zn(1\text{-}APT)_6](ClO_4)_2$ (**30**)		183	9	80	422	Def.	Def.	—
$[Cu(1\text{-}APT)_6](ClO_3)_2$ (**31**)	65	151	2.5	24	>1500	Def.	Def.	Def.
$[Cu(1\text{-}NET)_6](ClO_3)_2$ (**32**)	54	149	2	15	250	Def.	Det.	Combustion
$[Zn(1\text{-}NET)_6](ClO_3)_2$ (**33**)	71	154	7	14	750	Def.	Def.	Combustion
$[Cu(1\text{-}NET)_6](ClO_4)_2$ (**34**)	70, 117	165	3	25	480	Def.	Def.	Dec.
$[Cu(1,1\text{-}dtm)_3](ClO_4)_2$ (**35**)		230	<1	0.75	42	Def.	Def.	—
$[Cu(1,1\text{-}dte)_3](ClO_4)_2$ (**36**)		256	1.5	10	150	Def.	Def.	Def.
$[Mn(1,1\text{-}dtp)_3](ClO_4)_2$ (**37**)		243	3	216	300	Def.	Def.	—
$[Fe(1,1\text{-}dtp)_3](ClO_4)_2$ (**38**)		234	2	72	100	Def.	Def.	—
$[Cu(1,1\text{-}dtp)_3](ClO_4)_2$ (**39**)		231	1	54	330	Def.	Def.	—

[Cu(1,1-dtp)$_3$](ClO$_3$)$_2$ (**40**)	168	2	18	100	Def.	Def.	Det.
[Cu(1,1-idtp)$_3$](ClO$_4$)$_2$ (**41**)	235	4	20	150	Def.	Def.	Det.
[Cu(1,1-dtb)$_3$](ClO$_4$)$_2$ (**42**)	230	12.5	120	250	Def.	Def.	Def.

a) Endothermic events such as melting, dehydration, and phase transition.
b) Decomposition temperature (DTA or DSC; 5 °C min^{-1}).
c) Impact sensitivity.
d) Friction sensitivity.
e) Electrostatic discharge sensitivity.
f) Behavior when compound on a spatula is ignited by a Bunsen burner.
g) Hot needle test, compound fixed with adhesive tape and irradiated with a red glowing needle.
h) Behavior upon laser irradiation at the LMU setup (parameters: current $I = 7$–12 A; voltage $U = 4$ V; theoretical maximal output power $P_{max} = 45$ W; theoretical energy range $E = 0.17$–111 mJ; wavelength $\lambda = 915$ nm; pulse length time $\tau = 0.1$–20 ms).
—: Not tested; Dec.: decomposition; Deflag.: deflagration; Det.: detonation.

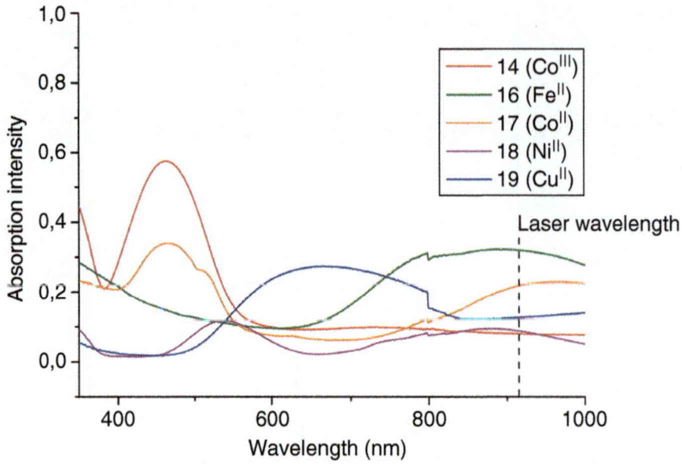

Figure 4.32 Solid-state UV–vis–NIR spectra of selected **1-MTZ** ECCs. Laser irradiation wavelength (915 nm) is shown with the black dotted line.

4.7 UV–Vis Spectroscopy of Energetic Coordination Compounds

The solid-state UV–vis–NIR spectra (in the range of 350–1000 nm) of many ECCs were also collected in order to get a better understanding of the laser ignition process. Spectra of selected ECCs, containing **1-MTZ** ligand – [Ni(1-MTZ)$_6$](ClO$_4$)$_2$ (**14**); [Zn(1-MTZ)$_6$](ClO$_4$)$_2$ (**16**), [Cu(ClO$_3$)$_2$(1-MTZ)$_4$] (**17**), [Cu(1-ETZ)$_6$](ClO$_4$)$_2$ (**18**), and [Cu(1-ETZ)$_6$](ClO$_3$)$_2$ (**19**), are shown in Figure 4.32.

The observed absorptions of aforementioned ECCs in the UV–vis–NIR region are quite typical to coordination compounds of the corresponding transition metals. These transitions can be assigned to the characteristic d–d transitions and correlate with the complementary colors. The d^5 Mn(II) and d^{10} Zn(II) ECCs didn't show significant absorption in the measured range, where all laser ignition experiments with these ECCs were not successful under the examined conditions, raising issues related to the molecular- and crystal-level mechanisms of a successful laser initiation.

4.8 Studies of Ignition Mechanism

General laser initiation mechanisms are still not fully understood. Many different reaction pathways, including photothermal and others, could be conceived. While measuring the UV–vis spectra of laser-ignitable ECC's, Klapötke and his team found a correlation between the absorption of the studied complexes at the laser wavelength of 915 nm (Table 4.1). Yet, some of the examined compounds, as for example [Co(1-MTZ)$_6$](ClO$_4$)$_2$ (**13**), which clearly absorbs in the irradiated region, showed no reaction upon exposure to laser irradiation. For **1-MTZ** ECCs of Fe(II) it is well known that the crystal structure, as well as the counterion identity (e.g. ClO_4^-, BF_4^-, PF_6^-, or $CF_3SO_3^-$), can have a profound effect on the spin crossover properties [67]. It is possible that not only the absorption properties, but also subtle changes in the geometry or crystal packing, would have an influence of the susceptibility of the complexes to the laser ignition. The underlying initiation processes are likely to depend on multiple different parameters. Based on the reported results, the observed general tendency was that compounds are more prone to laser-induced deflagration/detonation if they possess high sensitivity toward mechanical impact. Due to the fact that the ligands influence on the laser initiation process, as well as the process itself, are not determined yet, research focusing on a better understanding of the laser initiation mechanism of energetic materials is of a significant importance. Several laser ignition mechanistic studies have been conducted by various investigators. At the end of the 1960s, the first published laser initiation experiments were conducted by two research groups of Brish [5], and Menichelli and Yang [6]. Brish successfully initiated samples of PETN and LA by laser irradiation. These initiation experiments were conducted with a neodymium glass laser (1060 nm), operating in a Q-switched mode. LA was initiated by a 10 MW laser (0.1 μs pulse of 0.5 J), with a radiant flux density of 8 MW cm^{-2} at the materials surface, increasing the power density at the explosive surface by focusing the beam, which was found to be critical for the successful initiation of PETN under the explored conditions. In light of the successful initiation of these known explosives, Brish further investigated the mechanism of ignition using various laser sources. His investigations led to four different initiation mechanisms hypotheses: (i) light impact mechanism, (ii) electrical break-down mechanism, (iii) photothermal mechanism, and (iv) photochemical mechanism. During Brish's mechanistic research it was found that the exerted light pressure needed for successful initiation of the explosives such as LA and PETN should be several orders of magnitude lower than the required pressures for shock initiation. Hence, it is not likely that a light impact mechanism of ignition could be the answer for the laser initiation mechanism enigma.

When pressuring the electrical breakdown mechanism hypothesis, no differences in the initiation of the explosives were observed for dielectrics and metal powders. Thus, the latter hypothesis is also doubtful. Although it could not be fully excluded, these observations didn't fit the core of the hypothesis and hence didn't provide a suitable mechanistic explanation for the initiation phenomena that was observed. The dependency between the laser wavelength (694 or 1064 nm) and the initiation threshold serves as the core of the photochemical initiation mechanism hypothesis. Basically, this hypothesis suggests that there is a correlation between the laser wavelength, the explosives absorbance, and the initiation threshold. Therefore, changing the wavelength of the laser source to fit the explosives absorbance should lower the energy threshold for initiation. Furthermore, this hypothesis could be explained by a multi-quantum photoelectric effect. Yet, a dependency between the laser wavelength (694 or 1064 nm) and the initiation threshold was not observed. Consequently, it is unlikely that the photochemical initiation mechanism is a plausible mechanism for the laser initiation. The photothermal initiation mechanism hypothesis states that

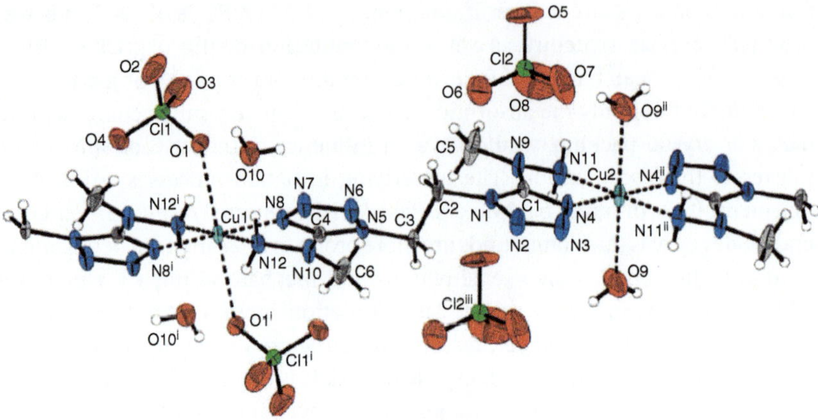

Figure 4.33 Structure of $(\{[Cu(ClO_4)_2(\mu\text{-bmhte})]\cdot 2H_2O\}\{[Cu(\mu\text{-bmhte})(H_2O)_2](ClO_4)_2\})_n$ (**2**).

simple heating of the examined explosive leads to this explosive decomposition. Basically, the laser radiation rapidly heats the explosive, the produced heat leads to a fast increase of the internal pressure. The rapid heating and rise in internal pressure results in an explosive decomposition that could be described as a light-to-energy transformation. The experimental results supported this mechanism. As previously mentioned, the initiation processes of energetic materials under laser irradiation most likely depend on multiple factors. Klapötke noticed a general correlation between the tendency to deflagrate/detonate under laser irradiation and a high mechanical sensitivity. During laser ignition experiments (operating at 940 nm, and τ of 100 μs of a single-pulse) of $(\{[Cu(ClO_4)_2(\mu\text{-bmhte})]\cdot 2H_2O\}\{[Cu(\mu\text{-bmhte})(H_2O)_2](ClO_4)_2\})_n$ (**2**), $\{[Co(\mu\text{-bmhte})(H_2O)_2](ClO_4)_2\}_n$ (**3**), and $\{[Ni(\mu\text{-bmhte})_{1.5}](ClO_4)_2\cdot H_2O\}_n$ (**4**) Klapötke and Joas proposed that the initiation mechanism might be of a thermal nature (Figure 4.33) [69]. The UV–vis–NIR spectra of these compounds showed different absorption at the laser irradiation wavelength (940 nm) (**3** > **4** > **2** absorption intensity) (Figure 4.34). In contrast, all three compounds exhibited a deflagration-to-detonation transition with no selectivity, as might be expected in a photoelectronic process [70]. Furthermore, it was concluded that the strong light absorption at this wavelength implies a strong energy absorption, and consequently a heating of a sample, suggesting a likelihood of the photothermal ignition mechanism.

Ilyushin investigated a mechanism of a laser initiation of perchlorate complexes containing Cu^{2+}, Cd^{2+}, Ni^{2+}, and Co^{2+} and 3-hydrazineyl-4*H*-1,2,4-triazol-4-amine (**HATr**) ligand (Figure 4.35) [45]. Initially, these coordination compounds were irradiated by a single pulse laser ($\lambda = 1064$ nm, $\tau = 30$ ns, $d = 0.48$ mm) and the minimum energies required their detonation were found (Table 4.2).

The results indicated that the sensitivity of the aforementioned compounds to the laser irradiation correlates to the oxidizing ability of the central metal ion (expressed as the sum of two ionization potentials, [I1+I2]). Basically, it was experimentally found that the greater the oxidizing ability of the central metal ion, the higher is the sensitivity of the compound toward the laser radiation. According to these results it

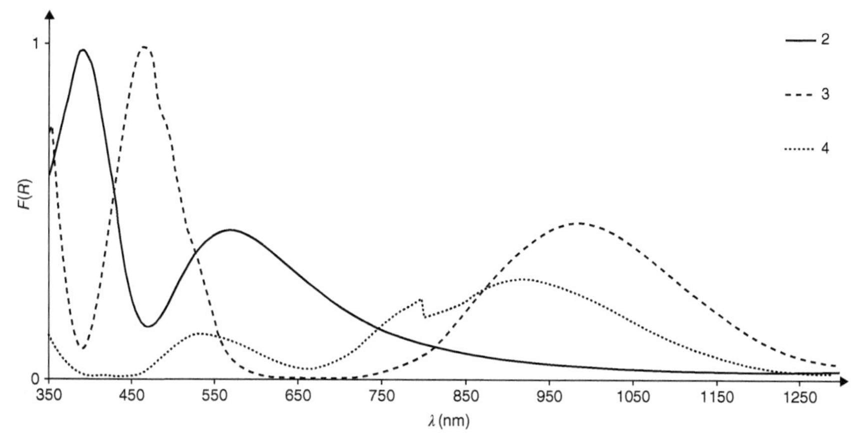

Figure 4.34 UV–vis–NIR spectra of compounds **2–4** in the range of 350–1300 nm.

Figure 4.35 Metal perchlorate complexes of **HATr** [M = Cu^{2+} (**5**), Cd^{2+} (**6**), Ni^{2+} (**7**), and Co^{2+} (**8**)].

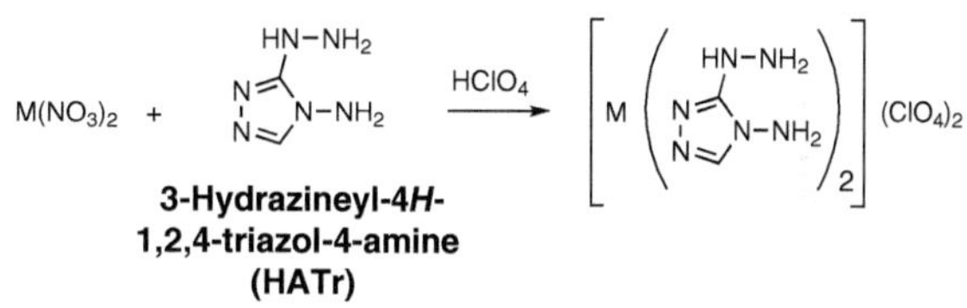

Table 4.2 Laser initiation resulted in detonation of metal complexes **5-8**.

Complex	Detonation threshold energy (J)	Ionization potentials of metals [I1+I2] (eV)
$[Cu(HATr)_2](ClO_4)_2$	$1.1 \cdot 10^{-5}$	28.02
$[Cd(IIATr)_2](ClO_4)_2$	$5.03 \cdot 10^{-4}$	25.90
$[Ni(HATr)_2](ClO_4)_2$	$5.75 \cdot 10^{-4}$	25.78
$[Co(HATr)_2](ClO_4)_2$	$1.36 \cdot 10^{-3}$	24.92

was found that the synthesized $[Cu(HATr)_2](ClO_4)_2$ complex exhibited higher sensitivity to the laser irradiation with the lowest laser detonation threshold (1.1×10^{-5} J). Furthermore, the fact that the inner sphere ligand oxidation was affected by the metal ion identity suggests that the reaction of the ligand with the metal ion is the one of the first steps of the initiation process. Hence, the following oxidation mechanism was suggested:

$$2ClO_4^- + M^{2+} \rightarrow ClO_4 \cdot + M^+ + ClO_4^- \quad (4.1)$$

$$ClO_4^- + M^+ \rightarrow ClO_4 \cdot + M^\circ \quad (4.2)$$

According to the suggested mechanism, the essence of the ignition mechanism lies an oxidative step that involves the metal cation and the perchlorate anion and results

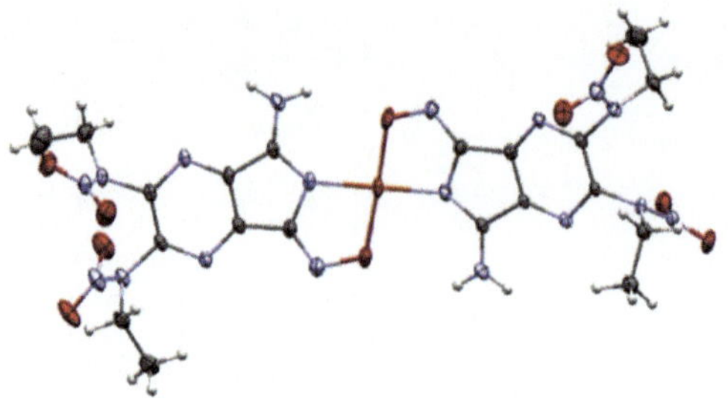

Figure 4.36 Structure of laser-ignitable Cu(II) complex **DS4**.

Table 4.3 Results of single-pulse laser ignition experiments.

	Pulse duration (τ) (ms)	Pulse intensity ($kW\,cm^{-2}$)	Pulse energy (mJ)
1	1	14.00	4.4
2	5	9.20	14.5
3	10	8.30	26.0
4	15	6.00	28.5
5	20	5.40	34.0
6	25	4.80	38.8
7	35	2.20	24.5
8	40	0.95	12.0
9	50	0.76	12.0
10	70	0.54	11.9
11	100	0.13	4.0

in the formation of a highly active perchlorate free radical that participates in the ligand oxidation [45].

To further investigate the nature of the laser ignition mechanism, Klapötke and Stierstorfer performed a mechanistic ignition research on a Cu^{2+} complex of pre-ligand 5,6-bis(ethylnitroamino)-N'^2,N'^3-dihydroxy-pyrazine-2,3-bis-(carboximidamide) (**DS4**) (Figure 4.36) [34]. As previously mentioned, laser ignition experiments were conducted on a custom-made laser ignition setup that can operate in a single and multiple pulse modes (915 nm laser, coupled with a fast camera filming at 5900 fps). To investigate the type of mechanism, whether it is of a photothermal or a photochemical nature, investigators conducted a series of single- and dual-pulse ignition experiments. The single-pulse ignition experiments were conducted to define the ignition energy threshold (Table 4.3).

The results of these various length single-pulse ignition experiments revealed two trends. First, it was evident that an increase in the pulse energy required for ignition directly correlates to the increasing pulse duration, up to 25 ms. Above this pulse duration they found that the required ignition energy decreases with the progressively increased τ (Figure 4.37a). However, the relationship between the single-pulse

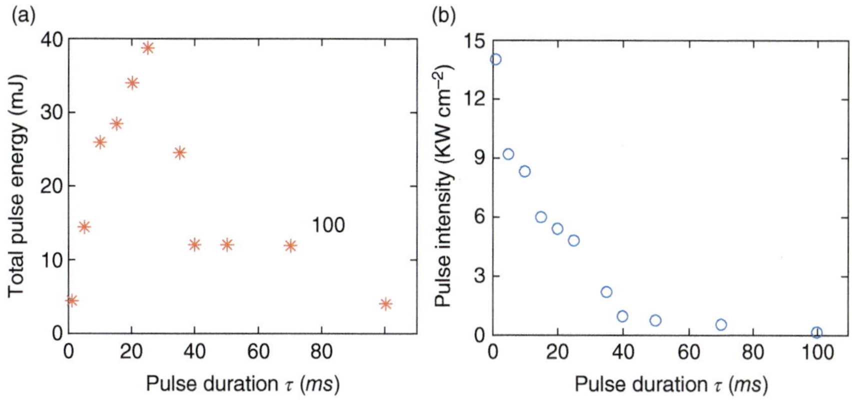

Figure 4.37 Results obtained in various durations of single-pulse laser ignition experiments of the complex **DS4**. (a): Pulse energy versus pulse duration; (b): pulse intensity versus pulse duration time.

duration and the pulse intensity revealed a clear trend of decay in the intensity required for ignition, as the pulse duration increases (Figure 4.37b).

Theoretically, in the photoexcitation ignition mechanism, the ignition threshold should depend on the excited-state population of the irradiated sample. Accordingly, it is expected that the minimum pulse intensity, required for ignition, would be constant and independent of changes in pulse duration. Furthermore, in a photothermal ignition mechanism, the laser light is partially absorbed by the sample at the irradiated region and a hotspot is created, transducing the heat to the rest of the sample. The aforementioned single-pulse ignition results showed that as τ increased the ignition threshold intensity decreased, in light of the aforementioned theory and the analysis of the fast camera results (Figure 4.38). This enabled the identification of the hotspot formation ($t = 4.67$ ms) and led the authors believe that the ignition mechanism of the synthesized compound **DS4** has a photothermal nature. Furthermore, the authors emphasized that the thermal conductivity of the irradiated material and its ability to diffuse the heat outside of the irradiated region as a critical parameter in photothermal ignition mechanism. Hence, the hypothesis that an efficient photothermal ignition could be achieved by reducing the heat dissipation by reducing the pulse duration was suggested and supported by the experimental results with $\tau \leq 25$ ms.

To support the suggested hypothesis, the authors conducted dual-pulse ignition experiments with long enough time intervals between the pulses that ensure that there is no increase in the excited-state population. The narrative of these experiments was to build up heat at the hotspot until ignition is achieved with two consecutive pulses with an energy output lower than the single-pulse ignition threshold at the same pulse duration ($\tau = 10$ ms, $E = 18$ mJ per pulse; at this τ, the single-pulse ignition threshold energy was found to be 26 mJ). The results of these experiments revealed that a minimum time interval of 190 ms between the two pulses was required for the successful ignition of the sample. Above this time interval, no ignition was observed, likely due to heat dissipation from the

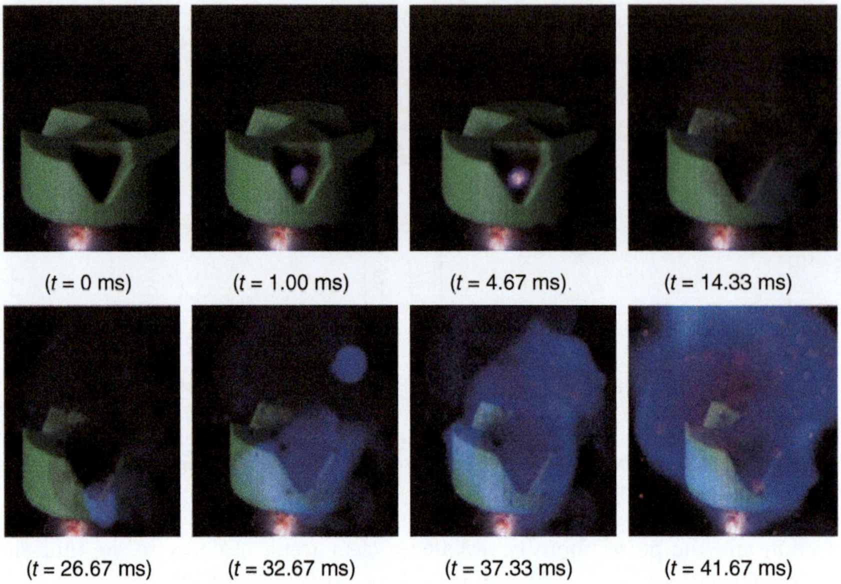

Figure 4.38 High-speed camera images of deflagration process of complex **DS4**.

hotspot area that cooled the hotspot below a certain critical ignition temperature, validating the photothermal ignition mechanism that was proposed for the complex **DS4**.

4.9 Conclusions

In this chapter, we reviewed the concept of laser ignition of energetic materials, with a focus on ECCs. We aimed to outline the research and development workflow in this area and thereby to provide a methodology for future research projects. We briefly reviewed the laser ignition research history, the different setup construction, the synthesis and characterization of the ECCs, and the different mechanistic research approaches of some of known laser-ignitable ECCs. Energetic coordination complexes containing nitrogen-rich ligands with low or non-toxic metals (such as, Mn, Fe, Cu, and Zn), stand out from other competing strategies and fulfill many of desired criteria for laser-initiating compounds. One of the biggest advantages of this promising approach is that a simple adjustment of one of the three building blocks (metal, ligand, and anion) allows the synthesis of a tailor-made compound that could be calibrated, from the perspective of the physico-chemical properties to the end application, whether it is a primer cap replacement, a model compound for mechanistic research, or a propellant ignitor. Another important benefit of Fe- and Cu-based ECCs, making them highly promising and suitable candidates for potential future applications, is their ignitability by optical methods like laser irradiation.

References

1 Maiman, T.H. (1960). *Nature* 187: 492–494.

2 Marx, J.L. (1975). *Science* 188: 821–823.

3 Tkachuk, A.M., Ivanova, S.E., Isaenko, L.I. et al. (1999). *J. Opt. Technol.* 66: 460–462.

4 Gottfried, J.L., Klapötke, T.M., and Witkowski, T.G. (2017). *Propellants Explos. Pyrotech.* 42: 353–359.

5 Brish, A.A., Galeeva, I.A., Zaitsev, B.N. et al. (1966). *Combust. Explos. Shock Waves* 2 (3): 81–82.

6 Yang, L.C. and Menichelli, V.J. (1971). *Appl. Phys. Lett.* 19: 473–475.

7 Ahmad, S.R. and Cartwright, M. (2015). *Laser Ignition of Energetic Materials*. Chichester: Wiley.

8 Tver'janovich, Y., Tver'janovich, A., Aver'yanov, A. et al. (2019). *Progress in Photon Science, Recent Advances*. Cham: Springer Nature Switzerland.

9 Leeuw, M.W., Rooijers, A.J.T., and van der Steen, A.C. (1986). *Advances in Chemical Reaction Dynamics*. Dordrecht: D. Reidel Publishing Company.

10 Moulard, H. (1999). Detonateur optique a deux etages et a transition choc-detonation. EP 1067357B1, filed 06 July 1999 and issued 05 November 2003, France.

11 Joas, M. and Deisenroth, U. (2016). An ignition material, in particular for use in an optical detonator and ignition means with such an ignition material. DE 102016120539A1, filed 27 October 2016 and issued 03 May 2018, Germany.

12 Sinditskii, V.P. and Serushkin, V.V. (1996). *Defence Sci. J.* 46: 371–383.

13 Kunz, S.C. and Salas, F.J. (1988). *13th International Pyrotechnics Seminar*, 505–523.

14 Merson, J.A., Salas, F.J., and Harlan, J.G. (1994). *19th International Pyrotechnics Seminar*, 191–206.

15 Zhilin, Y., Ilyushin, M.A., Tselinskii, I.V., and Brykov, A.S. (2001). *Russ. J. Appl. Chem.* 74: 99–102.

16 Talawar, M.B., Agrawal, A.P., and Asthana, S.N. (2005). *J. Hazard. Mater.* A120: 25–35.

17 Tver'janovich, S., Aver'yanov, A.O., Ilyushin, M.A. et al. (2015). *Russ. J. Appl. Chem.* 88: 226–231.

18 Ilyushin, M.A., Tver'yanovich, A.S., Tver, yanovich, Y.S. et al. (2018). *Glass Phys. Chem* 44: 120–122.

19 Tver'janovich, S., Abdrashitovh, G.O., Aver'yanovh, A.O. et al. (2016). *Int. J. Energetic Mater. Chem. Propul.* 15: 113–122.

20 Ilyushin, M.A., Kotomin, A.A., and Dushenok, S.A. (2019). *Russ. J. Phys. Chem. B* 13: 119–138.

21 Danzeisen, R., Williams, D.L., Viegas, V. et al. (2020). *Toxicol. Sci.* 174: 311–325.

22 Szimhardt, N. and Stierstorfer, J. (2018). *Chem. Eur. J.* 24: 2687–2698.

23 Kuzmenko, V., Tver'janovich, A.S., Ilyushin, M.A., and Tver'yanovich, Y.S. (2019). *J. Energetic Mater.* 37: 420–432.

24 Ilyushin, M.A., Ugryumov, I.A., Dolmatov, V.Yu., and Veretennikova, M.V. (2004). Initiating explosive composition. RU 2309139C2, filed 23 November 2004 and issued 27 October 2007, Russia.

25 Myers, T.W., Bjorgaard, J.A., Brown, K.E. et al. (2016). *J. Am. Chem. Soc.* 138: 4685–4692.

26 Sun, Q., Li, X., Lin, Q., and Lu, M. (2019). *J. Mater. Chem. A* 7: 4611–4618.

27 Xu, J.-G., Lin, S.-J., Li, X.-Z. et al. (2020). *Chem. Eng. J.* 390: 124587.

28 Steinhauser, G., Evers, J., Jakob, S. et al. (2008). *Gold Bull.* 41: 305–317.

29 Klapötke, T.M., Krumm, B., Galvez-Ruiz, J.-C., and Nöth, H. (2005). *Inorg. Chem.* 44: 9625–9627.

30 Ilyushin, M.A., Tselinsky, I.V., Ugryumov, I.A. et al. (2003). *6th New Trends in Research of Energetic Materials Seminar*, 146–152.

31 Chavez, D.E., Hanson, S.K., Scharff, R.J. et al., Photoactive energetic materials. US 20160031767 A1, filed 15 May 2014 and issued 27 February 2018, USA.

32 Wurzenberger, M.H.H., Gruhne, M.S., Lommel, M. et al. (2020). *Inorg. Chem.* 59: 17875–17879.

33 Wurzenberger, M.H.H., Lommel, M., Gruhne, M.S. et al. (2020). *Angew. Chem. Int. Ed.* 59: 12367–12370.

34 Shem-Tov, D., Petrutik, N., Wurzenberger, M.H.H. et al. (2021). *Inorg. Chem.* 60 (15): 10909–10922.

35 Lisitsyn, V.M., Tsipilev, V.N., Damamme, G., and Malis, D. (2011). *Combust. Explos. Shock Waves* 47: 591–600.

36 Renlund, M., Stanton, P.L., and Trott, W.M. (1989). *Proceedings of the 9th International Detonation Symposium*, 1118–1127.

37 Paisley, D.L. (1989). *Proceedings of the 9th International Detonation Symposium*, 1110–1117.

38 Aduev, B.P., Nurmukhametov, D.R., Furega, R.I. et al. (2013). *Russ. J. Phys. Chem. B* 7: 453–456.

39 Scholles, H., Schirra, R., and Zoellner, H. (2016). *42th International Pyrotechnics Seminar*, 422–428.

40 Gerasimov, S.I., Ilyushin, M.A., and Kuz'min, V.A. (2015). *Tech. Phys. Lett.* 41: 338–340.

41 Gerasimov, S.I., Ilyushin, M.A., Shugalei, I.V. et al. (2017). *Russ. J. Gen. Chem.* 87: 3156–3159.

42 United Nations Economic Commission for Europe www.unece.org (March 2021).

43 NATO Allied Ordnance Publication 7, 2, 2003.

44 Zhilin, Y., Ilyushin, M.A., Tselinskii, I.V. et al. (2003). High-energy-capacity cobalt (III) tetrazolates. *Russ. J. Appl. Chem.* 76: 572–576. https://doi.org/10.1023/A:1025778918488.

45 Ugryumov, I.A., Ilyushin, M.A., Tselinskiy, I.V., and Kozlov, A.S. (2003). *Russ. J. Appl. Chem.* 76: 439–441.

46 Ilyushin, M.A., Tselinskii, I.V., Bachurina, I.A. et al. (2007). *10th New Trends in Research of Energetic Materials Seminar*, 165–182.

47 Chernai, V., Sobolev, V.V., Chernai, V.A. et al. (2003). *Combust. Explos. Shock Waves* 39: 335–339.

48 Myers, T.W., Chavez, D.E., Hanson, S.K. et al. (2015). *Inorg. Chem.* 54: 8077–8086.

49 Sifain, E., Bjorgaard, J.A., Myers, T.W. et al. (2016). *J. Phys. Chem. C* 120: 28762–28773.

50 Myers, T.W., Brown, K.E., Chavez, D.E. et al. (2016). *Eur. J. Inorg. Chem.* 3178–3183.

51 Myers, T.W., Brown, K.E., Chavez, D.E. et al. (2017). *Inorg. Chem.* 56: 2297–2303.

52 Brown, K.E., Myers, T.W., Snyder, C., and Clarke, S.A. (2017). 20th Biennial Conference of the APS Topical Group on Shock Compression of Condensed Matter, July 9-14, 2017, St. Louis, Missouri. Bulletin of the American Physical Society, 100008.

53 Zhang, Q., Chen, D., and Li, H. (2020). Cadmium-based energetic metal organic framework directly initiated by 1064 nm laser, preparation method and application. CN 112280055 A, filed 30 October 2020 and issued 29 January 2021, China.

54 Wang, T., Zhang, Q., Deng, H. et al. (2019). *ACS Appl. Mater. Interfaces* 11: 41523–41530.

55 Huang, H., Zhang, T., Zhang, J., and Wang, L. (2010). *J. Hazard. Mater.* 179: 21–27.

56 Yan, Q.-L., Liu, P.-J., He, A.-F. et al. (2018). *Chem. Eng. J.* 338: 240–247.

57 Szimhardt, N., Wurzenberger, M.H.H., Klapötke, T.M., and Stierstorfer, J. (2018). Energetic materials, sythesis, processing, preformance. 49th international annual conference of the fraunhofer ICT June 26-29, 2018, published by Fraunhofer-Institut für Chemische Technologie ICT.

58 Szimhardt, N., Wurzenberger, M.H.H., Beringer, A. et al. (2017). *J. Mater. Chem. A* 5: 23753–23765.

59 Gaponik, P.N., Degtyarik, M.M., Lyakhov, A.S. et al. (2005). *Inorg. Chim. Acta* 358: 3949–3957.

60 Franke, P.L., Haasnoot, J.G., and Zuur, A.P. (1982). *Inorg. Chim. Acta* 59: 5–9.

61 Wurzenberger, M.H.H., Endraß, S.M.J., Lommel, M. et al. (2020). *ACS Appl. Energy Mater.* 3: 3798–3806.

62 Gaponik, P.N., Karavai, V.P., and Grigor'ev, Y.V. (1985). *Khim. Geterotsikl. Soedin.* 21: 1255–1258.

63 Ivashkevich, D.O., Lyakhov, A.S., Gaponik, P.N. et al. (2001). *Acta Crystallogr. E* 57: m335–m337.

64 Wurzenberger, M.H.H., Gruhne, M.S., Lommel, M. et al. (2019). *Chem. Asian J.* 14: 2018–2028.

65 Schneider, S., Hawkins, T., Rosander, M. et al. (2008). *Inorg. Chem.* 47: 3617–3624.

66 Müller, D., Knoll, C., Stöger, B. et al. (2013). A modified synthetic pathway for the synthesis of so far inaccessible N1-functionalized tetrazole ligands – synthesis and characterization of the 1D chain-type spin crossover compound $[Fe(3ditz)_3](BF_4)_2$. *Eur. J. Inorg. Chem.* 984–991. https://doi.org/10.1002/ejic.201201062.

67 Zeisel, L., Szimhardt, N., Wurzenberger, M.H.H. et al. (2019). *New J. Chem.* 43: 609–616.

68 Szimhardt, N., Wurzenberger, M.H.H., Klapötke, T.M. et al. (2018). *J. Mater. Chem. A* 6: 6565–6577.

69 Joas, M. and Klapötke, T.M. (2015). *Propellants Explos. Pyrotech.* 40: 246–252.

70 Aluker, E.D., Krechetov, A.G., Mitrofanov, A.Y. et al. (2011). *J. Phys. Chem. C* 115: 6893–6901.

5

Energetic 1,2,3,4-Tetrazines

Aleksandr M. Churakov, Michael S. Klenov, Aleksey A. Voronin, and Vladimir A. Tartakovsky

N.D. Zelinsky Institute of Organic Chemistry, Russian Academy of Sciences, 47 Leninsky Prospect, 119991 Moscow, Russian Federation

5.1 Introduction

This chapter discusses 1,2,3,4-tetrazine ring system. The parent member of the 1,2,3,4-tetrazine series, tetrazine **1**, has not yet been obtained (Figure 5.1). The first tetrahydro-tetrazines **2** were synthesized in 1972 [1], the first completely unsaturated annulated tetrazine **3** was obtained in 1988 [2], and benzotetrazine 1,3-dioxide **4** was first reported in 1990 [3]. To date, several reviews on 1,2,3,4-tetrazines have been published including a book series *Comprehensive Heterocyclic Chemistry I, II,* and *III* [4–6], a review on 1,2,3,4-tetrazines for 1994–2003 [7], a review on heterocyclic *N*-oxides where 1,2,3,4-tetrazines are mentioned [8], and a review on fused heterocycle-based energetic materials for 2012–2019, which also mentions 1,2,3,4-tetrazines [9].

This chapter does not aim to cover all 1,2,3,4-tetrazines known to date. We focus on those compounds that may be of interest as energetic compounds or their precursors. Some unstable 1,2,3,4-tetrazines are also considered in order to compare the thermal stability of different types of tetrazines. Methods for the synthesis of tetrazines, their reactivity, physicochemical and spectral properties are considered, and their energetic performance is assessed. Particular attention is paid to the thermal stability of these compounds, since it is crucial for the practical application.

Figure 5.2 shows the tetrazine frameworks discussed in this chapter. The substituents at the carbon atoms of the rings (hydrogen, carbon substituents, or functional groups) are not shown. The R groups at the nitrogen atoms are carbon substituents (R ≠ H). The list includes both partially unsaturated and fully unsaturated

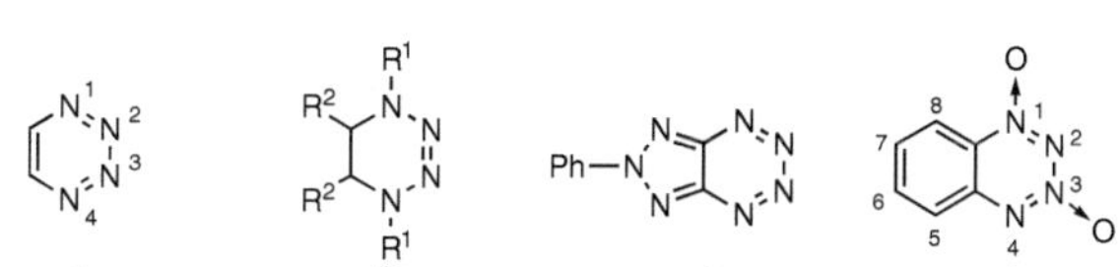

1 2 3 4

Figure 5.1 Parent 1,2,3,4-tetrazine **1** and tetrazines **2**, **3** and **4** which were synthesized very first.

Nitrogen-Rich Energetic Materials, First Edition. Edited by Michael Gozin and Leonid L. Fershtat.

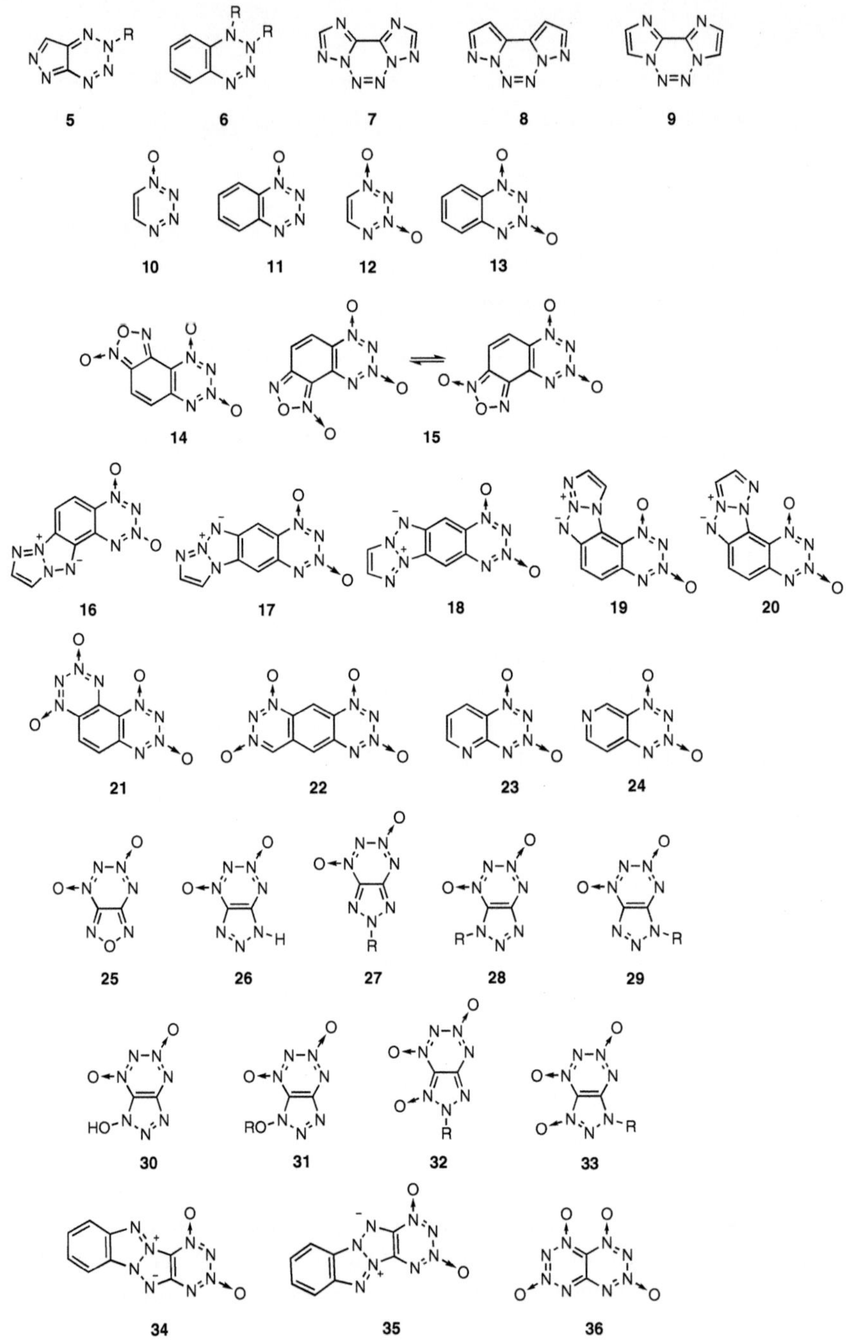

Figure 5.2 The frameworks of the heterocyclic systems discussed in this chapter.

compounds: tetrazines with one carbon substituent at the nitrogen atom **5**; with two carbon substituents at the nitrogen atoms **6**; tetrazines with a bridgehead (ring junction) nitrogen atom **7**, **8**, and **9**; completely unsaturated tetrazine 1-oxides **10** and **11**. The main group of compounds under consideration is 1,2,3,4-tetrazine 1,3-dioxides (TDOs). Among them are non-annulated TDOs **12**, benzotetrazine 1,3-dioxides (BTDOs) **13**, BTDOs annulated with the furoxan rings **14** and **15**, BTDOs annulated with the tetraazapentalene ring systems **16–20**, BTDOs annulated with the second TDO rings **21** and **22**, pyrido-annulated TDOs **23** and **24**, furazano-annulated TDO **25** (FTDO), 1,2,3-triazolo-annulated TDOs **26–33**, TDOs annulated with tetraazapentalene ring systems **34** and **35**, and tetrazino-tetrazine 1,3,6,8-tetraoxide **36** (TTTO).

5.2 Methods of Synthesis and Reactivity of 1,2,3,4-Tetrazines

5.2.1 1,2,3,4-Tetrazines with Carbon Substituents at Nitrogen Atoms

A series of substituted 5-aryl-3-methylpyrazolo-tetrazines **38** was synthesized by diazotization of arylazo-pyrazoles **37** with sodium nitrite in a mixture of sulfuric and acetic acids, followed by treatment of intermediate diazonium salts with sodium acetate and cyclization of intermediate diazo compounds (Scheme 5.1) [10]. Compounds **38** are quite stable (mp: 145–228 °C, depending on the substituents in the benzene ring). Note that this type of tetrazine does not always exist in cyclic form. Some of them exhibit tetrazine/azo-diazo equilibrium [7].

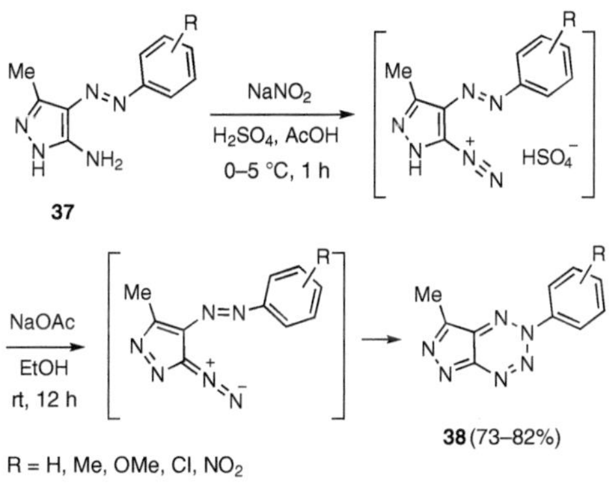

Scheme 5.1 Synthesis of *N*-aryl substituted tetrazines by coupling of azo and diazo groups.

A series of functionalized pyrazolo-benzotetrazin-3-ones **40** was synthesized in good yields (60–91%) [11, 12] using diazotization of 1-(2-aminophenyl)-pyrazol-3-ones **39** (Scheme 5.2) [13]. These compounds are also quite stable (mp: 143–238 °C, depending on the substituents in the benzene ring).

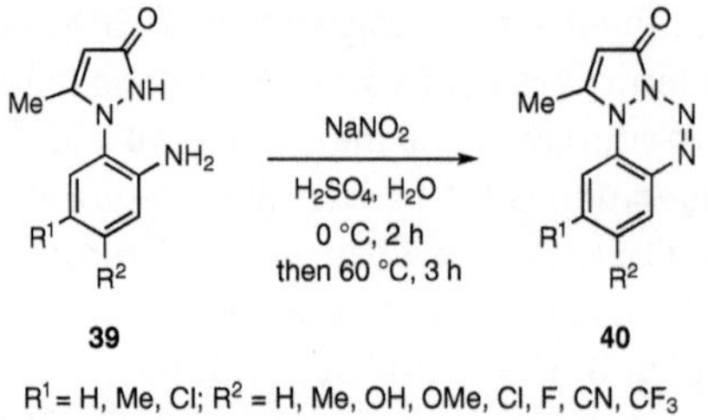

Scheme 5.2 Synthesis of di-*N*-substituted tetrazines by coupling of amino and diazonium groups.

5.2.2 Annulated 1,2,3,4-Tetrazines with Bridgehead Nitrogen Atoms

1,4-Dihydro-1,2,3,4-tetrazine **42** annulated with two nitrotriazole rings was obtained in high yield by oxidation of aminated bis-triazole **41** with *tert*-butyl hypochlorite (Scheme 5.3) [14].

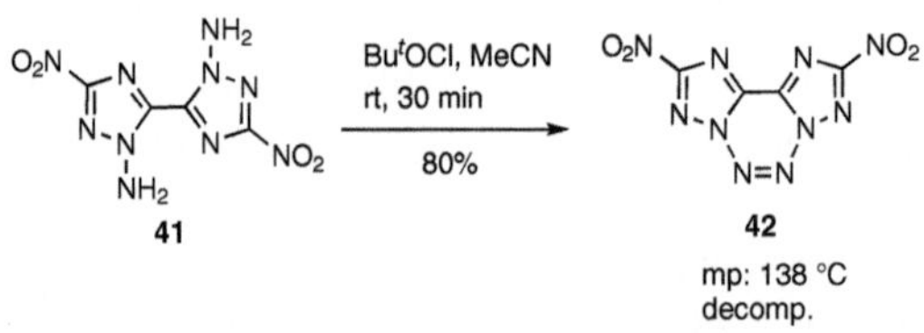

Scheme 5.3 Synthesis of tetrazine **42** with two bridgehead nitrogen atoms by oxidation of amino groups.

A similar synthetic approach was used to obtain 1,2,3,4-tetrazine **44** annulated with two dinitropyrazole rings (Scheme 5.4) [15]. Oxidation of aminated bis-pyrazole **43** with *tert*-butyl hypochlorite gave tetrazine **44** in 45% yield.

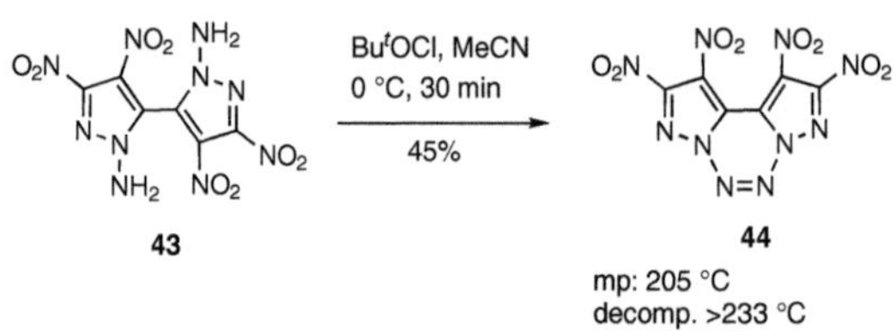

Scheme 5.4 Synthesis of tetrazine **44** with two bridgehead nitrogen atoms by oxidation of amino groups.

When aminated diimidazole **45** was oxidized with *tert*-butyl hypochlorite, dichloro azo compound **46** was obtained (Scheme 5.5). The following reaction with sodium azide in MeCN/H_2O resulted in a high yield of tetrazine **47** [16].

5.2.3 Fully Unsaturated 1,2,3,4-Tetrazines

5.2.3.1 Annulated 1,2,3,4-Tetrazines

The first fully unsaturated annulated tetrazine **3** was obtained by ring expansion of 1,2,3-triazole ring by an oxidation of amine **48** with lead tetraacetate (LTA)

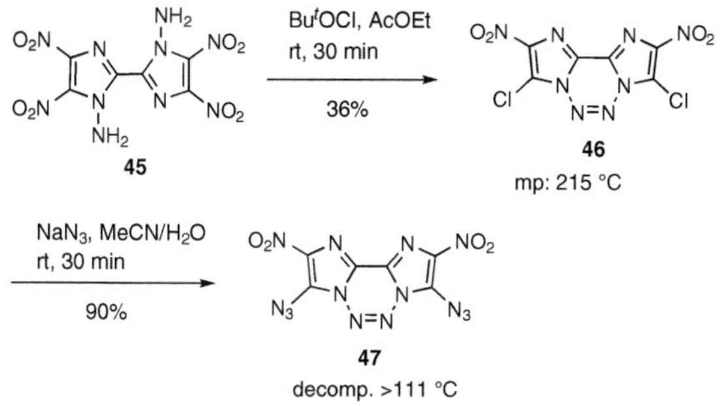

Scheme 5.5 Synthesis of tetrazine **47** with two bridgehead nitrogen atoms by oxidation of amino groups.

(Scheme 5.6) [2]. This tetrazine was unstable at room temperature and exploded on impact or heating. No further work has been done with this class of compounds.

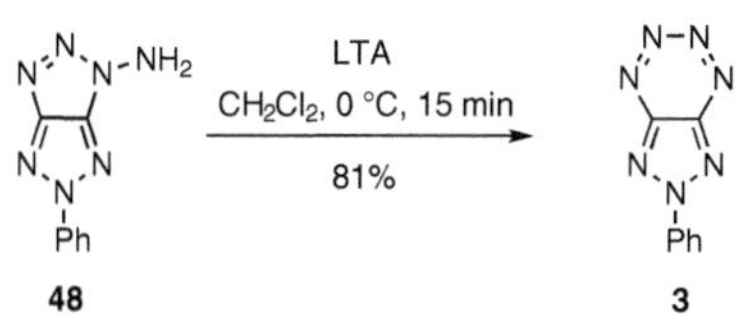

Scheme 5.6 Synthesis of the first fully unsaturated tetrazine.

5.2.3.2 1,2,3,4-Tetrazine 1-Oxides and Benzotetrazine 1-Oxides

Non-annulated 1,2,3,4-tetrazine 1-oxides and benzotetrazine 1-oxides (BTOs) are discussed here. BTOs **49** were synthesized from diazonium salts by intramolecular cyclization of the *tert*-butyl-*NNO*-azoxy group with the diazonium cation followed by elimination of the *tert*-butyl cation (Scheme 5.7) [17]. The thermal stability of BTOs depends on the substituents in the benzene ring. The most stable of them decomposed in eight days at room temperature (5-Br-BTO, 7,8-Me_2-BTO), whereas 5-NO_2- and 7-NO_2-BTOs were too unstable to be isolated [7]. The decomposition of BTOs involved ring opening to afford *ortho*-nitrosoazides, which easily cyclized to benzofurazans.

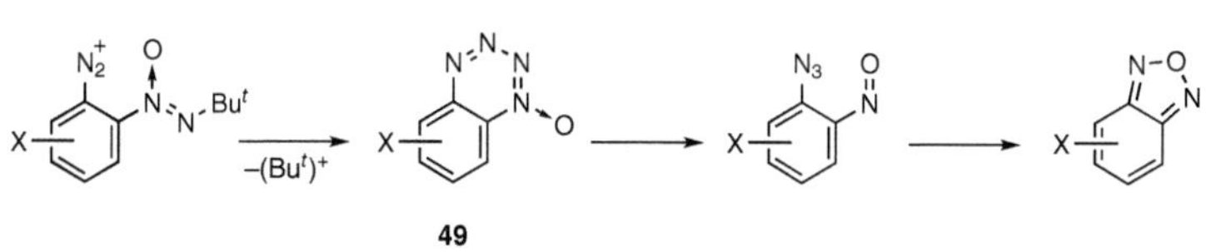

Scheme 5.7 Synthesis of benzotetrazine 1-oxides and their decomposition.

The first non-annulated 1,2,3,4-tetrazine 1-oxide **51** was obtained in three steps starting from 1-(benzyloxy)-1,2,3-triazole **50** (Scheme 5.8) [18], which was aminated with *O*-tosylhydroxylamine and oxidized with MnO_2. Presumably, nitrene was formed in the oxidation process, which participates in the ring expansion reaction to afford tetrazinium salt. The reduction in this salt with hydrogen in the presence of a Pd/C catalyst gave tetrazine 1-oxide **51**. It was not isolated individually, but was characterized by the ^{1}H, ^{13}C NMR, and mass spectra (EI). The thermal stability of this compound has not been determined.

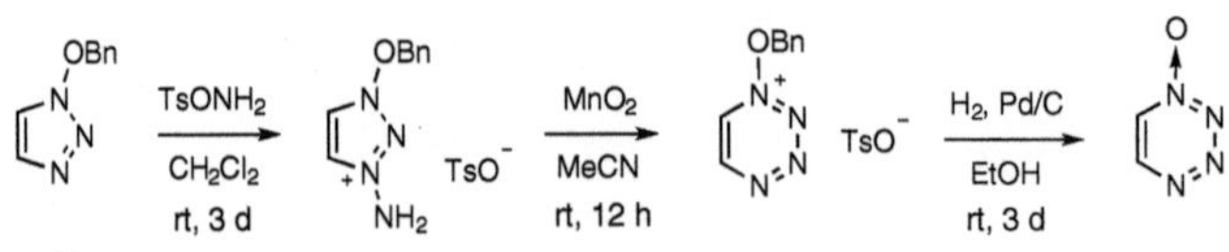

Scheme 5.8 Synthesis of the first tetrazine 1-oxide.

5.2.3.3 1,2,3,4-Tetrazine 1,3-Dioxides

The main approach to the synthesis of 1,2,3,4-tetrazine 1,3-dioxide cycle (TDO-cycle) is the intramolecular reaction of the *tert*-butyl-*NNO*-azoxy group with the intermediate oxodiazonium ion $[-N=N=O]^+$ followed by elimination of the *tert*-butyl cation (Scheme 5.9). The real existence of the oxodiazonium ion was confirmed by its trapping *via* an intramolecular reaction with the benzene ring to form cinnoline-*N*-oxide derivatives [19, 20].

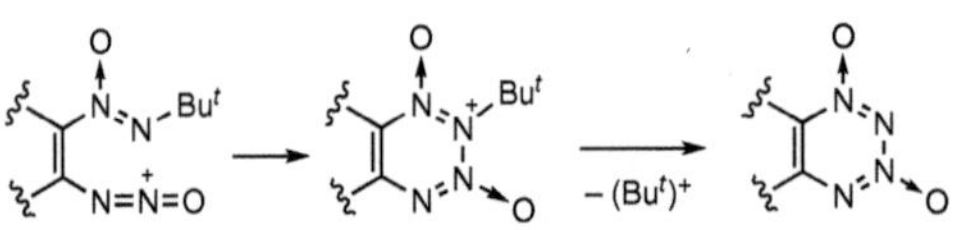

Scheme 5.9 Plausible mechanism for the 1,2,3,4-tetrazine 1,3-dioxide ring formation.

Methods for the synthesis of TDOs differ from each other depending on the generation of oxodiazonium ions. One of these methods includes the reaction of primary nitramines with nitrating reagents (so-called "nitration method"). Intermediate *O*-nitrated nitramines are assumed to dissociate in an acidic medium to afford the $[-N=N=O]^+$ ion (Scheme 5.10, Eq. (1)) [21]. In some cases, a "phosphorylation method" was used, which includes the treatment of nitramines with phosphorylating reagents (P_2O_5, PCl_5) [7]. The so-called "acetylation method," which includes the treatment of nitramines with acetic anhydride in the presence of small amounts of sulfuric acid, has recently become the most widespread. Presumably, intermediate *O*-acetylated nitramines dissociate in an acidic medium, giving the $[-N=N=O]^+$ ion (Scheme 5.10, Eq. (2)). Nitramine is usually generated *in situ* by nitrating the corresponding amine with HNO_3. This method is suitable for the synthesis of annulated TDOs. Aromatic and heteroaromatic compounds were used as starting compounds: benzenes [20], 1,2,5-oxadiazoles [22], 1,2,3-triazoles, and others, which contain adjacent amino and *tert*-butyl-*NNO*-azoxy groups.

For the synthesis of non-annulated TDOs from aliphatic precursors, another method for the generation of the oxodiazonium ion was used, which included the reaction of *O*-alkylated nitramines with boron trifluoride ether (Scheme 5.10, Eq. (3)) [23]. This method is also suitable for the synthesis of benzoannulated TDOs [24].

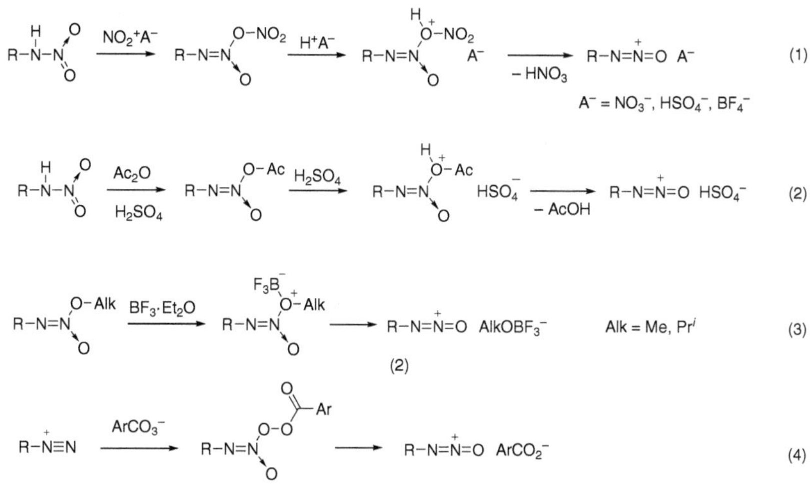

Scheme 5.10 Plausible mechanisms for oxodiazonium ion formation.

In addition, for the synthesis of BTDOs, the so-called "oxidation method" was developed, which involved generation of oxodiazonium ion by the reaction of diazonium salts with *m*-chloroperbenzoic acid in aqueous acetonitrile using sodium acetate as a base [21]. Presumably, azoperoxy intermediate first appeared, and then its ionic dissociation afforded oxodiazonium ion (Scheme 5.10, Eq. (4)). However, this method is not widely used.

5.2.3.3.1 Non-annulated 1,2,3,4-Tetrazine 1,3-Dioxides

The parent TDO has not yet been obtained.

Synthesis from Aliphatic Precursors The synthesis of non-annulated TDOs from aliphatic precursors involved the intramolecular reaction of alkenes containing *tert*-butyl-*NNO*-azoxy group and *O*-alkylated *N*-nitroamine group in adjacent positions. Cyclization took place under the action of boron trifluoride etherate (see Scheme 5.10, Eq. (5.3)). Non-annulated TDOs **56a** and **57a,b** were obtained from azoxyalkenes **54a** and **55a,b**, respectively (Scheme 5.11) [23, 25]. The latter were obtained from bis(*tert*-butyl-*NNO*-azoxy)acetonitrile **52** by reaction with alkyl thiols and subsequent nitration to give *N*-nitroimines **53a,b** followed by their *O*-alkylation. Oxidation of compounds **54a,b** led to sulfones **55a,b**. The best yields of TDOs were achieved precisely from these sulfones with electron-withdrawing alkylsulfonyl substituents. The relatively low yield of TDO **56a** (20%) can be explained by the fact that in this case the intermediate oxodiazonium cation is rather unstable and easily eliminates N_2O with the formation of a cation stabilized by the methylthio group. Alkylsulfonyl groups at the C-5 atom in TDOs **57a,b** were easily replaced by ammonia to form TDO **58**. The replacement of MeS group with ammonia in compound **56a** also gave TDO **58** but in a low yield (18–20%).

An attempt to perform the ring closure in compound **58** by a reaction with an excess of nitronium tetrafluoroborate in acetonitrile failed (for a successful ring

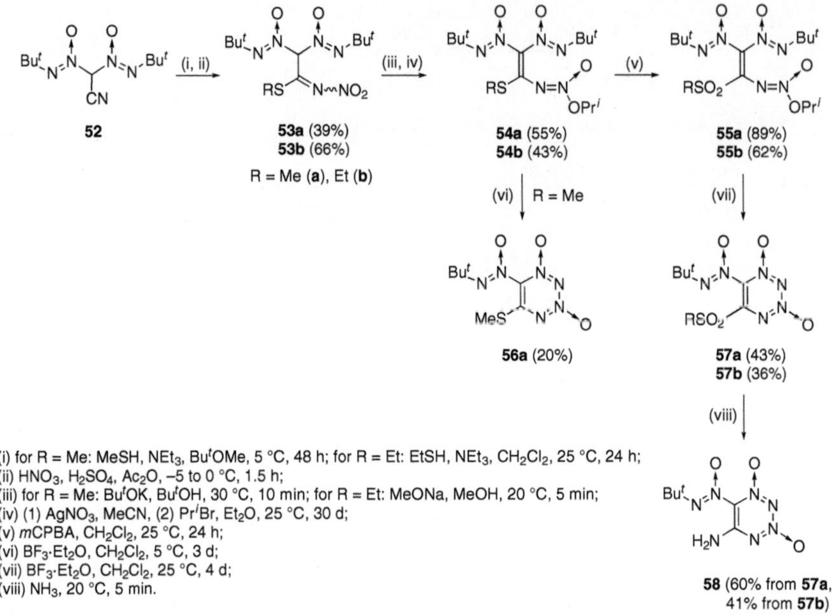

Scheme 5.11 Synthesis of BTDO **58**.

closure by "acetylation method" to obtain TTTO **36**, see Scheme 5.40). Instead of cyclization, 5-amino-6-nitro-TDO **59** (30%) was formed along with TDO **60** (45%) containing a nitro-*NNO*-azoxy group (Scheme 5.12) [26, 27]. Presumably, the formation of TDO **59** proceeds through electrophilic substitution of the *tert*-butyl-*NNO*-azoxy group at the carbon atom and includes the formation of cationic intermediate **A**. TDO **60** appears to be formed by electrophilic substitution of the *tert*-butyl group at the nitrogen atom *via* intermediate **B** (cf. lit. [28]).

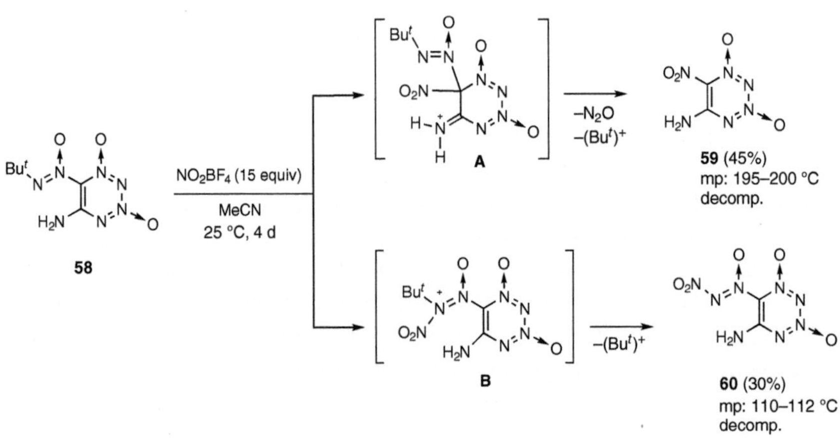

Scheme 5.12 Electrophilic replacement of the *tert*-butyl-*NNO*-azoxy or *tert*-butyl group with a nitro moiety.

Synthesis from Benzotetrazine 1,3-Dioxides Alternative methods for the synthesis of nonannulated TDOs are based on the benzene rings opening in BTDOs. The first method involves thermolysis of polyazido-substituted BTDOs.

5,7-Dinitro-6,8-dibromo-BTDO **62** was obtained by nitration of 6,8-dibromo-BTDO **61** (Scheme 5.13). The reaction of BTDO **62** with NaN_3 yielded di-, tri-, and tetraazido-substituted BTDOs **63**, **64**, and **65** depending on the conditions [29]. Heating a solution of 6,7,8-triazido-5-nitro-BTDO **64** in toluene at 100 °C for several minutes led to the cleavage of the benzene ring with the formation of non-annulated TDO **66** in a 35% yield. 5,6-Dicyano-TDO **67** was obtained in a 50% yield by heating 5,6,7,8-tetraazido-BTDO **65** in toluene at 80 °C for 1.5 hours.

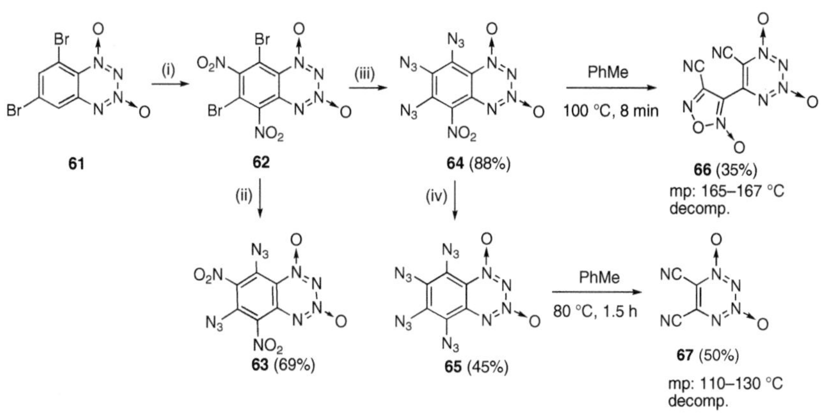

(i) HNO_3/oleum, 20 °C, 1 h and 90 °C, 1 h (58%); (ii) NaN_3 (2 equiv portionwise), acetone, 20 °C, 1 h; (iii) NaN_3 (3 equiv), acetone/H_2O, 20 °C, 5 min; (iv) NaN_3, DMF, 20 °C, 1.5 h.

Scheme 5.13 Synthesis of non-annulated TDOs by thermal benzene ring scission.

The second method for the benzene ring cleavage was ozonolysis of BTDOs bearing electron-donor substituents [30]. Ozonolysis of dimethoxy-BTDOs **68** and **71** in CH_2Cl_2 with subsequent treatment of primary ozonides with HCl in MeOH led to 5,6-di(methoxycarbonyl)-TDO **69** as the main product (Scheme 5.14). TDOs **70** and **72** were formed in these reactions as minor products. The reactivity of TDO **69** has not been studied.

5.2.3.3.2 Benzoannulated 1,2,3,4-Tetrazine 1,3-Dioxides

Benzotetrazine 1,3-Dioxides with Functional Groups Benzoannulated TDOs are the most studied class of TDOs. The first member of the series, unsubstituted BTDO **4**, was obtained by "nitration" (Scheme 5.15, yield 65%) [21, 31], "acetylation" (99%) [20], and "oxidation" (70%) [3] methods, as well as from *O*-alkylnitramines (99%) [24]. The "nitration method" involves the reaction of aniline **73** with nitrating reagents (N_2O_5, NO_2BF_4) in organic solvents (CH_2Cl_2, CH_3NO_2, CH_3CN, CF_3CO_2H) (Scheme 5.15). Under these conditions the benzene ring also underwent nitration, and the intermediate *C*-nitrated anilines were also involved in ring-closure reactions. As a result, along with BTDO **4**, BTDOs **74**, **75**, and **76** were

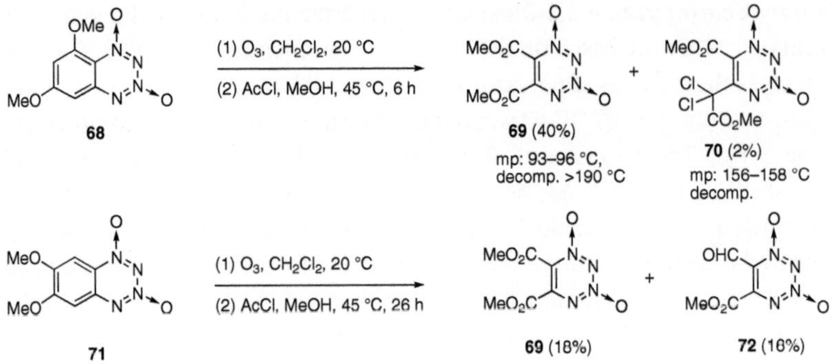

Scheme 5.14 Synthesis of non-annulated TDOs by ozonolysis.

obtained. The ratio of the reaction products depends on the nitrating reagent and the solvent. BTDO **4** itself is not nitrated under these reaction conditions.

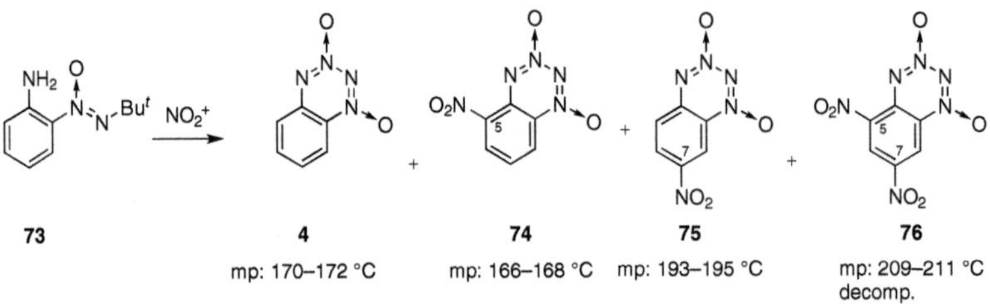

Scheme 5.15 Synthesis of parent BTDO **4** and nitro-substituted BTDOs **74**, **75**, and **76**.

A number of bromo- and chloro-BTDOs **78** were obtained by ring closure in aniline **77** by the "nitration" method from the corresponding bromo- and chloro-substituted anilines (yields: 27–82%) (Scheme 5.16) [7]. BTDOs **78** served as precursors for the preparation of a number of functionally substituted BTDOs.

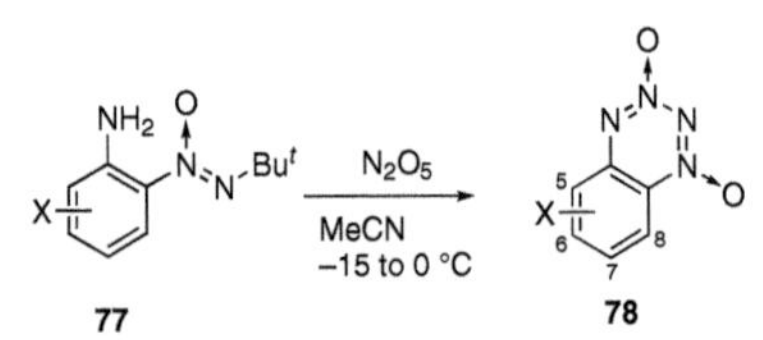

78: X = 6-Br, 7-Br, 8-Br, 5,7-Br_2, 6,7-Br_2, 6,8-Br_2, 7,8-Br_2, 5,6,7-Br_3, 6,8-Cl_2

Scheme 5.16 Synthesis of bromo- and chloro-substituted BTDOs.

The high resistance of the TDO ring to strong acids allows nitration and bromination of the benzene ring of BTDO. For example, nitration of BTDO **4** with a HNO_3/20% oleum mixture at 80 °C afforded 5,7-dinitro-BTDO **76** (yield: 90%) [31]. Nitration of BTDO **4** under milder conditions (HNO_3/H_2SO_4, rt) gives 5-nitro- and 7-nitro-BTDOs **74** and **75** in a ratio of ~1 : 2. In 2012, an optimized one-pot synthesis of 5,7-dinitro-BTDO **76** from aniline **73** was developed, which involved a tetrazine ring closure using N_2O_5 in CH_2Cl_2 as a solvent followed by exhaustive nitration with HNO_3 in 20% oleum (yield: 45%) [32].

Treatment of BTDO **4** with dibromoisocyanuric acid (DBI) in CF_3COOH containing H_2SO_4 (15–20%) allowed to obtain 5-bromo- and 7-bromo-BTDOs in a ratio of 47 to 53. Bromination of 6,8-dibromo-BTDO using DBI in conc. H_2SO_4 afforded tetrabromo-BTDO (yield: 96%) [7].

Generally, when BTDOs were nitrated or brominated, the reactivity of positions changed in the order $5 \approx 7 \gg 8 > 6$, though bromination is usually less selective [33]. Apparently, the difference in the rates of electrophilic substitution reactions at different positions of the BTDO ring can be explained by the higher thermodynamic stability of cationic σ-complexes **C** and **D** in comparison with cationic σ-complexes **E** and **F**, the positive charge in which is distributed less efficiently (by the example of nitration in Figure 5.3).

The difference in the thermodynamic stability of complexes **C** and **D** is not very significant and the difference in the rates of nitration at positions 5 and 7 is not large and depends on the electronic and steric structure of the substituents at positions 6 and 8.

The TDO ring in BTDO effectively activates the benzene ring to nucleophilic substitution, while the TDO ring itself is not affected in these reactions. This mode of reactivity made it possible to use bromo- and nitro-BTDOs as starting materials for the preparation of BTDOs containing amino, azido, methoxy, and other functional groups. When bromo or nitro groups in BTDOs were replaced by nucleophiles ($MeNH_2$, NaN_3, MeOK), the reactivity of positions changed in the order: $6 > 8 > 7 > 5$ [34]. Nucleophilic substitution (as exemplified by the replacement of bromine with a methoxy ion in Figure 5.4) probably follows the addition–elimination (AE) mechanism. In this case, the variation in the rate of substitution depending on the position of the substituent in the ring is associated with different thermodynamic stability of the intermediate anionic σ complexes **G–J**.

Better delocalization of the negative charge over the TDO ring in anions **G** and **H** makes them more thermodynamically stable than anions **I** and **J**. For this reason,

Figure 5.3 The plausible structures of cationic intermediates.

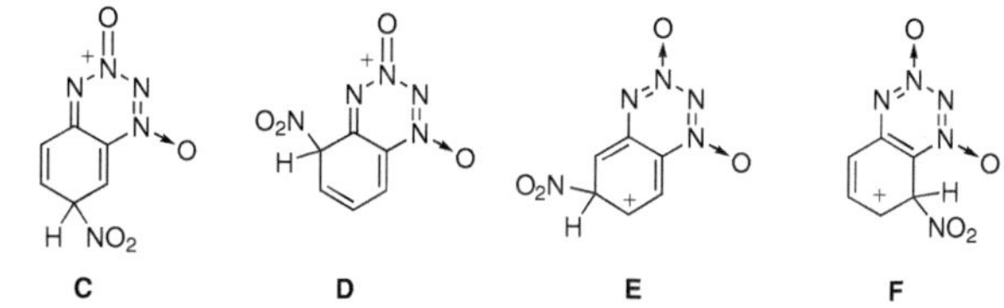

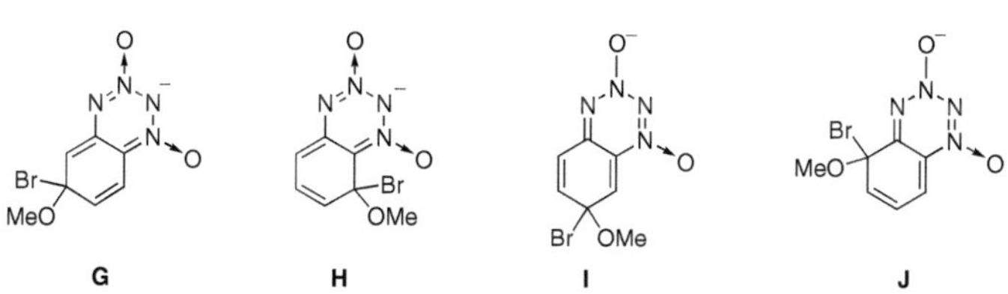

Figure 5.4 The plausible structures of the Meisenheimer-type anions.

positions 6 and 8 are preferable for substitution, though the 5- and 7-Br atoms can also be replaced rather easily. Thus, nucleophilic substitution in BTDOs is possible for any position of the benzene ring of BTDOs.

"Billiard" reaction. To obtain BTDOs annulated with tetraazapentalene systems, an attempt was made to synthesize BTDO **79** containing triazole moiety in position 5 by the reaction of BTDO **80** with 1,2,3-triazole in the presence of a base. However, instead of the expected 6-azido-5-triazolyl-BTDO **79**, an isomer with a reverse arrangement of substituents was formed, namely 5-azido-6-triazolyl-BTDO **81** (Scheme 5.17) [35].

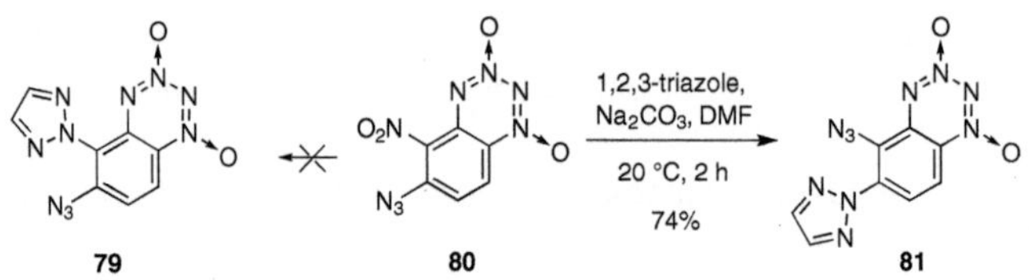

Scheme 5.17 Unusual "billiard" reaction.

The formation of BTDO **81** can be explained as follows (Scheme 5.18). At the first stage, the triazole anion attacks BTDO **80** at position 6 with the formation of the anionic σ-complex **K**. Further, the release of the azide anion occurs with the formation of BTDO **82**. However, the process does not stop at this stage. Now the azide anion itself acts as a nucleophile and attacks position 5 of BTDO **82** with the formation of the anionic σ-complex **L** followed by the release of the nitrite anion and formation of BTDO **81**.

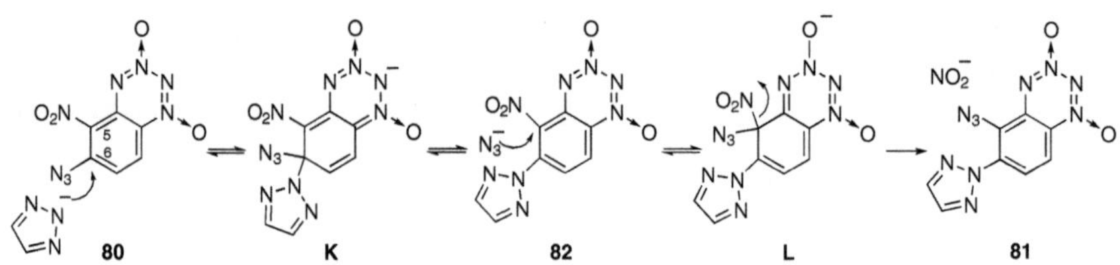

Scheme 5.18 Plausible mechanism of "billiard" reaction.

This unusual cascade of nucleophilic substitution reactions, in which a nucleofuge becomes a nucleophile, can be called a "billiard" process. The possibility of such a process is due to the fact that in BTDO nucleophilic substitution can occur both at positions 6 and at 5. Several more examples of "billiard" reactions are described in the BTDO series [35].

Benzotetrazine 1,3-Dioxide Annulated with One More Tetrazine 1,3-Dioxide Ring

The tetrazine ring closure in compound **83** using N_2O_5 resulted in BTDO **84** (Scheme 5.19). The treatment of the latter with ammonia afforded amines **85** and **86** in almost equal amounts. This mixture of amines was separated by chromatography and treatment of BTDO **85** with N_2O_5 yielded tetrazino-benzotetrazine tetraoxide (TBTTO) **87** [36]. The chlorine atom at position 6 of TBTTO **87** was easily replaced by nucleophiles. For example, reaction with methylamine afforded TBTTO **88**, while the TDO-rings remained unaffected.

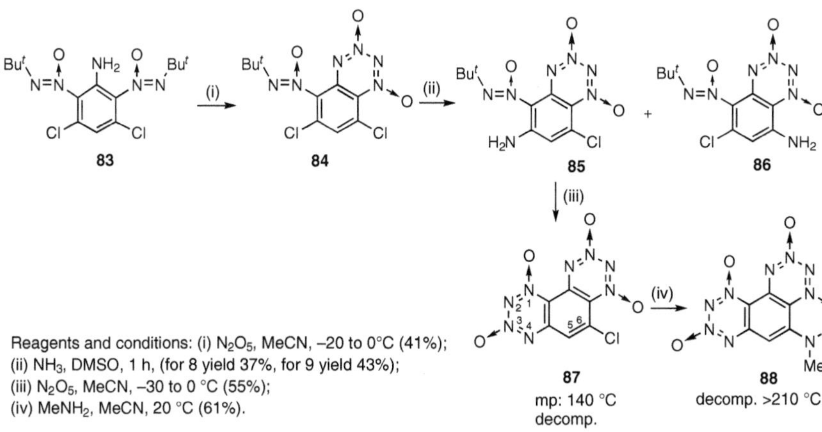

Scheme 5.19 The synthesis of phenanthrene-like TBTTOs.

TBTTOs **94a,b** were synthesized by simultaneous or sequential closure of two TDO-cycles (Scheme 5.20) [37]. In the first case, chloroaniline **89** was converted to phenylenediamine **93** under the action of NH_3 at 250 atm followed by cyclization into a mixture of TBTTOs **94a** and **94b** in a ratio close to 1 : 1. In the second case, BTDO **90** containing a labile chlorine atom was obtained from chloroaniline **89** under the action of N_2O_5. Replacement of chlorine with ammonia provided amino-BTDO **91**, which nitration with NO_2BF_4 afforded nitramine **95**. The cyclization of the latter under the action of P_4O_{10} in MeCN as a solvent allowed to obtain pure tetrazine **94a**. Treatment of BTDO **91** with an excess of N_2O_5 resulted in a mixture of BTDOs **94a** and **94b** in a ratio close to 1 : 2. Bromination of BTDO **91** followed by ring closure with N_2O_5 in MeCN as a solvent resulted in TBTTO **96**.

The TBTTOs obtained began to decompose without melting at 140–160 °C. These compounds including unsubstituted TBTTO **90a** were extremely sensitive to hydrolysis, most likely due to the easy formation of a thermodynamically stable anionic σ-complex **M** with a well-distributed negative charge (Figure 5.5). An attempt to replace the bromine atom in 5-bromo-TBTTO **96** with nucleophiles was unsuccessful, apparently due to the fact that the nucleophile attacks position 10 much easier than position 5.

Benzotetrazine 1,3-Dioxides Annulated with Tetraazapentalenes The starting compounds for the preparation of benzotetrazine 1,3-dioxides annulated with tetraazapentalenes were BTDOs incorporating the azide group and the 1,2,3-triazole fragment located in adjacent positions.

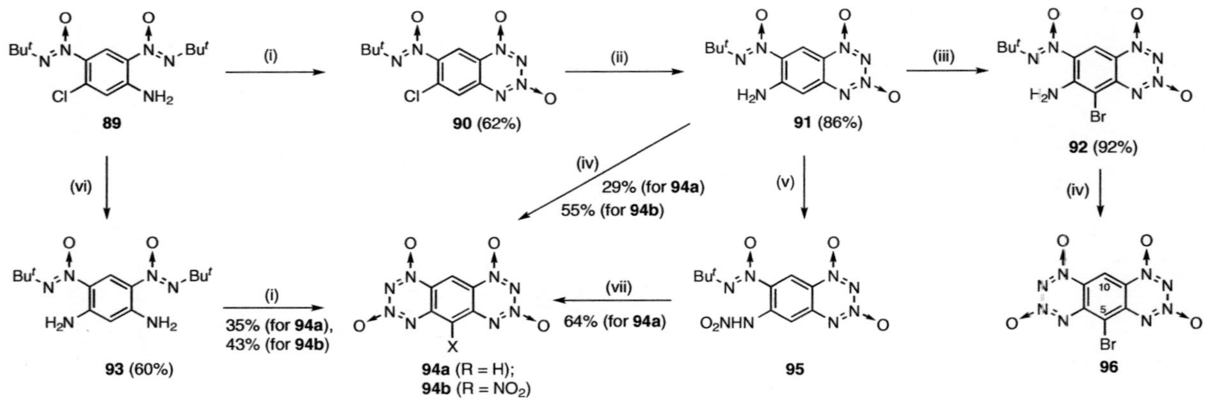

Reagents and conditions: (i) N_2O_5, CH_2Cl_2, −30 to 0 °C, 30 min; (ii) NH_3, DMSO, 60 °C, 3 h; (iii) Br_2, AcOH/AcONa, 20 °C, 5 min; (iv) N_2O_5, MeCN, −30 to 0°C, 30 min; (v) NO_2BF_4, MeCN, −20 to 0°C, 30 min; (vi) NH_3, toluene, 200 °C, 250 atm, 12 h; (vii) P_4O_{10}, MeCN, rt, 2 h.

Scheme 5.20 The synthesis of anthracene-like TBTTOs.

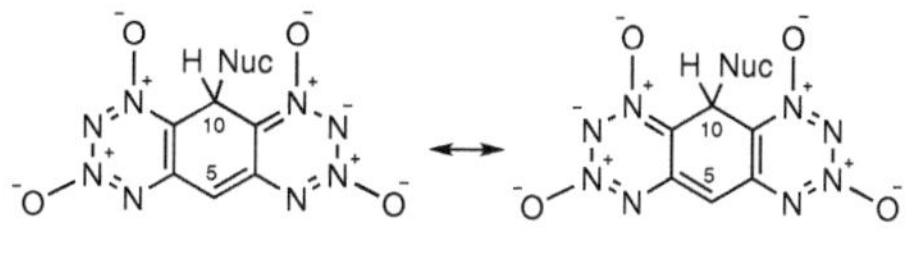

Figure 5.5 The plausible structure of the Meisenheimer-type anion as a result of an attack by Nuc⁻ at position 10 of TBTTO **94a**.

The precursor of BTDO **99** annulated with tetraazapentalene system at the C(5)—C(6) bond was 5-azido-6-(1,2,3-triazol-2-yl)-BTDO **81**. To obtain the latter compound, two consecutive nucleophilic aromatic substitution reactions were used (Scheme 5.21). First, in 6-bromo-5-nitro-BTDO **98**, obtained by nitration of 6-bromo-BTDO **97** [33], the bromine atom in position 6 was replaced by a 1,2,3-triazole fragment by treatment with 1,2,3-triazole in DMF in the presence of Na_2CO_3. Then, the nitro group in position 5 was replaced by the azido one by treating BTDO **82** with sodium azide in DMF. Heating BTDO **81** for eight hours at 140 °C afforded annulated BTDO **99** with a yield of 67% [38].

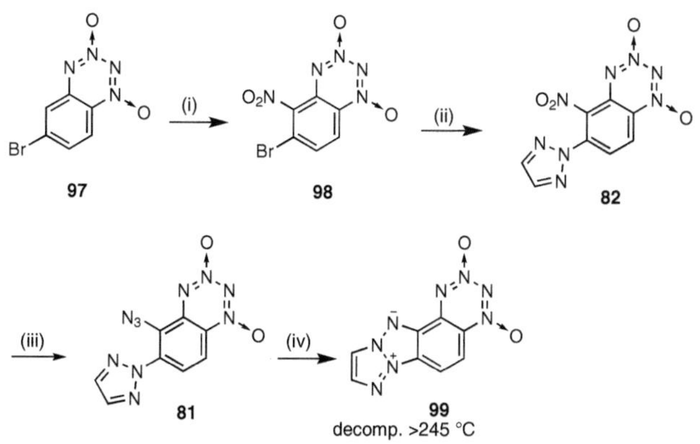

Reagents and conditions: (i) HNO_3/H_2SO_4; (ii) 1,2,3-triazole/Na_2CO_3, DMF, 20 °C, 3 h (47%); (iii) NaN_3, DMF, 20 °C, 30 min (63%); (iv) solvent-free heating, 140 °C, 8 h (67%).

Scheme 5.21 The synthesis of BTDOs annulated with the tetraazapentalene system at the C(5)−C(6) bond.

A similar approach was employed to obtain BTDOs **104** and **108** annulated with tetraazapentalene system at the C(6)—C(7) bond [39]. Interaction of 6,7-dibromo-BTDO **100** with 1,2,3-triazole in the presence of a base (Scheme 5.22) resulted in a selective replacement of the bromine atom in position 6. Triazol-1-yl-substituted isomer **106** was a minor product. The ratio of isomers **101** to **106** was 16 : 1. With an increase in the reaction time, the substitution of the bromine atom in position 7 also occurred to yield bis-triazol-2-yl-BTDO **103**. The triazol-1-yl-substituted isomer **106** was obtained in good yield by cycloaddition of acetylene to the azido group of BTDO **105**, which was prepared by the reaction of BTDO **100** with NaN_3. The bromine atom in position 7 of BTDO **106** was easily replaced by an azido group by reaction with NaN_3. Heating BTDO **107** without solvent at 140 °C afforded tetraazapentalene **108** in 90% yield.

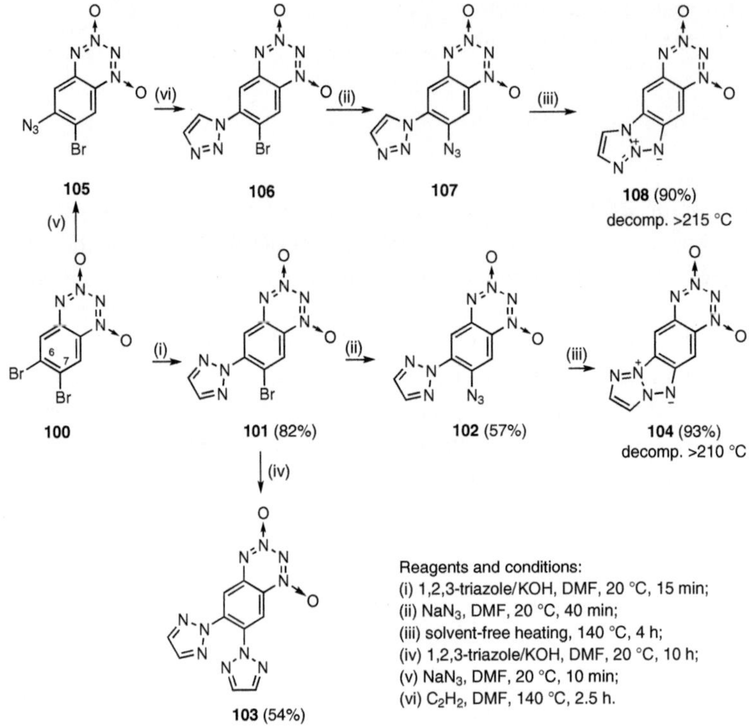

Scheme 5.22 Synthesis of BTDOs annulated with the tetraazapentalene system at the C(6)−C(7) bond.

A similar approach was used for the synthesis of BTDOs **112** and **115** annulated with tetraazapentalene system at the C(7)—C(8) bond [40]. In the reaction of 8-bromo-7-nitro-BTDO [33] **109** with 1,2,3-triazole in the presence of Na_2CO_3, bromine was rapidly replaced (Scheme 5.23). In this case, two isomeric triazol-2-yl and triazol-1-yl substituted isomers **110** and **113** were formed, respectively. When this reaction was carried out in acetone or DMF, the isomer ratio was 3 : 5. The isomer mixture was easily separated by extraction with benzene, in which the main isomer **113** was practically insoluble.

Despite the fact that position 8 in BTDO is activated to nucleophilic substitution reactions more strongly than position 7, the triazole fragment is not a good leaving group, and in the reaction of compounds **110** and **113** with NaN_3, the nitro group at position 7 was replaced by the azido group to form compounds **111** and **114**, respectively (Scheme 5.23). Heating these compounds without solvent at 140 °C led to tetraazapentalenes **112** and **115** in high yields.

Benzotetrazine 1,3-Dioxides Annulated with the Furoxan Ring The starting compounds for preparation of BTDOs annulated with a furoxan ring are BTDOs, in which the azide and nitro groups are located in adjacent positions. These compounds were prepared using electrophilic and/or nucleophilic substitution reactions.

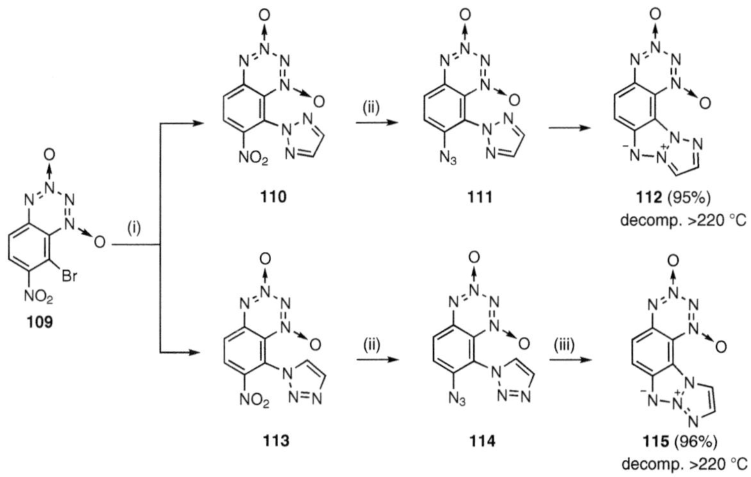

Reagents and conditions: (i) 1,2,3-triazole/Na_2CO_3, acetone, 20 °C, 15 h; (ii) NaN_3, DMF, 20 °C, 10 min; (iii) solvent-free, 140 °C, 5 h.

Scheme 5.23 Synthesis of BTDOs annulated with the tetraazapentalene system at the C(7)−C(8) bond.

5-Nitro-6-azido-BTDO **78** was obtained by nitration of 6-azido-BTDO **116** with 1 equiv of HNO_3 in H_2SO_4 (Scheme 5.24). Nitration takes place almost exclusively in position 5. This compound was also prepared by replacing the bromine atom with an azido group upon treatment of 5-nitro-6-bromo-BTDO **96** with sodium azide. The formation of the furoxan ring occurred upon heating of BTDO **78** at 100 °C without solvent or in AcOH at 90 °C. The ^1H and ^{13}C NMR spectra recorded at 297 K showed that BTDO **117** existed as an equilibrium mixture of two isomers **117a** and **117b** in a 1 : 1 ratio [41].

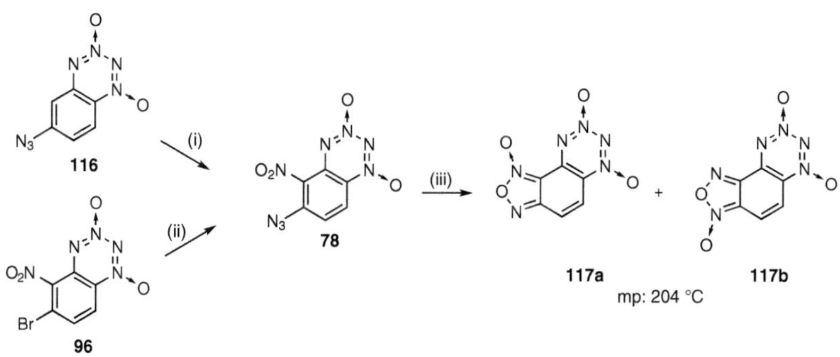

(i) HNO_3/H_2SO_4, 0 – 20 °C (82%); (ii) NaN_3, DMF, 20 °C, 10 min (85%); (iii) AcOH, 90 °C, 30 min (83%).

Scheme 5.24 Synthesis of BTDOs annulated with the furoxan ring at the C(5)−C(6) bond.

BTDO **119** was obtained by nitration of 6,7-dibromo-BTDO **100** followed by treatment of the resulting BTDO **118** with sodium azide (Scheme 5.25). The azido group was replaced exclusively in position 6. The cyclization of BTDO **119** proceeded under somewhat more drastic conditions: heating without solvent at 120–125 °C. The resulting bromo-substituted BTDO **120** was also an equilibrium mixture of the two isomers; however, in this case isomer **120a** was predominant with a ratio of isomers **120a** : **120b** being 9 : 1.

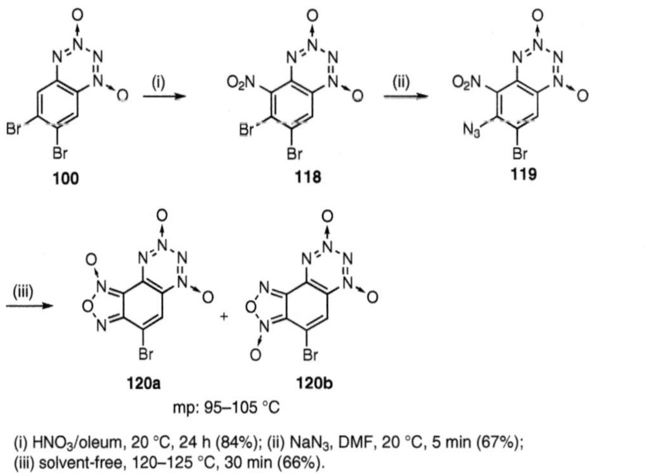

Scheme 5.25 Synthesis of brominated BTDOs annulated with the furoxan ring at the C(5)−C(6) bond.

Synthesis of BTDO **122** annulated with the furoxan ring at the C(7)—C(8) bond included replacement of the bromine atom for an azido group in 8-bromo-7-nitro-BTDO **108** and heating of 8-azido-7-nitro-BTDO **121** (Scheme 5.26) at 118 °C in AcOH. These conditions were somewhat harsher than those required for the cyclization of BTDO **78**. Compound **122** was not as thermally stable as BTDO **117**. It melted without decomposition at 74–76 °C, but began to decompose at temperatures above 90 °C. This, apparently, explains its relatively low yield [42].

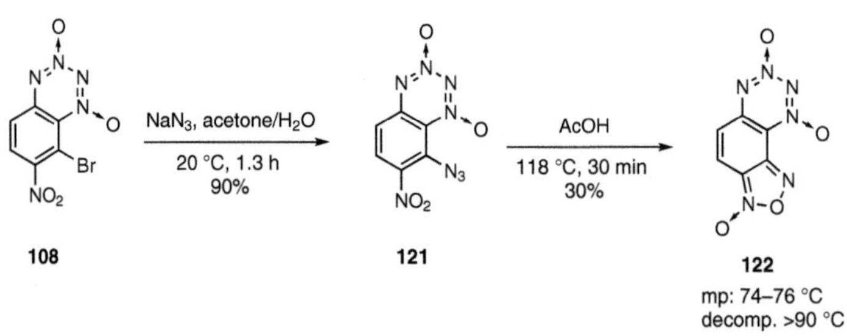

Scheme 5.26 Synthesis of BTDO annulated with the furoxan ring at the C(7)−C(8) bond.

In the reaction of 5,7-dibromo-BTDO **123** with NaN_3, the main product was 7-azido-5-bromo-BTDO **124** (yield: 29%) (Scheme 5.27). Isomeric 5-azido-7-bromo-BTDO was obtained in a yield of 12%. The nitration of BTDO **124** proceeded at position 8, giving BTDO **125**. However, heating this compound under various conditions resulted in only tarry products. Cyclization is difficult for steric reasons and probably insufficiently high thermal stability of target BTDO **126** [42].

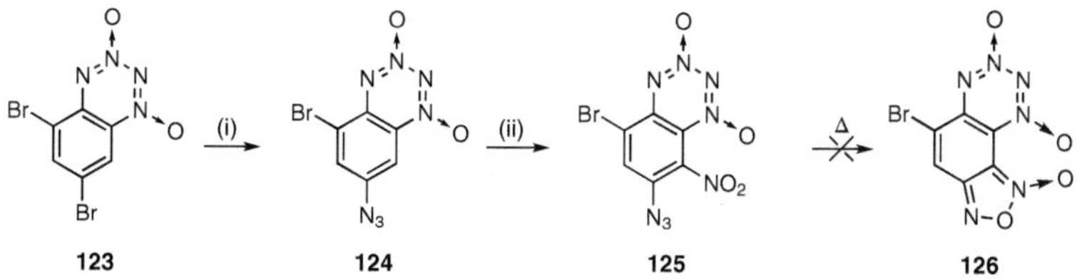

(i) NaN_3, DMF, 20 °C, 2 h (29%); (ii) HNO_3/oleum, 20 °C, 30 min (86%).

Scheme 5.27 Unsuccessful synthesis of BTDO **126**.

6-Azido-7-nitro-BTDO **128** was easily obtained by replacing the bromine atom with an azido group in BTDO **127** (Scheme 5.28). However, heating BTDO **128** under various conditions gave only tarry products [42]. Possibly, this is due to the quinoid structure of target **129**, which is prone to intermolecular reactions.

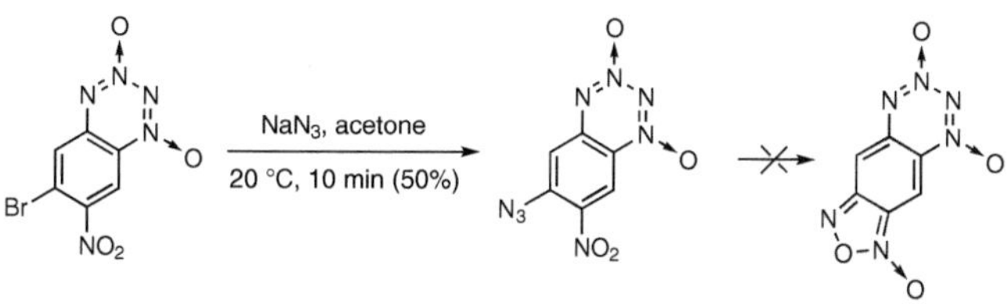

Scheme 5.28 Unsuccessful synthesis of BTDO **129** annulated with the furoxan ring at 6,7 position.

Heating BTDO **60** to obtain BTDO **130** with two furoxan rings was also unsuccessful (Scheme 5.29) [42].

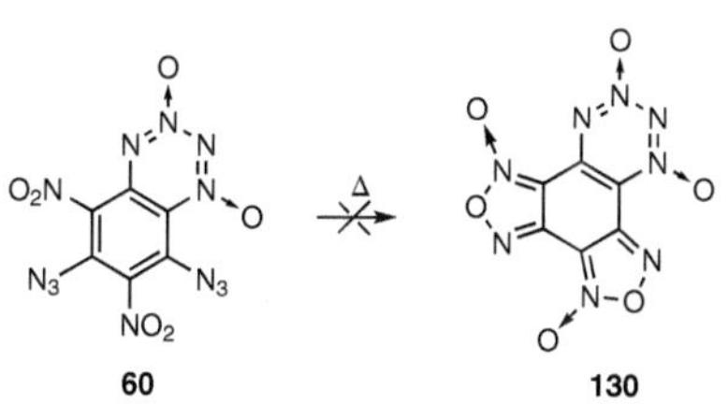

Scheme 5.29 Unsuccessful synthesis of BTDO **130**.

5.2.3.3.3 Heteroannulated 1,2,3,4-Tetrazine 1,3-Dioxides

Annulation with the Furazan Ring For the first time furazano-tetrazine 4,6-dioxide **25** (FTDO) was obtained from aminofurazan **131** by the "nitration" method [43]. Later, an improved two-step method was described, including the preparation of nitramine **132** by nitration of aminofurazan **131** with 70% HNO_3 and the subsequent closure of the TDO cycle by the "acetylation" method, that is, by treating nitramine with acetic anhydride in the presence of sulfuric acid (see Scheme 5.10) [22]. The conversion of aminofurazan **131** to TDO **25** was also carried out in one stage using the $HNO_3/H_2SO_4/Ac_2O$ reagent system, but the yield was lower than in the case of the two-stage method. At room temperature, TDO **25** was resistant to the action of water, but at 80 °C it completely hydrolyzed in two hours, giving triazolofurazan **133** (Scheme 5.30) [44].

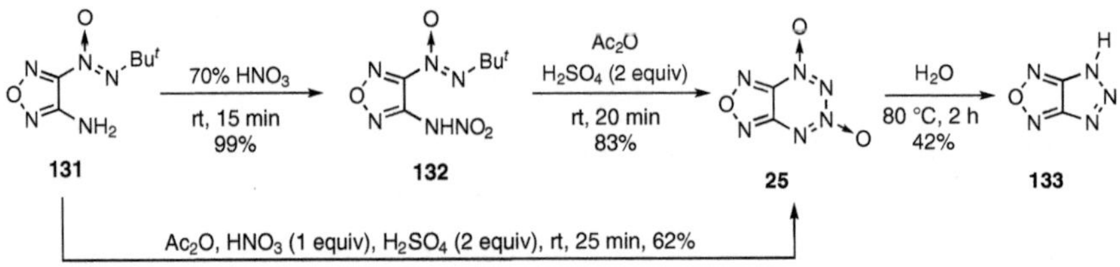

Scheme 5.30 Synthesis and hydrolysis of FTDO.

A plausible hydrolysis mechanism begins with water attack at the N-3 atom of the TDO-ring (Scheme 5.31). Note that benzo-TDOs are much more resistant to hydrolysis.

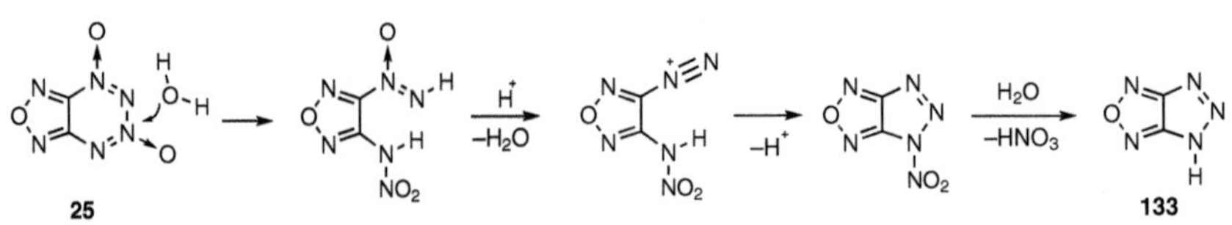

Scheme 5.31 Plausible mechanism of hydrolysis of FTDO.

Annulation with the 1,2,3-Triazole Ring TDOs **141, 142, 143**, and **30**, annulated with 2-alkyl-1,2,3-triazoles, 2-alkyl-1,2,3-triazole *N*-oxides, 1-methoxy-1,2,3-triazole and 1-hydroxy-triazole, were synthesized by the reaction of the corresponding amino-(*tert*-butyl-*NNO*-azoxy) triazoles **137–140** using the $HNO_3/H_2SO_4/Ac_2O$ reagent system (Scheme 5.32) [45, 46]. Aminotriazoles **138–140** were obtained from the same precursor **137** [47]. The latter was synthesized from aminofurazan **131** in several stages, including the furazan ring cleavage to cyanoxime **134**, conversion of the latter to amidoxime **135**, cyclization to furoxan **136** and treatment with alkylamines [48].

The reaction of amino-(*tert*-butyl-*NNO*-azoxy)triazoles **145a,b**, **151**, and **154** with a mixture of $HNO_3/H_2SO_4/Ac_2O$ made it possible to obtain TDOs **146a,b**, **152**, and **155** annulated with 1-aryl- and 2-aryl-1,2,3-triazoles (Scheme 5.33) [49, 50]. The key precursor of all these compounds was (*tert*-butyl-*NNO*-azoxy)acetonitrile **144** obtained by reductive amination of the sodium salt of cyanoxime **134** [51]. 1-Aryltriazoles **145a,b** were prepared by reacting nitrile **144** with the corresponding aryl azides in the presence of a base. The synthesis of 2-phenyltriazole **151** was based on the sequential conversion of nitrile **144** to hydrazone **147** and then to amidoxime

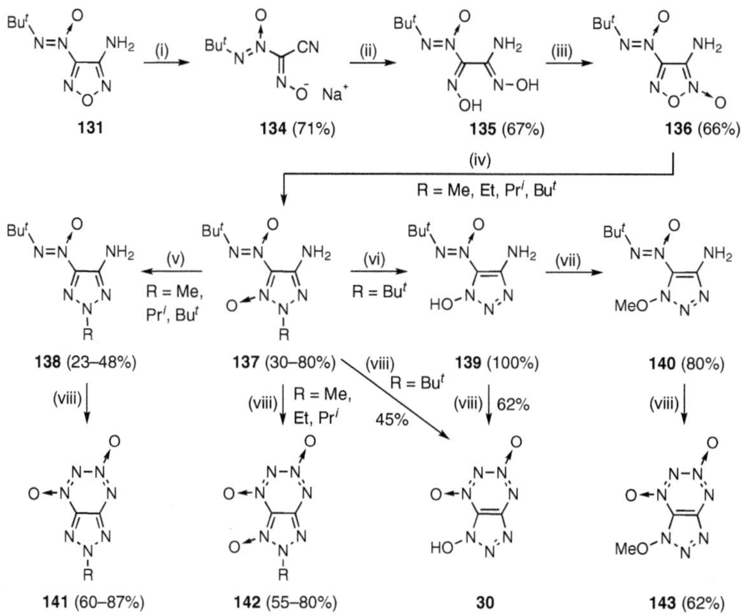

(i) $NaNO_2$, AcOH, Et_2O, 15 °C, 1.5 h; (ii) $NH_2OH{\cdot}HCl$, $NaHCO_3$, MeOH, reflux, 1.5 h; (iii) Br_2, HCl, H_2O, 1–4 °C, 10 min; (iv) RNH_2, MeOH; (v) Zn, AcOH, EtOH, 25 °C, 2 h; (vi) HCl, AcOEt, 25 °C, 30 min; (vii) CH_2N_2, Et_2O, MeCN, 5–10 °C; (viii) HNO_3 (1 equiv), H_2SO_4 (2 equiv), Ac_2O.

Scheme 5.32 Synthesis of *N*-substituted triazolo-TDOs.

149 with the subsequent closure of the 1,2,3-triazole ring upon heating at 150 °C in DMF in the presence of AcONa. Hydrazone **147** can be converted to triazole **151** in one-pot without isolation of intermediate amidoxime **149**. The method for the synthesis of 2-(2′-azido-phenyl)-triazole **154** includes the first two stages similar to the synthesis of triazole **149**, and then amidoxime **150** was converted to oxadiazole **153**. The following base-induced Boulton–Katritsky rearrangement and subsequent elimination of the formyl group in methanol in the presence of methane-sulfonic acid afforded triazole **154** [50].

The alkylation of TDO **30** and its Ag-salt **158** with various reagents (diazomethane, diazoacetone, bromoacetone, α-bromoacetophenone, methyl iodide, methyl vinyl ketone, *tert*-butanol in an acidic medium) was studied (Scheme 5.34). The reactions of TDO **30** with diazo compounds, as well as the reactions of salt **158** with halogenated derivatives, proceeded predominantly with the formation of *O*-alkylated TDOs **143, 159,** and **160** [46]. The reaction of TDO **30** with methyl vinyl ketone, as well as with *tert*-butanol in an acidic medium occurred at the nitrogen atom of the triazole ring to afford N-3 substituted TDOs **157** and **156**, respectively [45, 46].

To remove the hydroxy group from TDO **30**, several methods were developed that afforded triazolo-TDO **25** in high yield (77–94%): (i) reduction in 1-hydroxytriazolo-TDO **30** by treating with PCl_3 in acetonitrile, (ii) reduction in 1-methoxytriazolo-TDO **143** with $Na_2S_2O_4$, (c) reaction of 1-substituted triazolo-TDO **143, 159,** and **160** with Et_3N in methanol (Scheme 5.35). The mechanism of the latter nonconventional

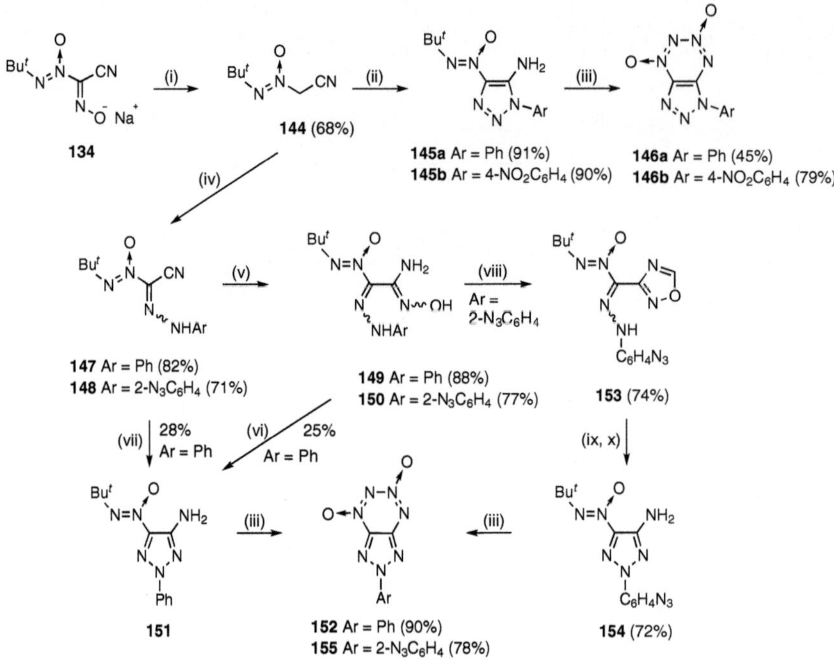

(i) $TsONH_2$, Et_4NBF_4, CH_2Cl_2, H_2O, 2–5 °C, 1 h; (ii) PhN_3 or $4\text{-}NO_2C_6H_4N_3$, DBU (0.5 equiv), MeCN, 25 °C, 10 min; (iii) HNO_3 (1 equiv), H_2SO_4 (2 equiv), Ac_2O, 25 °C, 1 h; (iv) $PhN_2^+Cl^-$ or $2\text{-}N_3C_6H_4N_2^+Cl^-$, AcONa, EtOH, H_2O, 0 – 25 °C, 1 h; (v) $NH_2OH{\cdot}HCl$, AcONa, MeOH, 35–40 °C, 3 h; (vi) AcONa, DMF, 150 °C, 30 min; (vii) $NH_2OH{\cdot}HCl$, AcONa, DMF, 150 °C, 30 min; (viii) $HC(OEt)_3$, $BF_3{\cdot}Et_2O$, DMF, 40 °C, 2 h; (ix) Bu^tOK, DMF, 0 °C, 0.5 h; (x) MeOH, $MeSO_3H$, 0 °C, 20 min.

Scheme 5.33 Synthesis of 1(2)-aryl-1,2,3-triazolo-TDOs.

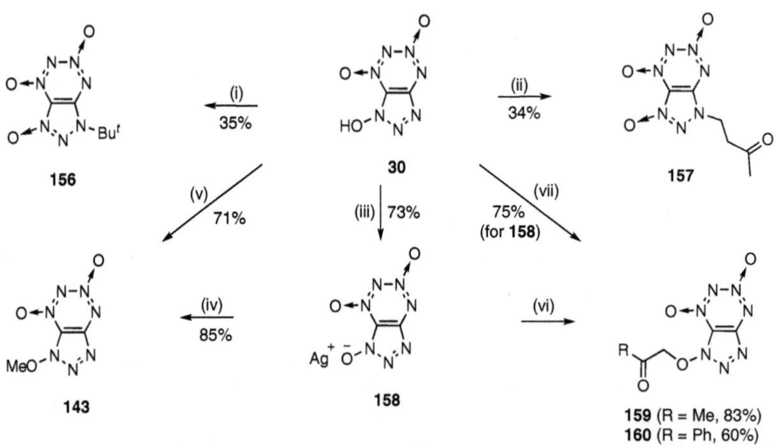

(i) Bu^tOH, H_2SO_4, CF_3CO_2H, rt, 5 min; (ii) $MeC(O)CH{=}CH_2$, $BF_3{\cdot}Et_2O$, MeCN, 25 °C, 1 h; (iii) $(NH_4)_2CO_3$, $AgNO_3$, H_2O, 25 °C, 10 min; (iv) MeI, MeCN, 25 °C, 1.5 h; (v) CH_2N_2, EtOAc, Et_2O, 25 °C; (vi) $MeC(O)CH_2Br$ or $PhC(O)CH_2Br$, MeCN, 25 °C, 5 h; (vii) $MeC(O)CHN_2$, $BF_3{\cdot}Et_2O$, MeCN, 25 °C, 10 min.

Scheme 5.34 Alkylation of hydroxytriazolo-TDO **30**.

reduction most likely involves the elimination of the acidic proton to afford the corresponding aldehyde, which in one case was isolated. The TDO-ring is not affected during reduction [52].

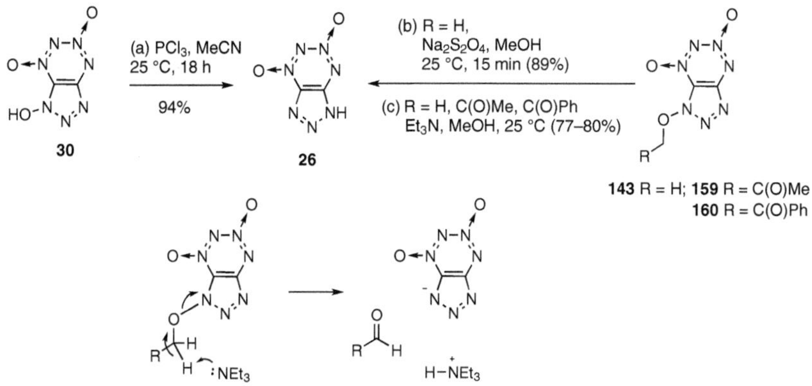

Scheme 5.35 Reduction of hydroxytriazolo-TDO **30**.

Benzo-annulated TDOs were easily reduced by the action of $Na_2S_2O_4$ to give benzotriazoles (Scheme 5.36) [53]. Under these conditions, the nitro group of the benzene ring is not affected, as, for example, in BTDO **75**. A plausible mechanism for the reduction in ^{15}N-labeled BTDO **75** includes the formation of tetrazine 2-oxide **N** in the first stage, followed by ring opening, recyclization to form nitrosotriazole **O**, and its subsequent hydrolysis with the release of ^{15}N-labeled HNO_2. It was suggested that the biological activity of BTDOs as thiol-dependent NO donors was due to their ability to release nitrosating species in the course of reduction.

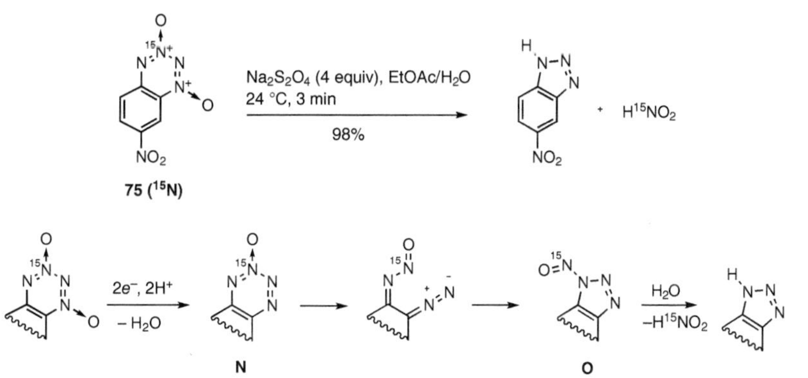

Scheme 5.36 Reduction of BTDO **75** and plausible mechanism of the reaction.

Methylation of triazolo-TDO **26** with diazomethane led to the formation of three isomers **162, 163,** and **164** in a ratio of 41 : 33 : 26 in a total yield of 34% (Scheme 5.37) [52]. The main product was isomer **162**. When Ag salt **161** was methylated with methyl iodide in acetonitrile, three isomers **162, 163,** and **164** were also formed in a ratio of 30 : 56 : 14 with a total yield of 83%. In this case, isomer **163** was the main product. In both reactions, the thermodynamically less stable isomer **164** was formed in the smallest amount.

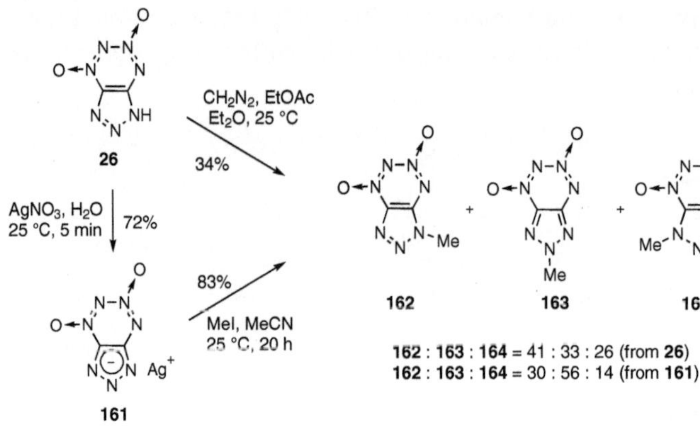

Scheme 5.37 Methylation of triazolo-TDO **26**.

Nitration of 2-azidophenyl-substituted triazolotetrazine **154** with concentrated nitric acid at room temperature led to mononitrated product **165** [50], and at 40 °C, to double nitrated product **166** (Scheme 5.38) [54]. Heating TDO **154** in 1,2-dichlorobenzene at 170 °C for 2.5 hours afforded a mixture of two isomeric tetraazapentalenes **169** and **170** in a 1 : 5.3 ratio. Cyclization of nitro-derivative **165** under the same conditions took 30 minutes, which gave a mixture of tetraazapentalenes **167** and **168** in a ratio of 1 : 1.1 in high yield. Tetraazapentalenes **168** and **170** were isolated in pure form [50]. The cyclization of dinitro-derivative **166** was also studied upon heating in toluene at 110 °C [54]. The azido group mainly reacted with the neighboring nitro group and, to a lesser extent, with the neighboring 1,2,3-triazole ring, which gave furoxan **173** as the major product, as well as isomeric tetraazapentalenes **171** and **172** as minor products. Tetraazapentalene **171** bearing dinitrophenyl substituent was synthesized by nitration of tetraazapentalene **168** and **170** with a HNO_3/H_2SO_4 mixture [50].

Annulation with the Pyridine Ring The reaction of pyridine **174** with an excess of N_2O_5 in CH_2Cl_2 gave pyrido-annulated TDO **175** and its nitro derivative **176** in the ratio of 1 : 2 (Scheme 5.39) [55]. The synthesis of isomeric pyrido-TDO **178** from pyridine **177** was carried out using nitronium tetrafluoroborate as a reagent for a tetrazine ring closure. The use of N_2O_5 in this case turned out to be ineffective.

Annulation with the 1,2,3,4-Tetrazine 1,3-Dioxide Ring: Synthesis of TTTO TTTO **36** was synthesized from TDO **56** using a mixture of $HNO_3/H_2SO_4/Ac_2O$ in a 22% yield (Scheme 5.40) [25]. This compound was quite thermally stable, but completely hydrolyzed in 50% aqueous EtOH at 20 °C for two hours, giving triazolo-TDO **26**.

The hydrolysis mechanism is apparently the same as the one of FTDO **25** (see Scheme 5.31), and begins with the attack of water molecule at the N-3 atom of the TDO cycle (Scheme 5.41). Thus, annulation of the TDO-ring by strongly electron-acceptor rings makes the TDO-fragment sensitive to hydrolysis.

In this regard, it is interesting to consider hypothetical compound **179**, in which two tetrazine dioxides are linked by a tetraazapentalene bridge (Figure 5.6). This compound could be sufficiently thermally stable (as similar in structure

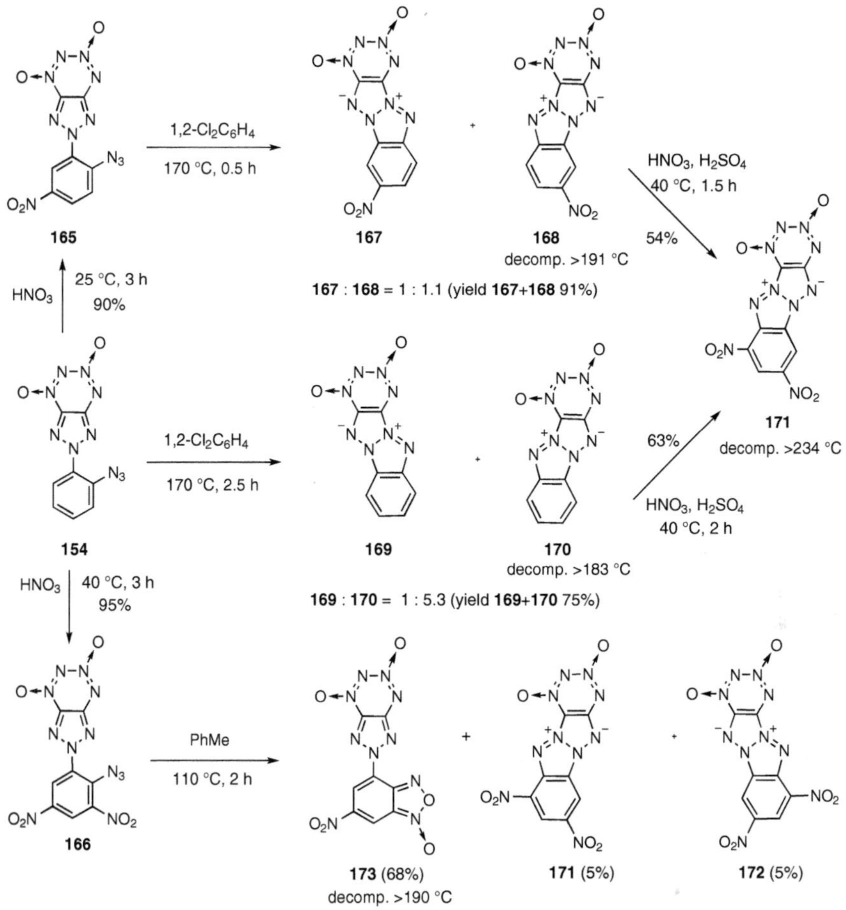

Scheme 5.38 Synthesis of TDOs annulated with the tetraazapentalene system.

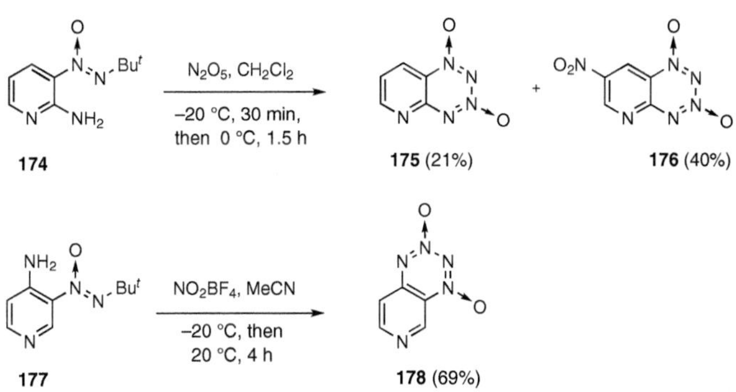

Scheme 5.39 Synthesis of pyrido-annulated TDOs.

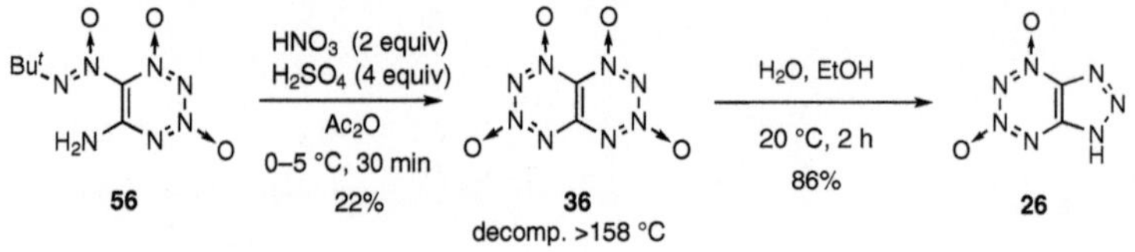

Scheme 5.40 The synthesis and hydrolysis of TTTO **36**.

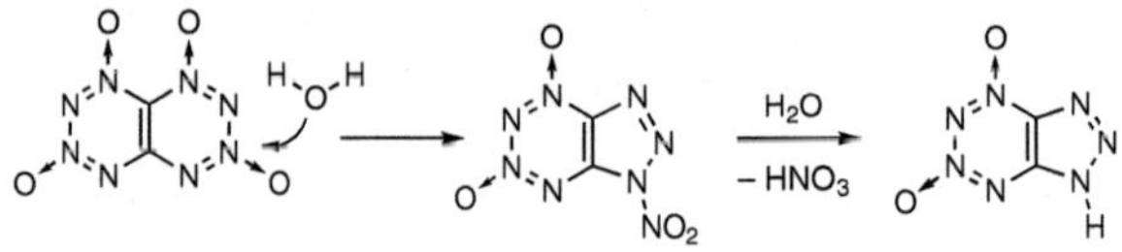

Scheme 5.41 Plausible mechanism of hydrolysis of TTTO.

to tetraazapentalene-annulated TDO **171**) and resistant to hydrolysis, since the tetraazapentalene system is much more electron-donor than the tetrazine dioxide or furazan systems. However, at present there is no practical synthetic approach to this system.

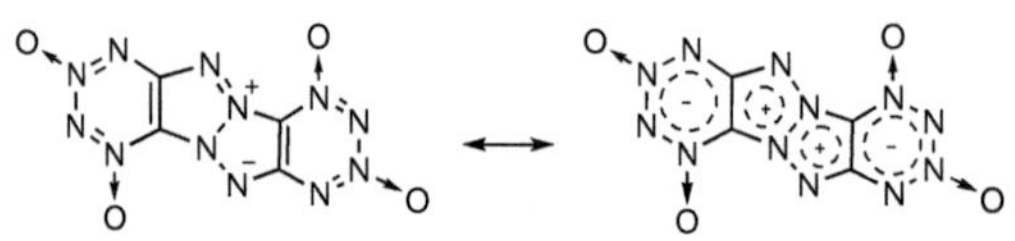

179

Figure 5.6 Hypothetical system **179**.

5.3 NMR and X-ray Studies

A number of methods have been used to identify tetrazines and their *N*-oxides, including X-ray analysis, IR and UV spectroscopy, mass spectrometry, and NMR spectroscopy [6, 7]. Nitrogen ^{14}N and ^{15}N NMR spectroscopy is an especially useful direct method to determine the tetrazine 1,3-dioxide structures.

The ^{15}N NMR of BTDO **4** (natural isotope content) (Figure 5.7), that showed four nitrogen signals, in combination with the selective polarization transfers (SPTs) from the H-5 and H-8 protons to the ^{15}N nucleus allowed the N-1 and N-4 nitrogen signals to be identified. The ^{15}N NMR of 7-NO_2-BTDO **75′** labeled with ^{15}N in position 3 made it possible to assign the remaining two signals [21].

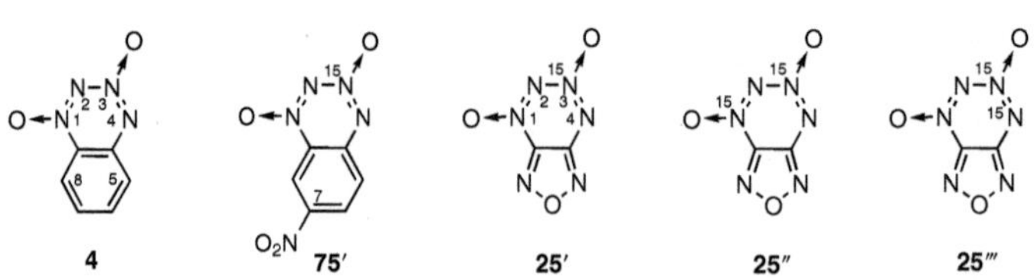

Figure 5.7 BTDO **4**, labeled BTDO **75′** and labeled FTDOs **25′**, **25″**, and **25‴**.

Isotope labeling method was used to unambiguously assign all nitrogen signals in FTDO **25** (FTDO was labeled with ^{15}N in position 3 (**25′**), positions 1 and 3 (**25″**), and in positions 3 and 4 (**25‴**)) [43]. In this case the N-3 signal ($\delta = -44$ ppm) is observed in nearly the same region as in BTDO **4**, whereas the N-1 signal ($\delta = -52.7$ ppm) was shifted upfield. These facts were used in assignment of ^{14}N NMR signals of TDOs.

The ^{14}N NMR spectrum of tetrazine 1,3-dioxide ring shows two narrow peaks attributed to the N→O nitrogen atoms. For instance, BTDO **4** shows two signals in the ^{14}N NMR spectrum, where the N-1 signal ($\delta = -40$ ppm, $\Delta\nu_{1/2} = 20$ Hz) is narrower than the N-3 signal ($\delta = -48$ ppm, $\Delta\nu_{1/2} = 30$ Hz). A similar pattern is observed for R-substituted BTDOs (R = Br, Cl, N_3, NMe_2, NO_2, OMe) [21, 31, 36] and BTDOs annulated with five-membered heterocycles at the benzene ring. Pyridoannulated and non-annulated TDOs show similar type of spectra. In these compounds, signals of N-1 are observed in the region of −39 to −45 ppm ($\Delta\nu_{1/2} = 30$–100 Hz) and for N-3 in the region of −42 to −51 ppm ($\Delta\nu_{1/2} = 45$–120 Hz) [55]. The ^{14}N NMR spectrum of FTDO **25** exhibits a narrow N-1 signal at −54 ppm ($\Delta\nu_{1/2} = 13$ Hz) and a wider N-3 signal at −39 ppm ($\Delta\nu_{1/2} = 32$ Hz). [43]. In triazole-annulated TDOs, the N-1 signals are observed in the region of $\delta = -50$ to −61 ppm ($\Delta\nu_{1/2} = 10$–50 Hz) and N-3 signals in the region of −37 to −44 ppm ($\Delta\nu_{1/2} = 45$–120 Hz) [49]. It is noteworthy that the spectra of TTTO **36** ($\delta = -37$ (N-3 and N-3′, $\Delta\nu_{1/2} = 32$ Hz), −54 (N-1 and N-1′, $\Delta\nu_{1/2} = 13$ Hz)) are similar to the spectra of FTDO **25** [25].

A distinctive feature of the ^{13}C NMR spectra (in $CDCl_3$ or acetone-d_6 as solvents) of TDO derivatives is a significant broadening of the signal from the carbon atom directly bonded to the *N*-oxide fragment of the tetrazine 1,3-dioxide ring which is a result of the spin–spin coupling of the ^{13}C and ^{14}N nuclei in particular. This feature should be taken into account in the assignment of signals in the ^{13}C NMR spectra.

Single-crystal X-ray and X-ray powder diffraction structural analyses of various 1,2,3,4-tetrazine derivatives have been repeatedly reviewed [6, 7]. In this chapter, we discuss the most prominent representatives of TDOs. The tetrazine ring in all compounds, both annulated and non-annulated, is almost planar. This planarity in combination with moderate alternation of endocyclic bond distances indicates an extensive delocalization of the π-electrons. The presence of cyclic conjugation in these molecules is probably one of the factors that stabilizes the system (see Section 5.4.1). A representative example of this is dicarboxylate **69** (mp: 93–96 °C, starts to decompose at 190 °C) having nearly planar TDO ring and the N—N bonds similar in length [30]. The presence of bulky substituents as in the case of TDO **70** can lead to the deviation from planarity (TDO ring has a boat conformation) causing a violation of cyclic conjugation [30], and/or as in the case of TDO **58** to alternation of endocyclic bond distances [25].

Tetrazines annulated with five-membered cycles are planar and as a rule have elongated N(2)—N(3) bond (more single-bond character) in the tetrazine ring as compared with the other two N—N bonds. The introduction of oxygen atoms in the tetrazine ring leads to a more pronounced bond length alternation (cf. triazolotetrazine **3** [2] and triazolotetrazin 1,3-dioxide **152** [49] in Table 5.1).

The same case is observed for benzene-annulated tetrazine 1-oxide **180** [56] and 1,3-dioxides **4** [57], **123** [58] and **76** [32]. These are flat structures with exocyclic

Table 5.1 Single-crystal X-ray diffraction studies of TDOs.

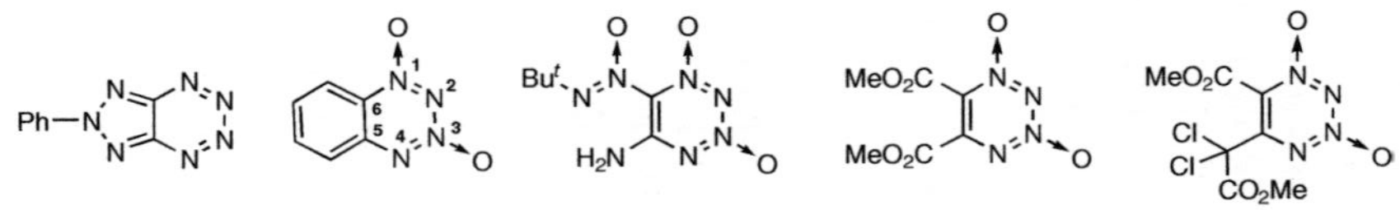

3 4 58 69 70

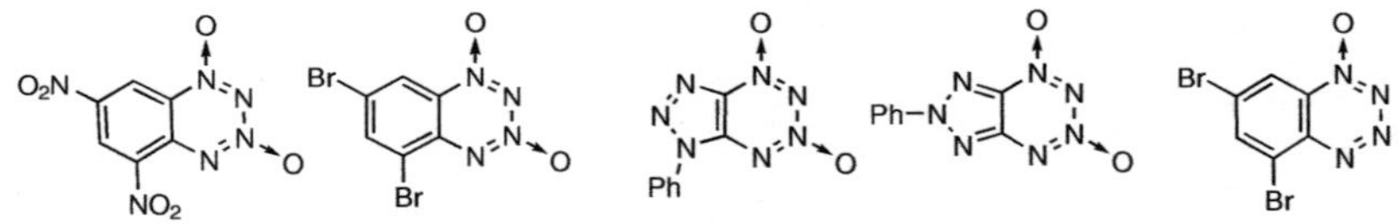

76 123 146a 152 180

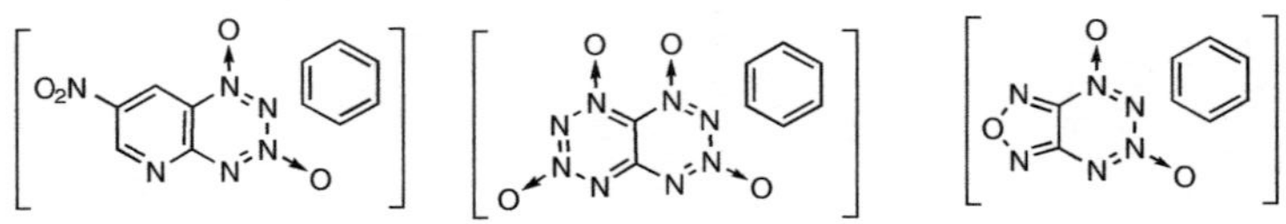

181 182 183

Compound	Ring	Bond length (Å)							
		C(5)–C(6)	C(6)–N(1)	C(5)–N(4)	N(1)–N(2)	N(2)–N(3)	N(3)–N(4)	N(1)–O(1)	N(3)–O(3)
3		1.383(9)	1.332(7)	1.354(8)	1.323(8)	1.388(8)	1.328(7)	—	—
4		1.386(4)	1.406(4)	1.371(4)	1.322(4)	1.380(4)	1.297(4)	1.238(4)	1.255(4)
58	**A**	1.3988(18)	1.3426(16)	1.3448(17)	1.3671(15)	1.3303(15)	1.3263(15)	1.2447(14)	1.2490(14)
	B	1.4089(17)	1.3399(16)	1.3461(17)	1.3658(15)	1.3380(15)	1.3256(15)	1.2529(14)	1.2416(14)

61		1.380(7)	1.414(8)	1.350(7)	1.315(7)	1.408(7)	1.287(8)	1.238(6)	1.247(6)
69	**A**	1.3717(16)	1.3721(15)	1.3397(15)	1.3445(14)	1.3558(14)	1.3363(14)	1.2468(12)	1.2430(13)
	B	1.1972(14)	1.3691(15)	1.3401(15)	1.3461(14)	1.3507(14)	1.3391(14)	1.2412(12)	1.2442(13)
70		1.3911(4)	1.3688(3)	1.3252(3)	1.3542(4)	1.3452(4)	1.3457(4)	1.2367(4)	1.2260(3)
76	**A**	1.41567	1.40279	1.34707	1.31171	1.39686	1.32591	1.23920	1.22843
	B	1.42311	1.40712	1.34480	1.30637	1.39092	1.31419	1.23911	1.21706
	A	1.412(4)	1.407(4)	1.347(4)	1.313(4)	1.391(4)	1.321(4)	1.237(3)	1.236(4)
	B	1.389(5)	1.405(4)	1.365(5)	1.317(4)	1.386(4)	1.311(4)	1.239(4)	1.238(4)
146a		1.380(3)	1.363(3)	1.340(3)	1.343(3)	1.385(2)	1.329(2)	1.241(2)	1.242(2)
152		1.388(3)	1.374(3)	1.363(3)	1.333(3)	1.405(3)	1.310(3)	1.244(3)	1.237(3)
180		1.401(13)	1.408(12)	1.363(14)	1.339(12)	1.347(13)	1.300(13)	1.239(11)	—
181		1.383(4)	1.391(4)	1.357(3)	1.321(3)	1.382(3)	1.308(3)	1.233(3)	1.2347(3)
182	**A**	1.375(3)	1.388(3)	1.348(2)	1.390(2)	1.328(3)	1.350(3)	1.231(2)	1.229(2)
	B	1.375(3)	1.389(3)	1.344(3)	1.387(3)	1.325(3)	1.356(3)	1.224(2)	1.230(2)
183		1.375	1.394	1.363	1.356	1.410	1.354	1.222	1.219

oxygen atoms located in the same plane. The distinctive feature of mono-oxide **180** as compared to BTDOs **4**, **123**, and **76** is the uniformity of the N(1)—N(2) and N(2)—N(3) bond distance and smaller alteration of bond distances in tetrazine cycle (Table 5.1).

In a number of cases, attempts to grow single crystals of TDOs were unsuccessful. In these cases, TDO-benzene solvates shed light on the structure of molecules (TTTO **36** [25] and pyrido-annulated TDO **176** [55]). In the event of FTDO **25**, attempts to determine the crystal structure from X-ray data failed due to problem of obtaining the parameters of the crystal lattice related to its plane-symmetrical structure that leads to a small number of responses on diffractogram. But the structure of FTDO-benzene solvate **183** was successfully decoded [59].

5.4 Thermal Stability

5.4.1 Theoretical Stability

One of the burning issues in the field of high nitrogen compounds is the question of the correlation between the structure of the molecule and its stability. Politzer and Murray discussed nitrogen catenation (directly linked nitrogen atoms) in five- and six-membered heterocyclic rings and suggested several reasons to explain why nitrogen catenation reduces molecular stability [60]. (i) N—N bond energies (single, double, or intermediate) are usually weaker than those of the corresponding C—N. Their formation increases the positive contribution to the heat of formation and the energy content of the molecule, and thereby destabilizes it. (ii) The presence of two linked doubly coordinated nitrogens may facilitate decomposition through the formation of the very stable N_2 molecule. (iii) Interactions between certain molecular orbitals that make destabilizing contributions (the instability of azobenzenes and the stabilizing effect of the N→O fragment using molecular orbitals theory is discussed in detail in work of Noyman, Zilberg, and Haas [61]). (iv) Repulsion between the lone pairs of adjacent doubly coordinated nitrogens.

It was demonstrated that there are strongly negative electrostatic potentials in the lone pair regions of the doubly coordinated nitrogens. This means that there is an electronic charge in these regions. So if such nitrogens are adjacent to each other and there is strong overlap between the negative regions, then a repulsive (destabilizing) effect can be anticipated. Possible solution to this problem is an introduction of N→O fragment in the molecule which "dilutes" the negative potentials associated with catenated doubly coordinated nitrogens, spreading them over larger areas and lessening the internal repulsion [62–64].

The effect of the N→O fragment on stability in a series of azines and other N-heterocycles has also been studied in detail [65, 66]. It was concluded that, as a rule, such introduction of an N→O fragment leads to destabilization of the system since elongation and weakening of the bonds adjacent to the N→O bond is induced. Elongation is clearly observed upon comparison of the corresponding bonds of compounds **3** and **152** (see X-ray Table 5.1). At the same time, the authors note that stabilization is possible if N→O fragments are located in the molecule, as in the case of TDOs [65, 66].

Aromaticity is considered as another factor of system stabilization. Partly for this reason, much attention is paid to the question of what happens to the aromaticity of the system in the case of sequential replacement of the CH moiety with a heteroatom in the conjugated hydrocarbon cycle (first of all in benzene) [67]. The main difficulty with aromaticity is that it is a multidimensional phenomenon, whose criteria are based on energy (increased stability), molecular geometry (very low alternation of bond lengths), magnetism (induction of diatropic ring current by external magnetic field), and reactivity (tendency to preserve the π-electron structure in chemical reactions). Aromaticity indices do not necessarily refer to the same aspects of aromaticity; in other words, aromaticity indices do not always "speak in the same voice" [68, 69]. So aromaticity remains one of the most controversial and debated issues in theoretical and computational chemistry. For example, based on magnetic and energy criteria, it is shown that the aromaticity of azines (pyridine, diazines, triazines, tetrazines, pentazine, and hexazine) remains almost unchanged [70–72]. Other researchers have shown that N-, P-, and As-substituted benzenes have aromaticity, which, however, decreases in the series from N- to As-substituted compounds, as well as with an increase in the number of heteroatoms in the system from one to six, with pentazine and hexazine being antiaromatic compounds [73, 74]. The trend of decreasing aromaticity for azabenzenes was noted in other works too [75–77]. However, it is worth noting that presence of aromaticity can lead to system stabilization, but it does not guarantee that this stabilization would be enough to achieve a relatively high stability of a molecule. Illustrative example of this is the drastic difference in the thermal stabilities of tetrazine **3** and TDO **152**, despite the fact that aromaticity of the tetrazine ring is significantly higher than that of the tetrazine 1,3-dioxide ring (according to geometric and magnetic criteria).

5.4.2 Heat Resistance

Heat resistance is one of the main characteristics of energetic materials that determines the possibility of their practical use. In this chapter, structural factors affecting thermal stability of 1,2,3,4-tetrazines are considered. Compounds with a decomposition temperature of at least 130 °C are considered potentially acceptable.

The thermal stability of tetrazines depends significantly on their structure. The obvious factors that reduce stability are a large energy storage in the molecule (high heat of formation), as well as the presence of a favorable degradation pathway of the molecule [60, 78]. At the same time, there are "islands of stability," which include substances with high heats of formation and relatively high stability due to the absence of favorable pathways for molecule decay. In this regard, we pay considerable attention to the degradation mechanisms of 1,2,3,4-tetrazines.

Tetrazine **3** decomposed at room temperature within two hours giving 2-phenyl-2*H*-1,2,3-triazole **184** and 5-cyano-2-phenyl-2*H*-tetrazole **185** through N_2 eliminations (Scheme 5.42) [2]. There are two possible decomposition pathways: pathway *b* is a concerned process that proceeds with the simultaneous release of two nitrogen molecules and triazolo-heteroaryne. This route is considered unlikely by the authors due to energetically unfavored structure of the latter. Pathway *a* is a stepwise process starting from the cleavage of the weakest bond in the molecule followed by denitrogenation to afford the biradical intermediate. The latter can abstract hydrogen from solvent to form a triazole or transform into a cyanotetrazole. The abstraction of the hydrogen atom from the solvent was confirmed by the decomposition in $CDCl_3$. As a result, a triazole with two deuterium atoms was formed. Note that the decay processes of high nitrogen compounds usually begin with the cleavage of the weakest bond. This fact is often used for initial stability assessment [79–84].

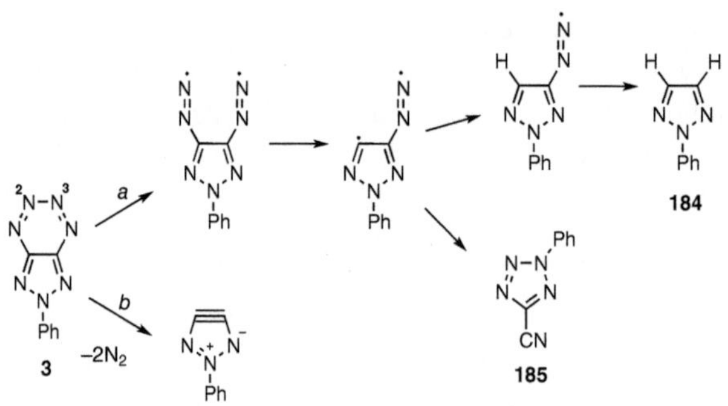

Scheme 5.42 Suggested mechanism of decomposition of tetrazine **3**.

Thus, although tetrazine **3** is an aromatic compound, aromatic stabilization itself is not sufficient to render the compound stable.

The experimental data on the stability of benzotetrazine 1-oxides were discussed in Section 5.2.3.2. Although these compounds are aromatic in character (see Section 5.3), they are rather unstable. The stability of benzotetrazine 1-oxides depends on the substituents in the benzene ring, and even the most stable of them decompose at room temperature [85]. The first decomposition step involves ring opening with low activation energy (ring-chain tautomerism) to afford 1-azido-2-nitrosobenzenes (Scheme 5.43).

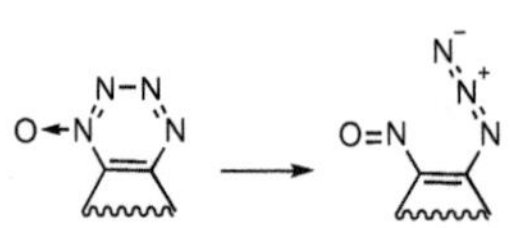

Scheme 5.43 Mechanism of BTO decomposition.

Stability of several hypothetical tetrazine *N*-oxides with different arrangements of *N*-oxide fragments should also be considered (Scheme 5.44). Tetrazine 2-oxide is predicted to be unstable since it could take part in ring-chain tautomerism with the formation of the diazo-nitrosoimino form. Tetrazine dioxide bearing *N*-oxide oxygen atoms at positions 2 and 3 may also undergo an easy ring opening to dinitrosoimine. Tetrazine dioxide bearing *N*-oxide oxygen atoms in positions 1 and 4 may easily eliminate the N_2 molecule with the formation of a thermodynamically stable dinitroso compound. Only in the case when the *N*-oxide oxygen atoms are located in positions 1 and 3 of the tetrazine ring, there are no favorable decomposition pathways for the molecule [58, 86]. The cleavage of the N(2)—N(3) bond with the formation of an azoxy radical occurs with difficulty (see below).

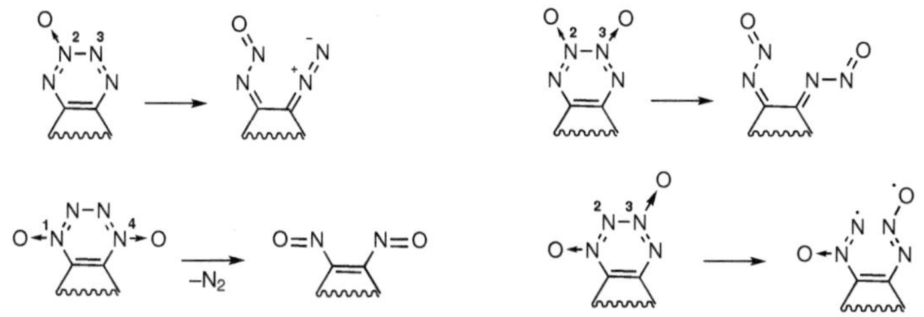

Scheme 5.44 Supposed decomposition pathways for hypothetic molecules: tetrazine 2-oxide, tetrazine 2,3-dioxide, and tetrazine 1,4-dioxide, and the first stage of plausible decomposition pathway for tetrazine 1,3-dioxide.

Comparison of the stability of compounds **3** and **152**, as well as compounds **123** and **180** clearly demonstrates the stabilizing effect of *N*-oxide oxygen atoms in positions 1 and 3 (Figure 5.8).

5.4.2.1 Non-annulated 1,2,3,4-Tetrazine 1,3-Dioxides (TDOs)

The stability of this class of tetrazines depends on substituents at carbon atoms (Figure 5.9). TDOs **66** [29] and **69** [30] bearing carbon substituents are rather stable. However, the stability of dicyano-TDO **67** is lower. 5-Amino-6-nitro-BTO **59**

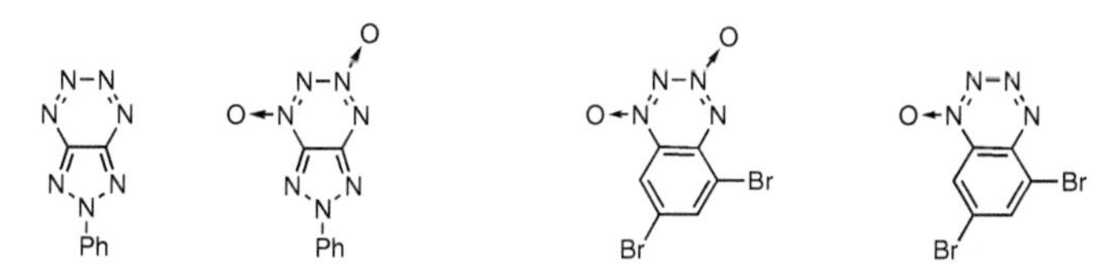

3 decomp. at rt; **152** decomp. >236 °C; **123** mp: 204–208 (decomp.); **180** decomp. at rt

Figure 5.8 Influence of *N*-oxide oxygen atoms on the thermal stability of 1,2,3,4-tetrazines.

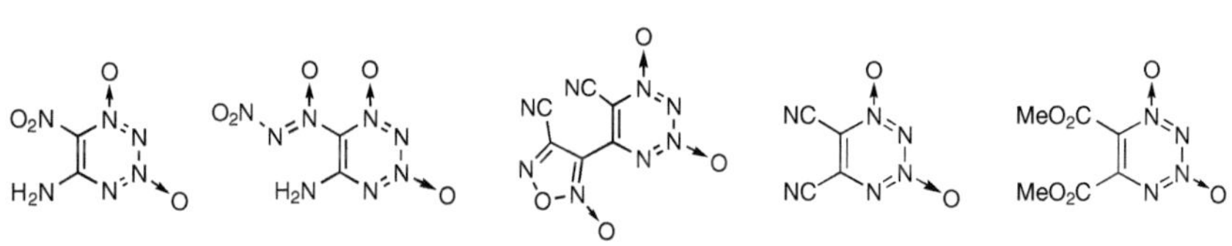

59 mp: 195–200 °C decomp.; **60** mp: 110–112 °C decomp.; **66** mp: 165–167 °C decomp.; **67** mp: 110–130 °C decomp.; **69** mp: 93–96 °C decomp. >190 °C

Figure 5.9 Thermal stability of non-annulated TDOs.

exhibits high thermal stability, with decomposition temperature above 190 °C [26]. Replacement of the nitro group with nitro-*NNO*-azoxy one leads to a decrease in heat resistance: 5-amino-6-(nitro-*NNO*-azoxy)-TDO **60** decays at 110–112 °C [27]. However, this is not due to the decomposition of the TDO cycle itself, but due to the disintegration of the nitro-*NNO*-azoxy group (cf. lit. [28]).

Note that TDO moiety is isoelectronic to nitro-*NNO*-azoxy group (Figure 5.10). However, the type of conjugation in these moieties is sharply different. In nitro-*NNO*-azoxy compounds, the angle between the planes of azoxy and nitro groups is almost 90°, demonstrating the absence of conjugation between these groups. As a result, the N(2)—N(3) bond is abnormally long (~1.480 Å) [87]. One would expect that the decomposition would start with the cleavage of this particular bond. However, an experiment with a ^{15}N labeled nitro group showed that this decomposition pathway is realized only to a small extent.

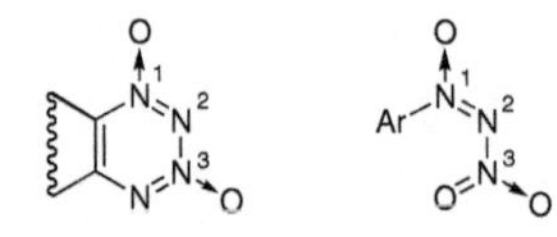

Figure 5.10 1,2,3,4-Tetrazine 1,3-dioxide and nitro-*NNO*-azoxy moieties.

(Nitro-*NNO*-azoxy)benzene **183** decomposed at 110 °C (octane, 4.5 hours) to afford nitrobenzene (69%) (Scheme 5.45). This is the first-order reaction. The gas phase consisted mainly of N_2O (63%).

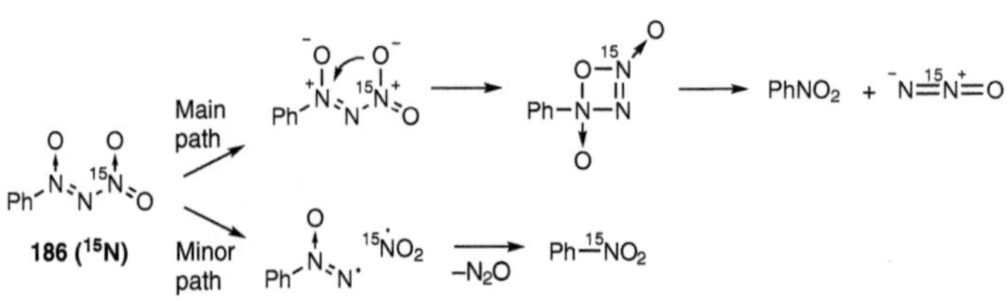

Scheme 5.45 Proposed mechanism for the decomposition of ^{15}N-labeled nitro-*NNO*-azoxybenzene.

During the thermal decomposition of (nitro-*NNO*-azoxy)benzene, the nitro group of which was enriched in the ^{15}N isotope, the label mainly moved to N_2O, and the proportion of labeled nitrobenzene is only 6% [58]. Thus, the main decomposition mechanism probably involves a four-membered cyclic transition state followed by N_2O elimination.

These data confirm that radical R–N(O)=N· is difficult to form. This is also confirmed by rather high stability of bromo-*NNO*-azoxy compounds (for example, bromo-*NNO*-azoxy-4-nitrobenzene melts with decomposition at 186 °C) [88].

Thus, apparently, the main reason for the higher stability of tetrazine 1,3-dioxides in comparison with isoelectronic nitro-*NNO*-azoxy compounds is the absence of favorable pathways for decomposition for the former. Cyclic conjugation, manifested in aligned lengths of N—N bonds in the tetrazine ring, evidently also stabilizes tetrazines, but is not a major factor.

5.4.2.2 Benzo-1,2,3,4-Tetrazine 1,3-Dioxides (BTDOs)

These are stable compounds that melt in the range of 170–250 °C (Figure 5.11). As a rule, they do not decompose below 200 °C. BTDO **4** melts without decomposition at

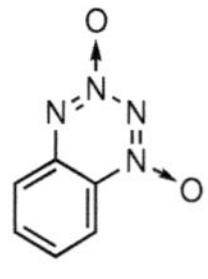
4
mp: 170–172 °C

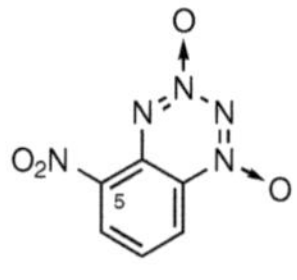
74
mp: 166–168 °C

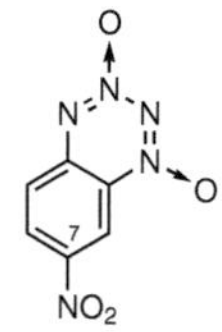
75
mp: 193–195 °C

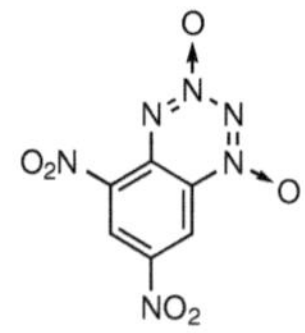
76
mp: 209–211 °C
decomp.

Figure 5.11 Melting points of nitro-BTDOs.

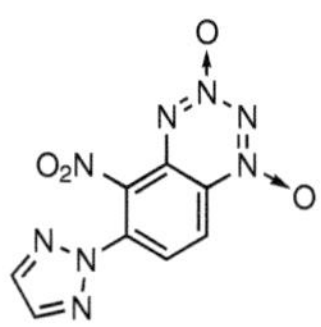
82
mp: 243–245 °C

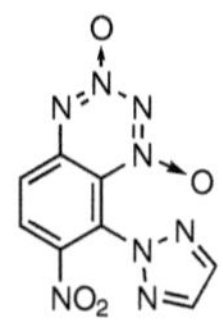
110
mp: 225–235 °C
decomp.

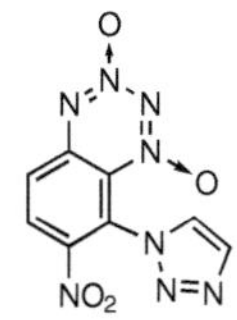
113
decomp. >260 °C

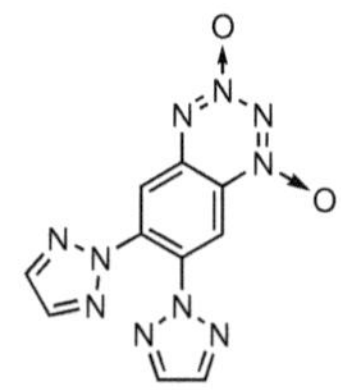
103
mp: 296–298 °C
decomp.

Figure 5.12 The most stable BTDOs.

170–172 °C; 5-nitro and 7-nitro-BTDOs **74** and **75** melts at 166–168 and 193–195 °C, respectively [21]. Exothermic decomposition of 5,7-dinitro-BTDO **76** according to differential scanning calorimetry (heating rate of 5 K min^{-1}) starts at 204 °C without melting [32].

The most stable high-melting BTDOs obtained are compounds **82**, **103**, **110**, and **113**, which begin to decompose at temperatures above 220 °C (Figure 5.12) [38–40].

5.4.2.3 BTDOs Annulated with TDO Ring

Currently, there are two types of BTDOs annulated with another TDO ring [36, 37] (Figure 5.13). Among phenanthrene-type compounds, the most stable is compound **88** with an electron-donor MeNH substituent (decompose without melting above

Figure 5.13 BTDOs annulated with another TDO ring.

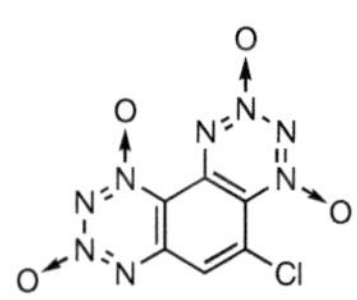
87
mp: 140 °C
decomp.

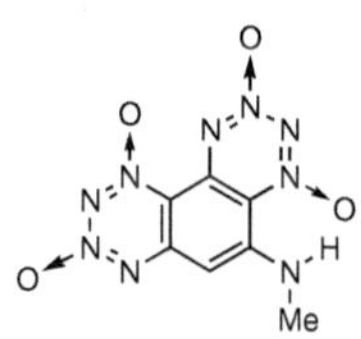
88
decomp. > 210 °C

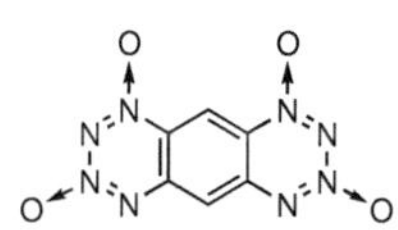
94a
decomp. > 160 °C

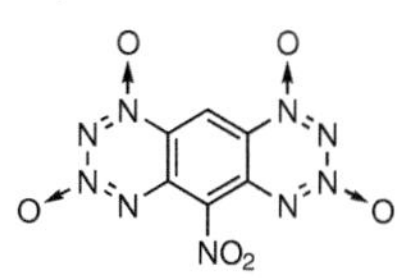
94b
decomp. > 150 °C

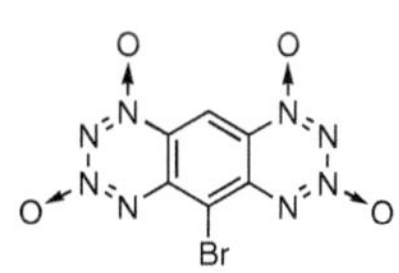
96
decomp. > 160 °C

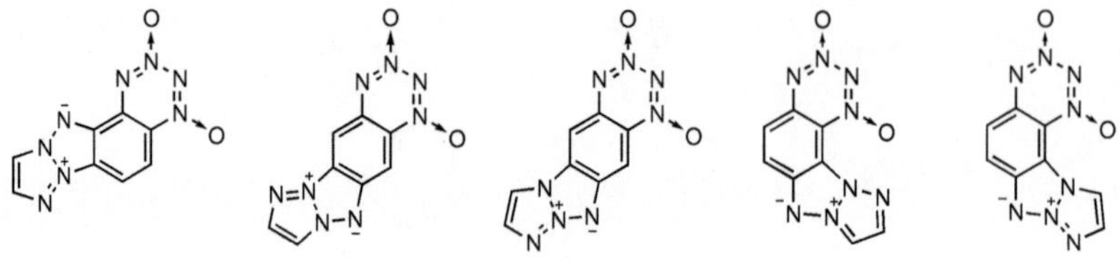

99 decomp. > 245 °C | 104 decomp. > 210 °C | 108 decomp. > 215 °C | 112 decomp. > 220 °C | 115 decomp. > 220 °C

Figure 5.14 BTDOs annulated with tetraazapentalene system.

210 °C), and the least stable is compound **87** with an electron-withdrawing Cl substituent (melts with decomposition above 140 °C). Anthracene-type compounds **94a,b** and **96** begin to decompose without melting at 150–160 °C.

5.4.2.4 BTDOs Annulated with Tetraazapentalene System

Thermal stability of BTDOs **99**, **104**, **108**, **112**, and **115** annulated with tetraazapentalene system was studied by differential scanning calorimetry (heating rate of 10 K min^{-1}) (Figure 5.14). All these compounds show good heat resistance and begin to decompose without melting at temperature of 210–245 °C. As a rule, the range between the beginning of decay and its intense point is ~20–40 °C. For example compound **99**, which proved to be the most stable in this series, starts to decompose at 245 °C, while an intensive decay starts at 280 °C [38]. Tetraazapentalenes **104**, **108**, **112**, and **115**, which are close in their stability, behave similarly when heated [39, 40].

5.4.2.5 BTDOs Annulated with Furoxan Ring

BTDO **117** annulated with a furoxan ring at the C(5)—C(6) bond exists in solutions as an equilibrium mixture of two isomers (Scheme 5.46). It melts without decomposition at 204 °C [41]. BTDO **122** annulated with a furoxan ring at the C(7)—C(8) bond

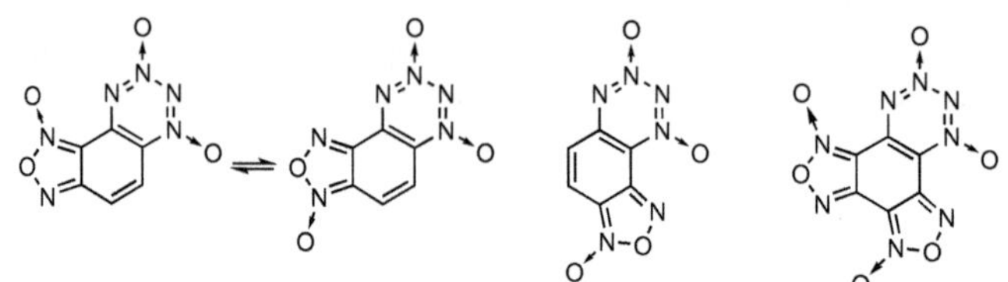

117a | 117b mp: 204 °C | 122 mp: 74–76 °C decomp. > 90 °C | 130 Hypothetic compound

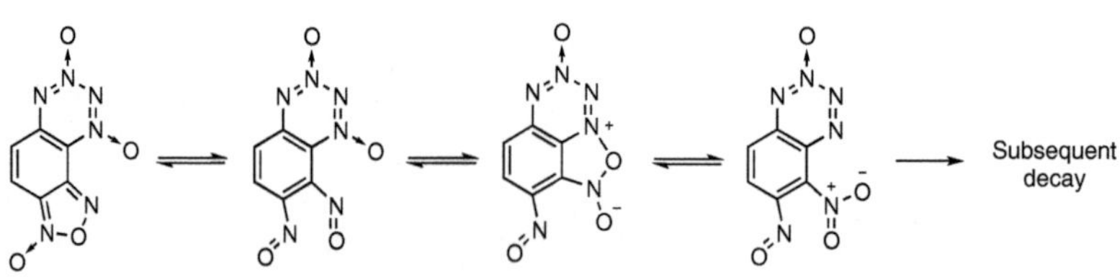

122

Scheme 5.46 BTDOs annulated with the furoxan ring. Proposed decomposition mechanism for BTDO **122**.

Figure 5.15 Melting points of TDO **26** and **30**.

26
mp: 150–152 °C
decomp.

30
mp: 182–192 °C
decomp.

is not so stable. It melts at 74–76 °C and begins to decompose above 90 °C. A possible reason for its instability is an intramolecular reaction of the *N*-oxide oxygen atom of the tetrazine 1,3-dioxide ring with the neighboring nitroso group followed by disintegration of 1,2,3,4-tetrazine 2-oxide ring [42]. The failure to obtain BTDO **130** is also associated with the low stability of this compound under conditions of thermal ring closure.

5.4.2.6 Tetrazine-1,3-Dioxides (TDOs) Annulated with Five-Membered Heterocycles

Thermal stability of these class of compounds depends on the type of annulating ring.

1,2,3-Triazolo- and 1-hydroxy-1,2,3-triazolo-TDOs **26** and **30** start to decompose at 150 and 182 °C, respectively (Figure 5.15). The thermal stability of these H-forms is comparable to that of their metal salts (Ag-salt of TDO **26** starts to decompose at 182 °C, K-salt of TDO **30** at 192 °C) [45, 52]. Apparently, temperature ~150–200 °C characterizes the stability of anions of these heterocyclic systems.

Me- and Ph-substituted TDOs **27–29** [49] show a ~50 °C increase in thermal stability as compared to H-form **26** (Figure 5.16). TDOs **32** with alkyl substituents at position 2 also exhibit increase in thermal stability as compared with H-form **30**. However, for TDOs **33** substituted at position 3 and for *O*-alkylated compounds **31** there is a slight decrease in stability [46].

FTDO **25** is isosteric and isoelectronic to TDOs **27** but is much less stable (mp: 113 °C, decomp.) [43]. Despite the fact that the thermal stability of FTDO is at the lower limit of acceptable values, the molecule was considered as a promising component of high-energy materials and rocket fuels (see Section 5.5).

5.4.2.7 TDOs Annulated with Tetraazapentalene Systems

TDOs **168**, **170–172** are similar in stability to TDOs annulated with triazole rings (Figure 5.17). They decompose without melting. Dinitro substituted TDO **171** is the most stable and had a thermal decomposition onset of 234 °C (differential scanning calorimetry (DSC), 10 °C min^{-1}) [50]. Isomer **172** is somewhat less stable (onset: 158 °C) [54], possibly due to a different decomposition mechanism.

5.4.2.8 TDO Annulated with Six-Membered Heterocycle

Pyrido-annulated compounds **175**, **176**, and **178** melt without decomposition in the range 179–228 °C [55] (Figure 5.18). TTTO **36** melts with decomposition at 183–186 °C (Kofler block) [25]. According to DSC measurements, onset temperature of TTTO **36** is 158 °C (heating rate: 5 °C min^{-1}), 165 °C (heating rate: 10 °C min^{-1}).

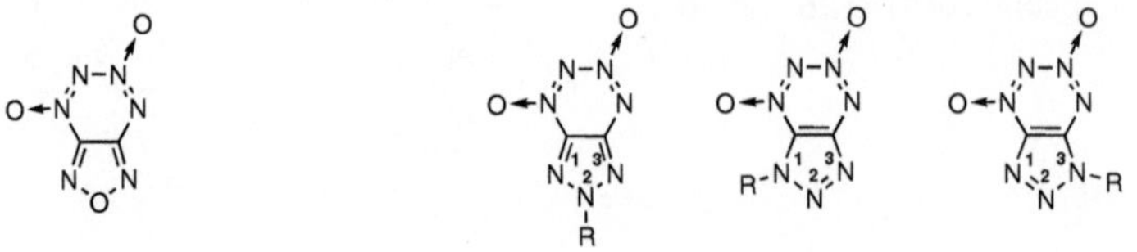

FTDO 25
mp: 110–112 °C
decomp.

	27	**28**	**29**
	mp, decomp. (°C)		
R = Me	198–199	200–201	213–214
R = Pr^i	173–175	—	—
R = Bu^t	151–152	—	—
R = Ph	decomp. > 236	—	233–236
R = 4-NO_2-C_6H_4	—	—	207–208

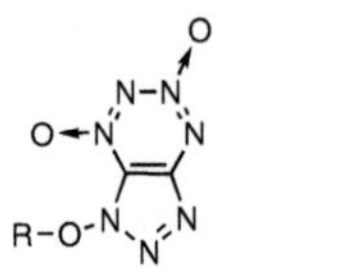

31

	mp, decomp. (°C)
R = Me	155–159
R = MeC(O)CH_2	136–138
R = PhC(O)CH_2	133–139

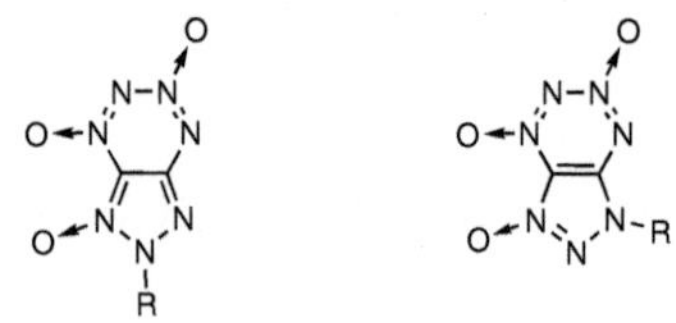

	32	**33**
	mp, decomp. (°C)	
R = Me	220–222	—
R = Et	208–209	—
R = Pr^i	231–232	—
R = Bu^t	—	176–178
R = $CH_3C(O)(CH_2)$	—	158–164

Figure 5.16 Melting points of TDOs annulated with five-membered heterocycles.

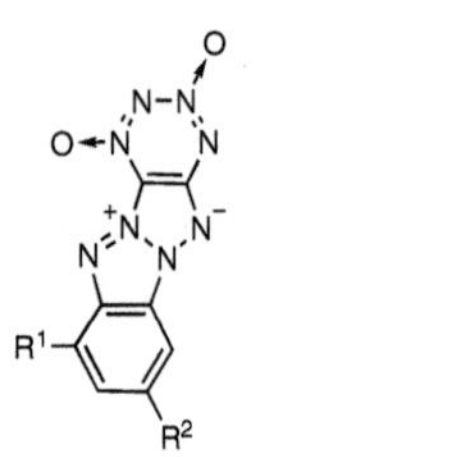

168: R^1 = H, R^2 = NO_2, decomp. >191 °C
170: R^1 = R^2 = H, decomp. >183 °C
171: R^1 = R^2 = NO_2, decomp. >234 °C

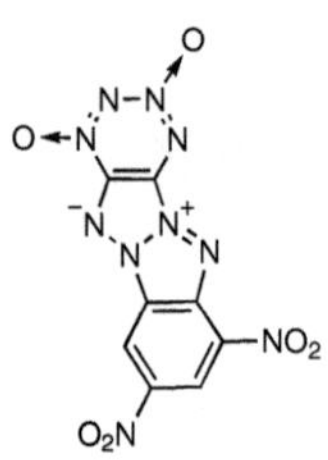

172
decomp. >158 °C

Figure 5.17 Melting points of TDOs annulated with tetraazapentalene systems.

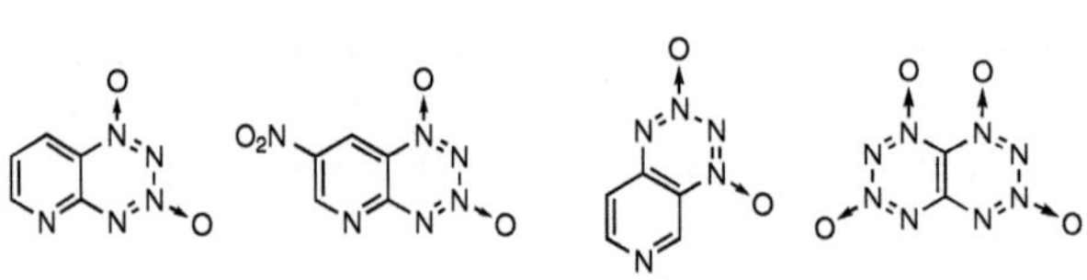

175 mp: 228–229 °C; **176** mp: 189–190 °C; **178** mp: 179–180 °C; **36** decomp. >158 °C

Figure 5.18 Melting points of TDOs annulated with six-membered heterocycle.

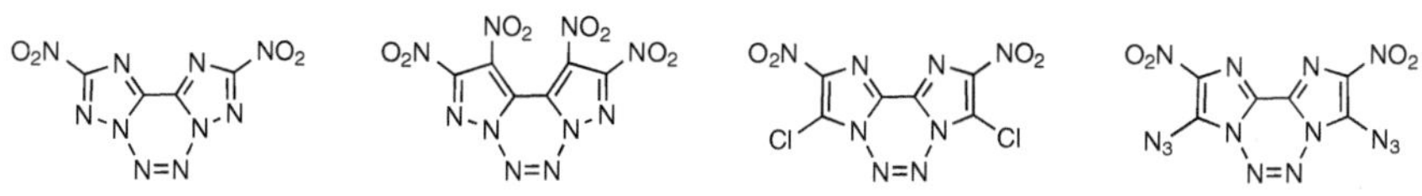

42
mp: 138 °C
decomp.

44
mp: 205 °C
decomp. 233 °C

46
mp: 215 °C

47
decomp. >111 °C

Figure 5.19 Melting points of 1,2,3,4-tetrazines with two bridgehead nitrogen atoms.

The peak of decomposition temperature is 183 °C (heating rate: 10 °C min^{-1}) (N.V. Murav'yev, Semenov Institute of Chemical Physics, Moscow, Russian Federation, unpublished results).

5.4.2.9 1,2,3,4-Tetrazines with Two Bridgehead Nitrogen Atoms

1,2,4-Triazole derivative **42** melts with decomposition at 138 °C [14] (Figure 5.19). Pyrazole derivative **44** is much more stable (mp: 205 °C, decomp. >233 °C, onset temperature at a heating rate of 5 °C min^{-1}). The authors explain the high thermal stability as the result of the conjugation of the fused backbone and the nature of the pyrazole ring [15]. Imidazole derivative **46** is also quite stable (mp: 215 °C) [16]. However, compound **47** bearing azido group begins to decomposes at temperature above 111 °C (onset temperature at a heating rate of 5 °C min^{-1}).

5.5 Applications

1,2,3,4-Tetrazine is considered as a promising scaffold for the construction of new high-energy materials, possessing high and in some cases record detonation performance. Moreover, the main focus is on compounds having TDO as a building block, modified by annulation and introduction of explosophoric moieties. In most cases, the articles dealing with tetrazines are theoretical, the primary goal of which is to select the most attractive structures for further synthesis.

Based on non-annulated 1,2,3,4-tetrazine and TDO, high-energy-density materials (HEDM) with covalent [79, 80, 89, 90] and ionic [91] structures were suggested for which the heat of formation, density, and detonation parameters were calculated. As substituents, both "standard" explosophore groups ($-NO_2$, $-NF_2$, $-ONO_2$, $-C(NO_2)_3$, tetrazole, tetrazole oxide, etc.), as well as non-annulated TDO cycle linked by a single bond, —NH— or —N=N— groups were considered.

Among polycyclic systems, attention was paid to 1,2,3,4-tetrazine and TDO annulated with five- and six-membered cycles. Energy compounds were proposed including furoxans [92, 93], tetraazapentalene [94, 95], nitro- and amino-substituted pyrazoles [96], imidazoles [97], oxazoles [97], and 1,2,3-triazoles [98], as well as their *N*-oxides. Among tetrazines annulated with six-membered cycles, compounds based on dinitrobenzene [31, 32], bis- and tris-annulated benzene [95, 99], pyridine [100], pyrazine [82, 94, 101], and piperazine [81, 82] were studied. As a rule, the calculated performance parameters of 1,2,3,4-tetrazines and TDOs annulated with five- and six-membered cycles are similar to or exceed the standard HEDMs, which

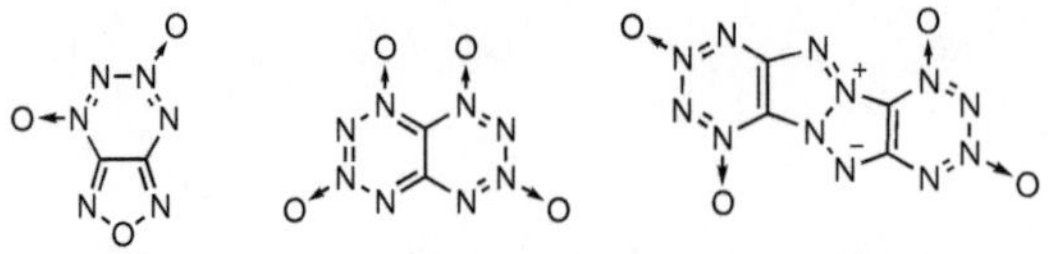

25 36 179

Figure 5.20 TDOs **25**, **36**, and **179**.

allows to consider them as a promising class of explosives. In most cases, these are high-density compounds ($d \geq 1.8\,\text{g cm}^{-1}$) with good oxygen balance, large positive heat of formation ($\Delta H_f \geq 500\,\text{kJ mol}^{-1}$), high detonation rate ($D \geq 9\,\text{km s}^{-1}$) and pressure, as well as acceptable sensitivity and thermal stability.

Individual representatives are unique compounds, such as TDOs **36** and hypothetical **179** (Figure 5.20). The latter according to calculations has record characteristics ($d = 2.047\,\text{g cm}^{-3}$, $\Delta H_f = 1479.9\,\text{kJ mol}^{-1}$, $D = 10.387\,\text{km s}^{-1}$, detonation pressure = 57.42 GPa), as well as low sensitivity coupled with high thermal stability. In turn TDO **36** has repeatedly been the subject of study by various research groups, which predicted excellent detonation performance ($\Delta H_f = 850\text{–}990\,\text{kJ mol}^{-1}$, $Q = 2000\,\text{cal g}^{-1}$, $D = 9.7\text{–}10.9\,\text{km s}^{-1}$, detonation pressure = 43–60 GPa) combined with impact sensitivity comparable to that of HMX [95, 100, 102–111]. A wide range of values is associated with the difficulty of calculating the crystal lattice and density of this compound, which, depending on the method [112–115], ranged from 1.9 to $2.45\,\text{g cm}^{-3}$.

Of all TDOs considered as a promising HEDM and rocket fuel components for which synthetic methods were developed, FTDO **25** is the most accessible, and therefore was actively studied (Figure 5.20). For this compound, heat of formation as well as enthalpy of sublimation was not only calculated [99, 116–119], but experimentally measured [120]. The data in combination with experimentally measured density allowed to refine the calculated values of the detonation performance of FTDO **25** [118, 120, 121]. Thermal decomposition of FTDO was studied in solid state, melt, and dinonyl phthalate solution by manometry and by thermogravimetry [122]. The kinetic parameters of thermal decomposition of FTDO and the (75/25) FTDO/dinitrazapentane system were experimentally investigated by differential scanning calorimetry [123]. It was shown that FTDO is highly sensitive to mechanical impact [124] and electrical impulses [125] and refers to initiating explosives. However, this did not prevent researchers from continuing to study the substance and its compositions [126, 127]. Its interaction with some organic solvents was investigated [128]. A FTDO–benzotrifuroxan complex in a ratio 3 : 1 has been synthesized and its properties have been investigated including thermal decomposition and sensitivity to impact and friction [129]. It was found that compositions based on the FTDO/2,4-dinitro-2,4-diazapentane system have interesting properties, which led to their detailed study [130–135].

Among synthesized high-nitrogen molecules compounds with the fused tricyclic 1,2,3,4-tetrazine ring system are considered as attractive primary and secondary explosives combining high density, a good thermal stability, and calculated detonation performance comparable to CL-20 [14–16].

References

1 Nelsen, S.F. and Fibiger, R. (1972). 1,4-Dimethyl-1,4,5,6-tetrahydro-1,2,3,4-tetrazine. Cyclic cis-2-tetrazene. *J. Am. Chem. Soc.* 94: 8497–8501.

2 (a) Kaihoh, T., Itoh, T., Yamaguchi, K., and Ohsawa, A. (1988). First synthesis of a 1,2,3,4-tetrazine. *J. Chem. Soc., Chem. Commun.* 24: 1608–1609. (b) Kaihoh, T., Itoh, T., Yamaguchi, K., and Ohsawa, A. (1991). Lead tetraacetate oxidation of 1- and 2-amino-5-phenyl[1,2,3]triazolo[4,5-*d*][1,2,3]triazoles; synthesis of a fused 1,2,3,4-tetrazine. *J. Chem. Soc., Perkin Trans. 1* (8): 2045–2048. (c) Yamaguchi, K., Kaihoh, T., Itoh, T., and Ohsawa, A. (1991). Structure of 2-phenyl-1,2,3-triazolo[4,5-*e*][1,2,3,4]tetrazine at 153 K. *Acta Crystallogr., Sect. C: Cryst. Struct. Commun.* 47: 590–592.

3 Churakov, A.M., Ioffe, S.L., and Strelenko, Y.A. (1990). 1,2,3,4-Tetrazine 1,3-dioxides: a new class of heterocyclic compounds. *Bull. Acad. Sci. USSR, Div. Chem. Sci.* 39: 639–640.

4 Neunhoeffer, H. (1984). Tetrazines and pentazines, Chapter 2.21. In: *Comprehensive Heterocyclic Chemistry* (ed. A.R. Katritzky and C.W. Rees), vol. 3 (ed. A.J. Boulton and A. McKillop), 531–572. Oxford: Pergamon.

5 Hurst, D.T. (1996). Other tetrazines and pentazines, Chapter 6.22. In: *Comprehensive Heterocyclic Chemistry II* (ed. A.J. Boulton, A.R. Katritzky, C.W. Rees, and E.F.V. Scriven), vol. 6 (ed. A.J. Boulton), 957–965. Oxford: Pergamon.

6 Benin, V. (2008). Tetrazines and pentazines, Chapter 9.13. In: *Comprehensive Heterocyclic Chemistry III* (ed. A.R. Katritzky, C.A. Ramsden, E.F.V. Scriven, and R.J.K. Taylor), vol. 9 (ed. K. Turnbull), 715–735. Oxford: Elsevier.

7 Churakov, A.M. and Tartakovsky, V.A. (2004). Progress in 1,2,3,4-tetrazine chemistry. *Chem. Rev.* 104: 2601–2616.

8 Chavez, D.E. (2017). Energetic heterocyclic *N*-oxides. In: *Heterocyclic N-Oxides, Topics in Heterocyclic Chemistry*, vol. 53 (ed. O.V. Larionov), 1–27. Cham: Springer.

9 Shreeve, J.M., Gao, H., and Zhang, Q. (2020). Fused heterocycle-based energetic materials (2012–2019). *J. Mater. Chem. A* 8: 4193–4216.

10 Yang, B., Lu, Y., Chen, C.-J. et al. (2009). The synthesis of 5-amino-4-arylazo-3-methyl-1*H*-pyrazoles and 5-aryl-3-methylpyrazole[3,4-*e*][1,2,3,4]tetrazines. *Dyes Pigm.* 83: 144–147.

11 Almerico, A.M., Mingoia, F., Diana, P. et al. (2005). 1-Methyl-3*H*-pyrazolo[1,2-*a*]benzo[1,2,3,4]tetrazin-3-ones. Design, synthesis, and biological activity of new antitumor agents. *J. Med. Chem.* 48: 2859–2866.

12 Mingoia, F., Di Sano, C., Di Blasi, F. et al. (2013). Exploring the anticancer potential of pyrazolo[1,2-*a*]benzo[1,2,3,4]tetrazin-3-one derivatives: the effect on apoptosis induction, cell cycle and proliferation. *Eur. J. Med. Chem.* 64: 345–356.

13 Almerico, A.M. and Boulton, A.J. (1985). An unusual substitution reaction in a benzotetrazine derivative. *J. Chem. Soc., Chem. Commun.* 4: 204–205.

14 Chavez, D.E., Bottaro, J.C., Petrie, M. et al. (2015). Synthesis and thermal behavior of a fused, tricyclic 1,2,3,4-tetrazine ring system. *Angew. Chem. Int. Ed.* 54: 12973–12975.

15 Tang, Y., Kumar, D., and Shreeve, J.M. (2017). Balancing excellent performance and high thermal stability in a dinitropyrazole fused 1,2,3,4-tetrazine. *J. Am. Chem. Soc.* 139: 13684–13687.

16 Tang, Y., He, C., Yin, P. et al. (2018). Energetic functionalized azido/nitro imidazole fused 1,2,3,4-tetrazine. *Eur. J. Org. Chem.* 19: 2273–2276.

17 Churakov, A.M., Smirnov, O.Y., Strelenko, Y.A. et al. (1996). Mechanism of benzo-l,2,3,4-tetrazine l-*N*-oxide formation. *Mendeleev Commun.* 6: 22–23.

18 Piercey, D.G., Chavez, D.E., Heimsch, S. et al. (2015). An energetic *N*-oxide and *N*-amino heterocycle and its transformation to 1,2,3,4-tetrazine-1-oxide. *Propellants Explos. Pyrotech.* 40: 491–497.

19 Klenov, M.S., Ratnikov, M.O., Churakov, A.M. et al. (2011). Generation of oxodiazonium ions 1. Synthesis of [1,2,5]oxadiazolo[3,4-*c*]cinnoline 5-oxides. *Russ. Chem. Bull. Int. Ed.* 60: 536–547.

20 Klenov, M.S., Zelenov, V.P., Churakov, A.M. et al. (2011). Generation of oxodiazonium ions 2. Synthesis of benzotetrazine 1,3-dioxides from 2-(*tert*-butyl-*NNO*-azoxy)-*N*-nitroanilines. *Russ. Chem. Bull. Int. Ed.* 60: 2040–2245.

21 Churakov, A.M., Smirnov, O.Y., Ioffe, S.L. et al. (2002). Benzo-1,2,3,4-tetrazine 1,3-dioxides – synthesis and NMR study. *Eur. J. Org. Chem.* 14: 2342–2349.

22 Zelenov, V.P., Lobanova, A.A., Sysolyatin, S.V. et al. (2013). New syntheses of [1,2,5]oxadiazolo[3,4-*e*][1,2,3,4]tetrazine 4,6-dioxide. *Russ. J. Org. Chem.* 49: 455–465.

23 Klenov, M.S., Anikin, O.V., Churakov, A.M. et al. (2015). Toward the synthesis of tetrazino-tetrazine 1,3,6,8-tetraoxide (TTTO): an approach to non-annulated 1,2,3,4-tetrazine 1,3-dioxides. *Eur. J. Org. Chem.* 28: 6170–6179.

24 Klenov, M.S., Churakov, A.M., Strelenko, Y.A. et al. (2011). Generation of oxodiazonium ions 4. Nitramine *O*-alkyl derivatives in the synthesis of benzotetrazine-1,3-dioxides. *Russ. Chem. Bull. Int. Ed.* 60: 2051–2056.

25 Klenov, M.S., Guskov, A.A., Anikin, O.V. et al. (2016). Synthesis of tetrazino-tetrazine 1,3,6,8-tetraoxide (TTTO). *Angew. Chem. Int. Ed.* 55: 11472–11475.

26 Guskov, A.A., Klenov, M.S., Churakov, A.M. et al. (2016). Synthesis of 5-amino-6-nitro-1,2,3,4-tetrazine 1,3-dioxide. *Russ. Chem. Bull. Int. Ed.* 65: 2763–2765.

27 Klenov, M.S., Leonov, N.E., Guskov, A.A. et al. (2019). Synthesis of 5-amino-6-(nitro-*NNO*-azoxy)-1,2,3,4-tetrazine 1,3-dioxide. *Russ. Chem. Bull. Int. Ed.* 68: 1798–1800.

28 Anikin, O.V., Leonov, N.E., Klenov, M.S. et al. (2019). An energetic (nitro-*NNO*-azoxy)triazolo-1,2,4-triazine. *Eur. J. Org. Chem.* 26: 4189–4195.

29 Tyurin, A.Y., Churakov, A.M., Strelenko, Y.A. et al. (2006). Synthesis of first nonannulated 1,2,3,4-tetrazine 1,3-dioxides. *Russ. Chem. Bull. Int. Ed.* 55: 1648–1653.

30 Klenov, M.S., Churakov, A.M., Strelenko, Y.A. et al. (2015). Ozonation of benzotetrazine 1,3-dioxides. Facile synthesis of di(methoxycarbonyl)-1,2,3,4-tetrazine 1,3-dioxide. *Tetrahedron Lett.* 56: 5437–5440.

31 Churakov, A.M., Ioffe, S.L., and Tartakovsky, V.A. (1991). The first synthesis of 1,2,3,4-tetrazine-1,3-di-*N*-oxides. *Mendeleev Commun.* 1: 101–103.

32 Klapötke, T.M., Piercey, D.G., Stierstorfer, J. et al. (2012). The synthesis and energetic properties of 5,7-dinitrobenzo-1,2,3,4-tetrazine-1,3-dioxide (DNBTDO). *Propellants Explos. Pyrotech.* 37: 527–535.

33 Smirnov, O.Y., Churakov, A.M., Strelenko, Y.A. et al. (2002). Nitration and bromination of benzo-1,2,3,4-tetrazine 1,3-dioxides. *Russ. Chem. Bull.* 51: 1841–1848.

34 Smirnov, O.Y., Churakov, A.M., Tyurin, A.Y. et al. (2002). Nucleophilic substitution in benzo-1,2,3,4-tetrazine 1,3-dioxides. *Russ. Chem. Bull. Int. Ed.* 51: 1849–1856.

35 Tyurin, A.Y., Smirnov, O.Y., Churakov, A.M. et al. (2009). An unusual cascade of S_N reactions of benzotetrazine 1,3-dioxide derivatives. *Russ. Chem. Bull. Int. Ed.* 58: 361–365.

36 Frumkin, A.E., Churakov, A.M., Strelenko, Y.A. et al. (1999). Synthesis of 1,2,3,4-tetrazino[5,6-*f*]benzo-1,2,3,4-tetrazine 1,3,7,9-tetra-*N*-oxides. *Org. Lett.* 1: 721–724.

37 Frumkin, A.E., Churakov, A.M., Strelenko, Y.A. et al. (2006). Synthesis of 1,2,3,4-tetrazino[5,6-g]benzo-1,2,3,4-tetrazine-1,3,7,9-tetraoxides. *Russ. Chem. Bull. Int. Ed.* 55: 1654–1658.

38 Tyurin, A.Y., Churakov, A.M., Strelenko, Y.A. et al. (2008). Benzo-1,2,3,4-tetrazine 1,3-dioxides annulated with tetraazapentalenes. 1. Annulation at the C(5)—C(6) bond. *Russ. Chem. Bull. Int. Ed.* 57: 193–196.

39 Smirnov, O.Y., Churakov, A.M., Strelenko, Y.A. et al. (2008). Benzo-1,2,3,4-tetrazine 1,3-dioxides annulated with tetraazapentalene systems. 2. Annulation at the C(6)—C(7) bond. *Russ. Chem. Bull. Int. Ed.* 57: 2180–2184.

40 Tyurin, A.Y., Churakov, A.M., Strelenko, Y.A. et al. (2009). Benzo-1,2,3,4-tetrazine 1,3-dioxides annulated with tetraazapentalene systems. 3. Annulation at the C(7)—C(8) bond. *Russ. Chem. Bull. Int. Ed.* 58: 212–215.

41 Smirnov, O.Y., Tyurin, A.Y., Churakov, A.M. et al. (2006). Annulated benzotetrazine 1,3-dioxides 1. [1,2,5]Oxadiazolo[3,4-*f*][1,2,3,4]benzotetrazine 2,4,7- and 2,4,9-trioxides. *Russ. Chem. Bull. Int. Ed.* 55: 137–140.

42 Tyurin, A.Y., Smirnov, O.Y., Churakov, A.M. et al. (2006). Annulated benzotetrazine 1,3-dioxides 2. [1,2,5]Oxadiazolo[3,4-*f*][1,2,3,4]benzotetrazine 1,3,7-trioxide. *Russ. Chem. Bull. Int. Ed.* 55: 351–356.

43 Churakov, A.M., Ioffe, S.L., and Tartakovsky, V.A. (1995). Synthesis of [1,2,5]oxadiazolo[3,4-*e*][1,2,3,4]tetrazine 4,6-di-*N*-oxide. *Mendeleev Commun.* 5: 227–228.

44 Zelenov, V.P., Lobanova, A.A., Lyukshenko, N.I. et al. (2008). Behavior of [1,2,5]oxadiazolo[3,4*e*][1,2,3,4]tetrazine 4,6-dioxide in various media. *Russ. Chem. Bull. Int. Ed.* 57: 1384–1389.

45 Voronin, A.A., Zelenov, V.P., Churakov, A.M. et al. (2014). Synthesis of 1,2,3,4-tetrazine 1,3-dioxides annulated with 1,2,3-triazoles and 1,2,3-triazole 1-oxides. *Tetrahedron* 70: 3018–3022.

46 Voronin, A.A., Zelenov, V.P., Churakov, A.M. et al. (2014). Alkylation of 1-hydroxy-1*H*-[1,2,3]triazolo[4,5-*e*][1,2,3,4]tetrazine 5,7-dioxide. *Russ. Chem. Bull. Int. Ed.* 63: 475–479.

47 Zelenov, V.P., Voronin, A.A., Churakov, A.M. et al. (2014). 2-Alkyl-4-amino-5-(*tert*-butyl-*NNO*-azoxy)-2*H*-1,2,3-triazole 1-oxides: synthesis and reduction. *Russ. Chem. Bull. Int. Ed.* 63: 123–129.

48 Zelenov, V.P., Voronin, A.A., Churakov, A.M. et al. (2013). Amino(*tert*-butyl-*NNO*-azoxy)furoxans: synthesis, isomerization, and rearrangement of *N*-acetyl derivatives. *Russ. Chem. Bull. Int. Ed.* 62: 117–122.

49 Voronin, A.A., Churakov, A.M., Klenov, M.S. et al. (2017). Synthesis of 1,2,3,4-tetrazine 1,3-dioxides annulated with 1(2)-aryl-1,2,3-triazoles. *Eur. J. Org. Chem.* 33: 4963–4971.

50 Konnov, A.A., Klenov, M.S., Churakov, A.M. et al. (2018). Synthesis of 1,2,3,4-tetrazine 1,3-dioxides annulated with 1,3*a*,4,6*a*-tetraazapentalene systems. *Asian J. Org. Chem.* 7: 2534–2543.

51 Klenov, M.S., Anikin, O.V., Guskov, A.A. et al. (2016). Serendipitous synthesis of (*tert*-butyl-*NNO*-azoxy)acetonitrile: reduction of an oxime moiety to a methylene unit. *Eur. J. Org. Chem.* 22: 3845–3855.

52 Voronin, A.A., Zelenov, V.P., Churakov, A.M. et al. (2015). Synthesis of 1*H*-[1,2,3]triazolo[4,5-*e*][1,2,3,4]tetrazine 4,6-dioxide and its methyl derivatives. *Russ. Chem. Bull. Int. Ed.* 64: 699–703.

53 Ratnikov, M.O., Lipilin, D.L., Churakov, A.M. et al. (2002). Release of nitrosating species in the course of reduction of benzo-1,2,3,4-tetrazine 1,3-dioxides. *Org. Lett.* 4: 3227–3229.

54 Shvets, A.O., Konnov, A.A., Klenov, M.S. et al. (2020). Synthesis of 2-(6-nitrobenzofuroxan-4-yl)-2*H*-[1,2,3]triazolo[4,5-*e*][1,2,3,4]tetrazine 4,6-dioxide. *Russ. Chem. Bull. Int. Ed.* 69: 739–741.

55 Tartakovsky, V.A., Filatov, I.E., Churakov, A.M. et al. (2004). Synthesis and structures of pyridoannelated 1,2,3,4-tetrazine 1,3-dioxides. *Russ. Chem. Bull. Int. Ed.* 53: 2577–2583.

56 Churakov, A.M., Smirnov, O.Y., Strelenko, Y.A. et al. (1994). Synthesis, structure and transformations of the first 1,2,3,4-tetrazine-1-*N*-oxide. *Mendeleev Commun.* 4: 122–124.

57 Yamaguchi, K., Takahashi, H., Kaihoh, T. et al. (1992). Structure of benzo-1,2,3,4-tetrazine 1,3-dioxide. *Acta Crystallogr., Sect. C: Cryst. Struct. Commun.* 48: 1237–1239.

58 Churakov, A.M. (2005). Cyclic conjugated poly-nitrogen systems of a new type. Dissertation for the degree of doctor of chemical sciences. N.D. Zelinsky Institute of Organic Chemistry (Moscow, Russian Federation).

59 Baraboshkin, N.M., Zelenov, V.P., Dzyabchenko, A.V. et al. (2019). X-ray study and computational model of the solid solvate of [1,2,5]oxadiazolo[3,4-*e*][1,2,3,4]tetrazine 4,6-dioxide (FTDO) with benzene and ab initio crystal structure prediction of pure FTDO. *J. Mol. Stuct.* 1190: 135–143.

60 Politzer, P. and Murray, J.S. (2017). Nitrogen catenation: polyazoles and polyazines. In: *PATAI'S Chemistry of Functional Groups* (ed. Z. Rappoport). Wiley Online Library https://doi.org/10.1002/9780470682531.pat0931.

61 Noyman, M., Zilberg, S., and Haas, Y. (2009). Stability of polynitrogen compounds: the importance of separating the σ and π electron systems. *J. Phys. Chem. A* 113: 7376–7382.

62 Politzer, P., Lane, P., and Murray, J.S. (2013). Computational analysis of relative stabilities of polyazine *N*-oxide. *Struct. Chem.* 24: 1965–1974.

63 Politzer, P., Lane, P., and Murray, J.S. (2014). Some interesting aspects of *N*-oxides. *Mol. Phys.* 112: 719–725.

64 Politzer, P. and Murray, J.S. (2019). *N*-Oxides of polyazoles and polyazines. In: *PATAI'S Chemistry of Functional Groups* (ed. Z. Rappoport). Wiley Online Library https://doi.org/10.1002/9780470682531.pat0936.

65 Yuan, J., Long, X., and Zhang, C. (2016). Influence of *N*-oxide introduction on the stability of nitrogen-rich heteroaromatic rings: a quantum chemical study. *J. Phys. Chem. A* 120: 9446–9457.

66 Lai, W., Lian, P., Ge, Z.-X. et al. (2016). Theoretical study of the effect of *N*-oxides on the performances of energetic compounds. *J. Mol. Model.* 22: 83.

67 Fabian, J. and Lewars, E. (2004). Azabenzenes (azines) – the nitrogen derivatives of benzene with one to six N atoms: stability, homodesmotic stabilization energy, electron distribution, and magnetic ring current; a computational study. *Can. J. Chem.* 82: 50–69.

68 Krygowski, T.M. and Szatylowicz, H. (2015). Aromaticity: what does it mean? *ChemTexts* 1: 12.

69 Feixas, F., Matito, E., Poater, J., and Solà, M. (2008). On the performance of some aromaticity indices: a critical assessment using a test set. *J. Comput. Chem.* 29: 1543–1554.

70 Wang, Y., Wu, J.I.-C., Li, Q., and Schleyer, P.R. (2010). Aromaticity and relative stabilities of azines. *Org. Lett.* 12: 4824–4827.

71 Sánchez-Sanz, G. (2015). Aromatic behaviour of benzene and naphthalene upon pnictogen substitution. *Tetrahedron* 71: 826–839.

72 Maria, M., Nisa, R.U., Hanif, M. et al. (2014). Aromaticities of azines relative to benzene; a theoretical approach through the dimethyldihydropyrene probe. *J. Phys. Org. Chem.* 27: 860–866.

73 Păuşescu, I., Medeleanu, M., Ştefănescu, M. et al. (2015). A DFT study on the stability and aromaticity of heterobenzenes containing group 15 elements. *Heteroat. Chem.* 26: 206–214.

74 Elguero, E., Alkorta, I., and Elguero, J. (2018). A theoretical study of the properties of ninety-two "aromatic" six-membered rings including benzene, azines, phosphinines and azaphosphinines. *Heteroat. Chem.* 29: e21441.

75 Pakiari, A.H. and Bagheri, N. (2011). A new approach for aromaticity criterion based on electrostatic field gradient. *J. Mol. Model.* 17: 2017–2027.

76 Dey, S., Manogaran, D., Manogaran, S., and Schaefer, H.F. III, (2018). Quantification of aromaticity of heterocyclic systems using interaction coordinates. *J. Phys. Chem. A* 122: 6953–6960.

77 Pandey, S.K., Manogaran, D., Manogaran, S., and Schaefer, H.F. III, (2016). Quantification of aromaticity based on interaction coordinates: a new proposal. *J. Phys. Chem. A* 120: 2894–2901.

78 Wilson, K.J., Perera, S.A., Bartlett, R.J., and Watts, J.D. (2001). Stabilization of the pseudo-benzene N_6 ring with oxygen. *J. Phys. Chem. A* 105: 7693–7699.

79 Wu, Q., Zhu, W., and Xiao, H. (2013). Molecular design of trinitromethyl-substituted nitrogen-rich heterocycle derivatives with good oxygen balance as high-energy density compounds. *Struct. Chem.* 24: 1725–1736.

80 Wu, Q., Pan, Y., Xia, X. et al. (2013). Theoretic design of 1,2,3,4-tetrazine-1,3-dioxide-based high-energy density compounds with oxygen balance close to zero. *Struct. Chem.* 24: 1579–1590.

81 Pan, Y., Zhu, W., and Xiao, H. (2013). Theoretical studies on the structures, heats of formation, energetic properties and pyrolysis mechanisms of nitrogen-rich difurazano[3,4-*b*:3′,4′-*e*]piperazine derivatives and their analogues. *Struct. Chem.* 24: 1071–1087.

82 Lai, W., Lian, P., Liu, Y. et al. (2014). Design and theoretical study of 15 novel high energy density compounds. *J. Mol. Model.* 20: 2479.

83 Lai, W.P., Lian, P., and Yu, T. (2014). Theoretical study on the structure and stability of [1,2,5]oxadiazolo[3,4-*e*][1,2,3,4]-tetrazine-4,6-di-*N*-dioxide (FTDO). *J. Mol. Model.* 20: 2343.

84 Zhang, X. and Gong, X. (2015). Theoretical investigations on stability of pyridylpentazoles, pyridazylpentazoles, triazinylpentazoles, tetrazinylpentazoles, and pentazinylpentazole searching for a replacement of phenylpentazole as N_5-source. *J. Mol. Model.* 21: 318.

85 Lipilin, D.L., Smirnov, O.Y., Churakov, A.M. et al. (2002). A New cyclization involving the diazonium and *ortho*-(*tert*-butyl)-*NNO*-azoxy groups – synthesis of 1,2,3,4-benzotetrazine 1-oxides. *Eur. J. Org. Chem.* 20: 3435–3446.

86 Churakov, A.M. and Tartakovsky, V. A. (1998). New high nitrogen heterocycles with the alternation of charges: stability and strategy of synthesis. *Energetic Materials. Production, Processing and Characterization*. 29th International Annual Conference of ICT, Karlsruhe, Germany; DWS Werbeagentur und Verlag GmbH: Karlsruhe, Germany; V7 (June 30–July 3 1998).

87 Churakov, A.M., Ioffe, S.L., and Tartakovsky, V.A. (1996). Synthesis of l-aryl-2-nitrodiazene 1-*N*-oxides. *Mendeleev Commun.* 6: 20–22.

88 Churakov, A.M., Tyurin, A.Y., Goneharova, E.L. et al. (1995). Synthesis of 1-aryl-2-alkenyldiazene 1-oxides. *Russ. Chem. Bull. Int. Ed.* 44: 897–900.

89 Wu, Q., Zhu, W., and Xiao, H. (2013). Molecular design of tetrazole- and tetrazine-based high-density energy compounds with oxygen balance equal to zero. *J. Chem. Eng. Data* 58: 2748–2762.

90 Xiang, D., Zhu, S., Qian, H. et al. (2019). Environmental degradability of 1,2,3,4-tetrazine-1,3-dioxide-functionalized FOX-7 derivatives with high energy and low sensitivity: a computational evaluation. *Struct. Chem.* 30: 327–340.

91 Xiang, F., Wu, Q., Zhu, W., and Xiao, H. (2013). A comparative theoretical study of heterocycle-functionalized tetrazolate-and tetrazolate-1-oxide-based dianionic salts. *Can. J. Chem.* 91: 1233–1242.

92 Wang, T., Zhang, T., Xu, L. et al. (2014). Theoretical studies on vicinal-tetrazine compounds: furoxano-1,2,3,4-tetrazine1,3,5-trioxide (FTTO-α) and furoxano-1,2,3,4-tetrazine-1,3,7-trioxide (FTTO-β). *J. Mol. Model.* 20: 2516–2527.

93 Khakimov, D.V., Zelenov, V.P., Baraboshkin, N.M., and Pivina, T.S. (2019). The unusual combination of beauty and power of furoxano-1,2,3,4-tetrazine 1,3-dioxides: a theoretical study of crystal structures. *J. Mol. Model.* 25: 107.

94 Molchanova, M.S., Pivina, T.S., Arnautova, E.A., and Zefirov, N.S. (1999). Computer-aided search for high-density energetic compounds among hydrogen-free heterocycles. *J. Mol. Struct. THEOCHEM* 465: 11–24.

95 Tan, B., Huang, M., Huang, H. et al. (2013). Theoretical investigation of several 1,2,3,4-tetrazine-based high-energy compounds. *Propellants Explos. Pyrotech.* 38: 372–378.

96 Wang, T., Zheng, C., Gong, X. et al. (2015). Theoretical studies of the structure, stability, and detonation properties of vicinal-tetrazine 1,3-dioxide annulated with a five-membered heterocycle. 2. Annulation with a pyrazole ring. *J. Mol. Model.* 21: 269.

97 Zheng, C., Wang, T., Wang, F. et al. (2019). Theoretical studies on a new series of 1,2,3,4-tetrazine 1,3-dioxide annulation with an imidazole ring or oxazole ring. *J. Mol. Model.* 25: 36.

98 Wang, T., Zheng, C., Liu, Y. et al. (2015). Theoretical studies of the structure, stability, and detonation properties of vicinal-tetrazine 1,3-dioxide annulated with a five-membered heterocycle. 1. Annulation with a triazole ring. *J. Mol. Model.* 21: 201.

99 Upadhyay, M.K., Sengupta, S.K., and Singh, H.J. (2015). Nitro and dinitroamino *N*-oxides of octaazaanthracene as high energy materials. *J. Mol. Model.* 21: 18.

100 Jorgensen, K.R., Oyedepo, G.A., and Wilson, A.K. (2011). Highly energetic nitrogen species: reliable energetics via the correlation consistent composite approach (ccCA). *J. Hazard. Mater.* 186: 583–589.

101 Wang, T., Zheng, C., Yang, J. et al. (2014). Theoretical studies on a new high energy density compound 6-amino-7-nitropyrazino[2,3-*e*][1,2,3,4] tetrazine 1,3,5-trioxide (ANPTTO). *J. Mol. Model.* 20: 2261.

102 Ye, C.-C., An, Q., Goddard, W.A. et al. (2015). Initial decomposition reaction of di-tetrazine-tetroxide (DTTO) from quantum molecular dynamics: implications for a promising energetic material. *J. Mat. Chem. A* 3: 1972–1978.

103 Shechter, H. (2004). Synthesis of high energy 1,2,3,4-tetrazine 1,3-di *N*-oxides and pentazine poly-*N*-oxides. Report AFRL-SR-AR-TR-05-0032. Los Alamos National Laboratory, Los Alamos, NM, USA.

104 Östmark, H., Wallin, S., and Goede, P. (2007). High energy density materials (HEDM): overview, theory and synthetic efforts at FOI. *CEJEM* 4: 83–108.

105 Christe, K.O., Haiges, R., Wagner, R.I., and Jones C.J. (2009). Synthesis of new high-oxygen carriers and ditetrazinetetroxide (DTTO). Final report for the office of naval research 2008–2009, Under Contr., N00014-08-1-0590.

106 Song, X., Li, J., Hou, H., and Wang, B. (2009). Extensive theoretical studies of a new energetic material: tetrazino-tetrazine-tetraoxide (TTTO). *J. Comput. Chem.* 30: 1816–1820.

107 Politzer, P., Lane, P., and Murray, J.S. (2013). Computational characterization of two di-1,2,3,4-tetrazine tetraoxides, DTTO and iso-DTTO, as potential energetic compounds. *CEJEM* 10: 37–52.

108 Dzyabchenko, A.V., Khakimov, D.V., and Pivina, T.S. (2017). Electrostatic model of crystal structure of tetrazinotetrazintetroxide benzene solvate. *Combust. Explos.* 10: 104–108.

109 Christe, K.O., Dixon, D.A., Vasiliu, M. et al. (2015). Are DTTO and iso-DTTO worthwhile targets for synthesis? *Propellants Explos. Pyrotech.* 40: 463–468.

110 Tan, B., Huang, M., Long, X. et al. (2016). Computational assessment of several hydrogen-free high energy compounds. *J. Mol. Graphics Modell.* 63: 85–90.

111 Tan, B., Huang, M., Long, X. et al. (2011). Basic characteristics for estimation polynitrogen compounds efficiency. *CEJEM* 8: 233–247.

112 Rice, B.M., Hare, J.J., and Byrd, E.F.C. (2007). Accurate predictions of crystal densities using quantum mechanical molecular volumes. *J. Phys. Chem. A* 11: 10874–10879, and works cited there.

113 Ammon, H.L. (2008). Updated atom/functional group and atom_code volume additivity parameters for the calculation of crystal densities of single molecules, organic salts, and multi-fragment materials containing H, C, B, N, O, F, S, P, Cl, Br, and I. *Propellants Explos. Pyrotech.* 33: 92–102.

114 Mendoza-Cortes, J.L., An, Q., Goddard, W.A. et al. (2015). Prediction of the crystal packing of di-tetrazine-tetroxide (DTTO) energetic material. *J. Comput. Chem.* 37: 163–167.

115 Khakimov, D.V., Dzyabchenko, A.V., and Pivina, T.S. (2020). Computer simulation of the crystal structure of tetrazino-tetrazine tetraoxide (TTTO) isomers with one and two independent molecules in the unit cell. *Russ. Chem. Bull. Int. Ed.* 69: 212–217.

116 Kiselev, V.G., Gritsan, N.P., Zarko, V.E. et al. (2007). Calculation of the formation enthalpy of [1,2,5]oxadiazolo[3,4-*e*][1,2,3,4]-tetrazine 4,6-di-*N*-oxide using modern multilevel quantum chemical methods. *Combust. Explos.* 43: 77–81.

117 Kiselev, V.G., Gritsan, N.P., Zarko, V.E. et al. (2007). Multilevel quantum chemical calculation of the enthalpy of formation of [1,2,5]oxadiazolo [3,4-*e*][1,2,3,4]tetrazine-4,6-di-*N*-dioxide. *Combust. Explos. Shock Waves* 43: 562–566.

118 He, L., Dong, L.-L., Zhang, G.-Q. et al. (2012). Structure and properties of furazano[3,4-*e*]-1,2,3,4-tetrazine-1,3-dioxide. *Chin. J. Energetic Mater.* 20: 693–696.

119 Lempert, D.B., Nechiporenko, G.N., and Soglasnova, S.I. (2004). Dependence of the specific impulse composite propellants containing oxidizing agents based on C, N, and O atoms. *Russ. J. Phys. Chem. B* 23: 75–81.

120 Pepekin, V.I., Matyushin, Y.N., and Gubina, T.V. (2011). Enthalpy of formation and explosive properties of 5,6-(3,4-furazano)-1,2,3,4-tetrazine-1,3-dioxide. *Russ. J. Phys. Chem. B* 5: 97–100.

121 Frem, D. (2018). The specific impulse as an important parameter for predicting chemical high explosives performance. *Z. Anorg. Allg. Chem.* 644: 235–240.

122 Nedel'ko, V.V., Zakharov, V.V., Korsunskii, B.L. et al. (2013). Thermal decomposition of [1,2,5]oxadiazolo[3,4-*e*][1,2,3,4]-tetrazine-4,6-di-*N*-oxide. *Russ. J. Phys. Chem. B* 7: 113–117.

123 Murav'ev, N.V., Monogarov, K.A., Zarko, V.E. et al. (2019). Kinetic parameters of thermal decomposition of furazano-1,2,3,4-tetrazine-1,3-dioxide and a binary solution based on it. *Combust. Explos. Shock Waves* 55: 629–631.

124 Teselkin, V.A. (2009). Mechanical sensitivity of furazano-1,2,3,4-tetrazine-1, 3-dioxide. *Combust. Explos. Shock Waves* 45: 632–633.

125 Oleshko, V.I., Zarko, V.E., Lysyk, V.V. et al. (2015). Explosion initiation of furazanotetrazine dioxide by a high-current electron beam. *Tech. Phys. Lett.* 41: 519–521.

126 Kvasov, A.A. (2010). Study of physical and chemical characteristics of condensed systems on the basis of furasanotetrazine dioxide and peculiarities of their laser initiation. *Gorenie i Plazmokhimiya* 8: 162–168.

127 Shastin, A.V. and Lempert, D.B. (2014). Energy potential of some triazine derivatives. *Russ. J. Phys. Chem. B* 8: 716–719.

128 Zelenov, V.P., Kalmykov, P.I., and Kalashnikov, A.I. (2013). Principles of interaction of furazano-1,2,3,4-tetrazine-1,3-dioxide with organic solvents. *Russ. J. Phys. Chem.* 87: 767–771.

129 Zelenov, V.P., Baraboshkin, N.M., Khakimov, D.V. et al. (2020). Time for quartet: the stable 3 : 1 cocrystal formulation of FTDO and BTF – a high-energy-density material. *CrystEngComm* 22: 4823–4832.

130 Simonenko, V.N., Kalmykov, P.I., Kiskin, A.B. et al. (2014). Study of combustion of model compositions based on furazano tetrazine dioxide and dinitro diazo pentane. Part I. Binary systems. *Combust. Explos. Shock Waves* 50: 68–77.

131 Zarko, V.E., Simonenko, V.N., Kalmykov, P.I. et al. (2009). Laser initiation of crystallized mixture of furazanotetrazine dioxide and dinitrodiazapentane. *Combust. Explos. Shock Waves* 45: 131–134.

132 Kalmykov, P.I., Zarko, V.E., Sidel'nikov, A.A. et al. (2011). Specific features of the crystal and phase structure of binary systems 5,6-(3′,4′-furazano)-1,2,3,4-tetrazine-1,3-dioxide-2,4-dinitro-2,4-diazapentane. *Russ. J. Appl. Chem.* 84: 248–255.

133 Zharkov, A.S., Chukanov, N.V., Kalmykov, P.I. et al. (2014). Phase equilibria and structural phase transformations in the furazano[3,4-*e*]tetrazine-4,6-dioxide-2,4-dinitro-2,4-diazapentane system. *Russ. Chem. Bull. Int. Ed.* 63: 1785–1800.

134 Kalmykov, P.I., Zakharov, V.V., Chukanov, N.V. et al. (2015). Crystallization and phase homogeneity of [1,2,5]oxadiazolo[3,4-*e*][1,2,3,4]tetrazine-4,6-dioxide–2, 4-dinitro2,4-diazapentane molecular complex. *Russ. J. Phys. Chem. B* 9: 714–720.

135 Zakharov, V.V., Kalmykov, P.I., Chukanov, N.V., and Korsounskii, B.L. (2015). Thermal transformations of a molecular complex of [1,2,5]oxadiazolo[3,4-*e*][1,2,3,4]tetrazine-4,6-di-*N*-oxide with 2,4-dinitro-2,4-diazopentane. *Russ. J. Phys. Chem. B* 9: 357–362.

6

Recent Advances in Chemistry of Nitrogen-Rich Energetic Polymers and Plasticizers

Michael Gozin[1,2,3] *and Leonid L. Fershtat*[4]

[1] *Tel Aviv University, School of Chemistry, Faculty of Exact Sciences, Tel Aviv 6997801, Israel*
[2] *Tel Aviv University, Center for Nanoscience and Nanotechnology, Tel Aviv 6997801, Israel*
[3] *Tel Aviv University, Center for Advanced Combustion Science, Tel Aviv 6997801, Israel*
[4] *N.D. Zelinsky Institute of Organic Chemistry, Russian Academy of Sciences, Laboratory of Nitrogen Compounds, Moscow 119991, Russian Federation*

6.1 Introduction

While much of this book has focused on new energetic molecules to be used as the main energetic fill in an energetic composition, overall performance is not solely dictated by this main energetic component. Energetic formulations often contain inert binder and plasticizer materials, which ultimately decrease performance compared to the pure energetic material, while they remain vital for the processing, handling, and loading of the energetic into the energetic system. For example, polymer-bonded explosive PBXN-107, used in the BGM-109 Tomahawk cruise missile, contains 86% of 1,3,5-trinitroperhydro-1,3,5-triazine (RDX), bonded with a polyacrylate binder. This inert binder offers an opportunity for replacement by an energetic binder, which has the capability to increase the performance of the overall energetic system. Energetic plasticizers and binders have been an area of a significant focus in the energetic materials community. In this chapter, we present a review of advancements in the field of energetic polymers and plasticizers since 2010, categorized by their energetic functional groups, such as high-nitrogen heterocycles, organic nitro, nitrato, and nitramine compounds, organic azides and their combinations.

6.2 Heterocyclic Energetic Polymers and Plasticizers

A number of energetic polymeric ionic liquids incorporating the imidazole subunit were recently synthesized through sequential transformations of the commercially available 1-vinyl-imidazole **1**. In particular, reaction of 1-vinyl-imidazole **1** with HNO_3, followed by radical polymerization with 2,2′-azobis(2-methylpropionitrile)

Nitrogen-Rich Energetic Materials, First Edition. Edited by Michael Gozin and Leonid L. Fershtat.

(AIBN), afforded poly(vinyl-imidazolium nitrate) **2** energetic poly(ionic liquid) (E-PIL) (Scheme 6.1) [1].

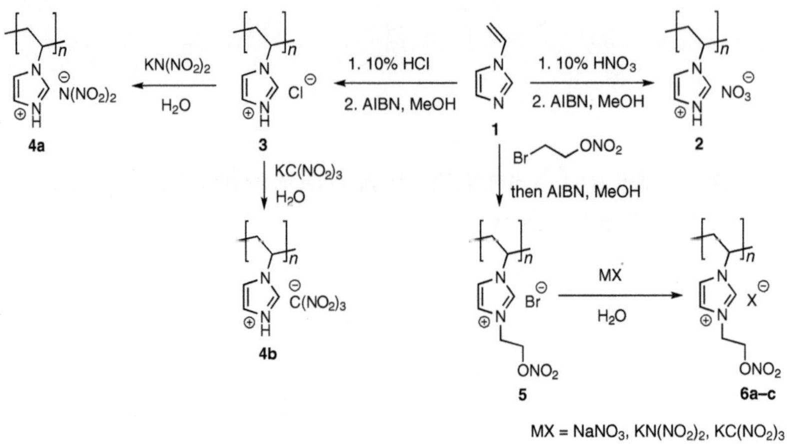

Scheme 6.1 Synthesis of imidazolium-based poly(ionic liquids).

Analogous procedure involving HCl instead of HNO_3 resulted in a formation of poly(vinyl-imidazolium chloride) **3**. Metathesis of the chloride anion in polymer **3** with potassium dinitramide or nitroformate provided E-PILs **4a,b** with corresponding explosophoric anions. Another route to such polyionic materials involved a preliminary alkylation of 1-vinyl-imidazole **1** with 2-bromoethyl nitrate, with subsequent radical polymerization to form poly(3-[2-(nitro-oxyethyl)]-1-vinyl-imidazolium bromide) **5**. Anion exchange with alkali metal salts provided a direct access to a series of highly energetic imidazolium-based poly(ionic liquids) **6a–c** (Scheme 6.1). Synthesized energetic polymers **6a–c** have high nitrogen-oxygen content (57–68%) and were calculated to exhibit good velocities of detonation (VODs: 7000–8000 m s^{-1}), exceeding the VODs of poly(glycidyl azide) (GAP) (5040 m s^{-1}) and poly(glycidyl nitrate) (6750 m s^{-1}). However, among studied materials, only poly(vinyl-imidazolium nitrate) **6a** showed an acceptable thermal stability, with an onset decomposition temperature (T_{decomp}) of 200 °C, while other poly(ionic liquids) were found to be less thermally stable, limiting their practical application.

Energetic mono- and bis(vinyl-imidazolium) perchlorates **7a** and **7b–d** were synthesized in high yields *via* a two-step alkylation/metathesis procedure, starting from 1-vinyl-imidazole **1** (Scheme 6.2) [2]. Obtained monomers were photo-polymerized, using phenyl-bis(2,4,6-trimethylbenzoyl) phosphine oxide as a photo-initiator. Prepared energetic polymers exhibited excellent thermal stability above 340 °C. In addition, it was demonstrated that the synthesized vinyl-imidazole monomers **7a–d** were able to be used as substrates for the 3D printing of energetic polymers with very low post-production shrinkage. The resulting polymeric materials were found to be compatible with RDX, ammonium perchlorate (AP), and ammonium dinitramide (ADN) fillers, and exhibited low sensitivity to mechanical impact, showing their excellent application potential to be used in various energetic formulations.

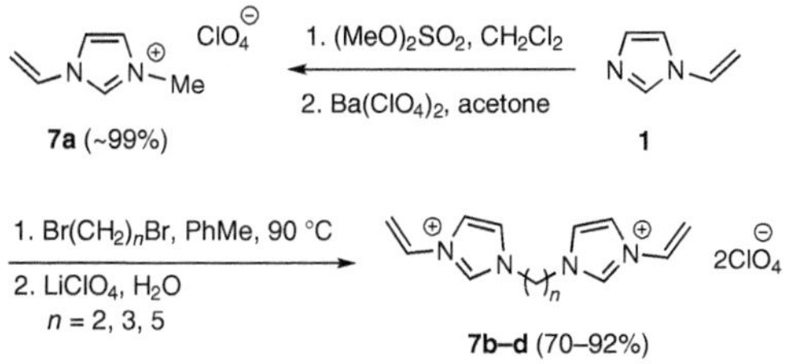

Scheme 6.2 Synthesis of vinyl-imidazolium perchlorate monomers.

A series of energetic heterocyclic polymers incorporating a 1,2,3-triazole subunit was synthesized through [3+2] cycloaddition of azide-functionalized polymers with corresponding acetylenes. An energetic nitro-ester plasticizer **10** was used as a dipolarophile in Cu-free click reaction with azide-functionalized polymer **11** to afford hybrid copolymer **12**, incorporating both 1,2,3-triazole moieties and azide groups (Scheme 6.3) [3]. Thermal degradation of copolymer **12** occurred in three steps, starting from 200 °C, while the T_g of alkyne-terminated polymer (−66.5 °C) and polymer grafted with GAP (−50.7 °C) were both lower than the T_g of the pure GAP (−47.8 °C). Analogous reaction of GAP with trifunctional alkynyl-terminated poly(ethylene-oxide-tetrahydrofuran) afforded corresponding poly(triazole-polyethylene-oxide-tetrahydro-furan) (PTPET) copolymer with an increased thermal stability (T_{decomp} of 240 °C). Thermal analysis of the latter copolymer showed that it has two mass-loss peaks related to the decomposition of azide groups and the main chain [4].

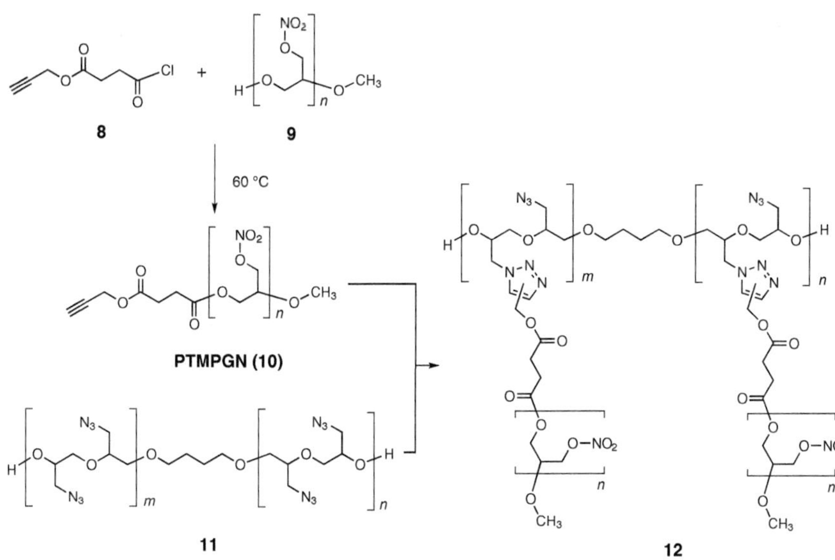

Scheme 6.3 Preparation of hybrid nitro-ester-triazole GAP-*g*-PGN copolymer **12**.

Bis(propargyl)hydroquinone (BPHQ) **13** was also used as an isocyanate-free curative for GAP to form a triazole linked polymeric network **14** (Scheme 6.4). This process was successfully performed under neat conditions and did not require any additional catalysts. BPHQ **13** has several advantages as a curation agent over other crosslinkers, including its good compatibility with oxidizers, such as ADN and hydrazinium nitroformate (HNF). In addition, curation with BPHQ **13** is not affected by moisture and the formed poly(triazole) **14** exhibited high T_{decomp} of above 240 °C [5].

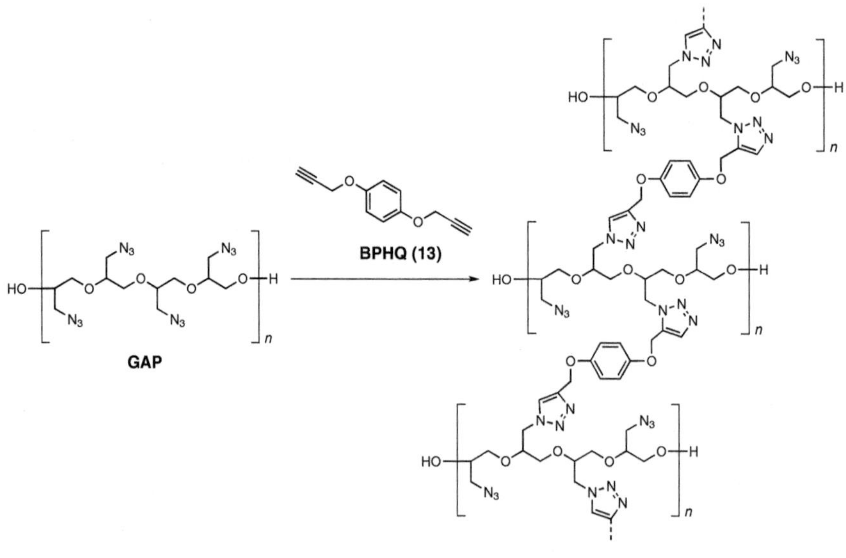

Scheme 6.4 Synthesis of hydroquinone-linked bis(triazole-azide) copolymer **14**.

The performance of solid propellants could be tailored by incorporating energetic binders and ballistic modifiers. Metal oxide nanoparticles are known for their capability to modify burning rates of AP based composite propellants. Nano iron oxide and poly(1-(3-nitrophenyl)-1*H*-1,2,3-triazol-4-yl) acrylate are incorporated in the composite propellant by partly replacing coarse AP, as well as hydroxyl-terminated poly(butadiene) (HTPB), respectively. Analysis of the mechanical properties of new composites revealed that by increasing the percentage of polymer in the propellant composition (by partially replacing HTPB), an increase in tensile strength with decrease of elongation percentage took place. The data of ballistic properties showed that incorporation of nano iron oxide in the propellant enhances the burning rate, while on partial replacement of HTPB with poly(1-(3-nitrophenyl)-1*H*-1,2,3-triazol-4-yl) acrylate led to a decrease in burning rate [6].

The elastomeric properties of the GAPs are normally provided by a urethane-based reticulation process, involving isocyanate compounds. Azides containing polymers can be also cured with dialkynes by 1,2,3-triazole formation, called a "click" reaction. The latter reaction kinetics of low-molecular-weight GAP with five different dialkynes allowed to determine the reactivity parameters of these dialkynes and to predict the thermal behavior of such reaction mixtures, which is a critical information for a safe control of exothermic curing process of GAP (Scheme 6.5) [7, 8].

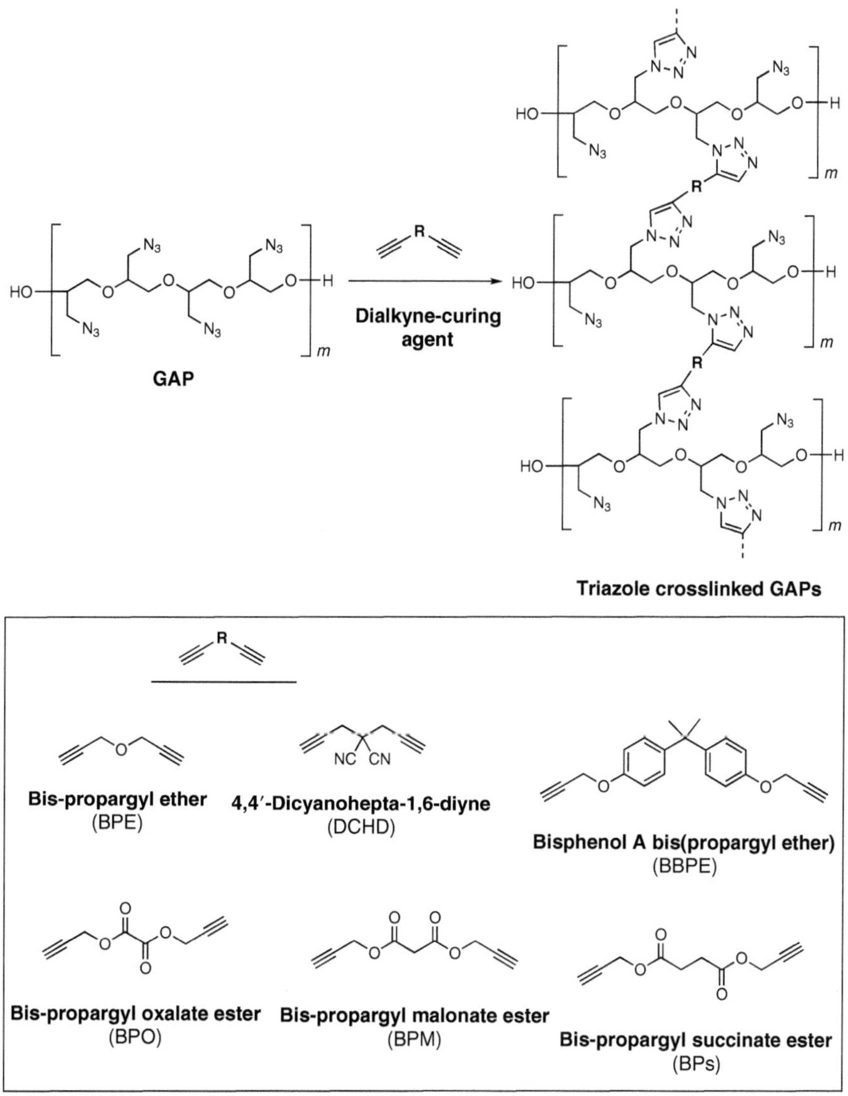

Scheme 6.5 Preparation of various dialkyne-cured GAPs.

Although ADN holds a great promise as a green replacement of the toxic AP, ADN's compatibility with many polymer binder systems has hindered the development of its formulations. Therefore, a polymer cure system based on hyperbranched copolymers of 3-ethyl-3-(hydroxymethyl)oxetane (TMPO) and tetrahydro-furan

(THF) was prepared and evaluated [9]. Some of these materials showed excellent compatibility with ADN, and rapid curing with GAP was demonstrated through 1,3-dipolar cycloaddition (Scheme 6.6).

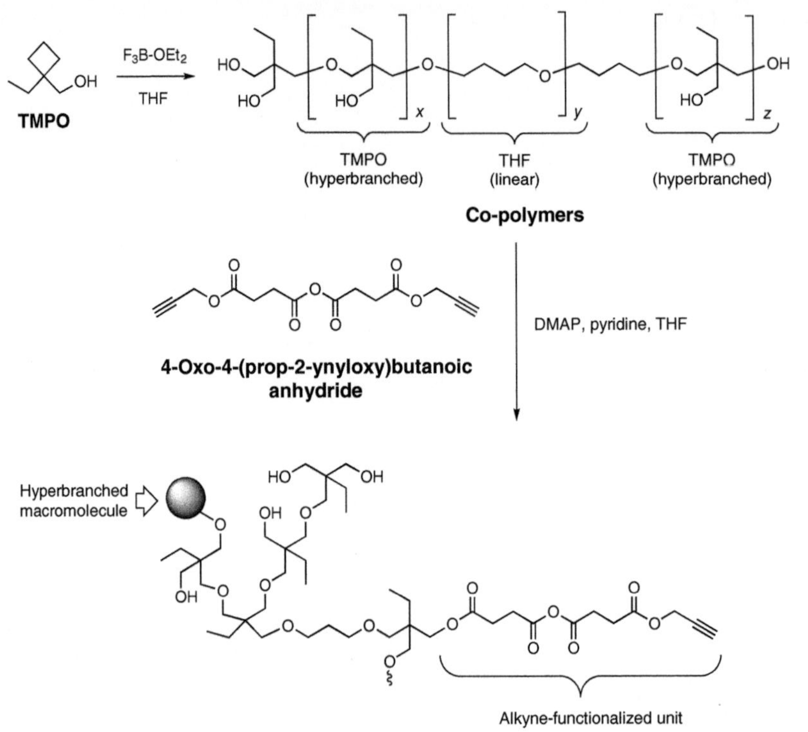

Scheme 6.6 Preparation of hyperbranched poly(TMPO)-*co*-poly(THF) copolymers *via* cationic ring-opening polymerization. Some of the branches were functionalized with alkynes. In the scheme, branching points were omitted for clarity (a majority of the TMPO units are dendritic).

Nucleophilic substitution of the chloride substituent in a copolymer **15**, comprised poly(vinyl chloride) and poly(vinyl-tetrazole) building blocks, with nitro-1,2,3-triazolate anion, afforded high-nitrogen energetic polymer **16**, which could not be obtained by a direct polymerization (Scheme 6.7, *left*) [10]. Poly(vinyl-tetrazole) **17** could be further functionalized with various acyclic and heterocyclic functional groups *via* an alkylation with corresponding alkyl chlorides (Scheme 6.7, *right*). In the latter cases, the degree of alkylation in the resulting polymers **18** did not exceed 85% [11].

Similar alkylation/acylation procedures were also employed for the preparation of a variety of energetic heterocyclic polymers incorporating tetrazole, 1,2,3- and 1,2,4-triazole subunits. Glycidyl triazolone, glycidyl nitrotriazolone polymers, and glycidyl triazolone-GAP copolymers were synthesized by nucleophilic substitution of poly(epichlorohydrin) (PECH) with 1,2,4-triazol-5-one and 3-nitro-1,2,4-triazol-5-one heterocycles or azide anions (Scheme 6.8) [12]. Inserted

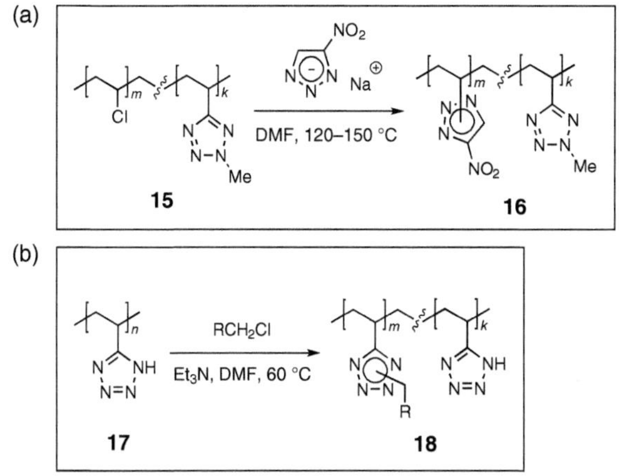

R = $N(NO_2)Me$, 5-azido-1,2,4-triazolyl, 5-nitro-1,2,4-triazolyl, 4-nitro-1,2,3-triazolyl

Scheme 6.7 Synthesis of substituted triazole-tetrazole- (a) and bitetrazole- (b) containing energetic polymers.

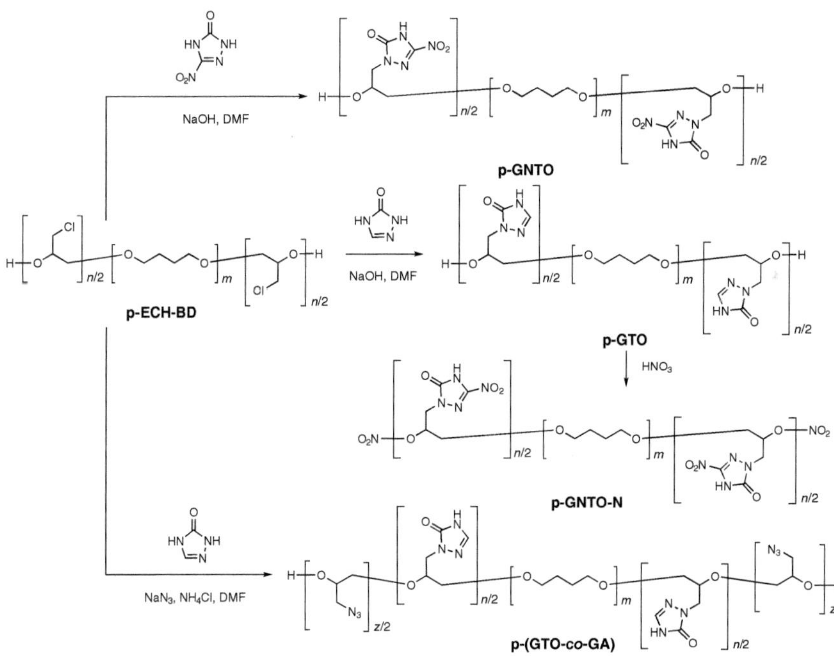

Scheme 6.8 Synthesis of GAP grafted with 1,2,4-triazol-5-one, 3-nitro-1,2,4-triazol-5-one functional groups.

triazolone heterocycles into the polymeric chain provided the resulting material with higher thermal stability and a much higher density in comparison to GAP homopolymer. The 1,3-dipolar cycloaddition demonstrated that spatially branched energy-rich polymeric binders bearing 1,2,4-triazol-5-one and 1,2,3-triazole heterocycles could be synthesized starting from *p*-(GTO-*co*-GA) azide copolymers.

Energetic monomers incorporating a tetrazole ring linked to a C≡C bond were prepared through a sequential transformation of vinyl-tetrazole **19** (Scheme 6.9) [13]. Bromination of the substrate **19** afforded dibromo derivative **20A**, which underwent regioselective alkylation at the N(2) position of the tetrazole ring followed by double dehydrobromination resulting in a formation of the C≡C bond. 5-Ethynyl-2*H*-tetrazole **20B** can serve as a valuable cross-coupling substrate, which was demonstrated by its dimerization to the corresponding butadiyne **21**. Copper nanocomposite (Cu-Pol) that was obtained *via* thermolysis of Cu(II) poly(5-vinyl-tetrazolate) (PVT) showed very good catalytic performance in this homocoupling reaction.

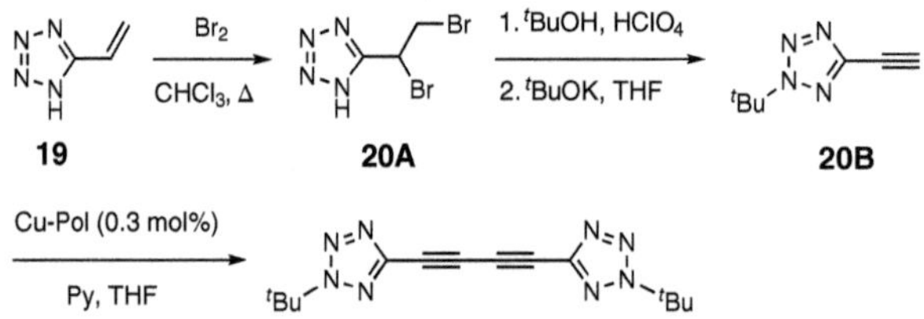

Scheme 6.9 Synthesis of tetrazole-based monomers.

Free radical copolymerization of 4-(chloromethyl)styrene with acrylonitrile afforded poly([*p*-(chloro-methyl)-styrene]-*co*-acrylonitrile) (PCSA), which was reacted with NaN_3 under microwave irradiation conditions. This protocol resulted in an azidation of PCSA and simultaneous [3+2] cycloaddition of the nitrile moiety in PCSA with the azide anion, providing an efficient route to the valuable energetic polymer – poly([*p*-(azidomethyl)styrene]-*co*-(5-vinyltetrazole)) (PAST) (Scheme 6.10) [14]. The yield of the final step was found to be dependent on the irradiation power. The maximum yield (96%) was obtained at 800 W, while analogous procedure at 200 W resulted in a formation of PAST in a lower yield (83%). Both synthesized copolymers demonstrated excellent thermal stability. Exothermic

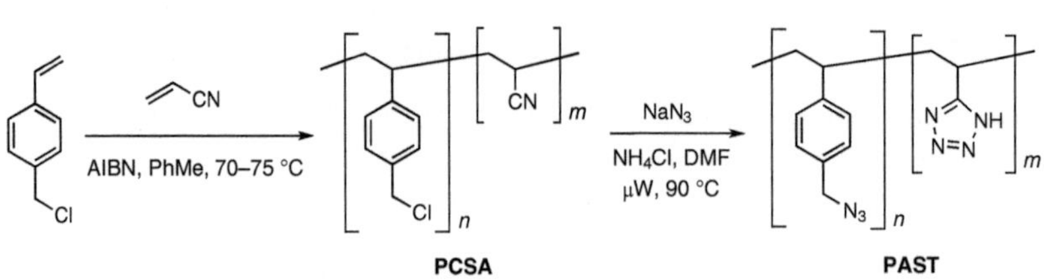

Scheme 6.10 Preparation of PAST copolymer.

degradation of PAST started at 200 °C, reaching maximum level at 253 °C, while PCSA was found to be thermally stable up to 280 °C.

Analogous cycloaddition of the nitrile moiety in poly(acrylonitrile) with azide anion afforded formation of poly(vinyl-tetrazole) (PVT) homopolymer (Scheme 6.11, *left*) [15]. PVT is not only a known energetic polymer, but it was also shown to exhibit excellent adsorption properties of heavy metal ions.

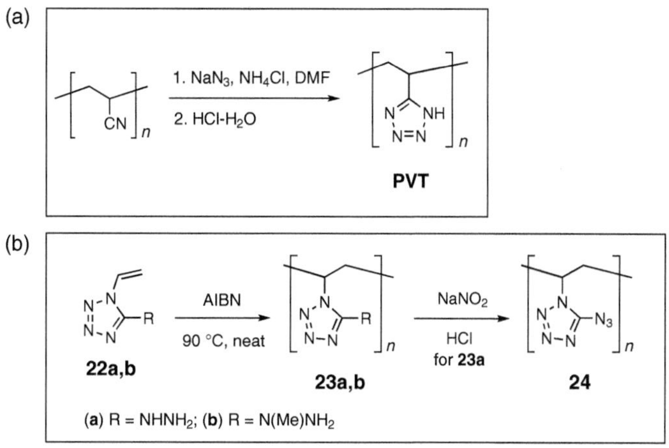

Scheme 6.11 (a) Synthesis of poly(vinyl-tetrazole) (PVT) homopolymer from poly(acrylonitrile); (b) synthesis of poly(vinyl-azido-tetrazoles).

Free radical polymerization of vinyl-tetrazoles **22a,b** bearing hydrazine functionalities afforded corresponding poly(vinyl-tetrazoles) **23a,b** (Scheme 6.11, *right*) [16]. Interestingly, polymer **23a**, bearing monosubstituted hydrazine functionality, was obtained in a moderate yield of 19%, while its *N*-methylated analogue **23b** was prepared in a much higher yield of 87%. In addition, polymer **23b** contained longer chain ($n = 100$) than the polymer **23a** ($n = 10$). These effects can be attributed to the NH proton of the hydrazine functionality in vinyl-tetrazole **22a**, which can act more efficiently as a radical interceptor. To increase the nitrogen content, polymer **23a** was subjected to nitrosation resulting in a formation of the corresponding azido-tetrazole polymer **24**. Energetic polymers **23a,b** and **24** have high-nitrogen content (50–60%) and showed good thermal stability (174–276 °C). Moreover, poly(vinyl-tetrazoles) **23a,b** exhibited much lower sensitivity to impact (35–45 J) in comparison to GAP (7 J). Aforementioned properties indicate that these polymers have a high potential for application in new gas-generating compositions.

Acylation of 5-aminotetrazole **25** with acryloyl or methacryloyl chlorides, followed by radical polymerization of formed intermediates, led to formation of poly(carbamoyltetrazoles) **26a,b**, which were further nitrated to the corresponding nitramides **27a,b** (Scheme 6.12) [17]. All synthesized polymers **26a,b** and **27a,b** demonstrated good thermal stability (>200 °C) and were completely insensitive to impact and friction. Moreover, poly(nitramidotetrazoles) **27a,b** were calculated to have detonation velocities (5450–5600 m s^{-1}), which are higher in comparison to GAP (5050 m s^{-1}), making them suitable as environment-friendly explosives or gas generators.

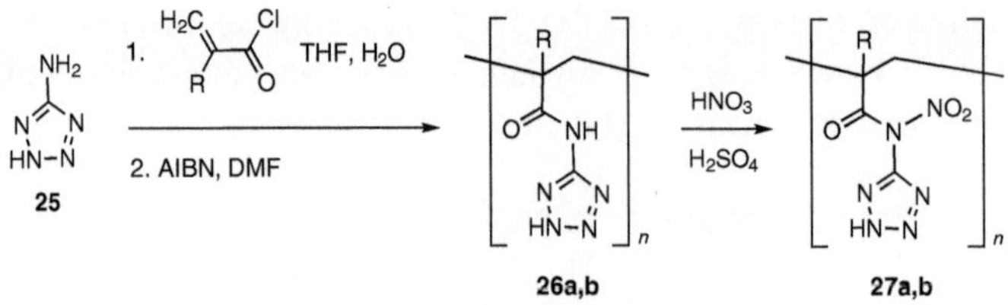

Scheme 6.12 Synthesis of poly(nitramidotetrazoles).

Michael addition of acrylonitrile to hydroxy groups of cellulose, followed by [3+2] cycloaddition of nitrile moieties with NaN_3, afforded tetrazolyl-ethyl cellulose ether **28**, which was further methylated using dimethyl sulfate (Scheme 6.13) [18, 19]. An incorporation of the *N*-methyl-tetrazole functional group instead of the NH-tetrazole moiety resulted in a loss of electrolytic properties and enhancement of organophilic properties, which led to the appearance of film-formation ability. This feature enables potential usefulness of *N*-methylated tetrazolyl-ethyl cellulose ether **29** as an effective binder for condensed energetic compositions.

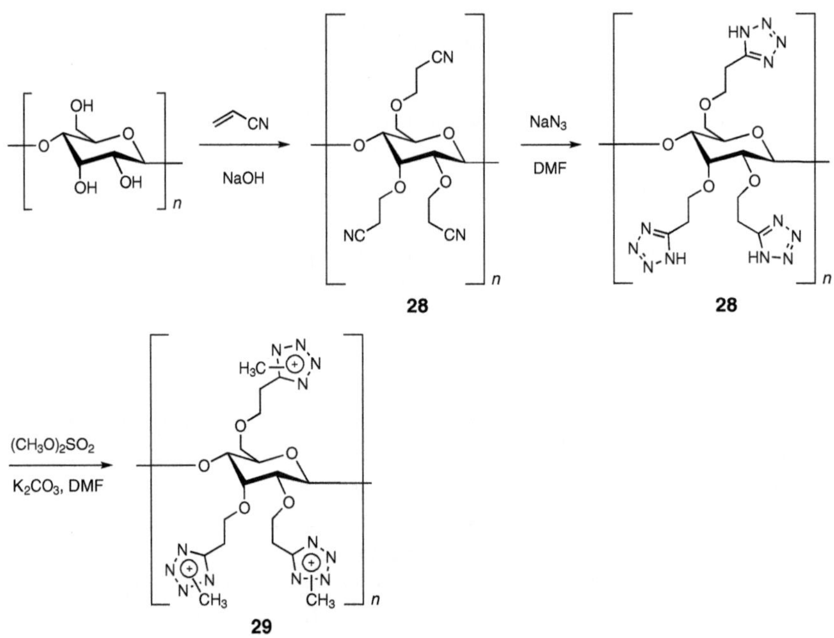

Scheme 6.13 Preparation of tetrazole-functionalized cellulose.

Tetrazole-acetate-functionalized nitrocellulose derivatives were found to be promising nitrogen-rich energetic polymers. These materials were found to be more thermally stable and were calculated to possess higher detonation properties than nitrocellulose. An incorporation of the tetrazole-acetate functional moiety resulted in a significant decrease of mechanical sensitivity of these energetic polymers, making them as potential candidates for replacement of nitrocellulose in high-energy formulations [20].

An interaction of bis(5-monomethylhydrazinyl-1*H*-tetrazolyl)alkanes **30a–c** with hexamethylene diiso-cyanate resulted in a formation of 1,1-methyltetrazolylhydrazine containing polymers **31a–c** (Scheme 6.14) [21]. All prepared macromolecular compounds exhibited high thermal stability around 240 °C and were completely insensitive to mechanical *stimuli*. Detonation velocities of the synthesized polymers were comparable to GAP and nitrocellulose.

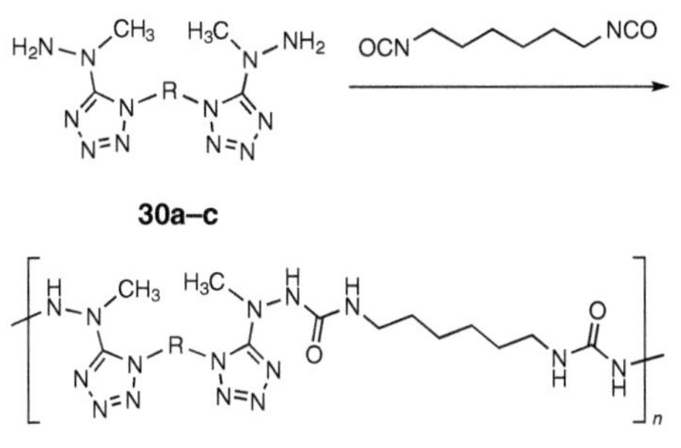

30a–c

31a–c

R = $(CH_2)_2$, $(CH_2)_4$, CH(Me)CH_2

Scheme 6.14 Synthesis of 1,1-methyltetrazolylhydrazine-containing polymers.

An incorporation of six-membered nitrogen heterocycles in structures of energetic polymers was less explored in recent years. Nevertheless, a series of 1,3,5-triazine-functionalized HTPBs **32a–c** were prepared from HTPB (Scheme 6.15) [22, 23]. An interaction of modified HTPBs **32a–c** with various diisocyanates was followed by treatment with dibutyltin dilaurate (DBTDL) and afforded target poly(urethanes) which were additionally cured at 70 °C for five days [24]. Obtained polymers were shown to form compact mass fractals of hard segment domains which resulted in high tensile strength. In addition, these polymers were found to be highly thermally stable with an onset decomposition temperature above 280 °C.

Reactions of 3,6-dihydrazinyl-1,2,4,5-tetrazine **33** with aliphatic and aromatic diisocyanates provided a series of tetrazine-based energetic polymers **34**, **35** (Scheme 6.16) [25]. Synthesized macromolecular materials have high nitrogen content (>40%) and thermal stability above 200 °C. However, detonation velocities of polymers **34** and **35** were lower than those of GAP and poly(glycidyl nitrate).

A three-step synthetic route for efficient preparation of 3,3-bis(azidomethyl) oxetane (BAMO), which is a monomer of an important energetic binder poly(bis (azidomethyl)oxetane) (poly(BAMO)) [26], was developed (Scheme 6.17) [27]. This methodology included initial tosylation of pentaerythritol to yield the tritosyl derivative **36**, which subsequently underwent base-induced intramolecular

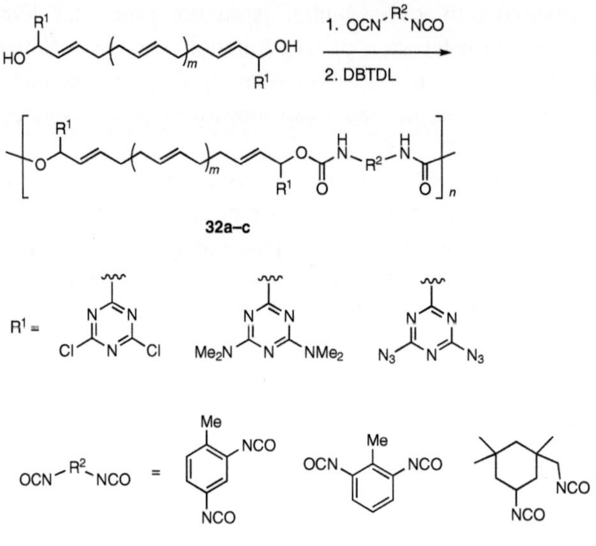

Scheme 6.15 Synthesis of 1,3,5-triazine-functionalized energetic polymers.

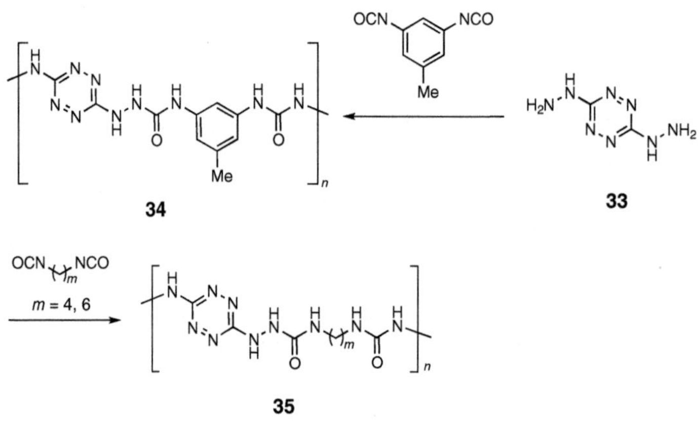

Scheme 6.16 Synthesis of tetrazine-containing polymers.

cyclization into 3,3-bis(tosyl-methyl)oxetane (BTMO). The final step was based on nucleophilic substitution of both tosyloxy functional groups with azide anion, which quantitatively afforded the target BAMO monomer.

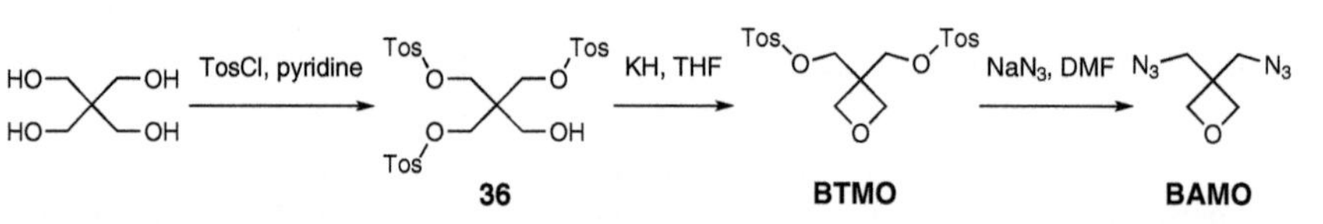

Scheme 6.17 Synthesis of BAMO.

6.3 Nitrogen-Rich Energetic Polymers Lacking Traditional Explosophoric Groups

While this review has focused on energetic polymers with traditional explosophores including nitrato, azido, and azoles, any material with high nitrogen content has the possibility of being energetic. A series of novel energetic nanomaterials comprising a triaminoguanidine-glyoxal polymer TAGP was synthesized using a polycondensation approach (Scheme 6.18) [28]. The prepared polymeric material has a layered 2D network-type structure formed by repeating 42-membered macrocyclic rings. TAGP was found to form stable complexes TAGP-M with various transition metal ions (M = Co, Ni, Cu, and Ag). All prepared energetic materials were found to be almost insensitive to mechanical *stimuli* and possessed higher nitrogen content than currently used energetic polymers, such as GAP and poly(glycidyl nitrate).

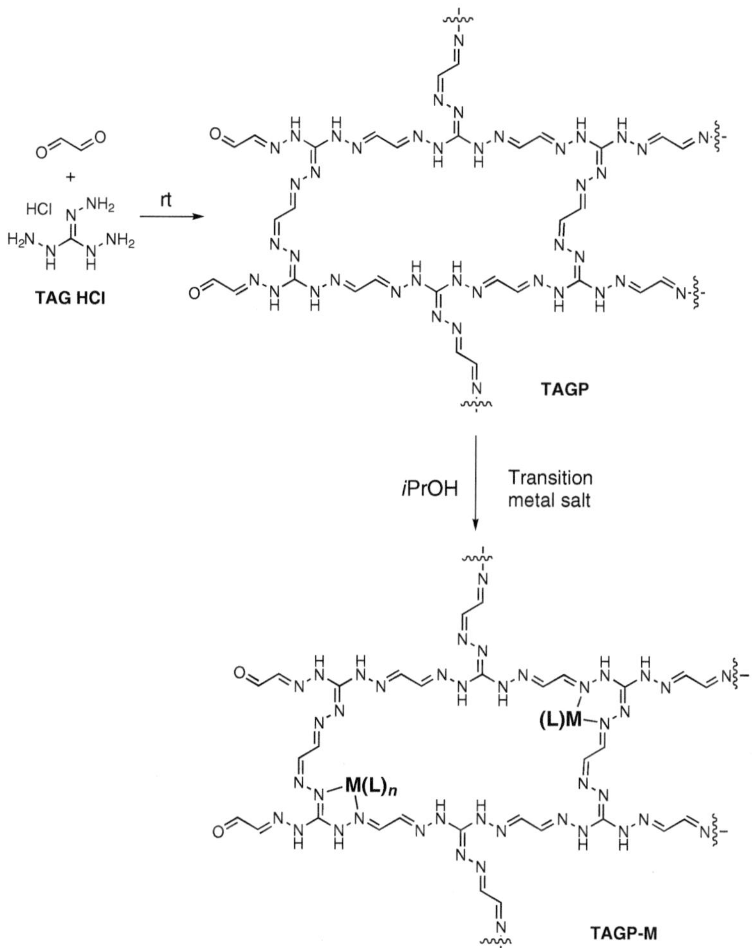

Scheme 6.18 Synthesis of TAGP polymer and corresponding TAGP-M complexes.

Synthesized compounds were calculated to exhibit a good detonation performance, exceeding that of GAP, and enabling their potential use in diverse civil and defense applications as new-generation energetic binders and combustion catalysts.

6.4 Azido-Rich Energetic Polymers and Plasticizers

During the last decade, a significant progress was made in chemistry of azide-containing energetic polymers and plasticizers, including copolymers of GAP [29–33]. Synthesis of energetic block copolymer of poly(glycidylazide-*b*-poly (azidoethylmethacrylate)) (GAP-*b*-PAEMA) was achieved by use of PECH-Br macro-initiator, which was obtained by reacting 2-bromoisobutyryl bromide with hydroxyl-terminated poly(epichlorohydrin) (PECH-OH). The atom-transfer radical polymerization (ATRP) of chloroethyl methacrylate was followed by azidation to complete the synthesis of block copolymer GAP-*b*-PAEMA (Scheme 6.19) [34].

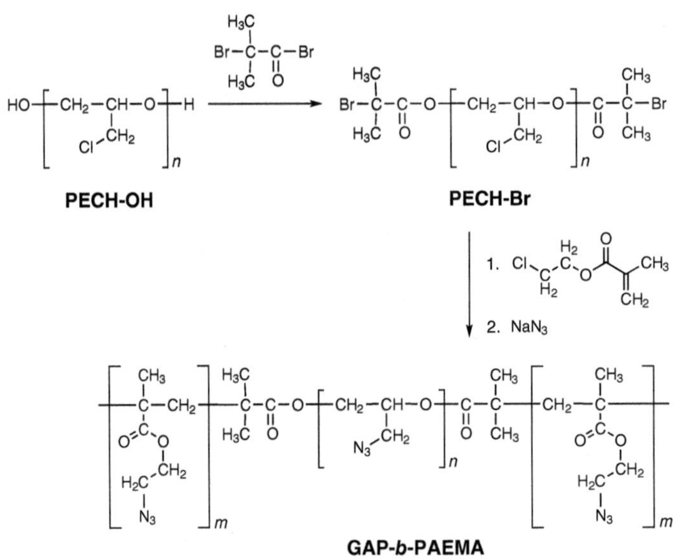

Scheme 6.19 Synthesis of GAP-*b*-PAEMA.

Gel permeation chromatography (GPC) of the resulting GAP-*b*-PAEMA copolymer showed an increase in the molecular weight over starting PECH-Br polymer, which indicated a formation of block copolymer, and FTIR spectroscopy allowed straightforward detection of substitution of chlorides with azide functional groups. Thermogravimetric (TG) analysis of GAP-*b*-PAEMA showed two-step degradation process that included fast decomposition of the azide groups, followed by progressive scission of the polymer backbone. Differential scanning calorimetry (DSC) analysis of GAP-*b*-PAEMA revealed two T_gs at −18 and +36 °C for this copolymer, suggesting that it is a thermoplastic elastomer.

In attempt to produce a new energetic binder for solid rocket propellants, a random copolymer containing glycidyl azide (GA) and BAMO monomeric units was synthesized *via* initial preparation of halogen precursor copolymer epichlorohydrin/3,3-bis(bromomethyl-oxetane) (ECH/BBrMO) that was further converted into an energetic polymer *via* azidation reaction with NaN_3 [35]. The introduction of BAMO units into the GAP chain has the advantage of an increased number of azide groups and consequently the energetic content of the resulting material. However, the introduction of the BAMO units should be done to limited extent to preserve the amorphous properties of the final copolymer.

A series of four energetic copolymers, poly(vinyl-*p*-nitrobenzal-acetal)-g-poly(glycidylazides) (PVPNB-g-GAPs), was prepared by cross-linking of four different molecular weights GAPs with poly(vinyl-*p*-nitrobenzal-acetal), using toluene diisocyanate as the cross-linking agent (Scheme 6.20) [36, 37].

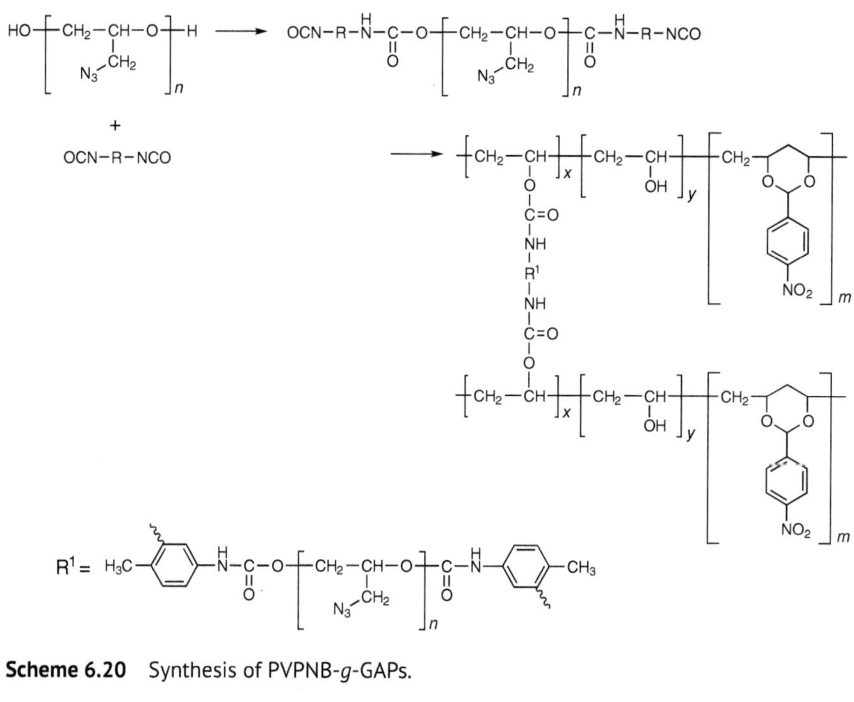

Scheme 6.20 Synthesis of PVPNB-*g*-GAPs.

The DTA and thermogravimetric analysis (TGA)–DTG thermograms of PVPNB-g-GAPs indicated that these copolymers have an excellent resistance to thermal degradation up to 200 °C, and begin to decompose gradually above 230 °C. The DSC results of PVPNB-g-2#GAP, PVPNB-g-3#GAP, and PVPNB-g-4#GAP showed two distinct T_gs, whereas PVPNB-g-1#GAP showed only one. Sensitivity and compatibility tests revealed that PVPNB-g-4#GAP is an insensitive material, which has good compatibility with TNT, HMX, and TATB explosives, but not with RDX. It was proposed that PVMNB-GAPs have a potential to be used as new energetic plasticizers for melt-castable explosives. Another energetic polymer that was evaluated for melt-castable explosives was polyether polyurethane (PU) azide elastomer, which was synthesized in a three-step process, starting by coupling of 2,6-dinitro-1,4-diphemethyol with ECH, followed by azidation of the resulting polyether, and completing the synthesis by cross-linking these soft segments with 4,4′-diphenylmethane diisocyanate (MDI), as hard segments (Scheme 6.21) [38].

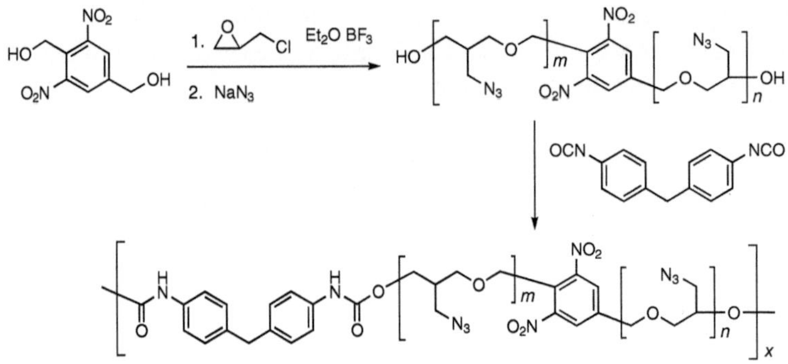

Scheme 6.21 Synthesis of polyether-polyurethane azide polymer.

By controlling the reaction conditions, polyether polyurethane azide elastomers could be prepared with average molecular weights (M_n) in the range of 8.68–15.6 × 10^3 and with polydispersity index (M_n/M_w) in the range of 1.3–1.73. Measurements of mechanical properties of melt-castable explosives containing RDX and 1.0% of polyether polyurethane azide elastomers showed that compressive loading, compressive strength, and compressive deformation were increased by 7.1%, 9.8%, and 22.5%, respectively, indicating that the latter elastomers could be good additives to improve the properties of RDX-based explosives. Polymer blends on a basis of GAP-containing energetic thermoplastic elastomers (ETPEs) and nitro-cellulose (NC), GAP-ETPE/NC, were prepared by solution-blending process, and their compositions were characterized by FTIR and wide X-ray diffraction (WXRD) (Scheme 6.22) [39].

The thermal and mechanical properties of these blends were investigated by DMA, shore durometer, and TGA, showing typical features of azido-containing polyether-based polyurethane and nitrocellulose. Yet, in comparison with NC, the thermostability of the GAP-ETPE/NC blends was found to be better. The degree of crystallinity and orientation in these blends was increased with increased NC content. Correspondingly, an increase in the strength and modulus of blends was observed. Significant improvements in elongation and low-temperature mechanical properties of polymer blends were found with the increase of the GAP-ETPE content. When the mass fraction between GAP-ETPE and NC decreases from 5/5 to 3/7, the tensile strength of GAP-ETPE/NC blends increased from 20.7 to 39.2 MPa, while breaking elongation decreased from 141% to 40.6%.

The triblock energetic copolymer poly(lactic-acid)-*block*-poly(glycidylazide)-*block*-polystyrene (PLA-*b*-GAP-*b*-PS) was synthesized by atom-transfer radical polymerization (ATRP) of styrene and ring-opening polymerization of D,L-lactide.

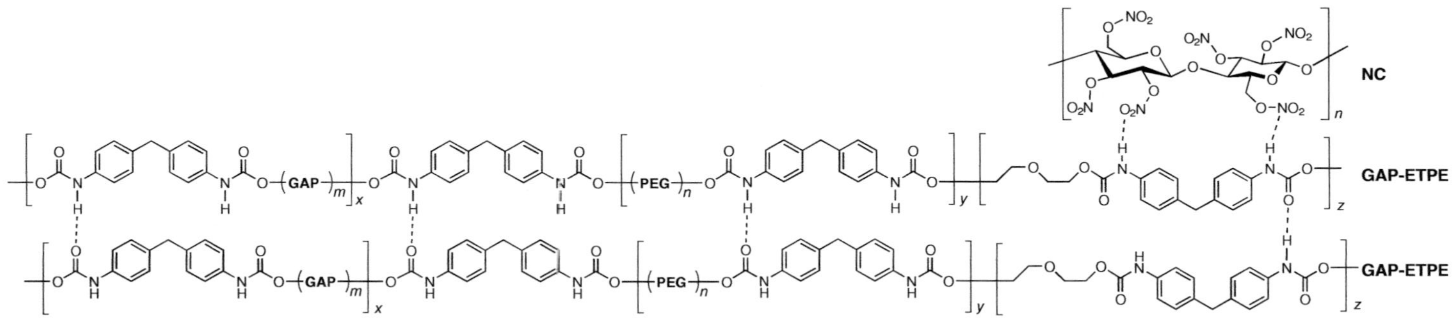

Scheme 6.22 Schematic structure of GAP-ETPE/NC polymer blends.

The energetic macro-initiator GAP-Br, made from reacting of GAP with α-bromoisobutyryl bromide, initially triggered the ATRP of styrene with its bromide group, and then the hydroxyl group on the GAP end of the resulting diblock copolymer participated in the polymerization of lactide in the presence of stannous octoate (Scheme 6.23) [40].

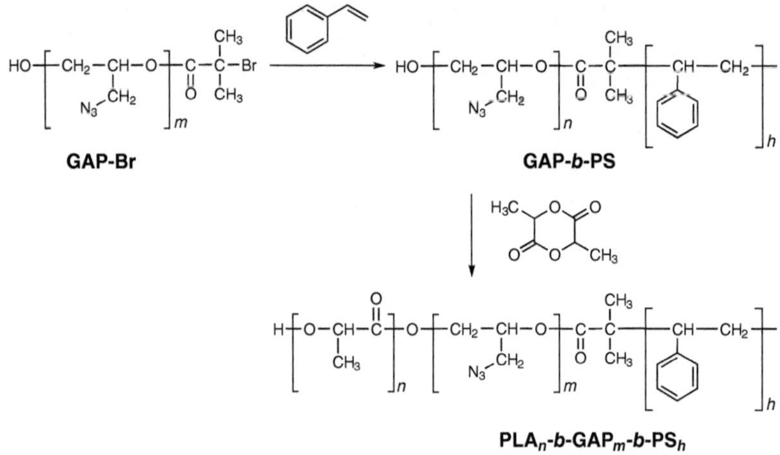

Scheme 6.23 Synthesis of PLA-*b*-GAP-*b*-PS.

This triblock copolymer had a narrow distribution of molecular weight, where the PS block was solvo-philic in toluene and improved the stability of the structure, the PLA block was solvophobic in toluene, serving as a sacrificial component for the preparation of porous materials, and GAP block was the basic and energetic material. The three blocks of the PLA-*b*-GAP-*b*-PS copolymer were fundamentally thermo-dynamically immiscible, which led to the self-assembly of the block copolymer in solution. Studies showed that the concentration and solubility of the copolymer and the polarity of the solvent significantly affected the morphology and size of the micelles generated from the self-assembly of PLA-*b*-GAP-*b*-PS. According to TEM results, the copolymer formed spherical aggregates in organic solvents, but generated irregular spherical aggregates when water was present as a co-solvent. When toluene was used as the solvent, the tri-block copolymer formed a vesicular structure with three distinct layers, with the diameter of the micelles increasing from 60 to 250 nm, as the concentration of triblock polymers in toluene increased from 5 to 15 mg ml^{-1}. The micelles prepared in 1 : 1 acetone/toluene were smaller than those from 1 : 1 DMF/toluene, which indicated that the presence of poor solvent could reduce the size of the micelles because of the excluded volume effect. The micelles generated in organic solvents at 10 mg ml^{-1} copolymer concentration were spherical, but became irregular, when water was used as a co-solvent.

An interesting molecular design of intrinsically plasticized copolymers, with decreased T_gs, was oriented to limit the use of external plasticizers for the formulation of GAP-based composite propellants. Based on this design, non-energetic *n*-alkyl substituted monomers were incorporated into the backbone of GAP, using cationic ring-opening polymerization of ECH and 1,2-epoxyhexane (EpH) with BF_3·THF, as a catalyst and completing the reaction by azidation of halogenated intermediate copolymer P(ECH-*co*-1,2-EpH) with NaN_3 in DMSO (Scheme 6.24) [41].

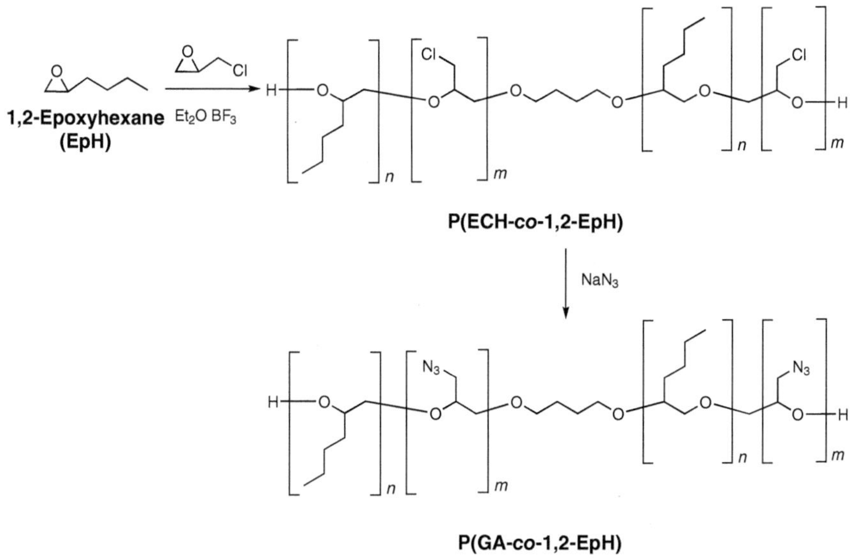

Scheme 6.24 Synthesis of P(GA-*co*-1,2-EpH) copolymer.

It was found that the introduction of pendant *n*-butyl chains by incorporation of EpH into the GAP backbone acted as an internal plasticizer, lowering the T_g of the resulting P(GA-*co*-1,2-EpH) copolymers in comparison to the commercial GAP. In a similar molecular weight range, P(GA-*co*-1,2-EpH) copolymers showed lower viscosity versus GAP, indicating the advantages of such copolymers as energetic binders for cast-cure applications. DSC and TGA studies of the P(GA-*co*-1,2-EpH) copolymers exhibited decomposition temperature of 248 °C, with the similar thermal behavior to GAP. By decreasing the nitrogen (azide) content in P(GA-*co*-1,2-EpH) versus GAP, the sensitivity toward friction and impact was reduced, giving P(GA-*co*-1,2-EpH) copolymers an advantage over GAP in fabrication of insensitive formulations.

PTPET is an energetic elastomer developed as a new binder for solid propellants. PTPET could be prepared by linking GAP with trifunctional alkynyl-terminated poly (ethyleneoxide-tetrahydrofuran) (ATPET) *via* click-chemistry between alkynyl and azide functional groups present in its building blocks [4, 42–44]. To explore properties and performance of PTPET, effects of energetic plasticizers A3, comprising a mixture of bis(2,2-dinitropropyl)-acetal (BDNPA) with bis(2,2-dinitropropyl)-formal (BDNPN), and *N*-butyl-*N*-(2-nitroxyethyl)nitramine (Bu-NENA), on the curing reaction between ATPET and GAP were investigated. In one study,

GAP with M_n of 4000 g mol^{-1} was used to replace GAP with M_n of 427 g mol^{-1} in the synthesis of PTPET with these plasticizers. A diffusion-ordered NMR spectroscopy (DOSY-NMR) technique was used to monitor the change in the diffusion coefficient of cross-linked PTPET. The change in the diffusion coefficient of PTPET with A3 plasticizer was found to be significantly higher than that of PTPET with Bu-NENA. Viscosity analysis also emphasized the difference between A3 and Bu-NENA plasticizers in the curing process of PTPET, showing that the curing curve of PTPET with A3 can be divided into two stages, while for the PTPET with Bu-NENA only a single stage could be observed. These methods, in conjunction with GPC analysis, revealed distinct effects of A3 and Bu-NENA on the curing process of PTPET. While XPS analysis showed that Bu-NENA has a little effect on the valence of copper catalyst, TG analysis indicated that Bu-NENA helps to improve the thermal stability of the catalyst. It was proposed that the formation of molecular cages between Bu-NENA and copper could be the reason for the above-mentioned differences. In another work, GAP-grafted tetrafunctional isocyanate (N100-g-GAP) was synthesized and used for the preparation of PTPET elastomer. TGA, DMA, FTIR, and mechanical properties analyses showed that such cross-linked materials can enhance mechanical strength of the resulting composite propellants and show a significant improvement in the interface interactions between the RDX and AP particles and the binder matrix.

GAP-based energetic thermoplastic polyurethane elastomers (ETPUs) were prepared by solution polymerization of GAP/poly(caprolactone) (PCL), as a soft segment, and methylene-bis(phenylisocyanate) (MDI) extended 1,5-pentane-diol, as a hard segment [45]. DSC and TG analyses showed that ETPU decomposition process could be divided into two stages, with different activation energies. The first main weight loss step was suggested to correspond to the elimination of N_2 due to the decomposition of azide functional groups in GAP segments, while the second weight loss step was attributed to the decomposition of the polymeric backbone. Additional energetic tri-block copolymer PCL–GAP–PCL (M_n = 1794) was synthesized by a ring-opening polymerization of ε-caprolactone with GAP (M_n = 1006 g mol^{-1}), as initiator, using DBTDL, as catalyst [46, 47]. Thermal analysis revealed that the resulting tri-block copolymer has greater thermal stability than GAP homopolymer. The T_g for PCL–GAP–PCL was influenced by the PCL blocks (T_g = −64 °C) and was found to be lower than that of the GAP homopolymer (T_g = −48 °C).

In addition to the linear-type azide-containing energetic polymers, hyperbranched poly(3-azidomethyl-3-hydroxymethyloxetane) (HBPAMHMO) polymers were synthesized and examined [48]. The results of the property studies of HBPAMHMO showed that the molecular weight of this material increased with the increasing ratio of r ([monomer]$_0$/[initiator]$_0$), while the degree of branching increased with the increase in polymerization reaction temperature. In comparison to linear-type azide-containing polymers, having similar structures, HBPAMHMO has lower T_g, lower sensitivity to mechanical impact and friction, higher combustion heat and better mechanical properties, suggesting a great potential of hyperbranched energetic polymers for broad range of applications. Using hyperbranched poly(3-ethyl-3-[hydroxymethyl]oxetane) core (PEHO-c) and linear poly (glycidylazide)-based shell (GAP-s), a series of multi-arm azide copolymers (POGs) were synthesized by sequential cationic ring-opening polymerization (CROP), followed by azidation reaction [49]. The molecular weight of POG having GAP-s and PEHO-c with a molar ratio of 14.95 : 1 ($R_{s/c}$) was found to be in the range of 31 000 g mol^{-1}, which is significantly higher than that of the linear GAP (in a range of 4000 g mol^{-1}). Thermal analyses of the obtained POGs showed that these materials have T_g in a range between −51 and −23 °C and exhibit thermostability up to 220 °C, whereas bomb calorimetry measurements demonstrated that the enthalpy of formation of POGs was higher than that of linear GAPs and comparable to branched GAPs.

In order to reduce the T_g of branched, star-type energetic thermoplastic elastomers and enhance their flexibility, A_nB star-type energetic thermoplastic elastomers (TSETPEs) were prepared by urethane reaction of functional pre-polymers mono-functionalized poly(3,3-bis-azidomethyloxetane) (UPBAMO), as hard blocks, and GAP-modified poly(tetrahydrofuran) (APP), with the average molecular mass (M_n) of 4000–6000 g mol^{-1}, as soft blocks, and toluene-2,4-diisocyanate (TDI), as crosslinking agent (Scheme 6.25) [50]. The average molecular mass (M_n) of synthesized TSETPE was in a range of 15 000–19 000 g mol^{-1}, the T_g decreased from −16.9 to −24.3 °C, and the elongation (20 °C) improved from 44.5% to 652–919%, in comparison with similar average molecular mass SETPE. The application study shows that the TSETPE has good compatibility with commonly used energetic materials, such as RDX, aluminum powder, and nitroglycerine. The melt cast plastic-bonded explosive (PBX) explosive based on the TSETPE was calculated to have heat of detonation of 7411 J g^{-1}.

Another family of ETPEs was prepared by using GAP and isocyanates with 2,2-bis(azidomethyl)propane-1,3-diol (DAP) chain extender [51, 52]. It was shown that DAP could be a promising building block for the synthesis of ETPEs, offering an opportunity to get elastomers with tailored hard-block segments and certain mechanical properties for specific applications. Controlling the ratios of DAP, GAP, and choice of isocyanates, a broad variety of ETPEs with consistencies ranging from sticky paste to rubbery flakes were prepared. Using selected ETPE as energetic binder, PBX formulations containing HMX, and 4,10-dinitro-2,6,8,12-tetraoxa-4,10-diazatetra-cyclo-[5.5.0.05,9.03,11]-dodecane (TEX) were prepared, evaluated, and compared to similar formulations in which paraffin wax was used as an inert binder. The comparison studies of the explosive properties revealed that although ETPE-based PBXs showed comparable detonation velocity to paraffin wax-based formulations, the ETPE formulations showed significantly increased detonation pressure. Additional chain extenders that were evaluated in preparation of ETPEs were 1,4-butanediol (BDO), *N*-(2-cyano-ethyl)-di-ethanolamine [53]. It was shown that the ability of ETPEs to form hydrogen bonds was weakened, when the percentage of *N*-(2-cyanoethyl)-diethanolamine was increased, which led to the decrease of maximum stress of ETPEs and increase in its elongation at break. On the other hand, the interfacial interactions between energetic fillers and ETPEs were enhanced due to the higher nitrile content. The RDX/ETPE propellants were synthesized, and their mechanical properties showed that the ETPE containing 50% of *N*-(2-cyano-ethyl)-diethanolamine can effectively prevent the de-wetting

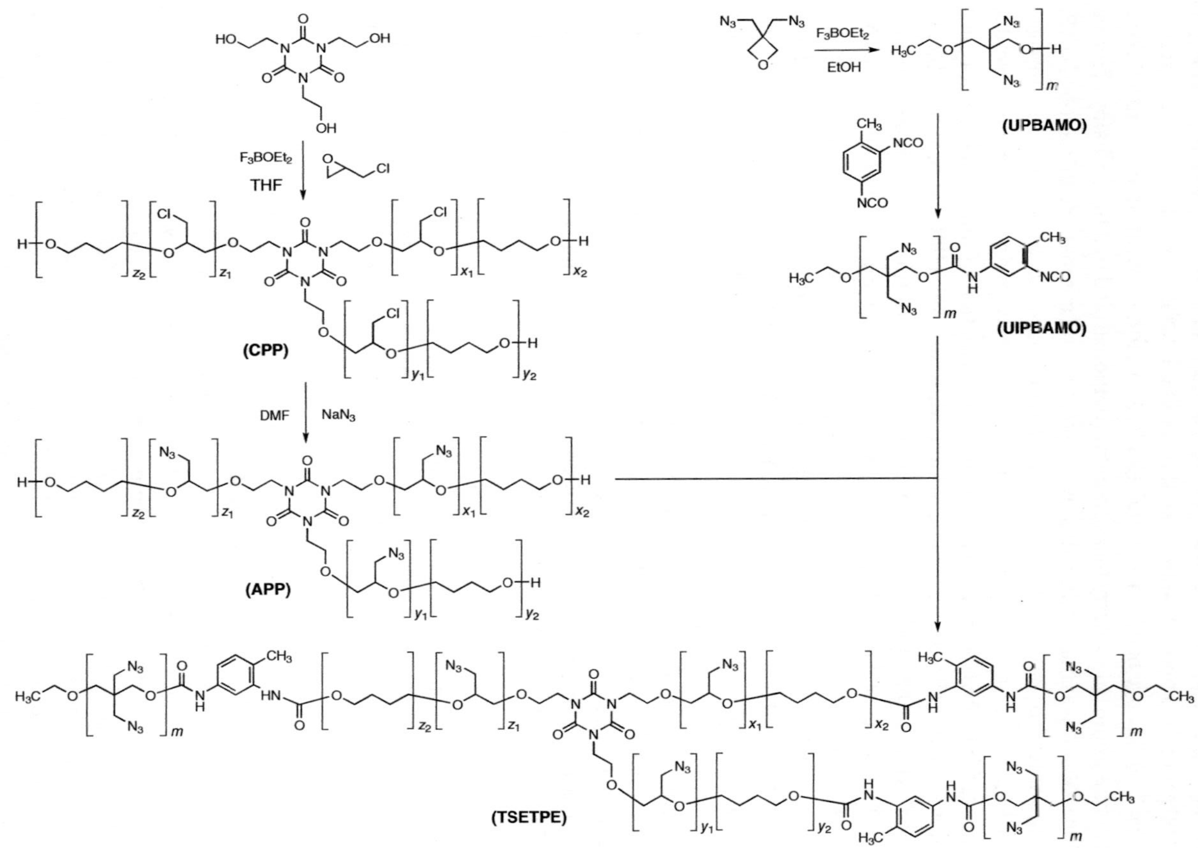

Scheme 6.25 Synthesis of TSETPE dendro-polymer.

and improve the mechanical properties of the corresponding propellants. The poly(3,3-bis(azidomethyl)-oxetane) (PBAMO)/GAP random block ETPE was synthesized by a solution polymerization of PBAMO, as hard segment pre-polymer, GAP as soft segment pre-polymer, BDO as chain extender, and TDI as curing agent [54]. GPC (gel permeation chromatography) results for this ETPE copolymer showed that the number average molecular weight was above 34 000 g mol^{-1}. FTIR analysis indicated that the imino groups of carbamate in this elastomer form hydrogen bonds with azido group, while the X-ray diffractometry (XRD) revealed that the crystallinity of PBAMO in copolymer was 16.6%.

Since GAP has poorer mechanical properties than HTPB, and HTPB-GAP blends are distinctly biphasic, the preparation of block copolymers could be an attractive methodology to obtain materials with advantages of both these homopolymers. The HTPB-GAP block copolymers were synthesized from the corresponding homopolymers by using two chain extenders: hexamethylene diisocyanate (HDI) and adipoyl chloride (AdCl) [55]. Both extenders lead to formation of homogeneous and stable polymeric materials. However, while in the case of the HDI, a gelation during reaction could be observed, AdCl was found to be a better reagent to be used as a binder and as a compatibilizer in GAP-HTPB mechanical blends. Another design approach included a direct functionalization of HTPB with azide groups. Two synthetic methodologies were explored for synthesis of azide-functionalized hydroxyl-terminated poly(butadiene) (azide-HTPB) (Scheme 6.26) [56]. Analysis of azide-HTPB obtained by azidation of 10% of HTPB double bonds resulted in formation of polymer with viscosity of 11 Pa s and T_g of −66 °C, while increasing nitrogen content by azidation of 20% of HTPB double bonds lead to soft gelatinous material with T_g of −34 °C.

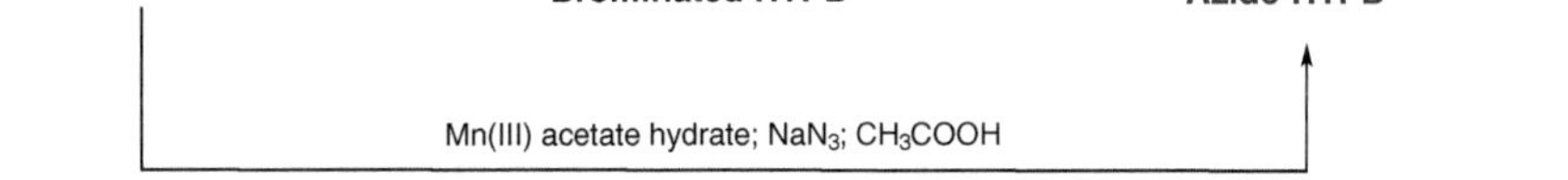

Scheme 6.26 Synthesis of azide-HTPB energetic binder.

Alternative to GAP, poly(vinyl-azido-acetate) (PVAA) energetic polymer was synthesized by initial polymerization of vinyl chloroacetate [57], or by chloroacetylation of poly(vinylalcohol) [58], followed by the azidation reaction (Schemes 6.26 and 6.27).

By controlling the polymerization reaction conditions of vinyl chloroacetate, the PVAA was prepared with an average molecular weight in the range of $1.54–4.02 \times 10^4$ and polydispersity index in the range of 2.0–4.0. Thermal analysis of PVAA showed

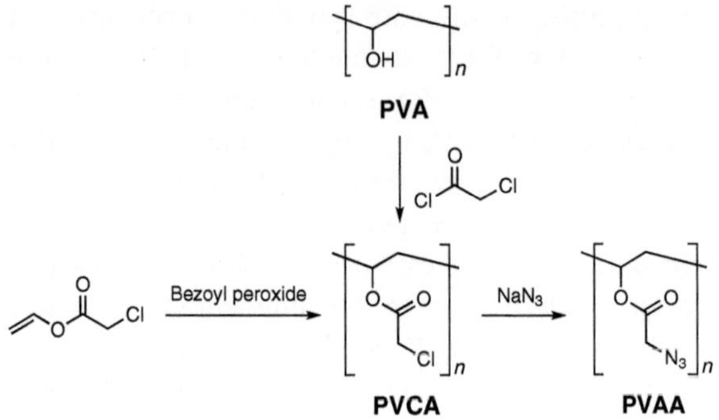

Scheme 6.27 Synthesis of poly(vinyl-azido-acetate) (PVAA) energetic polymer.

that it has a good thermostability up to 200 °C, with a single T_g transition at 24.6 °C. The results of the sensitivity and compatibility testing indicated that PVAA is relatively insensitive binder that can be safely used in melt cast explosives, such as TNT.

A molecular design of another azide-containing energetic polymer, poly(azidoaminoethers) allowed versatility in tailoring of certain properties of the latter polymers, including their endothermicity, oxygen balance, mechanical flexibility, toughness, and crystallinity, as well as making these types of polymers back-compatible with diisocyanate-based curing chemistry. Such characteristics provide poly(azidoaminoethers) a significant potential in meeting the demands for an energetic insensitive propellant binder. A series of bio-degradable energetic co-poly(ester/ether) elastomers were synthesized from PECH and sebacoyl chloride, using pyridine as a Lewis-base catalyst [59] (Scheme 6.28).

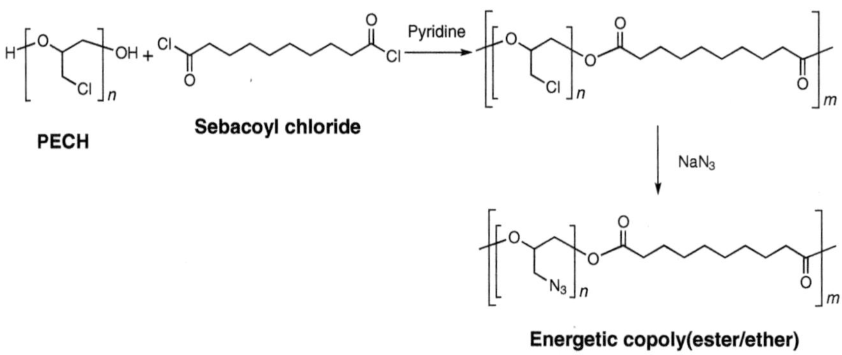

Scheme 6.28 Synthesis of energetic co-poly(ester/ether) elastomers.

In 2019, Tarchoun reported on the synthesis of normal and microcrystalline cellulose nitrate from *Posidonia oceanica* brown algae [60]. The preparation of NC based on this material is an improvement in both cost and sustainability of nitrocellulose production as the brown algae is an abundant and renewable resource in many countries surrounding the Mediterranean Sea. The authors found that the algae-derived nitrocellulose had comparable properties to traditionally produced NC and had great potential application in energetic materials formulation. Nitrocellulose-derived polymers continue to be of importance beyond their use as pure nitrocelluloses, but also as blends. Hu studied blends of nitrocellulose-GAP polymers [39]. In comparison with the pure nitrocellulose, it was found that such blends have improved thermostability and as the nitrocellulose content increased, the strength and crystallinity of these blends were increased. As GAP content increased, low-temperature properties were found to be improved. Thermal behavior of a series of energetic cellulose-based polymers, including azido-deoxy cellulose mononitrate (AC), cellulose dinitrate acetate (CAN), and azido-deoxy cellulose dinitrate (ACN), initially prepared by Vigo and Roberts [61, 62], was reinvestigated by Shamsipur (Scheme 6.29) [63].

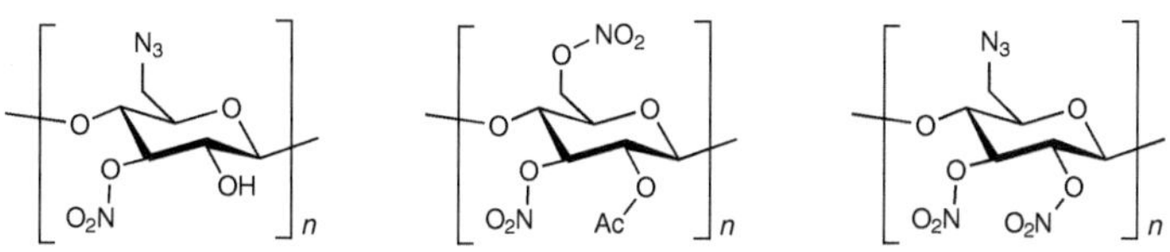

Scheme 6.29 Structures of azido-deoxy cellulose mononitrate (AC), cellulose dinitrate acetate (CAN), and azido-deoxy cellulose dinitrate (ACN) energetic polymers.

Study of the effects of azide and other functional groups on thermal behavior of cellulose-derived energetic polymers showed that ACN decomposes exothermally, with three continuous steps, in a broad temperature range between 180 and 500 °C. In contrast, AC and CAN undergo a single-step exothermic decomposition at 206 and 208 °C, respectively. Based on non-isothermal DSC experiments and obtained thermo-kinetic data, the relative thermostability of these prospective binders was found to follow AC > CAN > ACN order.

Using 2,2-bis(azidomethyl)propane-1,3-diol (BAMP) and 2,2-dinitropropane-1, 3-diol (DNPD) as energetic building blocks, four energetic polyurethanes were synthesized *via* polymerization reactions. In these polymerization reactions hexamethylene diisocyanate (HMDI) and diisocyanatoethane (DIE) were utilized (Scheme 6.30) [64].

Chemical composition, energy content, and combustion performance, as well as thermal and safety properties of the resulting poly(hexamethylene(2,2-bis (azidomethyl)propylene)carbamate) (HMDI-BAMP) and poly(hexamethylene(2,2-dinitropropylene)-carbamate) (HMDI-DNPD), and their ethylene analogues were studied and compared to GAP. The new energetic polyurethanes showed good physical stability, thermal stability in a range of 170–210 °C, and were found to be insensitive to friction (>360 N), less sensitive to impact (40 J) than GAP. Using EXPLO5 software, the prepared polymers were predicted to have VODs between 5000 and 7000 m s^{-1} and a detonation pressure between 64 and 176 kbar.

Another type of azido binder, poly(glycidylazide-*r*-3-azido-tetrahydrofuran) (PGAAT) was synthesized by cationic copolymerization of 3-methylsulfonyloxytetrahydrofuran with ternary cyclic ethers (Scheme 6.31) [65].

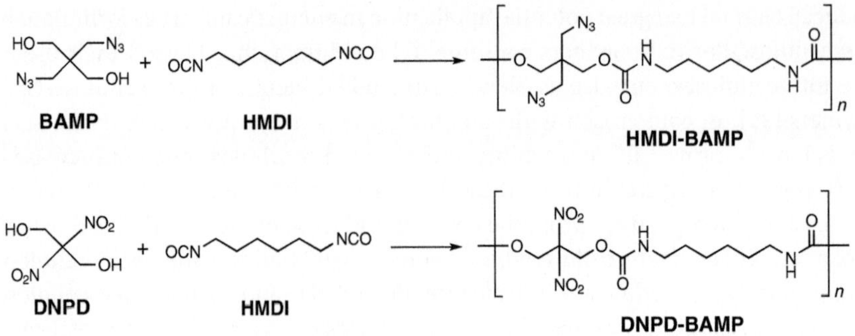

Scheme 6.30 Synthesis of poly(hexamethylene(2,2-bis(azidomethyl)propylene)carbamate) (HMDI-BAMP) and poly(hexamethylene(2,2-dinitropropylene)carbamate) (HMDI-DNPD).

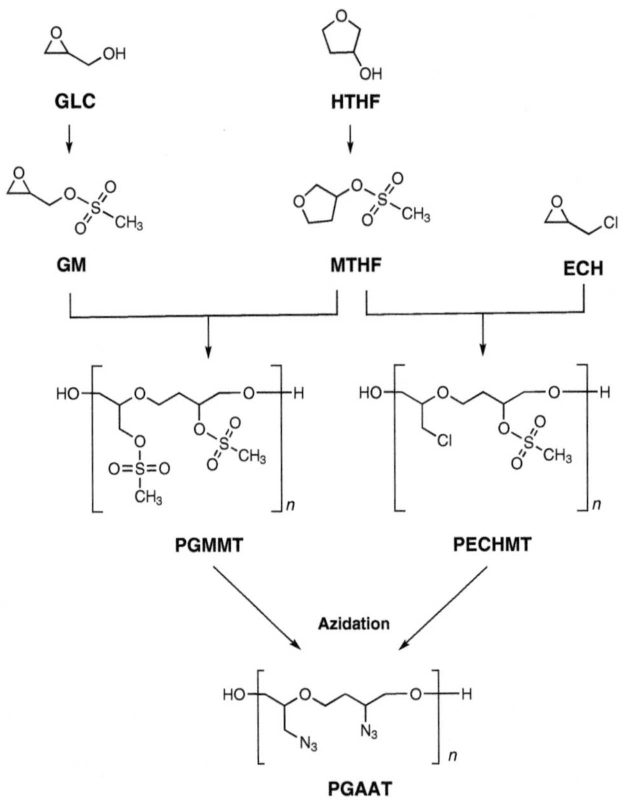

Scheme 6.31 Synthesis of poly(glycidyl-azide-*r*-3-azido-tetrahydrofuran) (PGAAT).

The thermal properties of PGAAT (having molecular weight of about 3811 g mol^{-1}, and polydispersity of 1.65) showed this polymer has low T_g (−60 °C), lower than that of GAP (−48 °C), and comparable to the T_g of poly(glycidyl-azide-*co*-tetrahydrofuran) (−64 °C). The exothermic decomposition of PGAAT was observed between 180 and 280 °C, and the exothermic enthalpy was found to be 1798 J g^{-1}, which is comparable to GAP (1828 J g^{-1}) and higher than that of poly(glycidyl-azide-*co*-tetrahydrofuran) (1166 J g^{-1}). These data indicate that introduction of the 3-azidotetrahydrofuran building block into the structure of PGAAT has a significant influence on the T_g of this material, without markedly affecting its energy content.

Study of the influence and molecular structure–properties relationships of azido-oxetane polymers, such as the location, length, and number of the pendant chains, on these polymers' densities, enthalpies of formation, T_g, and mechanical properties showed that the densities, enthalpies of formation, and elastic *moduli* decrease monotonously, as the number and length of the pendant chains increase [66, 67]. Also, the densities, elastic *moduli*, and the enthalpies of formation increase with the introduction of azido groups. The azido-oxetane polymers have best flexibility when the length of the pendant chain is one or two, leading to a minimal T_g and higher Poisson ratio. T_g of these polymers can be at best reduced by 17 °C, without compromising the energetic properties and mechanical performances.

An energetic polynorbornene with pendant bis(azidoacetyloxymethyl) (PNBAA) was designed, by combining energetic moieties, potential good mechanical property, compatibility with plasticizer, and polynorbornene in single polymeric structure. A green synthesis of PNBAA was achieved in four steps, starting with solvent-free Diels–Alder reaction, followed by chloroacetylation of norbornene-dimethanol in water, solvent-free catalytic ring opening metathesis polymerization (ROMP), and completed with azidation in aqueous 3-pentanone, with the tandem ROMP and azidation in one pot (Scheme 6.32) [68].

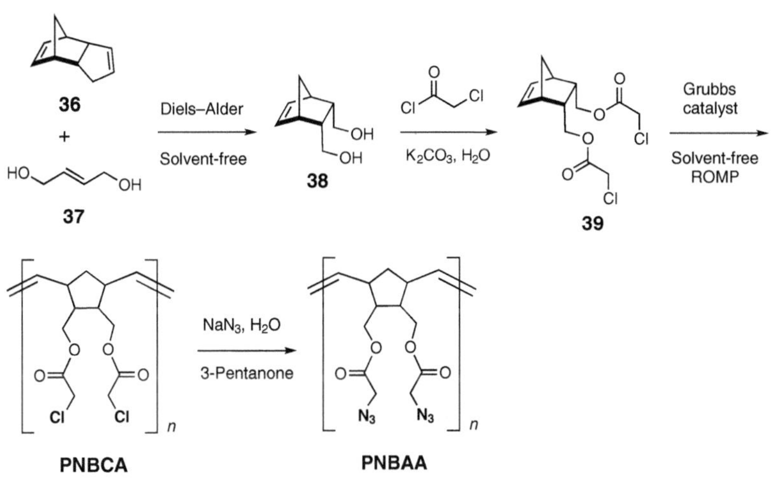

Scheme 6.32 Synthesis of PNBAA.

A methodology for preparation of two energetic azido polycarbonates, poly(2,2′-bisazidomethyl-1,3-propyl-carbonate) poly(BAMPC) and poly(3-azido-1,2-propyl-carbonate), included a two-step procedure starting with the bulk polymerization of suitable diols with diphenyl carbonate, utilizing La(acetylacetonate)$_3$ as a catalyst.

The synthesis was completed by azidation of the resulting bromo-polycarbonate in cyclohexanone (Scheme 6.33) [69].

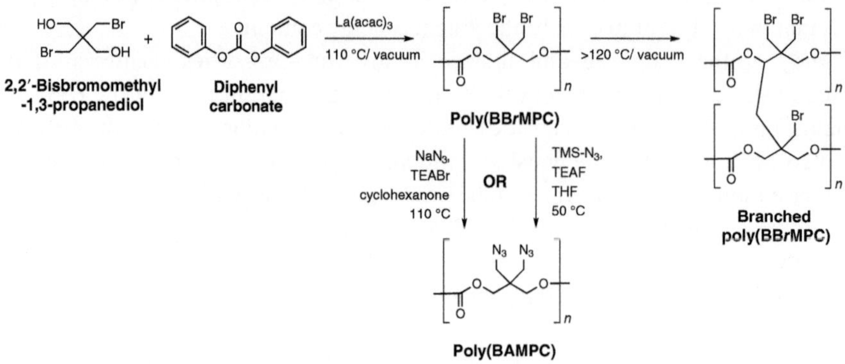

Scheme 6.33 Synthesis of branched poly(BB*r*MPC) and poly(BAMPC).

Alternative reaction pathway for the azidation of halogenated polyesters under mild conditions utilizes trimethylsilyl azide (TMS-N_3), as a less-polar azide source, and tetraethylammonium fluoride (TEAF) as a phase-transfer catalyst in tetrahydrofuran. Yet, the latter method suffers from drawbacks, which include partial TMS-N_3 hydrolysis due to the presence of hydroxyl end-groups in polymer chains, forming silane end groups that are no longer suitable for post-polymerization modifications. Furthermore, hydrolysis of $TMSN_3$ also leads to formation of highly toxic and potentially explosive hydrazoic acid. Poly(BAMPC) was found to have T_g of −46 °C, temperature of melting at 41 °C, and decomposition onset at 208 °C.

6.5 Azido Fluoropolymers

To improve the mechanical properties of GAP-based polyurethane binders, a fluorinated GAP, 2,2,2-trifluoro-ethoxymethyl-epoxy-*r*-glycidyl azide copolymer poly(TFEE-*r*-GA) was synthesized by cationic copolymerization of ECH and 2,2,2-trifluoroethoxymethyl-epoxy, followed by azidation (Scheme 6.34) [70]. The T_g and decomposition temperatures of poly(TFEE-*r*-GA) were found to be −49.5 and 230 °C (onset), respectively. The co-polyurethane networks were further synthesized by cross-linking poly(TFEE-*r*-GA) with trimethylol-propane, as a chain extender agent and using isophorone diisocyanate, as a cross-linking agent. In comparison with GAP, the poly(TFEE-*r*-GA) based co-polyurethane networks exhibited better mechanical properties, which had a tensile strength of 5.52 MPa, and an elongation at break of 162.8%. The density of poly(TFEE-*r*-GA) was determined and compared with GAP. Due to the presence of CF_3 groups, a density of poly(TFEE-*r*-GA) (1.36 g cm^{-3}) was found to be somewhat higher than that of the pure GAP (1.28 g cm^{-3}). The viscosity of poly(TFEE-*r*-GA) (2.66 Pa s at 25 °C) was also measured to be higher than that of the pure GAP (11.9 Pa s at 25 °C). In terms of safety, in comparison to pure GAP (H_{50} = 42.2 cm), poly(TFEE-*r*-GA)

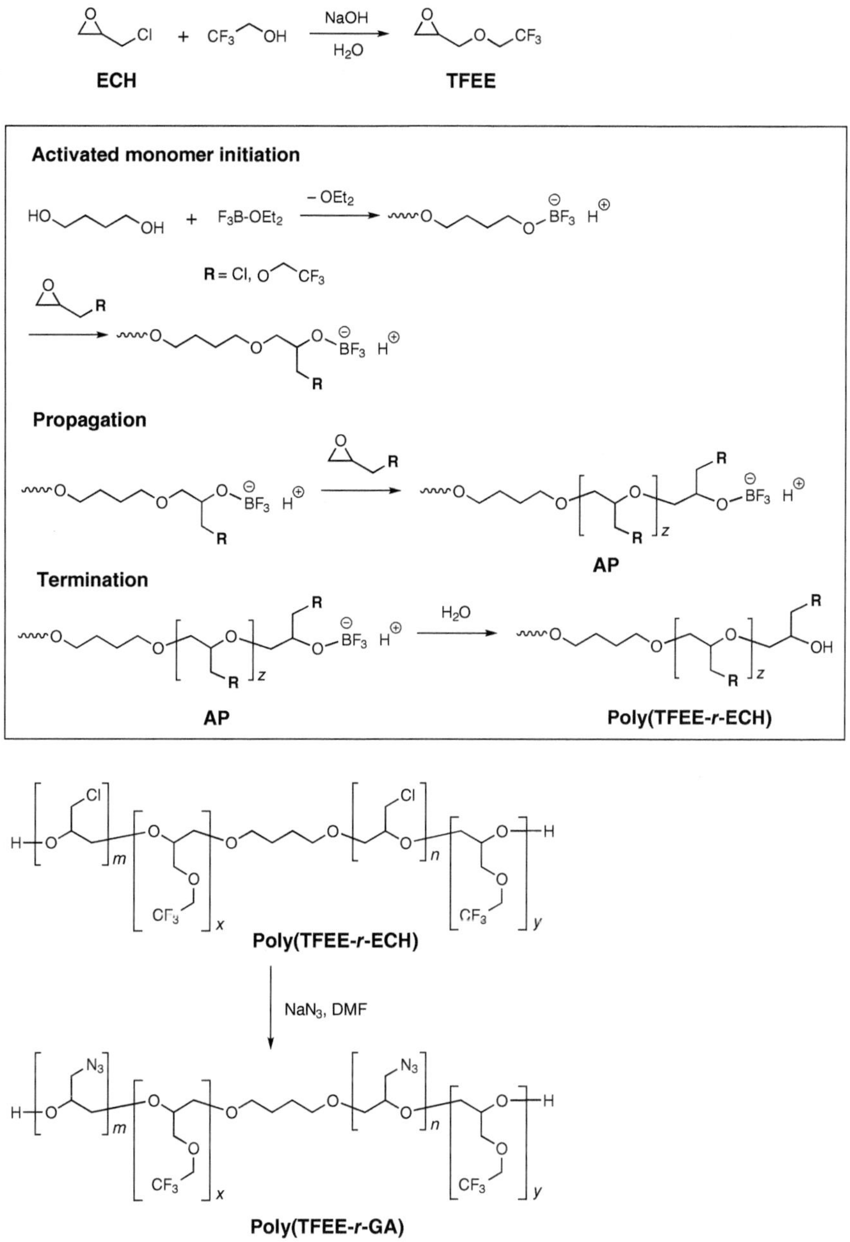

Scheme 6.34 Synthesis of TFEE monomer, formation of poly(TFEE-*r*-ECH) *via* cationic polymerization, azidation of poly(TFEE-*r*-ECH) into poly(TFEE-*r*-GA).

was significantly less sensitive to impact ($H_{50} > 129$ cm), which could be attributed to lower azide group content in poly(TFEE-*r*-GA) than in GAP.

Additional glycidyl azide fluorine-containing polymer (FGAP) was also synthesized by a cationic copolymerization of ECH and 1,1,1-trifluoro-2,3-epoxypropane that was followed by azidation (Scheme 6.35) [71]. Analysis of the resulting energetic FGAP polymer showed that it has molecular weight of about 2845 g mol^{-1}, and the T_g and decomposition temperature of −47.8 and 253 °C, respectively. FGAP-based polyurethane networks were further prepared using triphenyl-methane-4,4,4-triisocyanate (N-100), as the crosslinking agent. In comparison with GAP, FGAP-based poly(urethane) networks exhibited better mechanical behavior, having a tensile strength of 1.5 MPa and a highest elongation at break of 81.6%. FGAP showed good compatibility and could be safely used with HMX, RDX, and Al powder.

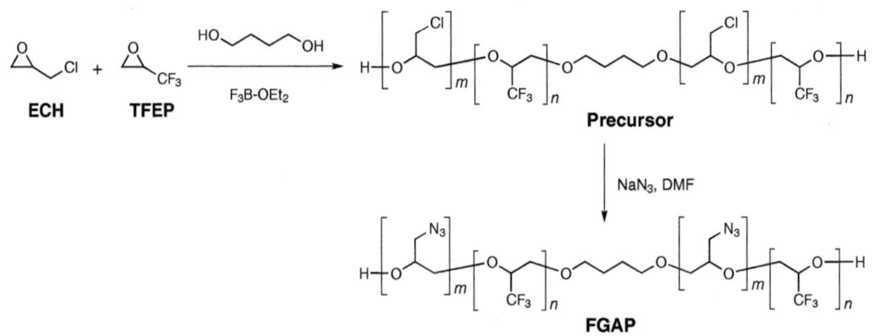

Scheme 6.35 Synthesis of glycidyl azide fluorine-containing polymer (FGAP).

Copolymer named PDA, based on 3-difluoroaminomethyl-3-methyloxetane (DFAMO) and 3-azido-methyl-3-methyloxetane (AMMO) energetic precursors, was prepared by cationic polymerization, using BDO as an initiator and BF_3-etherate (TFBE) as a catalyst (Scheme 6.36) [72, 73]. The resulting PDA copolymer had mole content of DFAMO of 45.66% and AMMO of 54.34%. It showed good thermostability and exhibited a three-step mass-loss process, with the first two steps (226.5 and 263.0 °C, at heating rate of 10 °C min^{-1}) belonging to the thermal decomposition of difluoroamino and azido groups, respectively. The third mass-loss step was attributed to the thermolysis of monofluoro-imino, cyano, imine intermediates, and the polymer chain.

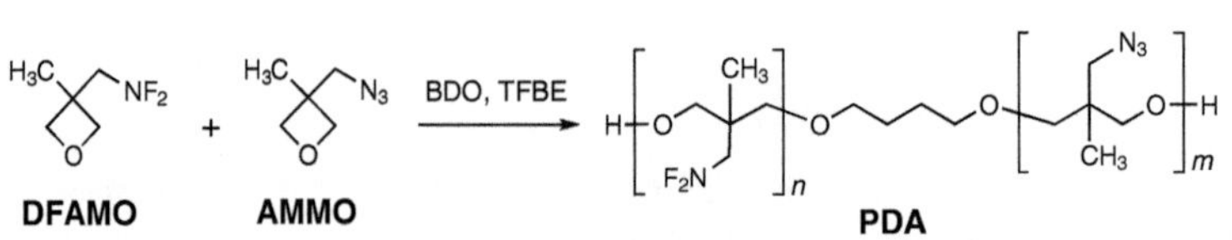

Scheme 6.36 Synthesis of PDA.

In extensive compatibility DSC-based studies several materials, including RDX, TNT, TATB, PETN, NH_4ClO_4, NH_4NO_3, KNO_3, Al powder, Al_2O_3 powder, 2-nitro-diphenylamine, and 1,3-diethyl-1,3-diphenyl urea were found to be compatible with PDA.

6.6 Azido Plasticizers

An energetic plasticizer, 1,3-diazido-2-ethyl-2-nitropropane (DAENP) **42**, was synthesized in three steps, starting with condensation of 1-nitropropane with formaldehyde to yield 2-ethyl-2-nitro-propane-1,3-diol **40**. The resulting 2-ethyl-2-nitro-1, 3-propane diol **40** was tosylated to give 2-ethyl-2-nitropropane-1,3-diyl-bis(4-methylbenzenesulfonate) **41**, which was finally azidated with NaN_3 in DMF to yield the target DAENP **42** (Scheme 6.37) [74].

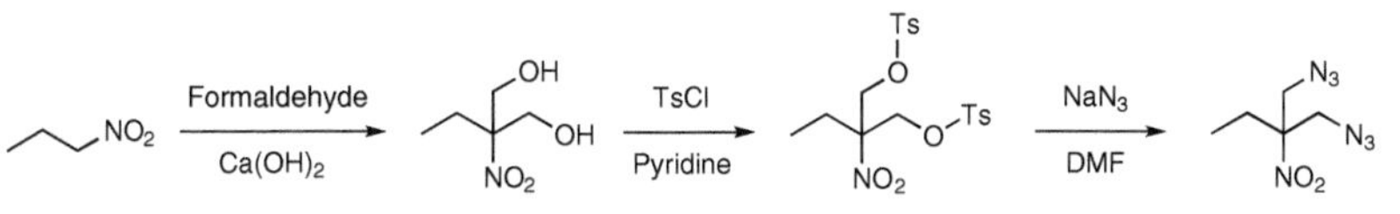

40 **41** **42**

Scheme 6.37 Synthesis of 1,3-diazido-2-ethyl-2-nitropropane (DAENP).

Thermal analysis showed that DAENP **42** was found to be stable up to 196 °C, had high-energy output of $2056\,J\,g^{-1}$, and a very low T_g (−96.76 °C) (Table 6.1).

DAENP **42** found to be compatible with energetic binders like GAP, 3,3-bis (azidomethyl)oxetane-tetra-hydroxyl-furan (BAMO-THF) copolymer, and poly-3-nitratomethyl-3-methyloxetane (polyNIMMO or PLN) (Tables 6.2 and 6.3).

Tetra-azido energetic plasticizers, bis(1,3-diazidoprop-2-yl)-malonate **43** and bis(1,3-diazidoprop-2-yl)-glutarate **44**, were prepared in two synthetic steps, starting from malonic and glutaric acids, respectively (Scheme 6.38) [75].

Both energetic plasticizers **43** and **44** showed good thermostability and exhibited decomposition temperatures of 233.5 and 232.6 °C and had densities of 1.25 and $1.27\,g\,cm^{-3}$, respectively. T_g of both compounds was measured to be −69 and −68 °C, respectively. Thermal analysis of energetic polymer-energetic plasticizer blends showed reduction of T_g in comparison to the T_g of pure polymers, while

Table 6.1 Thermal, physical, and sensitivity data of DAENP.

	Properties	**Data**
1.	DSC, T_{max} (°C)	245.5
2.	Heat release, ΔH ($J\,g^{-1}$)	2056.13
3.	Glass transition temperature T_g (°C)	−96.73
4.	Melting temperature (°C) (by DSC)	−5.85
5.	TGA weight loss	89.03% (179–389 °C)
6.	Oxygen balance (%)	−101
7.	Calculated heat of formation ΔH_f ($kJ\,mol^{-1}$)	454
8.	Density ($g\,cm^{-3}$) calculated.	1.32
9.	Impact sensitivity h_{50} (cm)	>170
10.	Friction sensitivity (kg)	>36

Table 6.2 Effect of plasticizer on thermal properties of binders.

Composition	Decomposition temperature T_{max} (°C)	Glass transition temperature T_g (°C)
GAP-2000	253.2	−45.0
GAP-2000 + 20% DAENP	252.6	−59.7
BAMO-THF	257.3	−49.0
BAMO-THF + 20% DAENP	256.2	−63.8
PLN	216.2	−30.0
PLN + 20% DAENP	215.6	−45.9

Table 6.3 Effect of plasticizer on rheological properties of binders.

	Compositions	Viscosity at 30 °C (cPs)	LVR (Pa)	G′ (Pa)	G″ (Pa)	Phase angle (degree)
1.	GAP	1 097	400	0.1239	7.059	89.9
2.	GAP + DAENP	359	300	0.0364	2.664	89.4
3.	PLN	13 590	875	0.7977	50.08	89.6
4.	PLN + DAENP	1 433	400	0.3866	9.141	89.2
5.	BAMO-THF	18 000	925	0.1651	62.58	89.1
6.	BAMO-THF + DAENP	2 347	450	0.4757	14.80	89.2

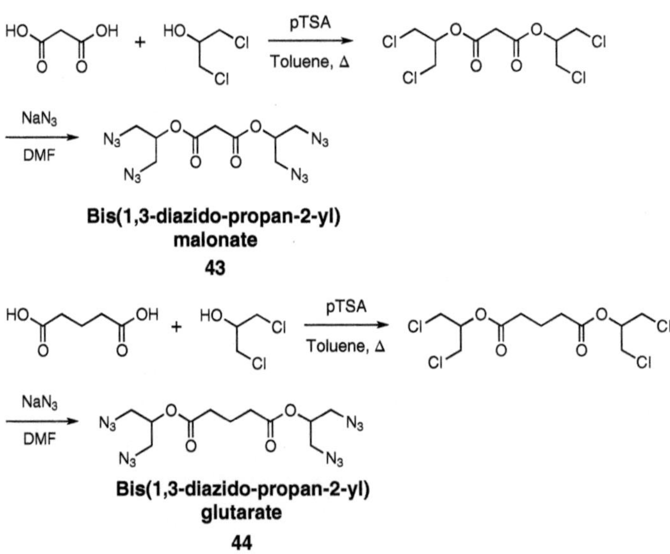

Scheme 6.38 Synthesis of bis(1,3-diazido prop-2-yl)-malonate **43** and bis(1,3-diazido-prop-2-yl)glutarate **44**.

Table 6.4 Effect of both plasticizers on thermal properties of three different energetic binders.

	Composition	Decomposition temperature T_{max} (°C)	Glass transition temperature T_g (°C)
1.	GAP	254.67	−45.00
2.	GAP + 20% of **43**	254.67	−64.07
3.	GAP + 20% of **44**	252.80	−62.02
4.	BAMO-THF	247.77	−49.00
5.	BAMO-THF + 20% of **43**	252.70	−65.38
6.	BAMO-THF + 20% of **44**	252.80	−66.07
7.	PLN	217.63	−30.00
8.	PLN + 20% of **43**	217.23	−56.28
9.	PLN + 20% of **44**	216.50	−53.29

single temperature decomposition strongly suggested formation of single-phase homogeneous systems with examined energetic polymers (Table 6.4). DFT calculations on energetic plasticizers **43** and **44** predicted positive heats of formations for these compounds, and their expected contribution to the energy content of the resulting blends.

α,α-Diazidated azido ester plasticizers (**47**, **50**, and **53**), containing up to six azido moieties, were synthesized on a basis of malonic acid (Scheme 6.39) [76]. The azido ester precursors **46**, **49**, and **53** were prepared *via* coupling of malonic acid with azido-alcohols and further reacted with NaN_3 and iodine, in the presence of $NaHCO_3$ base, to achieve the geminal diazidation of the malonic ester moiety.

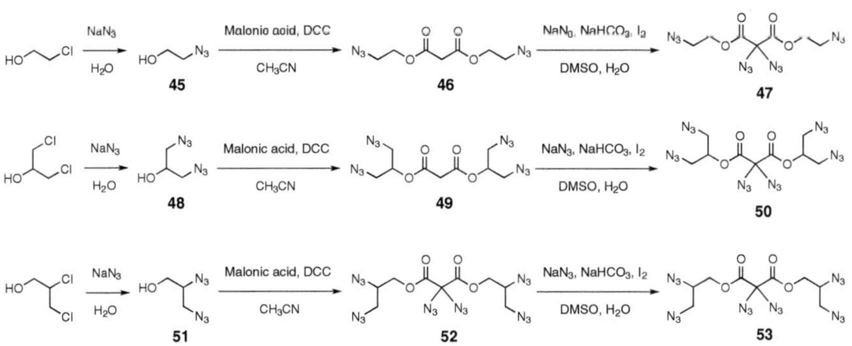

Scheme 6.39 Synthesis of energetic plasticizers **47**, **50**, and **53**.

High-nitrogen content energetic plasticizers **47**, **50**, and **53** were evaluated for their density, thermal stability, and safety properties. Heat of formation and detonation performance energetic properties of these materials were calculated (Table 6.5).

Additionally, four azido groups-containing plasticizers **56**, **59**, **62**, and **65** were synthesized based on various lengths terminal diols (Scheme 6.40) [77]. The plasticization effect of these energetic plasticizers on GAP was investigated by measurements

Table 6.5 Physicochemical, sensitivity, and energetic properties of plasticizers **46**, **47**, **49**, **50**, **52**, and **53**.

	bAEM 46	**DA-bAEM 47**	***b*-1,3-DAPM 49**	**DA-*b*-1,3-DAP 50**	***b*-2,3-DAPM 52**	**DA-*b*-2,3-DAP 53**
Formula	$C_7H_{10}N_6O_4$	$C_7H_8N_{12}O_4$	$C_9H_{12}N_{12}O_4$	$C_9H_{10}N_{18}O_4$	$C_9H_{12}N_{12}O_4$	$C_9H_{10}N_{18}O_4$
M (g mol^{-1})	242	324	352	434	352	434
$\Omega^{a)}$(%)	−99	−69	−90	−70	−91	−70
$T_{dec}{}^{b)}$(°C)	191	158	162	136	162	136
$\rho^{c)}$(g cm^{-3})	1.38	1.49	1.38	1.51	1.30	1.50
$\Delta H_f{}^{\circ d)}$(kJ mol^{-1})	−176	542	466	1186	440	1160
Impact (N)	>360	36	160	36	160	36
Friction (J)	>40	1	35	<1	35	<1
Explo 5 (V6.03)						
$-\Delta_{Ex}U^{\circ e)}$(kJ mol^{-1})	−2828	−4001	−3647	−4359	−3578	−4300
$T_{det}{}^{f)}$ (K)	2150	3016	2769	3227	2636	3207
$P_{CJ}{}^{g)}$(kbar)	118	170	160	181	119	178
$V_{det}{}^{h)}$(m s^{-1})	6192	6974	6576	7147	6252	7087
Vo$^{i)}$ (l kg^{-1})	562	514	530	507	588	509

a) Oxygen balance assuming the formation of CO_2.
b) Onset decomposition temperature (DSC).
c) Density (RT).
d) Enthalpy and of formation (CBS-4M, Gaussian: 09.23).
e) Heat of combustion.
f) Detonation temperature.
g) Detonation pressure.
h) Detonation velocity.
i) Volume of gaseous products.

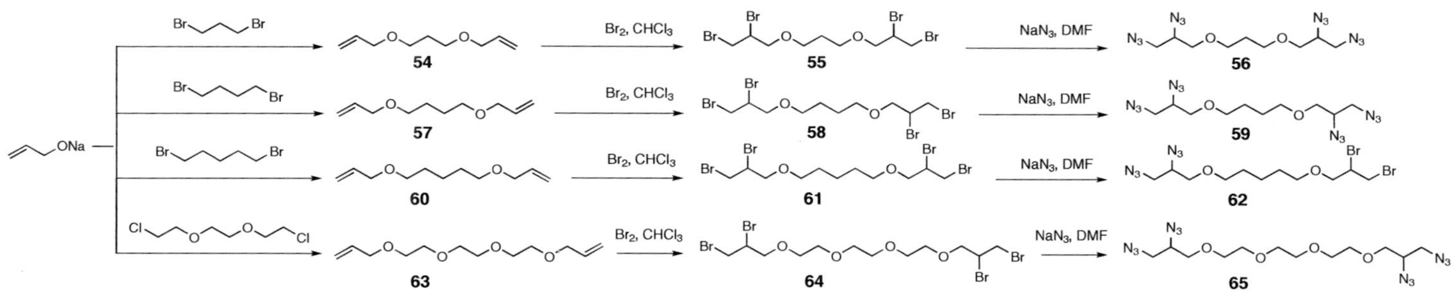

Scheme 6.40 Synthesis of azido energetic plasticizers **56**, **59**, **62**, and **65**.

Table 6.6 Thermal properties of compounds **56**, **59**, **62**, and **65**.

Compound	T_g (°C)	$T_{d, onset}$ (°C)	$T_{d, max}$ (°C)	$\Delta H_{decom.}$ (J g^{-1})	Weight loss steps	Total weight loss (%)
56	−101.3	169.8	230.4	1382.4	122.2–252.8 °C, 58.3% 252.8–550.0 °C, 11.3%	69.6
59	−92.3	166.8	232.3	1617.2	121.1–251.3 °C, 57.4% 251.3–550.0 °C, 13.5%	70.9
62	−96.4	169.6	232.1	1616.6	121.1–253.6 °C, 57.3% 253.6–550.0 °C, 16.2%	73.5
65	−86.0	167.6	234.9	1585.6	119.5–259.0 °C, 57.4% 259.0–550.0 °C, 11.3%	68.7

of viscosity and thermal analysis of the pre-polymer-plasticizer blends and plasticized binders (Table 6.6).

The plasticized blends of compounds **56**, **59**, **62**, and **65** with GAP were cured, and T_g and decomposition temperatures of the cured poly(urethane) binders were measured. Thermal analysis of the pre-polymer-plasticizers and cured polymer mixtures showed that the plasticizers **56**, **59**, **62**, and **65** were found to be completely compatible with the GAP binder and have a very good plasticizing effect (Tables 6.7 and 6.8).

A tetra-azido ether energetic plasticizer 1,2,8,9-tetraazido-4,6-dioxol-nonane (TADONA) was designed and synthesized through consecutive azidation reaction and condensation reaction, and its structure was characterized (Scheme 6.41) [78].

Table 6.7 Effect of synthesized plasticizers on T_g of GAP pre-polymer.

	Composition	Glass transition temperature (°C)	Average interaction parameter ($\bar{I}$)
1.	GAP pre-polymer	−53.7	—
2.	GAP + 35% **56** GAP + 50% **56**	−75.3 −86.5	−28.86
3.	GAP + 35% **59** GAP + 50% **59**	−69.6 −78.3	−15.85
4.	GAP + 35% **62** GAP + 50% **62**	−74.3 −79.6	−21.53
5.	GAP + 35% **65** GAP + 50% **65**	−69.4 −74.3	−18.56

Table 6.8 Effect of synthesized plasticizers on thermal properties of GAP polyurethane binder.

	Composition	Glass transition temperature (°C)	Decomposition temperature (°C)
1.	GAP polyurethane	−33.8	223.8
2.	GAP + 10% **56**	−43.8	222.0
3.	GAP + 10% **59**	−43.5	221.6
4.	GAP + 10% **62**	−42.6	221.3
5.	GAP + 10% **65**	−42.4	221.5

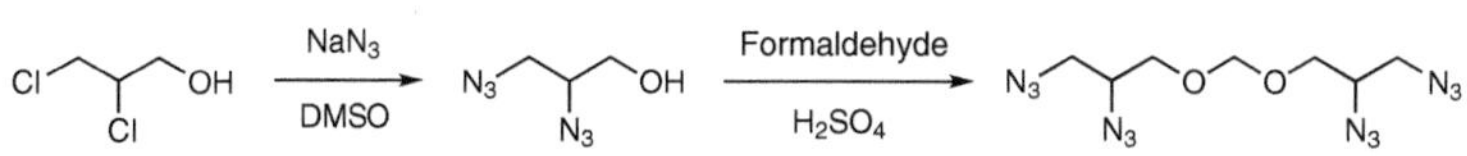

DClOH **DAPOH** **TADONA**

Scheme 6.41 Synthesis of energetic plasticizer 1,2,8,9-tetraazido-4,6-dioxol-nonane (TADONA).

Its impact sensitivity was measured as 25.2 J, and friction sensitivity was measured as 160 N, its T_{max} was 236.2 °C as well as its T_g was −88.9 °C, which indicated that TADONA had insensitive mechanical sensitivities, good thermostability, and excellent low-temperature properties. Theoretical calculations showed that TADONA has superior energetic properties in comparison to some reported azide plasticizers and good miscibility with GAP. Moreover, TADONA had notable effect on the adjustment of the viscosity of GAP, and the higher content of TADONA in GAP, the better adjustment effect was achieved. Furthermore, the T_gs of [TADONA + GAP] (1 : 10), (2 : 10), (3 : 10), and (4 : 10) mixtures were determined to be −54.4, −58.8, −61.9, and −65.8 °C, respectively, and are much lower than those of GAP (−49.1 °C). Therefore, TADONA exhibited excellent plasticizing effect on GAP, demonstrating considerable application potential in GAP-based propellants and explosives (Tables 6.9 and 6.10).

6.7 Nitro Group Containing Polymers

The nitro group is naturally one of the most quintessential functional groups seen in energetic materials; as a result, it is also found regularly in energetic polymers. The first energetic polymer was nitrocellulose, containing the -ONO_2, group and despite being discovered in the 1800s, energetic polymers containing this functional group and those that can be considered nitrocellulose derivatives are still being investigated. The energetic nitrato-containing polymers, poly(glycidyl nitrate) (PGN) were continuously investigated, especially copolymers of PGN with various other monomers, in order to obtain energetic polymers exhibiting desired tailored properties. Ghorbani prepared the PGN triblock copolymer poly(glycidyl nitrate-*block*-poly(butadiene)-*block*-glycidyl nitrate) (PGN–PB–PGN)

Table 6.9 Comparison of properties of TADONA with 2,2-bis(azido-methyl)propane-1,3-diyl diacetate (BAMP) [79], diethyleneglycol bis(azido-acetate) (DEGBAA) [80], and bis(2,3-diazido-propoxy)diethylene glycol (BDAP) [77].

	TADONA	BAMP	DEGBAA	BDAP
Formula	$C_7H_{12}N_{12}O_2$	$C_9H_{14}N_6O_4$	$C_8H_{12}N_6O_5$	$C_{10}H_{18}N_{12}O_3$
M (g mol^{-1})	296.26	270.25	272.22	354.34
IS[a] (J)	>25.2	>40	>10	N/A
FS[b] (N)	160	>360	160	>360
N[c] (%)	56.74	31.10	30.87	47.44
OB[d] (%)	−97.21	−124.33	−99.92	−117.40
ρ[e] (g cm^{-3})	1.28	1.21	1.00	N/A
T_{onset}[f] (°C)	212.1	N/A[m]	N/A[m]	167.6
T_d[g] (°C)	236.8	230.0	215.0	234.9
T_g[h] (°C)	−88.9	−80.0	−63.0	−86.0
ΔH_f[i] (kJ mol^{-1})	857.3	242.0	329.0	698.2
P[j] (GPa)	18.0	7.9	4.6	N/A
D[k] (m s^{-1})	6824	5420	4363	N/A
I_{sp}[l] (s)	230	175	169	215

a) Impact sensitivity.
b) Friction sensitivity.
c) Nitrogen content (mass%).
d) Oxygen balance (based on CO_2).
e) Density.
f) Onset decomposition temperature.
g) Peak decomposition temperature.
h) Glass transition temperature.
i) Heat of formation.
j) Detonation pressure.
k) Detonation velocity.
l) Specific impulse (isobaric combustion, chamber pressure: 7 MPa, equilibrium expansion).

Table 6.10 The viscosity test results for [TADONA + GAP] mixtures and GAP under different temperatures.

	Viscosity (mPa s)				
Sample	**30 °C**	**40 °C**	**50 °C**	**60 °C**	**70 °C**
GAP	4260	2241	1350	915	662
TADONA + GAP (1 : 10)	2055	1150	721	506	375
TADONA + GAP (2 : 10)	1111	652	426	306	229
TADONA + GAP (3 : 10)	660	402	267	197	150
TADONA + GAP (4 : 10)	418	263	179	133	103

by ring-opening polymerization of glycidyl nitrate in the presence of HTPB. This reaction was catalyzed by boron trifluoride etherate ($F_3B{\cdot}OEt_2$) [81]. It was found that this type of polymerization was extremely sensitive to water, and the reaction conditions must be kept dry to prevent otherwise preferential homopolymerization to PGN. T_g of the prepared PGN–PB–PGN block copolymer was determined to be −27 °C, and the authors proposed its use as an energetic propellant binder (Scheme 6.42).

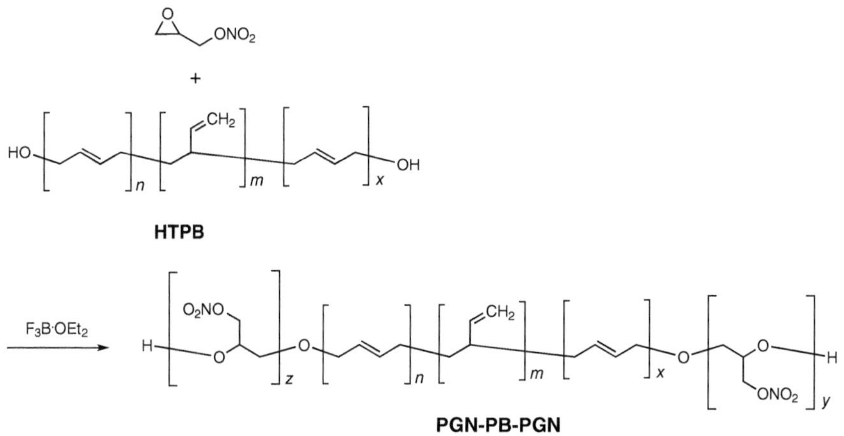

Scheme 6.42 Synthesis of poly(glycidyl nitrate-*block*-poly(butadiene)-*block*-glycidyl nitrate) (PGN-PB-PGN).

Another energetic binder was synthesized by catalytic ring-opening copolymerization of glycidyl nitrate with PCL, as a macro-initiator, using DBTDL or boron trifluoride etherate as catalysts. These catalytic reactions resulted in formation of a tri-block copolymer poly(glycidyl nitrate-*block*-poly(caprolactone)-*block*-poly (glycidyl nitrate)) (PGN–PCL–PGN) (Scheme 6.43) [82]. Effect of monomer concentration, type of a catalyst, reaction time, and solvent were investigated on this polymerization. The yields of the polymerization were solvent dependent and were found to be higher in chloroform than in less polar toluene. Use of $F_3B{\cdot}OEt_2$ as catalyst gave higher yield of the target copolymer, while increasing the concentration of monomer resulted in higher-molecular-weight polymerization products. Under optimized conditions, the obtained molecular weight of PGN–PCL–PGN polymer was 2900 g mol^{-1} with PDI of 2.2. Thermal analysis of the tri-block copolymer showed glass transition temperature (T_g) of −56.2 °C, which is lower than that of the PGN homopolymer (T_g of −35 °C), which resulted in a higher flexibility of PGN–PCL–PGN copolymer than PGN, making PGN–PCL–PGN copolymer a promising candidate for use as energetic propellant binder.

Nitrate-ester-containing polymers, such as poly(vinyl nitrate) (PVN), could be prepared by nitration of poly(vinylalcohol) (PVA). It was found that decreasing of the nitrogen content leads to improved thermal stability of such polymers [83]. An energetic binder was prepared through ring-opening copolymerization of

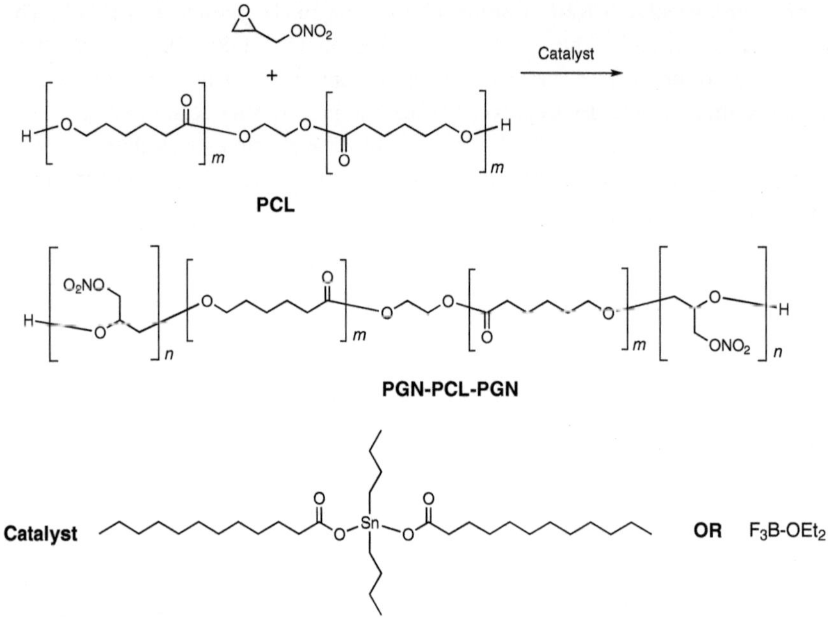

Scheme 6.43 Synthesis of poly(glycidyl nitrate-*block*-poly(caprolactone)-*block*-poly (glycidyl nitrate)) (PGN–PCL–PGN).

3-nitrato-methyl-3-methyloxetane (NIMMO) with HTPE, as a macro-initiator, to form tri-block copolymer poly(NIMMO)–(HTPE)–poly(NIMMO) (Scheme 6.44) [84]. This block copolymer exhibited two T_gs, due to its copolymer structure. The first transition was observed at −69 °C, and results from the HTPE segment, where the second transition was observed at −30 °C, originating from the terminal poly(NIMMO) segments. Poly(NIMMO)–(HTPE)–poly(NIMMO) was found to have a good thermostability exhibiting an exothermic peak at 215 °C.

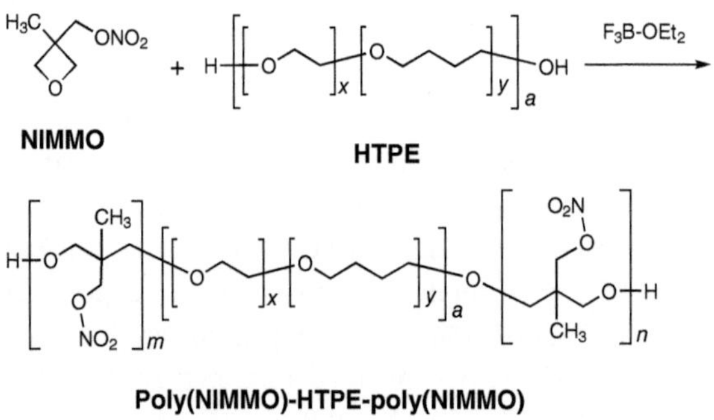

Scheme 6.44 Synthesis of tri-block copolymer poly(NIMMO)-(HTPE)-poly(NIMMO).

Xu prepared an energetic nitrate ester-containing block copolymer, poly(3,3-bis (2,2,2-trifluoroethoxy-methyl)-oxetane)-glycol-*block*-poly(3-nitratomethyl-3-methyloxetane) (PBFMO-*b*-PNMMO) (Scheme 6.45) [85]. This copolymer was prepared by reacting pre-polymers of PBFMO with well-known poly(NMMO) polymer, using toluene-diisocyanate as a coupling agent. The resulting elastomer is unique due to the intrinsic combination of energetic nitrate esters and fluorinated moieties in the same molecule, making this interesting copolymer of a significant potential for use in metallized energetic formulations. PBFMO-*b*-PNMMO copolymer exhibited T_g of −20.4 °C and had a decomposition temperature of over 200 °C. It was also found to be stable upon mixing with common propellant ingredients, such as HMX and aluminum powder.

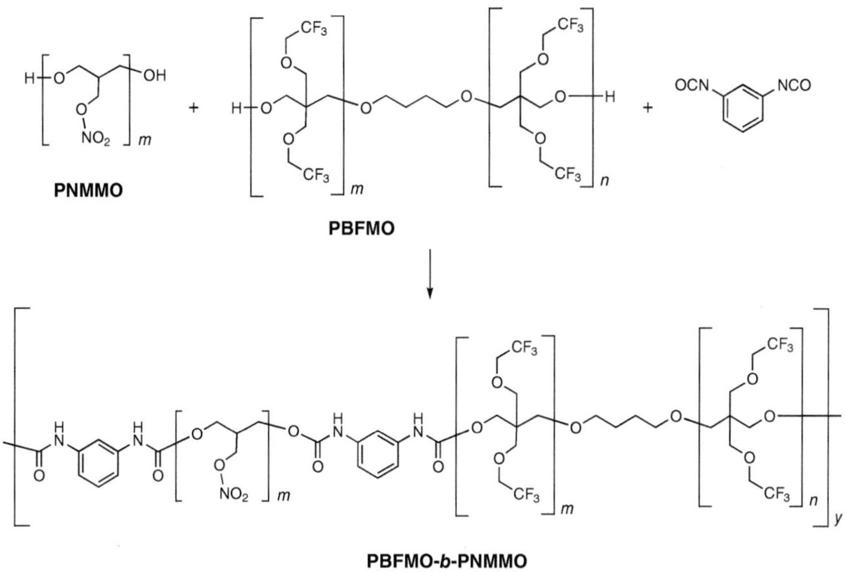

Scheme 6.45 Synthesis of poly(3,3-bis(2,2,2-trifuoroethoxy-methyl)-oxetane)-glycol-*block* poly(3 nitratomethyl-3-methyloxetane) (PBFMO-*b*-PNMMO).

Li prepared an energetic poly(3-difluoroaminomethyl-3-methyloxetane-3-nitratomethyl-3-methyl-oxetane) (PDN) copolymer containing both nitro and difluoroamino groups (Scheme 6.46) [86]. This copolymer was prepared by a cationic copolymerization of NIMMO and difluoroamino-methyl-3-methyl oxetane (DFAMO), using BDO and TFBE as an initiator and a catalyst, respectively. PDN showed low glass transition temperature and had good thermal stability, decomposing with the maximum temperature of 223.7 °C. PDN copolymer was found to be stable with RDX, HMX, TATB, and PETN and explosives. In contrast, PDN was sensitized and destabilized in its formulations with CL-20, nitroguanidine, TNT, nitrocellulose, and 2-nitrodiphenylamine. It was concluded that the NIMMO segments in PDN improved the compatibility of this copolymer with HMX, carbon black, and lead carbonate ($PbCO_3$), in comparison with compatibility of PDFAMO.

Straessler and coworkers prepared a series of aliphatic nitro substituted polymers by reacting a geminal dinitrodiol with various diacid chlorides and producing geminal dinitro polyesters (Scheme 6.47) [87]. The authors stated that they also obtained effective crosslinking by using diisocyanates.

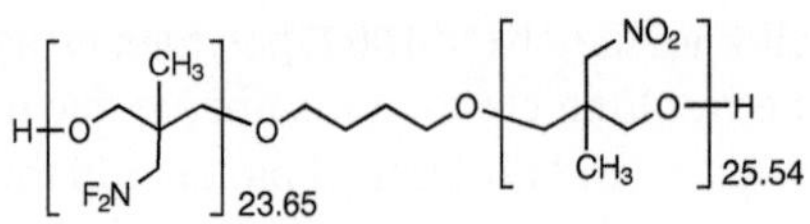

Scheme 6.46 Structure of poly(3-difluoroaminomethyl-3-methyloxetane-3-nitratomethyl-3-methyl-oxetane) (PDN) copolymer.

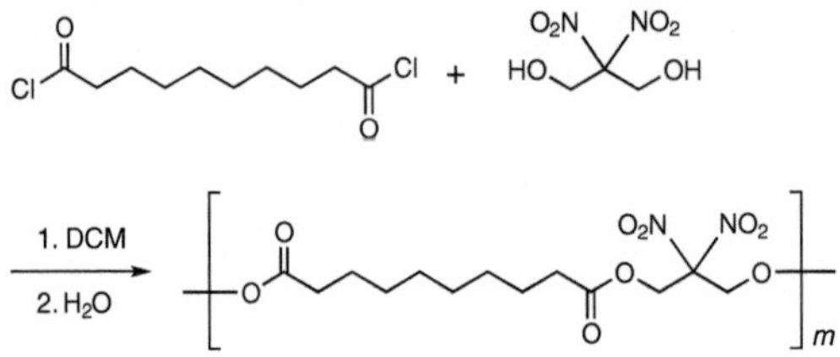

Scheme 6.47 Synthesis of energetic geminal dinitro polyester.

Additional energetic polymers containing aliphatic nitro groups were reported by Ashrafi, who developed a unique approach for the direct nitration of poly(butadiene) (PB) and HTPB (Scheme 6.48) [88]. To retain the unique physicochemical properties of low-molecular-weight PB it was nitrated to an extent of 10% of its double bonds, using sodium nitrite and iodine (*in situ* formed nitryl iodide, as an active nitrating agent), producing poly(nitro butadiene) (NPB). NPB was used in cured and unfilled nitro-hydroxyl-terminated poly(butadiene) (NHTPB) polymer. Isothermal thermogravimetric analysis (Iso-TGA) was used to determine the migration rate in cured and unfilled HTPB binder systems and compared to the dioctyl-adiphate (DOA) plasticizer. It was found that the exudation of the NPB plasticizer is slower than that of the DOA plasticizer. Thus, the energetic NHTPB/NPB binder system presents better mechanical properties than non-energetic HTPB/DOA, making the former binder/plasticizer combination to a promising energetic binder system for polymer-bonded explosives.

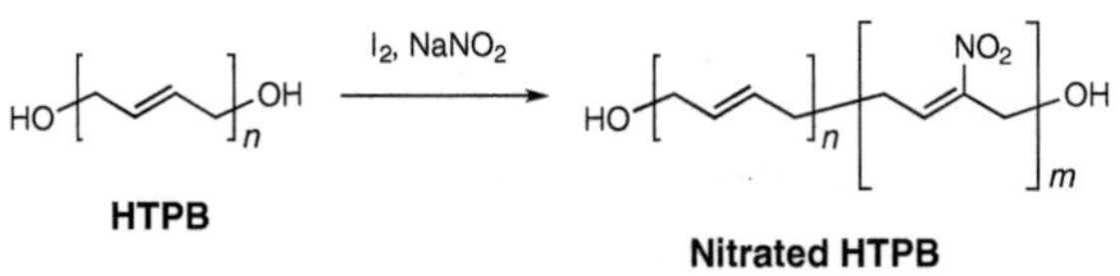

Scheme 6.48 Synthesis of nitrated-HTPB.

6.8 Aromatic C-NO_2 Containing Polymers

In addition to reports regarding aliphatic nitrate and nitrate ester-containing polymers, many new energetic polymeric materials, containing nitroaromatic (Ar-NO_2) moieties, were also investigated. Within the latter class of energetic polymers, Jin and coworkers performed numerous studies on energetic polymers substituted with nitrophenylacetal functional groups [36, 89, 90]. The authors prepared poly (vinyl-trinitro-nitro-benzal acetal) (PVPNB) by reacting PVA with trinitro-nitrophenylacetaldehyde (Scheme 6.49).

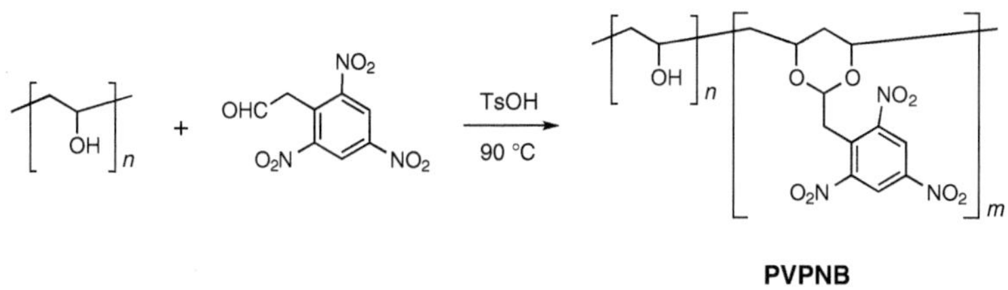

Scheme 6.49 Synthesis of PVPNB.

Subsequently, PVPNB was crosslinked with GAP, vinyl acetate azide, and isocyanates. The mechanical properties and thermal stability of the resulting binders were good and suitable for use in melt cast TNT-based formulations, due to the excellent compatibility of trinitrophenyl substituents in the molecular structures of these binders with TNT. Wu took a different approach to nitroaromatic substituted heat-resistant polymers [91]. Starting with GAP, the authors performed either a click reaction with 2,4,6-trinitro-*N*-(prop-2-yn-1-yl)-aniline or with 2,4-dinitro-*N*-(prop-2-yn-1-yl)aniline (Scheme 6.50). They also explored reducing the azide of GAP under Staudinger conditions, followed by the reaction with 2,4,6-trinitrochlorobenzene or 2,4-dinitro-chlorobenzene. The prepared polymers showed excellent thermal stabilities and the prepared polymer samples could be directly ignited without use of any additives. The authors proposed the use of their

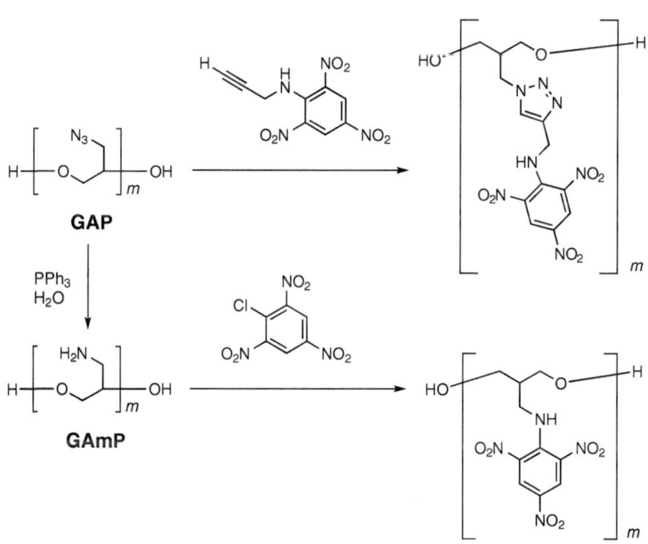

Scheme 6.50 Synthesis of GAP-derived energetic polymers containing aminotrinitrophenyl.

polymers in solid rocket propellants and PBXs, due to these polymers' impressive chemical and physical properties.

The Staudinger-reduced GAP precursor has been used by other groups as a synthetic precursor to synthesize a variety of additional energetic polymers. Glycidyl nitramine was prepared by Klapötke *via* the nitration of this same material after a protection step (Scheme 6.51) [92]. The authors found the poly(glycidyl nitramine) to have a better energy of explosion, detonation velocity, and detonation pressure compared to GAP, while possessing low sensitivity toward impact and friction and being stable up to 170 °C.

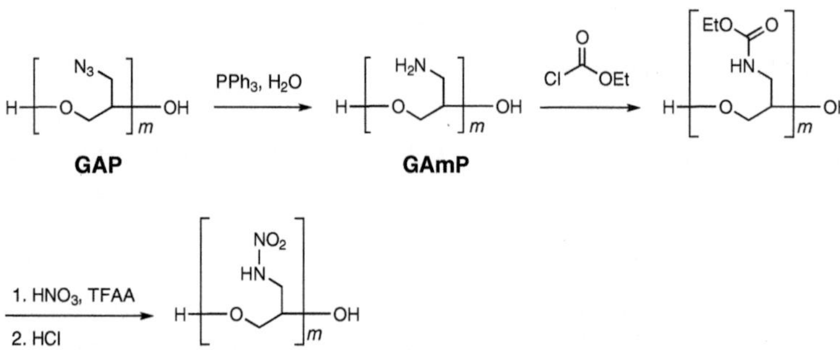

Scheme 6.51 Synthesis of poly(glycidyl nitramine).

Formulations of reactive energetic plasticizers (REPs) with hydroxy-telechelic poly(glycidylazide-*co*-tetrahydrofuran) (PGT)-based energetic polyurethane (PU) binders were investigated for use in solid propellants and PBXss. A motivation for the development of *gem*-dinitro REPs containing terminal alkyne groups was aimed to satisfy the demands for conventional energetic plasticizers, as well as to prevent the migration of plasticizers inside a solid energetic matrix. The *gem*-dinitro REPs with the terminal alkyne group were synthesized *via* Michael addition, hydrolysis reaction, and esterification reaction, while hydroxy-telechelic PGT pre-polymer was prepared by cationic ring-opening polymerization, followed by an azidation reaction (Schemes 6.52 and 6.53) [93].

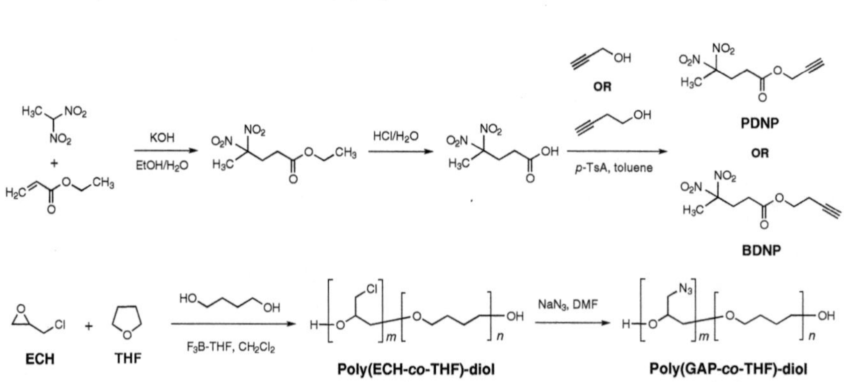

Scheme 6.52 Synthetic route to REP *via* PDNP and BDNP, and to poly(GAP-*co*-THF).

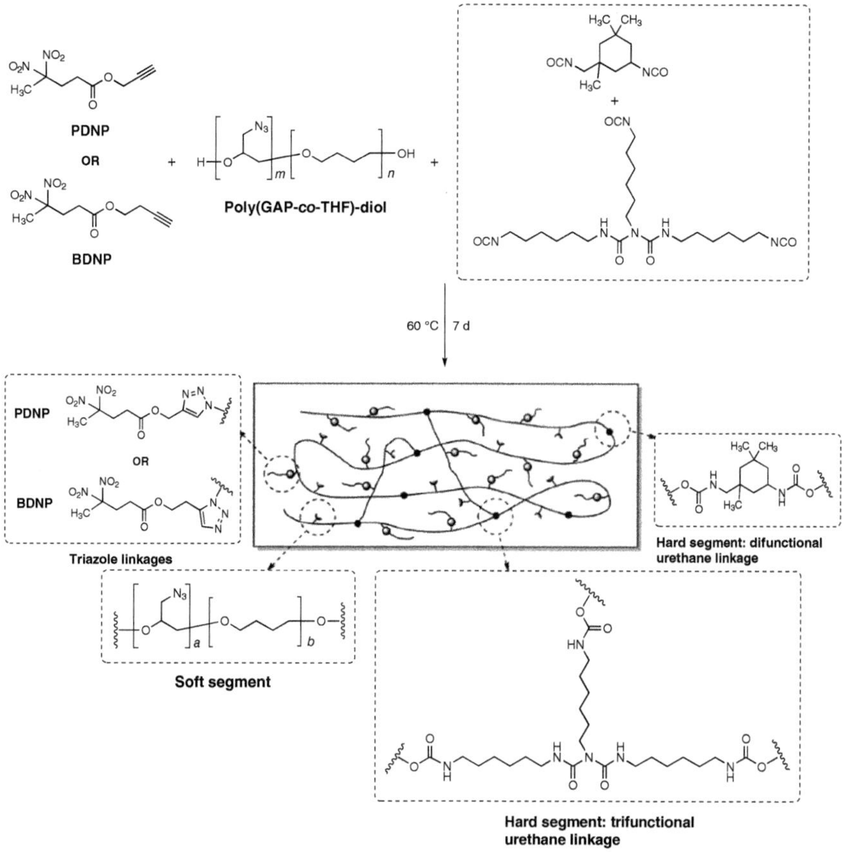

Scheme 6.53 Synthetic route to the PGT-based PU in the presence of REP.

The miscibility and rheological behavior of a binary mixture of PGT/REP with various REP fractions were determined. The kinetics on the reactivity of propargyl versus 3-butynyl species of REPs toward the azide groups of the PGT pre-polymer, in terms of Cu-free azide-alkyne-1,3-dipolar cycloaddition, was studied by ^{1}H NMR, where the activation energies (E_a) of 83.4 kJ mol^{-1} for PDNP/PGT and 91.6 kJ mol^{-1} for BDNP/PGT were obtained from the DSC measurements. The REPs could effectively reduce the viscosity of the PGT pre-polymer from 1485 to 350–450 cP (at 30 °C), with a REP loading of up to 50 wt%, which was much lower than that of the PGT/EP mixtures. The thermal stability of the finally cured energetic binders with the incorporation of REPs indicated that the thermal stability of the REP/PGT-based PUs was maintained independent of the REP content. TGA of all PUs showed a similar decomposition temperature of around 233 ± 4 °C, indicating that REP/PGT-based PUs maintained the thermal stability of the control PGT-based PU, and the REPs presented good thermal stability until the polymeric matrix decomposed. The tensile strength and modulus of the PUs increased with an increase in the REP content. In addition, the REP/PGT-based PUs presented higher tensile strength than

REP/GAP-based PUs at the same REP content, suggesting that the tetramethylene oxide soft segment in the REP/PGT-based PUs played a dominant role in enhancing the mechanical properties. The impact sensitivity of the REP/PGT-based PUs was also reduced by the converting the azide groups to the triazole groups and was also lower than the REP/GAP-based PUs for the same REP content. Compared to the pure REPs, REP triazoles showed better energetic performance in terms of energy of explosion, explosion temperature, detonation pressure, detonation velocity, and specific impulse.

6.9 Conclusions

In the recent decade, energetic nitrogen-rich polymers and plasticizers have gained rapid development in materials science. Great research and development efforts were directed on the study of various possible and prospective structural modifications of glycidyl azide polymer to increase its performance and to decrease its mechanical sensitivity to impact and friction. In this regard, a number of cross-linked polymers and copolymers, incorporating a glycidyl azide subunit and possessing optimized functional properties, was prepared and evaluated. At the same time, a broad array of new nitro substituted aliphatic and aromatic polymers and binders, as valuable components of solid propellants and plastic-bonded explosives, were also reported. In addition, the preparation of nitrogen heterocycle-based energetic polymers and plasticizers became an important modern trend in the design of next-generation high-energy materials. Heterocyclic polymers prepared *via* curation of glycidyl-azide polymer, or through sequential functionalization of the pre-installed heterocyclic rings, were evaluated as promising alternatives to known materials, due to the improved tensile strength and elongation of the new polymers. Moreover, a novel promising sub-class of imidazolium-based polymeric ionic liquids received much attention as low-sensitive energetic materials with high nitrogen and oxygen content. There is no doubt that the construction of novel nitrogen-rich polymers, plasticizers, and binders will be in the spotlight of energetic chemistry, materials, and engineering in upcoming decades and will provide new fundamental and applied perspectives for advanced materials in general.

References

1 Wang, B., Feng, Y., Qi, X. et al. (2018). *Chem. Eur. J.* 24: 15897–15902.
2 Sevilia, S., Yong, M., Grinstein, D. et al. (2019). *Macromol. Mater. Eng.* 304: 1900018.
3 Bodaghi, A. and Shahidzadeh, M. (2019). *Cent. Eur. J. Energetic Mater.* 16: 259–280.
4 Hu, J., Gong, L., He, J., and Yang, R. (2020). *Polym. Adv. Technol.* 31: 749–758.
5 Sonawane, S., Anniyappan, M., Athar, J. et al. (2017). *Propellants Explos. Pyrotech.* 42: 386–393.

6 Kumar, D., Singh, H., Maurya, M. et al. (2020). *Mater. Lett.* 280: 128585.
7 Landsem, E., Jensen, T.L., Kristensen, T.E. et al. (2013). *Propellants Explos. Pyrotech.* 38: 75–86.
8 Ma, M., Shen, Y., Kwon, Y. et al. (2016). *Propellants Explos. Pyrotech.* 41: 746–756.
9 Rahm, M., Malmström, E., and Eldsäter, C. (2011). *J. Appl. Polym. Sci.* 122: 1–11.
10 Kizhnyaev, V.N., Pokatilov, F.A., Vereshchagin, L.I. et al. (2009). *Russ. J. Appl. Chem.* 82: 1769–1775.
11 Kizhnyaev, V.N., Golobokova, T.V., Pokatilov, F.A. et al. (2017). *Chem. Heterocycl. Compd.* 53: 682–692.
12 Sukhanov, G.T., Bosov, K.K., Sukhanova, A.G. et al. (2021). *Propellants Explos. Pyrotech.* 46: 1526–1536.
13 Belousov, G.K. and Zuraev, A.V. (2019). *Tetrahedron Lett.* 60: 151217.
14 Kshirsagar, A.D., Hundiwale, D.G., Gite, V.V., and Mahulikar, P.P. (2017). *Chem. Heterocycl. Compd.* 53: 1090–1093.
15 Wang, Y., Chen, H., Xu, Y. et al. (2015). *J. Macromol. Sci. Part A Pure Appl. Chem.* 52: 707–712.
16 Klapötke, T.M. and Sproll, S.M. (2010). *Eur. J. Org. Chem.* 1169–1175.
17 Betzler, F.M., Boller, R., Grossmann, A., and Klapötke, T.M. (2013). *Z. Naturforsch.* 68B: 714–718.
18 Pokatilov, F.A., Kizhnyaev, V.N., Akamova, E.V., and Edel'shtein, O.A. (2020). *Polym. Sci. Ser. B Polym. Chem.* 62: 190–195.
19 Pokatilov, F.A. and Kizhnyaev, V.N. (2012). *Polym. Sci. Ser. A Polym. Phys.* 54: 894–899.
20 Tarchoun, A.F., Trache, D., Klapötke, T.M., and Khimeche, K. (2020). *Chem. Eng. J.* 400: 125960.
21 Klapötke, T.M. and Sproll, S.M. (2010). *J. Polym. Sci. Part A: Polym. Chem.* 48: 122–127.
22 Sankar, R.M., Saha, S., Meera, K.S., and Jana, T. (2009). *Bull. Mater. Sci.* 32: 507–514.
23 Sankar, R.M., Roy, T.K., and Jana, T. (2011). *Bull. Mater. Sci.* 34: 745–754.
24 Rao, B.N., Yadav, P.J.P., Malkappa, K. et al. (2015). *Polymer* 77: 323–333.
25 Cohen, A., Yan, Q.-L., Shlomovich, A. et al. (2015). *RSC Adv.* 5: 106971–106980.
26 Pisharath, S. and Ang, H.G. (2007). *Polym. Degrad. Stab.* 92: 1365–1377.
27 Maksimowski, P., Kasztankiewicz, A.B., and Kopacz, W. (2017). *Propellants Explos. Pyrotech.* 42: 1020–1026.
28 Yan, Q.-L., Cohen, A., Chinnam, A.K. et al. (2016). *J. Mater. Chem. A* 4: 18401–18408.
29 Pasquinet, E. (2012). *Nitrogen-Rich Polymers as Candidates for Energetic Applications* (ed. A. De Souza Gomes), 313–338. London. https://doi.org/10.5772/46245: New Polymers for Special Applications, IntechOpen.
30 Klapötke, T.M. (2012). *Propellants Explos. Pyrotech.* 37 (4): 510.
31 Kozuch, K. and Wojewodka, A. (2010). *Chemik* 64 (1): 15–20.
32 Nair, U.R., Asthana, S.N., Rao, A.S., and Gandhe, B.R. (2010). *Defence Sci.* 60 (2): 137–151.

33 Eroglu, M.S. and Bostan, M.S. (2017). *Org. Commun.* 10 (3): 135–143.

34 Zhang, Y., Zhao, J., Yang, P. et al. (2012). *Polym. Eng. Sci.* 52 (4): 768–773.

35 Keicher, T., Janitschek, W., Schaller, U., and Krause, H. (2014). Energetic materials: particles, processing, applications. 45th International Annual Conference of the Fraunhofer Institute of Chemical Technologies, Karlsruhe, Germany (24–27 June 2014).

36 Chen, C., Bo, B., Peng, R. et al. (2015). *J. Appl. Polym. Sci.* 132: 42126.

37 Shan, D., Jin, B., Peng, R., and Chu, S. (2016). *Polym. Sci. Ser. B Polym. Chem.* 58 (2): 194–204.

38 Wang, S., Huo, J., and Shu Y. (2015). Proceedings of the 18th International Seminar "New trends in research of energetic material, University of Pardubice, Pardubice, Czech Republic (15–17 April 2015) 293–300.

39 Hu, Y.-W., Zheng, Q.-L., Zhou, W.-L., and Xiao, L.Q. (2016). *Chin. J. Energetic Mater.* 24 (4): 331–335.

40 Li, G., Dong, H., Liu, M. et al. (2017). *Polym. Int.* 66 (7): 1037–1043.

41 Hafner, S., Keicher, T., and Klapoetke, T.M. (2018). *Propellants Explos. Pyrotech.* 43 (2): 126–135.

42 Hu, J., Liu, Y., Cong, K. et al. (2020). *Polymers* 12 (5): 1199.

43 Hu, J., Tang, W., Yonghui, L. et al. (2020). *Polymers* 12 (2): 278.

44 Hu, J., Chen, J., Cong, K. et al. (2021). *Mater. Sci. Eng., B* 272: 115335.

45 You, J.S., Kweon, J.O., Kang, S.C., and Noh, S.T. (2010). *Macromol. Res.* 18 (12): 1226–1232.

46 Mostafa, C. and Yadollah, B. (2018). *Cent. Eur. J. Energetic Mater.* 15 (2): 243–257.

47 Mostafa, C. and Yadollah, B. (2019). *Cent. Eur. J. Energetic Mater.* 16 (1): 33–48.

48 Zhang, C., Zheng, X., Li, Z. et al. (2017). *Propellants Explos. Pyrotech.* 42 (9): 1037–1043.

49 Zhang, G., Zhang, T., Li, J., and Luo, Y. (2018). *Polym. Int.* 67 (1): 68–77.

50 Lu, X.-M., Mo, H.-C., Ding, F. et al. (2016). *Chin. J. Energetic Mater.* 24 (10): 947–952.

51 Holler, M., Keicher, T., Gerber, P. et al. (2014). Energetic materials: particles, processing, applications. *45th International Annual Conference of the Fraunhofer Institute for Chemical Technology*, Congress Center, Stadthalle, Karlsruhe, Germany (24–27 June 2014). Fraunhofer Institute for Chemical Technology. Holler/1-Holler/14, 14 pp.

52 Holler, M.; Keicher, T.; Gerber, P.; Happ, A. (2015). DAP: A Promising New Energetic Chain Extender. 41st International Pyrotechnics Seminar (EuroPyro 2015), Toulouse, France (4–7 May 2015). 803–815.

53 Zhang, Z., Wang, G., Wang, Z. et al. (2015). *Polym. Bull.* 72: 1835–1847.

54 Li, B., Zhao, Y., and Li, X. (2015). *Chin. J. Energetic. Mater.* 23 (7): 624–628.

55 Filippi, S., Mori, L., Cappello, M., and Polacco, G. (2017). *Propellants Explos. Pyrotech.* 42 (7): 826–835.

56 Pant, C.S., Santosh Mada, S.S.N.M., Mehilal et al. (2016). *J. Energetic Mater.* 34 (4): 440–449.

57 Jin, B., Shen, J., Peng, R. et al. (2012). *J. Polym. Res.* 19 (10): 1–9.

58 Jin, B., Shen, J., Peng, R. et al. (2012). *Polym. Degrad. Stab.* 97 (4): 473–480.

59 Cossu, C., Heuzey, M.C., Lussier, L.S., and Dubois, C. (2011). *J. Appl. Polym. Sci.* 119 (6): 3645–3657.

60 Tarchoun, A.F., Trache, D., Klapötke, T.M. et al. (2019). *Macromol. Chem. Phys.* 220: 654–658.

61 Vigo, T.L. and Sachinvala, N. (1999). *Polym. Adv. Technol.* 10: 311–320.

62 Roberts, R.W. (1957). *J. Am. Chem. Soc.* 79: 1175–1178.

63 Shamsipur, M., Pourmortazavi, S.M., Hajimirsadeghi, S.S., and Atifeh, S.M. (2012). *Fuel* 95: 394–399.

64 Bellan, A.B., Hafner, S., Hartdegen, V.A., and Klapötke, T.M. (2016). *J. Appl. Polym. Sci.* 133 (40): 43991.

65 Dong, Q., Li, Y., Wu, F. et al. (2017). *Propellants Explos. Pyrotech.* 42 (10): 1143–1148.

66 Gaur, B., Lochab, B., Choudhary, V., and Varma, I.K. (2003). *J. Macromol. Sci.* C43 (4): 505–545.

67 Mohan, Y.M., Mani, Y., and Raju, K.M. (2006). *Des. Monomers Polym.* 9 (3): 201–236.

68 Jiang, K., Deng, H.L., Zhang, Q. et al. (2018). *Green Chem.* 20 (12): 2813–2820.

69 St-Charles, J.C. and Dubois, C. (2020). *Propellants Explos. Pyrotech.* 45 (6): 889–897.

70 Xu, M., Ge, Z., Lu, X. et al. (2017). *RSC Adv.* 7 (75): 47271–47278.

71 Xu, M., Ge, Z., Lu, X. et al. (2017). *Polym. Int.* 66 (9): 1318–1323.

72 Li, H., Pan, R., Wang, W., and Zhang, L. (2014). *Appl. Mech. Mater.* 628: 28–32.

73 Li, H., Pan, J., Wang, W. et al. (2018). *J. Macromol. Sci. Part A Pure Appl. Chem.* 55 (2): 135–141.

74 Ghosh, K., Athar, J., Pawar, S. et al. (2012). *J. Energetic Mater.* 30 (2): 107–123.

75 Kumari, D., Singh, H., Patil, M. et al. (2013). *Thermochim. Acta* 562: 96–104.

76 Claßen, M., Heimsch, S.B., and Klapötke, T.M. (2019). *Propellants Explos. Pyrotech.* 44: 1515–1520.

77 Baghersad, M.H., Habibi, A., and Heydari, A. (2017). *J. Mol. Struct.* 1130: 447–454.

78 Xu, R., Li, Z., Chen, Y. et al. (2020). *Dalton Trans.* 49 (26): 9016–9023.

79 Hafner, S., Hartdegen, V.A., Hofmayer, M.S., and Klapötke, T.M. (2016). *Propellants Explos. Pyrotech.* 41: 806–813.

80 Drees, D., Löffel, D., Messemer, A., and Schmid, K. (1999). *Propellants Explos. Pyrotech.* 24: 159–162.

81 Ghorbani, M. and Bayat, Y. (2015). *Polym. Sci. Ser. B Polym. Chem.* 57: 654–658.

82 Abrishami, F., Zohari, N., and Vida Zeynali, V. (2017). *Propellants Explos. Pyrotech.* 42: 1032–1036.

83 Kazemi, A., Hayaty, M., Mousaviazar, A. et al. (2015). *J. Therm. Anal. Calorim.* 119: 613–618.

84 Zhang, Q., Liu, N., Xu, M.H. et al. (2019). *Eur. Polym. J.* 119: 514–522.

85 Xu, M., Lu, X., Mo, H. et al. (2019). *RSC Adv.* 9: 29765–29771.

86 Li, H., Yang, Y., Pan, J. et al. (2018). *Cent. Eur. J. Energetic Mater.* 15 (1): 85–99.

87 Straessler, N.A., Killpack, M.O. and Doll, D.W. (2020). Methods of producing energetic polymers, energetic binders, and energetic compositions. US Patent 10,696,786, filed 11 December 2014 and issued 16 June 2016.

88 Ashrafi, M., Fakhraian, H., and Dehnavi, M.A. (2017). *Propellants Explos. Pyrotech.* 42: 269–275.

89 Jin, B., Shen, J., Gou, X. et al. (2016). *Polymers* 8 (1): 10.

90 Jin, B., Shen, J., Peng, R., and Dong, H. (2014). *Macromol. Res.* 22: 117–123.

91 Wu, Q., Ma, Q., Zhang, Z. et al. (2020). *Chem. Eng. J.* 399: 1385–8947.

92 Betzler, F., Hartdegen, V.A., Klapötke, T.M., and Sproll, S.M. (2016). *Cent. Eur. J. Energetic Mater.* 13: 289–300.

93 Ma, M. and Kwon, Y. (2018). *Polymers* 10 (5): 516.

7

Tetrazole Energetic Salts Based on Various Explosophores: Recent Overview of Synthesis and Energetic Properties

Saira Manzoor, Qamar-un-nisa Tariq, and Jian-Guo Zhang

Beijing Institute of Technology, State Key Laboratory of Explosion Science and Technology, School of Mechatronic Engineering, Beijing, No. 5, South Street, Zhongguancun, Haidian District, Beijing 100081, PR China

7.1 Introduction

In recent years, the design and synthesis of innovative high-energy-density materials (HEDMs) have gained prevalent attention [1–3]. The substantial interest in this essential and functional research area is owing to the dual civilian (space technologies, mining explosion, and welding) and military applications (propellants, explosives, pyrotechnics, and gas-generating agents) [4]. These military and civilian applications bring enlarged requirements to the preparation of novel HEDMs with advanced performance. On one hand, these requirements are connected with the exploration of innovative energetic materials that comprehends high density, detonation performance (detonation pressure and velocity), thermal stability, low sensitivity (impact and friction), and eco-friendly disintegration products. On the other hand, meticulous needs for their non-violent synthesis, separation, and handling should be considered [5]. However, the conflict between these necessities, particularly between energetic performance and safe handling, is frequently a challenging task, because the improved properties originate generally at the cost of molecular stability [6]. The classic chemical explosives like black powder, ammonium perchlorate, nitroglycerine, and 2,4,6-trinitrotoluene (TNT) symbolize the utmost significant materials in the past and attained sufficient consideration in chemical science. TNT was established to be significant due to the insensitivity toward mechanical stimulation; however, it is inferior in energetic properties in evaluation to the modern HEDMs such as 1,3,5-trinitro-1,3,5-triazine (RDX) and 1,3,5,7-tetranitro-1,3,5,7-tetrazocine (HMX), 2,4,6-triamino-1,3,5-trinitrobenzene (TATB), and hexanitrohexaazaisowurtzitane (CL-20) (Table 7.1). The preparation of TNT, RDX, and HMX suffers from harsh, less eco-friendly synthetic approaches [8], whereas the energetic compound CL-20 has been recommended to a restricted use due to the economic issues of synthesis and possession of high energy with greater sensitivity toward mechanical stimulation [9]. Additionally, the environmental concerns associated with extensive consumption of primary explosives such

Nitrogen-Rich Energetic Materials, First Edition. Edited by Michael Gozin and Leonid L. Fershtat.

Table 7.1 Physicochemical and energetic properties of typically used energetic compounds [7].

Compound	T_d a) (°C)	d b) (g cm^{-3})	D c) (m s^{-1})	P d) (GPa)	IS e) (J)	FS f) (N)	OB g) (%)	$\Delta H°_f$ h) (kJ mol^{-1})
TNT	295	1.65	7303	21.3	15	353	−74.0	−59.3
TATB	360	1.94	8544	32.1	50	360	−55.8	−154.2
RDX	210	1.80	8795	34.9	7.5	120	−21.6	70.7
HMX	280	1.91	9144	39.2	7	120	−21.6	74.8
CL-20	210	2.04	9706	45.2	4	94	−11.0	397.8

Note: In this review, all the quantity symbols are used in tables.
a) Decomposition temperature (onset).
b) Density.
c) Detonation velocity (calculated).
d) Detonation pressure (calculated).
e) Impact sensitivity.
f) Friction sensitivity.
g) Oxygen balance (based on CO_2) for $C_aH_bO_cN_d$, [1600($c - 2a - b/2$)/MW]; MW = molecular weight.
h) The heat of formation (calculated).
Source: Adapted from Gao et al. [7].

as lead azide and lead styphnate require urgent replacement. Therefore, the newly proposed HEDMs should possess high detonation performances and acceptable mechanical sensitivity in addition to the environmental friendliness to bring world safety [10].

The development of new HEDMs as propellants, explosives, and pyrotechnics in modern times proceeds with various emerging concepts and techniques, in particular, "energetic salt formation." [11–13] The concept of the synthesis of energetic salts originated from ionic liquids. They possess low vapor pressure and are considered as "green energetic materials" in comparison to conventional materials. Recently, a specific group of energetic nitrogen-rich salts has been introduced. These high-nitrogen energetic salts gain their energy from the high heat of formation (HOF) rather than from heat of combustion due to the presence of N—N and C—N bonds. Therefore, the nitrogen-rich cations and anions used for the synthesis of energetic salts can strengthen the value of HOF, density, thermal stability, and improve oxygen balance (OB) owing to the low hydrogen and carbon content [14–16]. The common methods to prepare nitrogen-rich energetic salts include neutralization or metathesis reaction approaches which can be optimized to advance the energetic properties by combining the different cations and anions. Therefore, to achieve these key requirements of HEDMs, modification of existing or new heterocyclic compounds with nitrogen-rich cations and anions leads to the inauguration of the new direction of success toward operational energetic materials [11].

Recently, heterocyclic compounds with high-nitrogen contents have been explored for energetic salts preparation. In this regard, various azoles, such as pentazole, tetrazole, triazole, pyrazole, and imidazole, have received considerable attention

Table 7.2 Properties of five-membered nitrogen-containing heterocyclic compounds (azoles) [14].

Azole	Pentazole	Tetrazole	1*H*-1,2,3-triazole	2*H*-1,2,3-triazole	1,2,4-triazole	Pyrazole	Imidazole
N (%)	98.6	79.8	60.8	60.8	60.8	41.2	41.2
HOF ($kJ\,mol^{-1}$)	—	320	271.7	255.5	192.7	177.4	129.5
Density ($g\,cm^{-3}$)	—	0.79	1.19	—	1.15	1.10	1.01

Source: Adapted from Kommu et al. [14].

[12, 13]. Some of the major properties of azoles are given in Table 7.2 [14]. From Table 7.2 it can be concluded that tetrazole and pentazole have excellent energetic properties as compared to other azoles, but pentazole is of unstable and very reactive nature [15]. Therefore, tetrazole drives additional importance for energetic compounds and salts investigation owing to the practical significance and potential for diverse properties [16–18]. The significant characteristics of the tetrazole ring are high HOF, ring strain, stability, and density, which are useful properties for high-performing explosives [19, 20].

In this review, we present the tetrazole-based energetic salts with a variety of energetic groups such as nitro, nitramino, amino, hydrazine, along with tetrazole *N*-oxide, bistetrazole, or bridged bistetrazole motifs since 2010 and their impact on energetic properties. The reported tetrazole-based compounds and their energetic salts were usually characterized by infrared, Raman, multinuclear NMR spectroscopy, elemental analysis, differential scanning calorimetry (DSC), mass spectrometry, and X-ray diffraction analysis. Moreover, the energetic properties were also calculated by using EXPLO5 [21, 22], and CHEETAH software [23, 24]. The values of impact and friction sensitivities were also determined by standard BAM drop hammer and friction tester practices [25]. The properties of these materials such as density, thermal decomposition temperature, HOF, detonation performance (detonation pressure and velocity), oxygen balance, impact, and friction sensitivity are discussed in detail. TNT, TATB, RDX, HMX, and CL-20 are used as standard HEDMs for comparison. It is anticipated that this review will give a summarized impression of the major methodologies used for the design and synthesis of tetrazole-based nitrogen-rich energetic salts and the effect of functional groups on the energetic properties more specifically on density, detonation properties, and thermal stabilities to fulfill the requirement of HEDMs. A comparison of these synthesized compounds with standard HEDMs will encourage further studies to design similar energetic compounds with improved properties.

7.2 Tetrazole-Based Energetic Salts

Nitrogen-rich heterocyclic compounds are one of the basic sources in a preparation of energetic salts. For the synthesis of energetic materials, the accessibility of raw material is important in addition to the easy and safe synthetic method, and tetrazole is one heterocycle that allows both. Tetrazole-based compounds attain high density and release extensive energy upon disintegration because of the flat structure and high-nitrogen content [26, 27]. These types of compounds also possess high HOFs due to a large number of N—N bonds, ring strain, and high density [28]. Additionally, this ring displays extraordinary stability during preparation from its heterocyclic derivatives in the presence of alkalis, acids, reducing and oxidizing agents [29]. Tetrazolates and tetrazolium salts are formed due to the deprotonating and protonation of the tetrazole ring, respectively. The former enhance thermal stability, and the latter decrease the decomposition temperature [30]. Moreover, the physicochemical and energetic properties of tetrazole derivatives can be upgraded by the introduction of different functional groups such as $-NO_2$, $-ONO_2$, $NHNO_2$, $-N_3$, –N=N–, and *N*-oxide [31]. Recently, many tetrazole-based energetic compounds and their salts are reported as primary and secondary explosives, sensitizers, and electric igniters and proved to be good due to their distinctive properties (high nitrogen content, density, HOF, thermal stability, and low sensitivity) [32–45]. Therefore, the synthesis of tetrazole-containing nitrogen-rich salts with a combination of different explosophores to achieve exciting energetic properties has appealed to many researchers [46–49].

7.2.1 Synthesis of Nitro, Nitramino, or Azido Tetrazole Energetic Salts

The nitro and nitramino explosophores along with organic heterocyclic frameworks such as tetrazole is one ideal combination for the synthesis of HEDMs. These kinds of explosophores play a vital role to increase the density, detonation performance, as well as oxygen balance of the energetic material. Further, the azido (N_3) group gained more interest in the field of energetic materials due to 100% nitrogen content, high enthalpy of formation, and strong explosive power [50]. According to the recent research, an introduction of one azido group into an organic compound increases its energy content in approximately $300\,kJ\,mol^{-1}$ and HOF in about $360\,kJ\,mol^{-1}$ [51, 52]. However, such energetic materials often suffer from low thermal stability and high sensitivities toward impact and friction; therefore, it is difficult to meet the essential requirement of HEDMs for practical applications. Additionally, the synthesis of fully nitro, nitramino, and azido functionalized tetrazoles and their energetic salts remains challenging to gain high-energy compounds. Therefore, synthetic chemists tried to introduce these energetic groups into the molecular skeleton of the compound and synthesized their energetic salts [53]. Qian Wang et al. prepared a series of energetic compounds by utilizing tetrazolotriazine building block as shown in Scheme 7.1 [54]. The stability and density of energetic compounds through hydrogen bonding were achieved by the introduction of a tetrazole group into the fused framework. Compound **1** was easily synthesized by diazotization of 5-amino-1*H*-tetrazole following by azo coupling with malononitrile in a weak base solution and the assembly of the tetrazole ring using sodium azide in hydrochloric acid. Further, a series of energetic salts (**1-1** to **1-3**) was prepared *via* metathesis reactions in good yields. The method for the preparation of these compounds was easy and cost effective as the starting materials were commercially available. Physicochemical and energetic properties of **1** and its energetic salts are given in Table 7.3. Energetic compound **1** and its salts retain impact and friction sensitivities up to 31–35 J and 300–330 N, respectively, showing their insensitive nature. The detonation properties (pressure and velocity) were in the range of 24.9–28.0 GPa and 8313–8679 m s^{-1}. Compound **1** and its energetic salts possess densities in the range of 1.73–1.79 g cm^{-3} which were greater than those of TNT. High density and good HOF were responsible for the better detonation performance of these energetic compounds and salts. Moreover, the thermal decomposition range was 216–230 °C, reachable to RDX only in the case of energetic salt **1-2**.

Recently, methylated polyazole energetic compounds prepared by combining the tetrazolyl and dinitromethyl groups with a 1,2,3-triazole ring were reported [55]. The synthetic pathway toward compounds **2** and **2-1** to **2-4** is illustrated in Scheme 7.1. The dinitromethyl-1,2,3-triazole potassium salt was prepared using 2-methyl-4,5-dicyano-1,2,3-triazole as a raw material through the diazotization and nitration reactions. The nitrated product was obtained using a mixture of trifluoroacetic acid anhydride (TFAA) and 100% HNO_3 in a form of colorless oil; this oily product immediately reacted with KI to afford the target potassium salt. After that, the neutral compound was attained through a click reaction with NaN_3 and acidification by 2 M HCl. The energetic salts **2-1** to **2-4** were obtained after the reaction of the neutral compound with nitrogen bases in ethanol. Salt **2-4** was obtained as a result of a metathesis reaction.

The energetic properties of prepared energetic salts **2-1** to **2-4** are given in Table 7.3. The thermal decomposition range was from 153 to 209 °C which was not reasonable as compared to standard HEDMs. The calculated density of target compounds was found to be in the range of 1.61–1.73 g cm^{-3} which was higher than TNT. All compounds have comparatively high positive HOFs (402.8–1144.9 kJ mol^{-1}) in comparison to traditional HEDMs. Prepared compounds were less sensitive than RDX toward impact and the measured oxygen balance (OB) value for **2-3** (−22%) is advanced than the values of TNT. Therefore, the hydroxylammonium (**2-3**) energetic salt has good energetic properties that were higher than TNT, close to RDX, and lower than for other standard HEDMs.

Honglei Xia and coworkers performed the synthesis of 3,6-dinitro-1,4-di(1*H*-tetrazol-5-yl)-pyrazolo[4,3-*c*]pyrazole (**DNTPP** (**3**)) by using an acid–base neutralization of 3,6-dinitropyrazolo[4,3-*c*]pyrazole (**DNPP**) with KOH followed by an introduction of the tetrazole group *via in situ* produced cyanogen azide (upon interaction of sodium azide and cyanogen bromide) as presented in Scheme 7.1 [56, 57]. Additionally, energetic salts (**3-1** to **3-5**) were prepared by reacting of **3** with various nitrogen-rich bases that were stable under air and can be stored for a long time. Prepared compounds were thermally stable in the range of 178–281 °C which fulfills the requirement of HEDMs, except in the case of hydrazinium energetic **3-2** (178 °C) (Table 7.3). The detonation pressures and velocities were in the range of 22.4–31.3 GPa and 8023–9082 m s^{-1}, respectively. This indicated that the detonation performance of diaminomethaniminium salt (**3-4**) was found to be close to RDX. Hence, the detonation performance of **DNTPP** (**3**) and their energetic

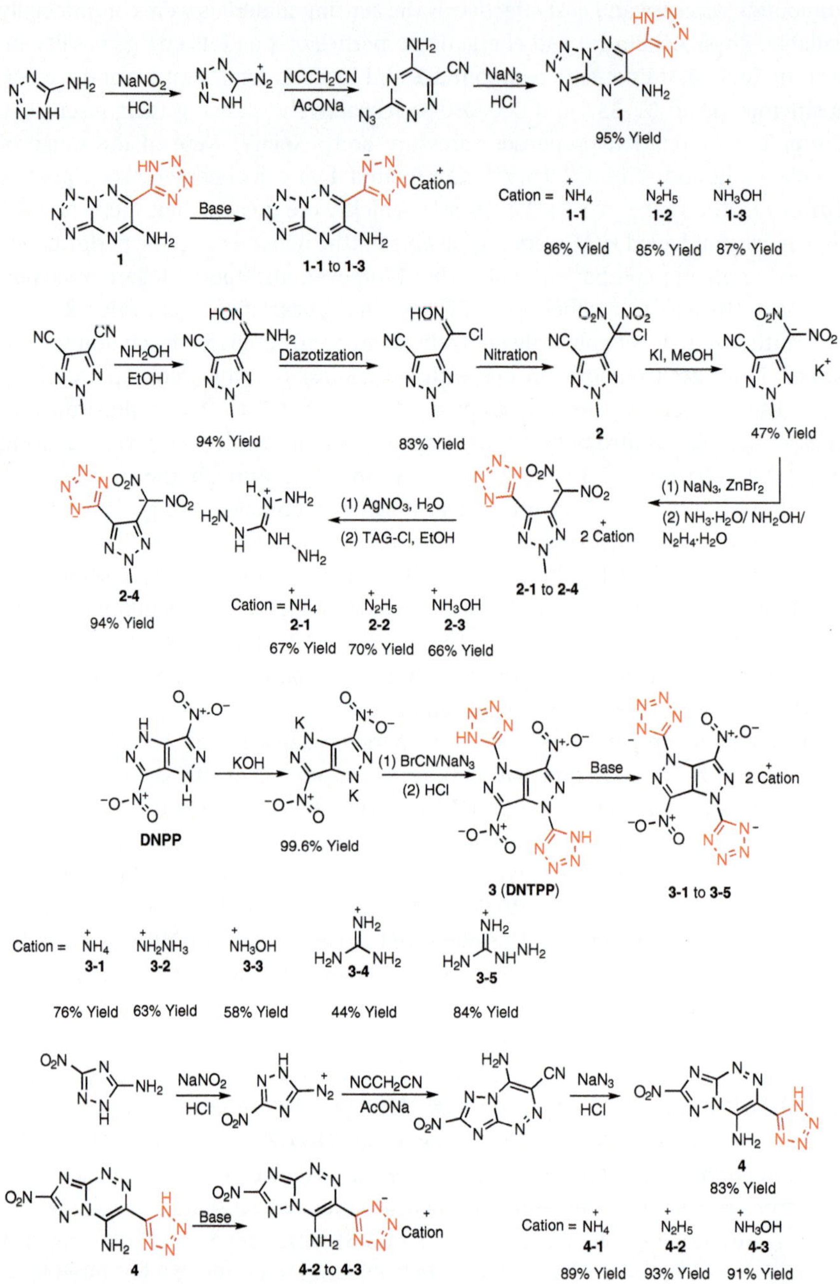

Scheme 7.1 Synthesis of nitrotetrazoles**2-4**, and their energetic salts and azidotetrazole compound **1** and its energetic salts.

Table 7.3 Physicochemical and energetic properties of hetaryltetrazoles**1-4** and their energetic salts.

Compound	T_d (°C)	d (g cm^{-3})	D (m s^{-1})	P (GPa)	IS (J)	FS (N)	OB (%)	$\Delta H°_f$ (kJ mol^{-1})
1	220	1.79	8367	25.6	31	300	−42	866.5
1-1	216	1.73	8313	24.9	33	320	−50	739.5
1-2	230	1.74	8679	27.6	33	310	−50	882.1
1-3	192	1.75	8618	28.0	35	330	−43	785.7
2-1	153	1.63	8178	24.4	19	>360	−35	402.8
2-2	209	1.65	8640	27.7	16	>360	−37	724.0
2-3	180	1.73	8811	30.9	18	>360	−22	509.9
2-4	176	1.61	8516	25.9	28	>353	−39	1144.0
3	281	1.79	8721	30.9	15	192	−14	1111.5
3-1	280	1.69	8404	26.2	19	>360	−26	916.8
3-2	178	1.61	8237	26.0	27.5	324	−28	1223.0
3-3	299	1.62	8023	22.4	>60	144	−16	1062.8
3-4	221	1.75	9082	31.3	12	>360	−35	926.9
3-5	255	1.64	8396	24.9	35	>360	−36	1143.3
4	181	1.87	9010	32.8	35	300	−41	721.6
4-1	185	1.81	8793	28.6	40	360	−42	625.3
4-2	196	1.86	8950	32.9	40	360	−42	781.2
4-3	200	1.86	9220	33.1	40	360	−36	677.4

derivatives **3-1** to **3-5** is greater than **DNPP** (literature comparison) due to the manifestation of two tetrazole rings. Moreover, the impact and friction sensitivities of prepared compounds were better than RDX and other standard HEDMs. Also, the prepared compound **3** and its energetic salts **3-1** to **3-5** exhibited high HOFs (916.8–1223.0 kJ mol^{-1}) and good oxygen balance (−14.4% to −36.5%) in comparison to traditional HEDMs.

An energetic compound **4** and its salts were prepared using triazolo-triazine building block following the same technique as for compound **1** and its energetic salts presented in Scheme 7.1 [54]. Physicochemical and energetic properties of **4** and its energetic salts are given in Table 7.3. The impact and friction sensitives were found up to 35–40 J and 300–360 N, respectively, showing their insensitive nature. The detonation pressures and velocities were in the range of 28.6–33.1 GPa and 8793–9220 m s^{-1}. The detonation velocity of hydroxylammonium salt **4-3** (9220 m s^{-1}) was greater than that of HMX, while the detonation pressure was 33.1 GPa comparable to RDX. Compound **4** and its hydrazinium salt **4-2** exhibited detonation performance comparable to RDX which was 9010 and 8950 m s^{-1} (velocity) and 32.8 and 32.9 GPa (pressure), correspondingly. Compound **4** and its energetic salts possess densities in the range of 1.81–1.87 g cm^{-3}. High densities and good HOFs were responsible for the better detonation performance of these energetic compounds. Therefore, the nitro tetrazole compound **4** and its hydrazinium (**4-2**) and hydroxylammonium (**4-3**) salts have slightly better densities and detonation performance than RDX, and detonation velocity was reachable to HMX in the case of salt **4-3**.

5-(Trinitromethyl)-2*H*-tetrazole has a density [58] of 1.92 g cm^{-3}; therefore, several HEDMs were designed and synthesized using this raw material [59, 60]. 5-(Trinitromethyl)-2*H*-tetrazole was alkylated with bromoacetone followed by nitration with various acids mixture to obtain compound **5** with additional nitro groups shown in Scheme 7.2. Further nitration of **5** leads to the formation of an unstable compound. In addition, a series of energetic salts **5A-1** to **5A-3** and **5B-1** to **5B-5** were prepared by reaction of **5A** and **5B** with selected nitrogen bases. 2,5-Bis(dinitromethyl)-2*H*-tetrazole (**6**) was synthesized from potassium salt **5B-1** after the acidification with sulfuric acid. The energetic salts of **5B-2** were obtained *via* the reaction of silver salt and related hydrochlorides. Prepared compounds thermally decomposed in the range of 73–121 °C which is moderate in comparison to standard HEDMs as illustrated in Table 7.4. However, the densities ranging between 1.75 and 1.92 g cm^{-3} were good as compared to standard HEDMs. The range of detonation pressures and velocities were from 31.4 to 42.4 GPa and 8511 to 9590 m s^{-1}, respectively. Among synthesized energetic salts, the dihydroxylammonium salt **5B-5** found to be the best due to excellent detonation velocity and pressure (9520 m s^{-1} and 42.4 GPa, respectively) which were higher than those of HMX. Dihydrazinium salt (**5B-4**) had a higher detonation velocity (9230 m s^{-1}) than that of HMX, while detonation pressure (37.9 GPa) and density (1.83 g cm^{-3}) were greater than those of RDX. Compound **6** also shows promising energetic performance (velocity: 9123 m s^{-1} and pressure: 37.9 GPa) and density (1.92 g cm^{-3})

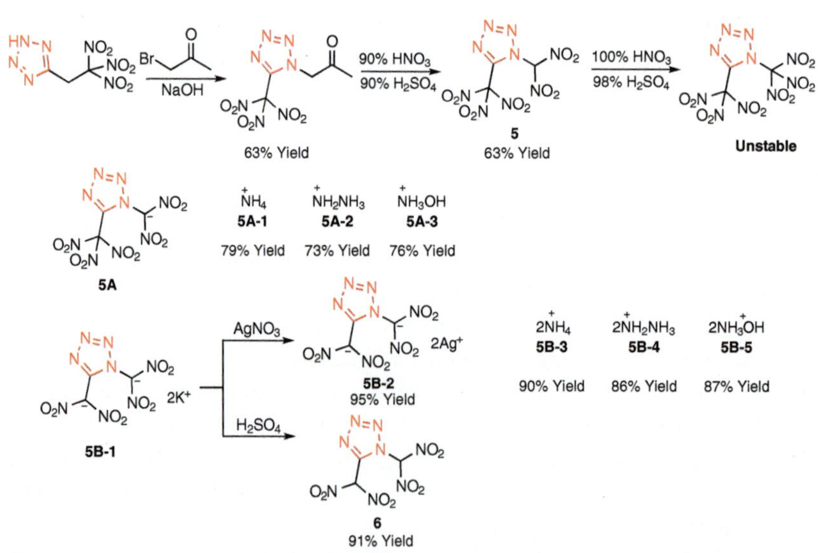

Scheme 7.2 Preparation of polynitromethyltetrazoles and their energetic salts.

Table 7.4 Physicochemical and energetic properties of compounds **5-8** and their nitrogenous salts.

Compound	T_d (°C)	ρ (g cm^{-3})	D (m s^{-1})	P (GPa)	IS (J)	FS (N)	OB (%)	H_f (kJ mol^{-1})
5	73	1.88	8511	31.2	2	80	32	296.67
5A-1	121	1.79	8617	31.6	1	120	23	247.90
5A-2	107	1.92	9054	36.3	1	20	27	288.74
5A-3	97	1.81	8920	34.6	1	10	20	399.62
6	88	1.92	9123	37.7	3	120	23	259.98
5B-3·H_2O	114	1.75	8655	31.8	8	240	5.1	17.21
5B-4	115	1.83	9230	37.9	1	120	0	338.41
5B-5	89	1.91	9590	42.4	5	80	13	107.42
7	-	1.83	8499	26.7	30	>360	−42	181.5
7-1	242	1.66	8446	25.2	40	>360	−54	255.4
7-2	275	1.80	9596	34.0	35	>360	−54	398.4
7-3	286	1.76	9016	30.4	30	>360	−50	71.1
7-4	120	1.73	8445	29.0	8	240	−33	354.5
8	272	1.76	8257	25.9	30	360	−41	407.7
8-1	251	1.70	8286	24.9	40	>360	−52	339.1
8-2	229	1.75	8801	28.9	35	360	−57	547.1
8-3	187	1.78	8818	29.9	32	360	−50	445.0
8-4	224	1.69	8244	23.9	40	>360	−50	485.5
8-5	287	1.72	8138	23.6	40	>360	−58	337.7

similar to HMX and higher than RDX. Other energetic salts (**5A-1** to **5A-3**) also exhibited detonation performance and densities better or similar to those of RDX. Moreover, the prepared compounds showed positive HOFs and displayed positive oxygen balances (+5.1% to 32.2%), whereas the oxygen balance of **5B-4** was zero. However, most of these compounds revealed low thermal decomposition temperatures and high sensitivities toward impact and friction as compared to standard HEDMs. Hence, polynitro groups along with a tetrazole ring play an important role to enhance the density and detonation performance of the target compounds.

Recently, by getting inspiration from the properties and structure of FOX-7-based compounds, new HEDMs bearing a tetrazole ring instead of one nitro group were synthesized [61]. 4,6-Dichloro-5 nitropyrimidine was used as the starting material shown in Scheme 7.3. The tetrazole derivative of FOX-7 like compounds **7** was obtained after treating the starting material with NaN_3 and a mixture solution of HCl in methanol [62]. Furthermore, energetic salts **7-1** to **7-3** of compound **7** were also obtained in good yields after treating with the nitrogen-rich bases. Neutral energetic compound**7-4** was obtained after nitration of **7**, although it was found to be unstable in organic solvents.

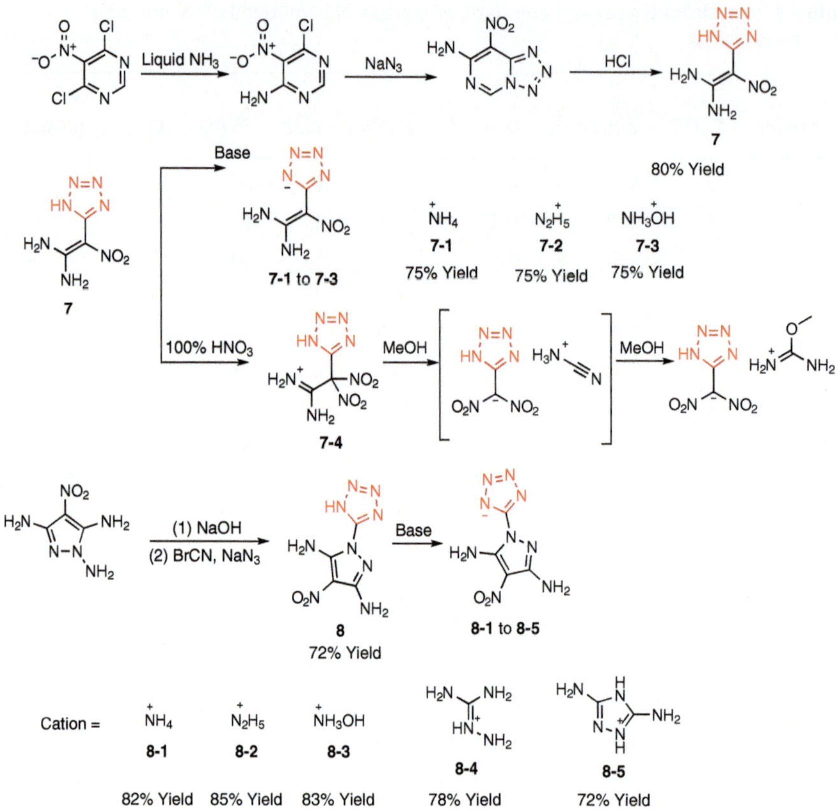

Scheme 7.3 Synthetic routes to the tetrazole-based energetic compounds **7**, **8**, and their energetic salts.

The energetic and physicochemical properties of prepared compounds are shown in Table 7.4. The decomposition temperatures were found in the range of 120–286 °C. The hydrazinium (**7-2**) and hydroxylammonium (**7-3**) energetic salts have higher thermal stability (275 and 286 °C, respectively) among all compounds, greater than RDX and lower than TNT, TATB, and HMX. The densities of neutral compound **7** and hydrazinium salt **7-2** (1.83 and 1.80 $g\,cm^{-3}$, respectively) were comparable to RDX, higher than TNT and lower than TATB and HMX as compared to other prepared compounds. Owing to the high density (1.80 $g\,cm^{-3}$) and HOF (398.4 $kJ\,mol^{-1}$) the energetic salt **7-2** has good detonation performance (9596 $m\,s^{-1}$ velocity and 35 GPa pressure) which is competitive to HMX and higher than RDX. The impact and friction sensitivities (35 J and >360 N) of the salt **7-2** were found to be acceptable among all studied derivatives, while HOFs of all prepared compounds were positive. Therefore, tetrazole-containing neutral compound with less number of nitro groups **7** and its energetic salts **7-2** and **7-3** may be considered as useful HEDMs for future work, especially concerning their high thermal stability.

Nitrogen-rich heterocyclic compounds consisting of five- or six-membered rings (tetrazole, triazole, pyrazole, furazan, tetrazine, etc.) have attained prevalent attention as frameworks for HEDMs. These building blocks are quite useful in a preparation of suitable derivatives with high energetic properties [63, 64]. In recent work, 3,5-diamino-4-nitropyrazole was used as a precursor to prepare various nitrogen-containing energetic salts as shown in Scheme 7.3 [65]. Compound **8** was prepared from the reaction of cyanogen azide with the 3,5-diamino-4-nitropyrazole sodium salt. Energetic salts (**8-1** to **8-5**) were synthesized by neutralization with nitrogen-rich bases. The prepared compounds were thermally stable in a range of 187–287 °C as shown in Table 7.4. The neutral compound **8** and 3,5-diamino-1,2,4-triazole energetic salt (**8-5**) decomposed at 272 and 287 °C, respectively. The densities were found to be between 1.69 and 1.78 g cm^{-3}. The calculated HOFs for tetrazole **8** and its salts **8-1** to **8-5** were positive. Ammonium and hydroxylammonium salts (**8-1** and **8-3**) have better detonation velocities (8818 and 8286 m s^{-1}) as compared to TNT and RDX. However, the detonation pressure of prepared energetic compounds was not higher than that of the standard HEDMs. The impact and friction sensitivities of **8** and salts **8-1** to **8-5** were lowered as compared to other synthesized compounds. However, the oxygen balance was negative in all cases. Therefore, these compounds with higher thermal stability and lower sensitivities were found to be promising in the area of energetic materials.

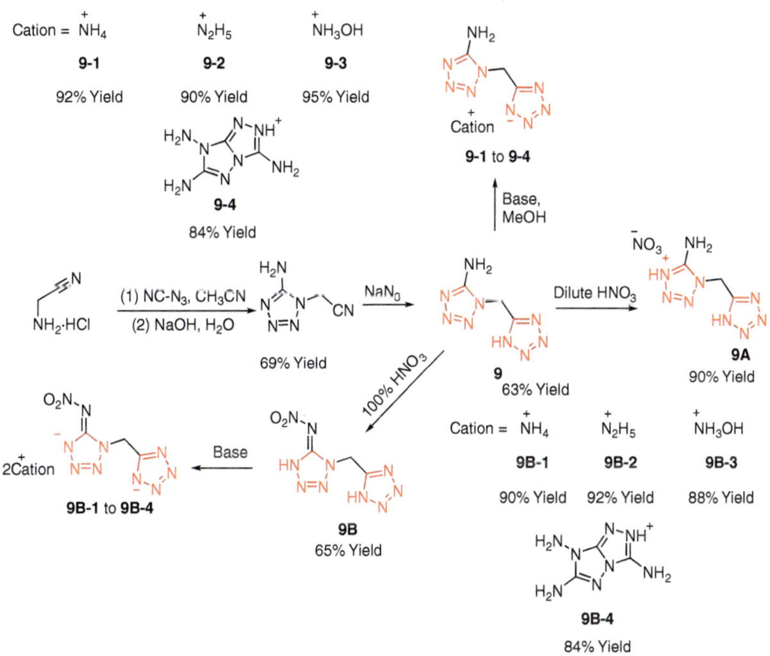

Scheme 7.4 Synthetic routes to tetrazole energetic compounds **9** and its energetic salts.

Methylenebistetrazole**9** (Scheme 7.4) [63] was easily obtained upon interaction of 2-(5-amino-1*H*-tetrazol-1-yl)acetonitrile with sodium azide. The nitro group was introduced to obtain energetic compounds **9A** and **9B** after the reaction of **9** with diluted and conc. nitric acid, respectively. Furthermore, the energetic salts (**9-1** to **9-4**

Table 7.5 Properties of nitrotetrazole energetic compounds **9**, **10**, **11**, and their energetic salts.

Compound	T_d (°C)	ρ (g cm^{-3})	D (m s^{-1})	P (GPa)	IS (J)	FS (N)	OB (%)	ΔH_f (kJ mol^{-1})
9	217	1.61	7780	20.9	>40	>360	−52	534.7
9A	156	1.79	8857	31.2	28	240	−31	449.5
9B	178	1.76	8652	29.6	>12	160	−30	615.9
9-1	239	1.49	7476	18.1	>40	>360	−60	504.2
9-2	244	1.60	8330	23.5	>40	>360	−60	649.1
9-3	172	1.59	8155	23.1	>40	>360	−51	550.5
9-4	246	1.62	7912	21.5	>40	>360	−57	1027.0
9B-1	229	1.60	8076	22.9	>40	>360	−45	387.6
9B-2	224	1.62	8654	26.9	>40	>360	−46	728.1
9B-3	186	1.60	8778	29.5	>40	>360	−34	509.3
9B-4	238	1.71	8499	26.5	>40	>360	−49	1634.6
10	279	1.86	9289	38.6	35	240	−36	856.4
10-1	299	1.79	8953	33.3	>40	168	−43	672.6
10-2	290	1.84	9364	37.0	>40	360	−43	819.8
10-3	296	1.84	9232	37.4	>40	216	−35	719.7
10-4	256	1.67	8264	25.9	>40	360	−53	648.8
10-5	216	1.72	8756	28.6	>40	32	−63	808.7
10-6	285	1.79	9066	32.5	>40	288	−53	844.9
10-7	287	1.81	9290	34.2	>40	84	−53	954.9
10-8	286	1.84	8951	32.2	>40	360	−56	840.2
10-9	261	1.82	9003	32.2	>40	360	−56	960.0

and **9B-1** to **9B-4**) of compound **9** and **9B** were synthesized through neutralization with nitrogen-rich bases. The reason to prepare energetic salts was to improve the thermal stability and sensitivity as compared to the parent compounds (**9** and **9B**).

The physicochemical and energetic properties of compounds **9**, **9A**, **9B**, and salts **9-1** to **9-4** and **9B-1** to **9B-4** are presented in Table 7.5. All the nitrogen-rich and nitrotetrazole compounds and their salts decompose between 156 and 246 °C. Except for compounds **9A**, **9B**, and hydroxylammonium salt **9-3** which have decomposition temperatures 156, 178, and 204 °C respectively, all other compounds have better thermal stability than RDX. However, the densities of prepared compounds were found in the range of 1.49–1.79 g cm^{-3} which is not better than for standard HEDMs. The calculated HOFs were found higher than TNT and RDX. The calculated detonation velocities were found in the range of 7476–8857 m s^{-1} and detonation pressures were between 18.1 and 31.2 GPa. These data reveal that the detonation performance of target compounds was not higher than that of standard HEDMs. However, the impact and friction sensitivities of these compounds were better in comparison to RDX, except in the case of **9A** and **9B**. As a result, a connection of more stable tetrazole ring with sensitive nitramino moiety and salt formation improved thermal stability and decreased sensitivity of prepared compounds.

The azole bridge tetrazole-containing compound **10** and energetic salts **10-1** to **10-9** were prepared according to Scheme 7.5 [66]. Compound **10** is a combination of a nitro group containing moiety, namely *N*-(3,5-dinitro-1*H*-pyrazol-4-yl)-1*H*-tetrazol-5-amine (TNP), and an additional tetrazole ring [67]. TNP was prepared through nitration, amination, and oxidation steps using 3,5-dinitro-4-chloropyrazole as a starting material [68]. TNP was converted to compound **10** through the reaction of 5-amino-1*H*-tetrazole (ATZ) in potassium hydroxide (4 equiv) aqueous solution. The subsequent raw product was acidified with 20% H_2SO_4 to yield a yellow solid product. Furthermore, the energetic salts **10-1** to **10-9** were prepared through the reaction of selected nitrogen-rich bases with compound **10**. The latter and energetic salts **10-1** to **10-9** exhibited good thermal stability (Table 7.5). The decomposition temperature of prepared compounds was between 216 and 299 °C. The neutral compound **10** decomposed at 279 °C, which is higher than that of RDX. However, ammonium and hydroxylammonium (**10-1** and **10-3**) energetic salts were the most stable among all (thermal decomposition at 299 and 296 °C, respectively), which is comparable to standard HEDMs. This result confirmed that the introduction of a tetrazole ring and further preparation of salts improved the stability as compared to LLM-116. The calculated densities were in the range of 1.67–1.86 g cm^{-3}, indicating the high density of neutral compound **10** in comparison to their salts. However, most of the compounds possess higher densities comparable to standard HEDMs. Further, the calculated HOFs were positive for all compounds and the highest value was for salt **10-9** (960 kJ mol^{-1}). Detonation pressures were in the range of 25.9 and

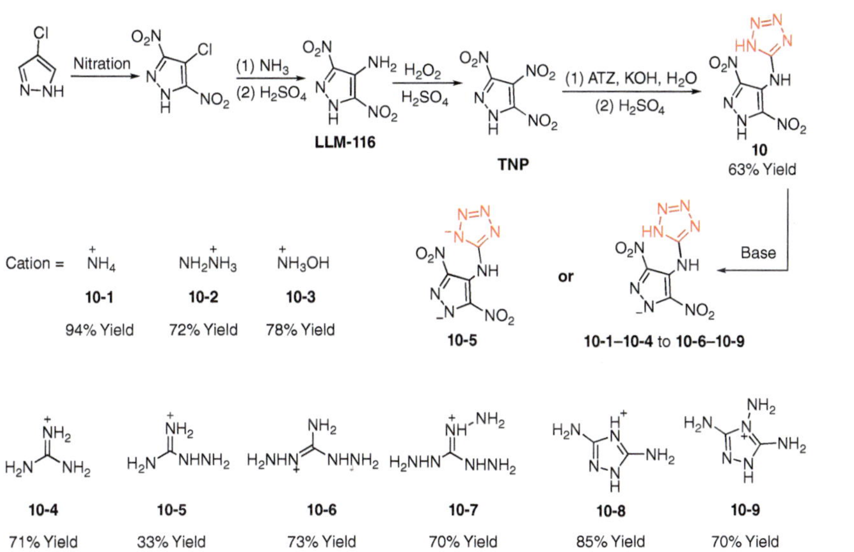

Scheme 7.5 Synthetic routes to energetic compound **10** and its energetic salts.

38.6 GPa, and detonation velocities were between 8264 and 9364 m s^{-1}. The highest detonation velocity was found for hydrazinium salt (**10-2**), i.e. 9364 m s^{-1} comparable to HMX. Therefore, prepared compounds have attractive properties to contribute to future applications of HEDMs.

3-Nitro-1-(2*H*-tetrazol-5-yl)-1*H*-1,2,4-triazol-5-amine (HANTT)**11** attained promising importance as HEDM due to good thermal stability and insensitivities toward impact and friction. The synthesis of HANTT (**11**) was achieved by an interaction of 3-nitro-1*H*-1,2,4-triazol-5-amine and cyanogen azide as depicted in Scheme 7.6 [69]. Additionally, the energetic salts **11-1** to **11-7** were obtained *via* metathesis reaction of Ba(ANTT) with one equivalent of the corresponding sulfate salt, whereas the energetic salt **11-8** was formed in a reaction of 3,4,5-triamino-1,2,4-triazole and HANTT (**11**). The decomposition temperature of HANTT (**11**) was 246 °C as listed in Table 7.6. Salt formation is considered commonly as a good method to improve the stability of the existing compound. Therefore, the salts of HANTT showed better thermal stability ranging from 264 to 321 °C, approaching the standard HEDMs. The measured density of HANTT was 1.77 g cm^{-3}, whereas the energetic salts possessed densities from 1.65 to 1.81 g cm^{-3}, which were greater than TNT but comparable to RDX in the case of hydroxylammonium salt **11-2** (1.81 g cm^{-3}). HANTT (**11**) and

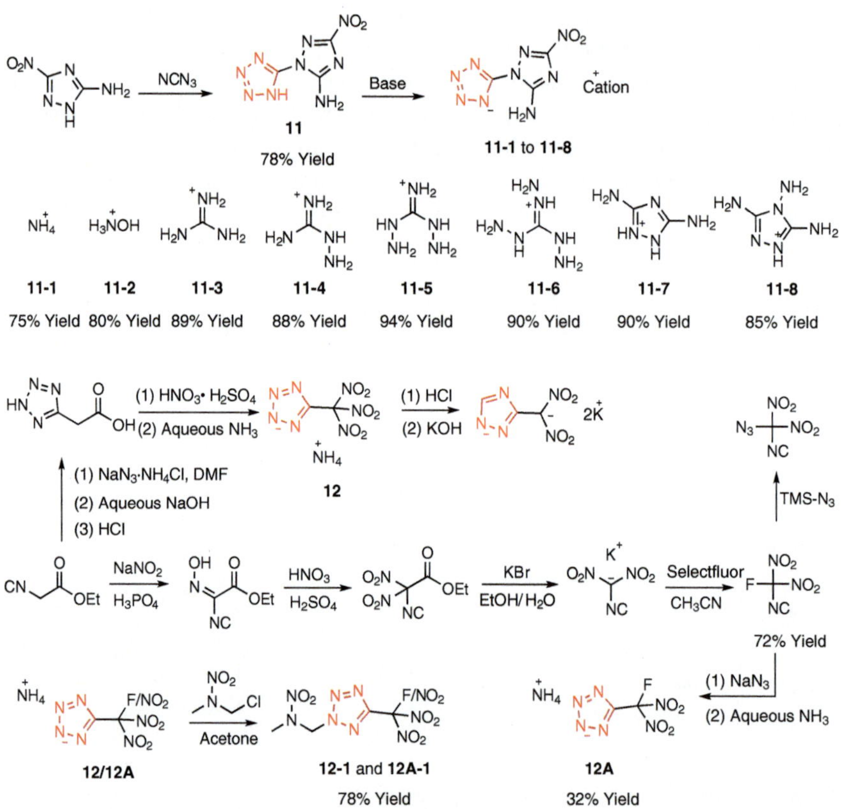

Scheme 7.6 Synthesis of energetic compounds **11**, **12**, **12A**, and their energetic salts.

Table 7.6 Properties of energetic compounds **11**, **12**, and their energetic salts.

Compound	T_d (°C)	ρ (g cm^{-3})	D (m s^{-1})	P (GPa)	IS (J)	FS (N)	OB (%)	ΔH_f (kJ mol^{-1})
11	246	1.77	8316	28.8	>40	>360	−44	487.2
11-1	281	1.79	8507	29.6	>40	>360	−52	409.1
11-2	264	1.81	8779	32.6	>40	>360	−41	460.8
11-3	321	1.65	7742	22.6	>40	>360	−62	392.2
11-4	269	1.72	8149	25.9	>40	>360	−61	486.0
11-5	—	1.69	8249	26.4	>40	>360	−61	595.5
11-6	—	1.69	8390	27.5	>40	>360	−61	707.9
11-7	277	1.68	7840	23.6	>40	>360	−64	595.5
11-8	301	1.74	8217	26.7	>40	>360	−64	711.2
12	128	1.78	9058	34.1	2	8	0.0	231.4
12A	166	1.70	8476	29.9	1	16	−15	6.7
12-1	134	1.75	8704	33.4	3	40	−13	313.9
12A-1	184	1.83	8773	34.2	2	36	−25	98.3

energetic salts (**11-1** to **11-8**) exhibited negative oxygen balance and positive values of HOF. The impact and friction sensitivities of compound **11** and its salts were greater than 40 J and 360 N, verifying them as insensitive HEDMs. The detonation pressure and velocity fell from 22.6 to 32.6 GPa and from 7742 to 8779 m s^{-1}, respectively. However, the hydroxylammonium salt **11-2** had greater detonation performance (velocity: 8779 m s^{-1} and pressure: 32.6 GPa) as compared to other compounds and similar to RDX. These prepared compounds are important due to their high thermal stability and insensitive behavior toward impact and friction.

The purpose of manufacturing polynitrotetrazole compounds **12** and **12A** was to maintain high thermal stability and low sensitivity toward impact and friction as represented in Scheme 7.6 [70]. The synthesis was based on a commercially available ethyl 2-cyanoacetate. To prepare compound **12**, decarboxylative nitration of ethyl 2-(1*H*-tetrazol-5-yl)acetate was performed. Ammonium 5-(fluorodinitromethyl)-2 *H*-tetrazolate (**12A**) was attained from potassium dinitroacetonitrile, and the addition of NaN_3. Compounds **12** and **12A** were synthesized in the form of ammonium salts due to their low stability. Further, these compounds were reacted with 2-nitro-2-azapropyl chloride to accomplish the synthesis of **12-1** and **12A-1** in acetone. The physicochemical and energetic properties of these prepared compounds are listed in Table 7.6. The obtained densities were in the range of 1.70–1.83 g cm^{-3}. The detonation performance was high ranging from 8476 to 9058 m s^{-1} (detonation velocity) and 29.9–34.2 GPa (detonation pressure). The polynitro groups containing ammonium salt **12** has high detonation velocity (9058 m s^{-1}) comparable to HMX and detonation pressure (34.2 GPa) similar to RDX. All compounds possess high oxygen balance and positive HOFs, but low thermal stability (128–184 °C). Moreover, the focused compounds correspond to very sensitive HEDMs owing to their high sensitivity toward impact and friction.

The nitrotetrazole compounds (**13**), namely 4-nitro-3-(1*H*-tetrazol-5-yl)furoxan (HTNF) and its salts (**13-1** to **13-9**), were prepared in search of HEDMs with high thermal stability, densities, and HOFs [71]. 4-Amino-3-cyanofuroxan were readily synthesized to achieve the production of HTNF (**13**) as articulated from Scheme 7.7.

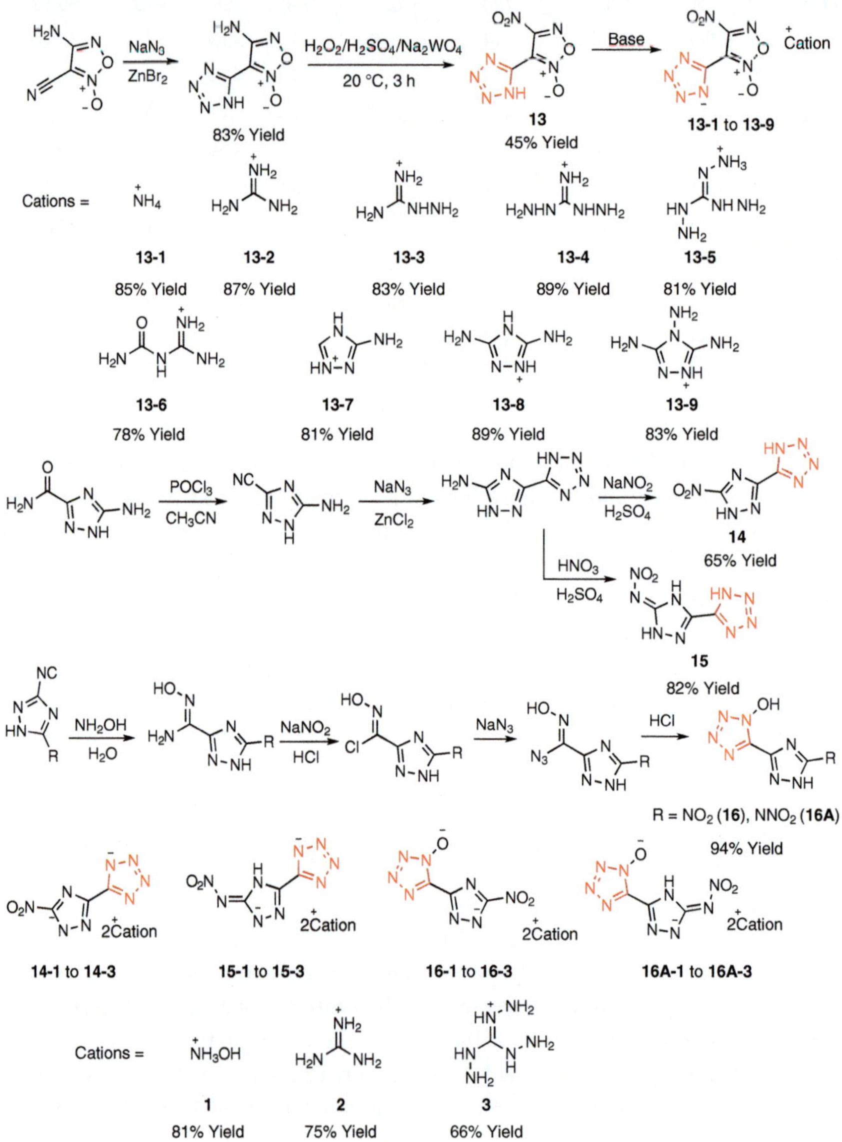

Scheme 7.7 Synthesis of different nitro- and nitraminotetrazole energetic compounds and their energetic salts.

The 4-amino-3-cyanofuroxan reacted with sodium azide in the presence of zinc bromide to construct the tetrazole ring [72]. Subsequent oxidation of the amino group at the furoxan ring using a mixture of 50% $H_2O_2/H_2SO_4/Na_2WO_4$ yielded HTNF (**13**). Additionally, energetic salts (**13-1** to **13-9**) were harvested by employing direct metathesis strategy. The degradation temperature of HNTF (**13**) was 169 °C whereas, for energetic salts, it was in the range of 122–213 °C (Table 7.7). These results indicate that only *N*-carbamoylguanidinium salt **13-6** has thermal stability similar to RDX. Moreover, the calculated densities were in a range of 1.63–1.84 g cm^{-3}, nearly the same for ammonium salt **13-1** and neutral compound **13** (1.84 and

Table 7.7 Physicochemical and energetic properties of different nitrotetrazole energetic compounds and their energetic salts.

Compound	T_d (°C)	ρ (g cm^{-3})	D (m s^{-1})	P (GPa)	IS (J)	FS (N)	OB (%)	ΔH_f (kJ mol^{-1})
13	169	1.81	8681	34.3	<2	—	4	499.5
13-1	142	1.84	8919	36.2	2	—	−7	427.4
13-2	159	1.67	8095	26.4	>40	—	−18	406.3
13-3	153	1.63	8103	26.0	10	—	−20	508.0
13-4	122	1.81	8882	33.6	12	—	−22	603.1
13-5	133	1.75	8766	31.8	2	—	−23	720.0
13-6	213	1.70	7934	25.5	7	—	−18	199.6
13-7	167	1.75	8331	29.2	>40	—	−19	640.7
13-8	182	1.72	8208	27.6	12	—	−21	608.3
13-9	180	1.75	8430	29.4	13	—	−23	690.1
14	211	1.70	7919	25.4	25	288	−43	518
15	215	1.70	8062	26.2	<1	18	−44	576
16	152	1.86	8655	33.7	35	360	−32	446
16A	116	1.85	8776	34.2	<1	60	−33	515
14-1	182	1.78	8473	29.1	40	360	−38	316
14-2	296	1.71	7682	21.9	40	>360	−74	234
14-3	190	1.68	8644	29	40	>360	−69	958
15-1	194	1.75	8500	29.3	40	160	−38	354
15-2	238	1.68	7709	22.1	40	>360	−74	261
15-3	191	1.66	8628	28.9	40	324	−69	978
16-1	175	1.81	8996	35.2	8	360	−30	356
16-2	269	1.72	7974	24.4	40	360	−65	260
16-3	181	1.70	8728	29.9	10	360	−63	977
16A-1	179	1.79	9014	34.8	8	288	−31	418
16A-2	212	1.70	7970	24.2	40	360	−65	304
16A-3	186	1.67	8628	28.8	10	288	−62	1009

1.81 g cm^{-3}). The imperative factor to determine the performance of HEDMs is oxygen balance which ranged from −7.4% to −23.7% for HNTF and its salts. It is higher than TNT for all compounds and even better than TATB in the case of ammonium salt **13-1** (−7.4%). The calculated HOFs were in the range of 499.5 and 308.4 kJ mol^{-1}, while impact sensitivities were in a range of 2–40 J, respectively. Only energetic guanidinium (**13-2**) and 3-amino-1,2,4-triazolium (**13-7**) salts were insensitive with impact sensitivity greater than 40 J, while the other was classified as sensitive in comparison to standard HEDMs. As indicated from Table 7.7, the detonation pressure and velocity were between 25.5 and 36.2 GPa and 7934 and 8919 m s^{-1}, respectively. The ammonium salt (**13-1**) attained higher detonation velocity (8919 m s^{-1}) and pressure (36.2 GPa) in comparison to RDX, while the diaminoguanidinium energetic salt **13-4** has comparable detonation performance (8882 m s^{-1} and 33.6 GPa). Moreover, the triaminoguanidinium salt **13-5** has slightly lower detonation velocity (8766 m s^{-1}) and pressure (31.8 GPa) as compared to RDX. This detonation performance was due to moderate HOFs, high density, and good oxygen balance.

Compounds **14-16** and **16A** in which triazole ring was connected with tetrazole or *N*-hydroxy-tetrazole were designed (Scheme 7.7) to enhance the energetic properties, due to presence of enormous energy content (nitro and nitramino explosophores) [73]. Further, to increase the stability and energetic performance, corresponding salts of these compounds were also prepared. The manufacturing of 5-(1,2,4-triazol-3-yl)tetrazoles (**14, 15**) and 5-(1,2,4-triazol-3-yl)tetrazol-1-oles (**16, 16A**) was achieved through 5-amino-1*H*-1,2,4-triazole-3-carbonitrile [74, 75]. The tetrazole ring was installed through a cycloaddition, followed by addition of the energetic group to attain compounds **14** and **15**, whereas in the case of compounds **16** and **17** the energetic groups were introduced first and then cyclization was performed in several steps. The respective 12 ionic nitrogen-rich salts were prepared in ethanol *via* the reaction of neutral compounds with 2 equiv of bases.

The ionic *N*-oxide compounds (**16-1** and **16-2** and **16A-1** and **16A-3**) possess lower decomposition temperature in contrast to non *N*-oxide (**14-1** to **14-3** and **15-1** to **15-3**) (Table 7.7). The thermal decomposition temperature ranged from 116 to 296 °C, highest values were for guanidinium salts **14-2, 16-2**, and**15-2** (296, 269, and 238 °C). The calculated HOFs were superior to RDX in the range of 234–1009 kJ mol^{-1}. Moreover, detonation parameters were from 21.9 to 35.2 GPa, and from 7682 to 9014 m s^{-1}. The detonation velocities and pressures for salts **15-1**, **16-1**, and **16A-1** were superior among all others and comparable to reference HEDMs. The triaminoguanidinium salt **16-3** and the hydroxylammonium salt **16A-1** exhibited higher detonation velocities (8996 and 9014 m s^{-1}) and pressures (35.2 and 34.8 GPa) similar to RDX. The impact and friction sensitivities were lower in the case of ionic salts as compared to their neutral compounds. Therefore, triaminoguanidinium salt **16-3** and the hydroxylammonium salt **16A-1** demonstrated superior energetic performance in terms of detonation performance, sensitivities, and thermal stability in comparison to RDX. However, the diaminomethaniminium salts **14-2** and **16-2** possessed higher decomposition temperatures (296 and 269 °C), but lower energetic performance.

For the development of efficient HEDMs, heterocyclic compounds **17**, **17A**, and **18** consisting of alkyl- and nitraminotetrazole moieties were synthesized in a recent work (Scheme 7.8) [76]. Furthermore, different salts **17A-1** to **17A-3** were also synthesized to enhance the thermal stability and sensitivity. The synthetic route to compounds **17** and **18** was the same with different starting materials, methyl-2-(5-amino-1*H*-tetrazol-1-yl) acetate and ethyl-2-(5-amino-2*H*-tetrazol-2-yl)acetate, respectively. The starting materials were treated with hydrazine monohydrate, *N*-methyl-*N*-nitroso-*N*′-nitroguanidine, and the final step included refluxing with aqueous sodium hydroxide solution with subsequent acidifying with conc. HCl. Compound **17A** was obtained after the nitration of **17** with conc. HNO_3.

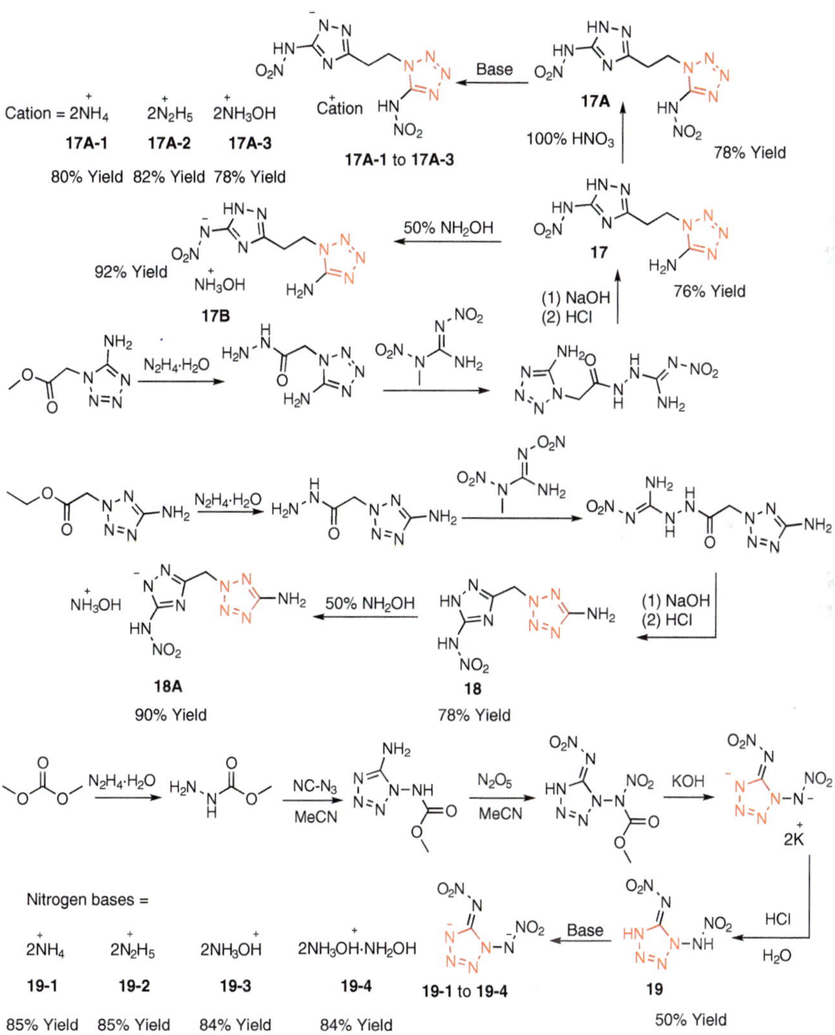

Scheme 7.8 Synthesis of nitraminotetrazole energetic compounds **17**, **17A**, **18**, **19**, and their energetic salts.

Moreover, the reaction of nitrogen-rich bases with compound **17A** formed the energetic salts **17A-1** to **17A-3**. Hydroxylammonium salts **17B** and **18A** were obtained through the reaction of aqueous hydroxylamine with **17** and **18**, respectively. The densities of prepared compounds were found in the range of 1.61–1.81 g cm^{-3} as shown in Table 7.8. Compound **17A** has a higher density (1.81 g cm^{-3}) as compared to other prepared compounds, and similar to RDX. The decomposition temperatures of the prepared compounds were between 159 and 218 °C, which is not appropriate

Table 7.8 Physicochemical and energetic properties of nitraminotetrazole energetic compounds **17**, **17A**, **18**, **19**, and their several energetic salts.

Compound	T_d (°C)	ρ (g cm^{-3})	D (m s^{-1})	P (GPa)	IS (J)	FS (N)	OB (%)	ΔH_f (kJ mol^{-1})
17	204	1.71	8187	24.9	15	240	−53	476.2
17A	159	1.81	8750	31.8	10	159	−42	285.0
17B	193	1.71	8737	29.0	15	240	−55	597.9
17A-1	187	1.63	8362	25.8	22	>360	−49	539.4
17A-3	167	1.74	8997	32.7	18	>360	−50	643.5
17A-2	199	1.71	9052	31.2	18	>360	−40	850.1
18	218	1.72	8213	25.2	18	360	−42	462.8
18A	193	1.61	8248	25.0	15	240	−43	597.9
19	240	2.18	10011	52.2	1	<5	−6	−112.4
19-1	170	1.73	9078	31.5	1.5	30	−7	212
19-2	150	1.78	9872	37.6	1	6	−12	541.3
19-3	130	1.85	9807	39.9	1.5	10	6	351.5
20	232	1.67	7752	25.4	5	240	−48	299.2
20-1	203	1.75	8535	31.8	6	108	−48	613.8
20-2	233	1.82	8826	34.7	5	112	−30	425.6
20-3	284	1.62	7117	21.1	>40	>360	−65	326.7
20-4	238	1.71	7686	25.4	>40	288	−64	538.3
20-5	233	1.66	7668	26.2	>40	240	−63	807.3
20-6	230	1.69	8174	28.5	29	160	−63	1040.6
21	116	1.66	8690	31.4	5	—	−24	504.3
21-1	179	1.60	8141	24.9	13	>360	−47	405.5
21-2	167	1.59	8210	25.2	4.1	89	−47	484.7
21-3	134	1.56	8187	24.7	4.7	51	−47	565.7
21-4	139	1.59	8479	27	8.2	20	−48	700.1
21-5	151	1.71	9065	34.2	5.5	13	−36	787.8
21-6	174	1.61	8107	25.1	4.4	83	−45	644.6
21-7	134	1.65	8408	27.7	4.6	59	−45	774.0
21-8	159	1.60	8013	24.1	11	102	−46	580.6

as for standard HEDMs. Impact and friction sensitivities of all compounds were acceptable. Prepared energetic salts (**17A-1** to **17A-3**) have impact sensitivities of 18, 18, and 22 J, respectively, and friction sensitivities were higher than 360 N. Furthermore, the detonation properties of the prepared compounds were not higher than standard HEDMs. The detonation velocity of neutral compound **17A** and its hydrazinium and hydroxylammonium (**17A-2** and **17A-3**) energetic salts were 8750, 9052, and 8997 m s^{-1}respectively, which is comparable to RDX.

The preparation of nitramino tetrazole-containing compound**19** was achieved through dimethyl carbonate, which was further treated with hydrazine hydrate to produce methyl carbazate as depicted in Scheme 7.8 [77]. Methyl carbazate further reacted with cyanogen azide to give *N*-methoxycarbonyl-1,5-diaminotetrazole which was further nitrated and treated with a solution of KOH to produce potassium salt. Acidification of potassium salt yielded compound **19**, which was further neutralized to prepare selected nitrogen-rich salts **19-1** to **19-4**. Salts **19-1** to **19-3** were synthesized in methanol under the action of a relevant base. However, hydroxylamine salt **19-4** was obtained as a covalent complex when hydroxylamine was used in excess. The decomposition temperature of neutral nitramino tetrazole compound **19** was 240 °C, while its salts **19-1** to **19-4** decomposed in the range of 130–150 °C as shown in Table 7.8. The detonation velocity of compound **19** was 10 011 m s^{-1}, while the detonation pressure was 52.2 GPa comparable to CL-20. Moreover, the detonation velocities of salts **19-2** and **19-3** were also higher than CL-20 (9872 and 9807 m s^{-1}) and pressure (37.6 and 39.9 GPa) higher than RDX. However, salt **19-1** has a detonation velocity (9078 m s^{-1}) higher than RDX and slightly lower than HMX and detonation pressure (31.5 GPa) lower than RDX. The density of neutral compound **19** was 2.18 g cm^{-3}. However, compound **19** and its different salts (**19-1** to **19-3**) were highly sensitive toward impact and friction; therefore, they can be used as a primary explosive.

The ammonium or barium salt (**20**) was prepared upon nitration of 3-nitramino-4-(tetrazolyl)furazan (NATF) with mixed acids (H_2SO_4 and HNO_3) and neutralization with ammonia or aqueous $Ba(OH)_2$ as indicated from Scheme 7.9 [78]. Further, energetic salts **20-1** to **20-6** were synthesized after the reaction of NATF cations with selected nitrogen-rich bases. The measured densities of synthesized compounds were in a range of 1.62–1.82 g cm^{-3} (Table 7.7). Thermal decomposition temperatures were in the range of 203–284 °C, higher than RDX except for hydrazinium salt **20-1** which decomposed at 203 °C. Moreover, HOFs of these NATF-based energetic salts were found positive ranging from 299.2 to 1040.6 kJ mol^{-1}. The detonation pressures and velocities were 21.1–34.7 GPa and 7117–8826 m s^{-1}, respectively, which was not much better than for standard HEDMs except for hydroxylammonium salt **20-2** (8826 m s^{-1} and 34.7 GPa). However, the impact and friction sensitivities were also lower for most of the prepared salts.

The synthesis of compound **21** was attained through intermediate 1-methoxy-5-aminotetrazole, which was obtained after the cyclization of cyanogen azide with methoxyamine as shown in Scheme 7.9 [79]. Further nitration of 1-methoxy-5-aminotetrazole led to the formation of nitraminotetrazole compound **21**. The energetic salts **21-1** to **21-4** were prepared by a metathesis reaction of silver

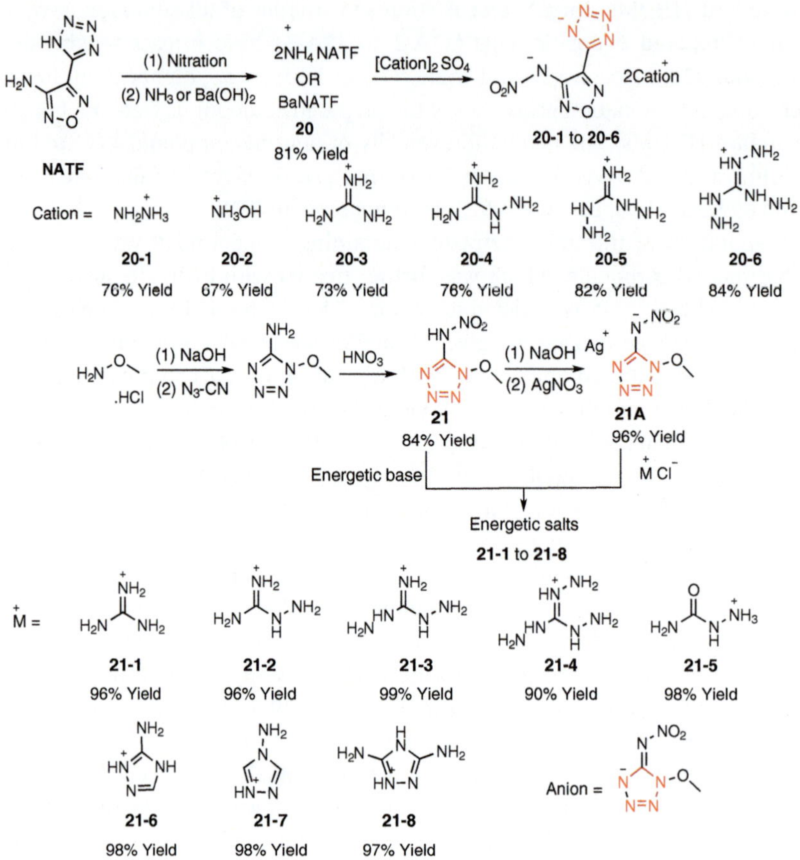

Scheme 7.9 Synthesis of nitraminotetrazole energetic compounds **20**, **21**, and their several nitrogen-rich energetic salts.

1-methoxy-5-nitroiminotetrazolate (**21**) with nitrogen-rich bases. Salts **21-5** to **21-8** were prepared *via* simple acid–base reaction in water. The prepared compounds thermally decomposed from 117 to 179 °C, which is not better than for standard HEDMs as shown in Table 7.8. The values of measured densities were in the range of 1.56–1.71 g cm^{-3}. Friction sensitivities were lower than for RDX for all compounds except for diaminomethaniminium salt **21-1**. The detonation pressures and velocities were relatively high from 24.1 to 34.1 GPa and 8013 to 9065 m s^{-1} respectively. The detonation values were greater than those of TNT but less than those of other standard HEDMs except for the detonation velocity and pressure (9065 m s^{-1} and 34.1 GPa) of the salt **21-5**. However, the calculated oxygen balance was negative for prepared compounds.

Furthermore, a series of nitroiminotetrazole derivatives (**22**, **23**, and **24**) and their salts were prepared to achieve good physical and energetic properties as shown in Scheme 7.10 [30]. The preparation of 1-methoxy-5-aminotetrazole (**22**) was successfully attained from cyanogen azide and methoxyamine with neutralization

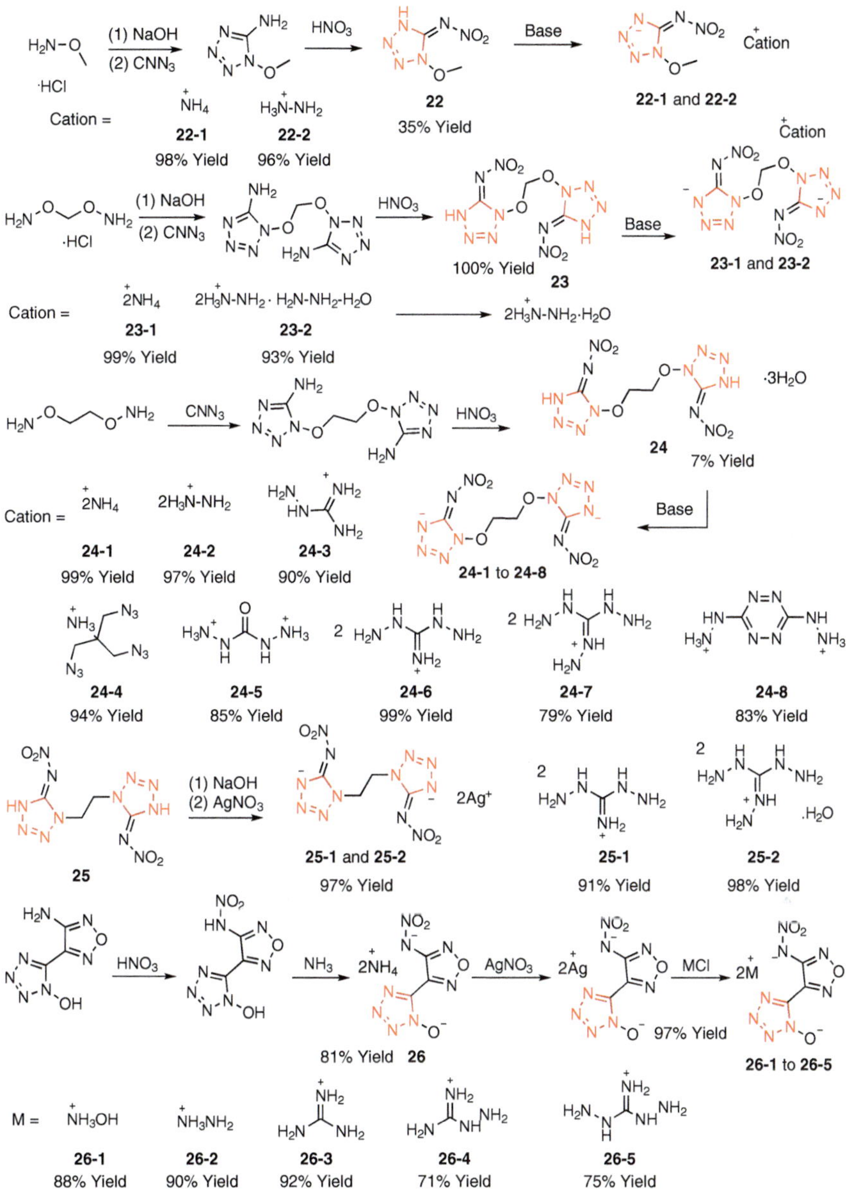

Scheme 7.10 Synthesis of nitraminotetrazole energetic compounds **22–26**, and their nitrogen-rich energetic salts.

with NaOH, and nitration using 100% HNO_3. Compound **22** was extremely sensitive; therefore, energetic salts **22-1** and **22-2** were prepared to attain more stability. Compounds **23**, **24**, and their energetic salts **23-1**, **23-2**, **24-1**, and **24-2** were obtained by analogous reactions. The physicochemical and energetic properties of these compounds are given in Table 7.9. These compounds were thermally

Table 7.9 Energetic properties of nitraminotetrazole energetic compounds **22–26** and their several nitrogen-rich energetic salts.

Compound	T_d (°C)	ρ (g cm^{-3})	D (m s^{-1})	P (GPa)	IS (J)	FS (N)	OB (%)	ΔH_f (kJ mol^{-1})
22	116	1.66	8660	31.5	5	—	−30	504.28
22-1	184	1.55	8448	27.7	1.5	—	−41	425.09
22-2	156	1.63	9036	32.3	4	—	−42	572.55
23	157	1.90	9867	46.7	1	—	−11	1088.82
23-1	167	1.71	8984	33.9	1	—	−24	804.94
24	134	1.81	9200	38.4	1.5	—	−25	1291.63
24-1	202	1.72	9014	33.3	2	—	−36	788.99
24-2	169	1.73	9305	35.5	1.5	—	−38	1116.17
24-3	200	1.602	8472	26.78	6	—	−51	1029
24-4	151	1.597	8377	26.56	<1	—	−63	2944
24-5	146	1.695	8789	32.20	6	—	−35	1134
24-5·2H$_2$O	156	1.716	—	—	2	—	−35	-
24-5·4H$_2$O	147	1.647	—	—	8	—	−35	-
24-6	164	1.675	8741	28.72	5	—	−52	1254
24-7	162	1.607	8840	29.36	4	—	−52	1487
24-8	147	1.780	9030	33.25	2	—	−42	1270
24-8·2H$_2$O	131	1.683	—	—	—	—	−42	—
25-1	220	1.577	8568	26.51	7	—	−62	1233
25-1·H$_2$O	218	1.639	—	—	>40	—	−62	—
25-2	195	1.601	8860	28.68	>40	—	−61	1458
26	250	1.74	8645	28.9	15	160	−19	43.3
26-1	204	1.84	9323	38.3	14	160	−5	398.9
26-2	254	1.72	9094	32.1	18	240	−23	631.1
26-3	269	1.70	8289	25.1	35	360	−33	336.9
26-4	251	1.74	8755	28.8	>40	360	−35	570.6
26-5	209	1.69	9033	30.4	35	360	−37	1063.2

less stable exhibiting decomposition temperatures in the range of 116–202 °C in comparison to standard HEDMs. The densities ranged from 1.55 to 1.90 g cm^{-3}. Compound **23** has the highest density (1.90 g cm^{-3}) comparable to RDX, HMX, and TATB. The impact sensitivities values were from 1 to 1.5 J indicating them as the sensitive class of HEDMs. Moreover, the detonation pressures and velocities were between 27.7 and 46.7 GPa and 8448 and 9305 m s^{-1}, respectively. Among all the nitroiminotetrazoles, compound **23** exhibited higher detonation velocity and pressure (9305 m s^{-1} and 46.7 GPa) higher than HMX, while the ammonium salt **23-1** has closer detonation performance (8984 m s^{-1} and 33.9 GPa) to RDX.

Although these compounds were thermally less stable and impact sensitive, their detonation performance was excellent. In enduring efforts with nitroiminotetrazole compound 1,1′-ethylenebis(oxy)bis(5-nitroiminotetrazole) (**24**), more energetic salts (**24-3** to **24-8**) were synthesized as expressed in Scheme 7.10 [80]. This work aimed to reduce sensitivity and enhance the thermal stability of compound **24** by salt formation. Furthermore, energetic salts **25-1** and **25-2** were prepared from 1,1′-ethylenebis(5-nitroiminotetrazole) (**25**). The energetic salts (**24-3** to **24-8**) were prepared *via* acid–base reaction, and consuming silver salts of **24** with nitrogenous bases followed by the metathesis method.

The synthesis of furazan-functionalized nitraminotetrazole compound **26** was attained as a result of the reaction of 5-(4-amino-1,2,5-oxadiazol-3-yl)-1-hydroxytetrazole with conc. HNO_3 as illustrated in Scheme 7.10 [81]. The resulted product was unstable at room temperature; therefore, it was directly converted to the corresponding ammonium salt and subsequently to silver salt. Furthermore, the energetic salts (**26-1** to **26-5**) were also prepared after the reaction of silver salt with selected nitrogenous bases. The thermal decomposition temperature of these furazan-functionalized nitraminotetrazole compounds and their salts was in a range of 204–291 °C as depicted in Table 7.9. The values suggested that most of the compounds possess comparable or higher decomposition temperature than RDX. The calculated densities were established between 1.65 and 1.85 $g\,cm^{-3}$. All these compounds have higher densities than TNT. The hydroxylammonium energetic salt **26-1** has an even higher density than RDX, i.e. 1.84 $g\,cm^{-3}$. The detonation velocities were found excellent fit as compared to standard HEDMs. Salts **26-1**, **26-2**, **26-5** have higher detonation velocities than RDX, i.e. 9323, 9094, 9033 $m\,s^{-1}$, respectively. The hydrazinium salt (**26-2**) has an even higher detonation velocity than HMX. The calculated detonation pressure of hydroxylammonium salt (**26-1**) was 38.3 GPa which is higher than RDX and TATB. Moreover, these compounds show low sensitivity toward impact and friction and high positive HOF.

From Table 7.9, it is observed that the thermal decomposition temperature is in the range of 131–220 °C not higher than for reference HEDMs. The utmost thermally stable was diaminoguanidinium salt **25-1** which decomposed at 220 °C. All synthesized compounds exhibited lower densities 1.58–1.78 $g\,cm^{-3}$, although they have better detonation performance. The detonation pressures (33.3 and 32.5 GPa) and velocities (9030 and 8789 $m\,s^{-1}$) were established highest for 3,6-dihydrazino-1,2,4,5-tetrazine and carbohydrazide salts (**24-8** and **24-5**) exceeding those of RDX. However, all energetic salts were sensitive toward impact sensitivity (2–8 J) excluding diaminoguanidinium (**25-1**) and triaminoguanidinium (**25-2**) salts which were insensitive. Consequently, the nitrogen-rich heterocyclic compounds showed greater HOFs ranging from 1134 to 2944 $kJ\,mol^{-1}$. However, the found oxygen balance was negative for all compounds.

7.2.2 Synthesis of Amino/Hydrazine-Tetrazole Energetic Salts

1-Aminotetrazol-5-one (ATO) (**27**) is an attractive compound due to good detonation performance and low impact sensitivity. Therefore, ATO (**27**) was obtained from benzaldehyde in six steps as shown in Scheme 7.11 according to a literature modified method [17]. 1-Phenyl-*N*-(1*H*-tetrazol-1-yl)methanimine was also prepared according to the reported method [82] and further treated with $Hg(OAc)_2$ and brominated with Br_2. The raw product was treated with NaOH and HCl to furnish the synthesis of target compound **27**. In addition to the synthesis of ATO, the energetic salts **27-1** to **27-3** were synthesized through the reaction of carbonate and hydroxide nitrogen bases with ATO, while the synthesis of salts **27A-1** to **27A-3** was accomplished using the silver salt of ATO and corresponding hydrochlorides in aqueous media. Prepared compounds decomposed in a range of 214.5–227.1 °C and were found to be thermally stable as RDX. However, the ammonium salt **27-1** was thermally unstable (Table 7.10). Moreover, all the compounds were insensitive toward impact sensitivity except salt **27-1**, though the friction sensitivity was not measured. The detonation

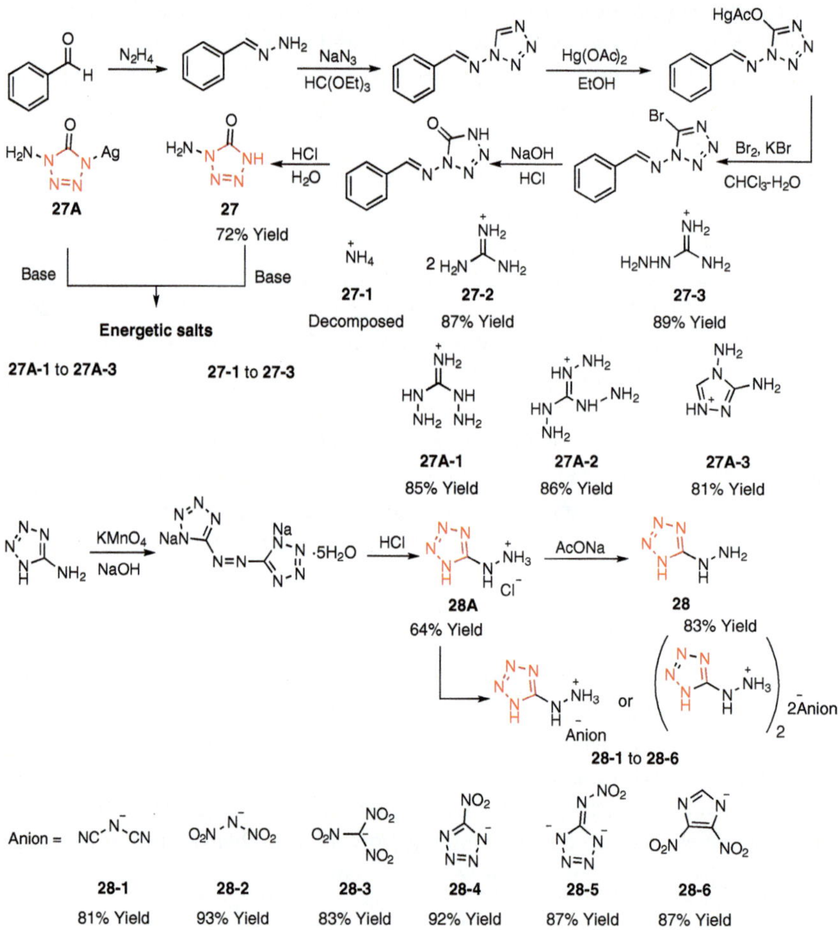

Scheme 7.11 Synthesis of ATO (**27**), HTZ (**28**), and their several nitrogen-rich energetic salts.

Table 7.10 Physicochemical and energetic properties of ATO (**27**), HTZ (**28**), and their several nitrogen-rich energetic salts.

Compound	T_d (°C)	ρ (g cm^{-3})	D (m s^{-1})	P (GPa)	IS (J)	OB (%)	ΔH_f (kJ mol^{-1})
27	227.1	1.80	8800	35	>40	−23	343.0
27-1	—	1.65	8260	28.7	—	−40	255.0
27-2	224.8	1.56	7830	25.0	>40	−50	286.5
27-3	220.6	1.60	8160	26.7	>40	−50	420.5
27A-1	220.7	1.53	8410	28.4	>40	−50	608.4
27A-2	214.5	1.57	8720	31.0	>40	−50	743.3
27A-3	220.5	1.63	7530	23.6	>40	−48	494.4
28	288.7	1.74	8226	27.4	>40	−47	590.5
28-1	150.6	1.52	7052	17.4	>40	−81	430.7
28-2	188.7	1.78	9626	46.8	6.8	−3.9	516.3
28-3	179.6	1.82	9206	39.2	—	−3.2	419.4
28-4	184.4	1.70	8885	32.5	4.3	−33	768.7
28-5	198.6	1.67	9227	35.1	28.4	−43	1424.3
28-6	182.5	1.73	8602	31.3	>40	−43	572.5

velocity and pressure were between 7530 and 8800 m s^{-1} and 23.6 and 35.0 GPa, respectively. These values revealed that only compound **27** has detonation performance (8800 m s^{-1} and 35.0 GPa) comparable to RDX.

Compounds **28** and **28A** were prepared from a cheap starting reagent 5-amino-1*H*-tetrazole (Scheme 7.11) [83]. 5-Amino-1*H*-tetrazole was readily oxidized by $KMnO_4$ in NaOH solution to produce sodium 5,5′-azotetrazolate pentahydrate, which was further reacted with diluted HCl resulting in compound **28A**. Compound **28** (HTZ or 5-hydrazinotetrazole) was produced after the neutralization with sodium acetate. Energetic salts (**28-1** to **28-6**) based on HTZ cations and nitrogen-rich anions were prepared to fulfill the necessity of high nitrogen content and good oxygen balance. The thermal stability of prepared salts (**28-1** to **28-6**) was 150.6–198.6 °C with decomposition temperature (Table 7.10). The thermal decomposition was lower in comparison to standard HEDMs. The value of calculated density ranged from 1.52 to 1.82 g cm^{-3}, which is reasonable for all compounds except for dinitramide (**28-2**) and nitroformate (**28-3**) salts which have higher densities (1.78 and 1.82 g cm^{-3}). The detonation velocities were found from 7052 to 9450 m s^{-1} and detonation pressures were between 17.4 and 46.8 GPa. The detonation performance (9450 m s^{-1} and 46.8 GPa) of dinitramide salt **28-2** was advanced than CL-20, and the detonation velocities (9206, 8885, and 9227 m s^{-1}) and pressures (39.2, 32.5, 35.1 GPa) of salts **28-3** to **28-5** were competitive to RDX.

7.2.3 Synthesis of Tetrazole *N*-oxide Energetic Salts

The synthesis of tetrazole *N*-oxide is a new trend in the advancement of tetrazole-based energetic materials and their nitrogen-rich salts. The main purpose of this technique is to improve the oxygen balance, energetic properties, and densities. The synthesis of compound **29** was achieved by using malononitrile as starting material to afford 4-amino-1,2,5-oxadiazole-3-carbohydroximoyl-chloride intermediate (Scheme 7.12) [84]. The synthetic process was completed after the treatment of intermediate compound with sodium azide followed by passing of HCl gas. The energetic salts **29-1** to **29-9** were prepared simply by the reaction of compound **29** with selected nitrogen-rich bases.

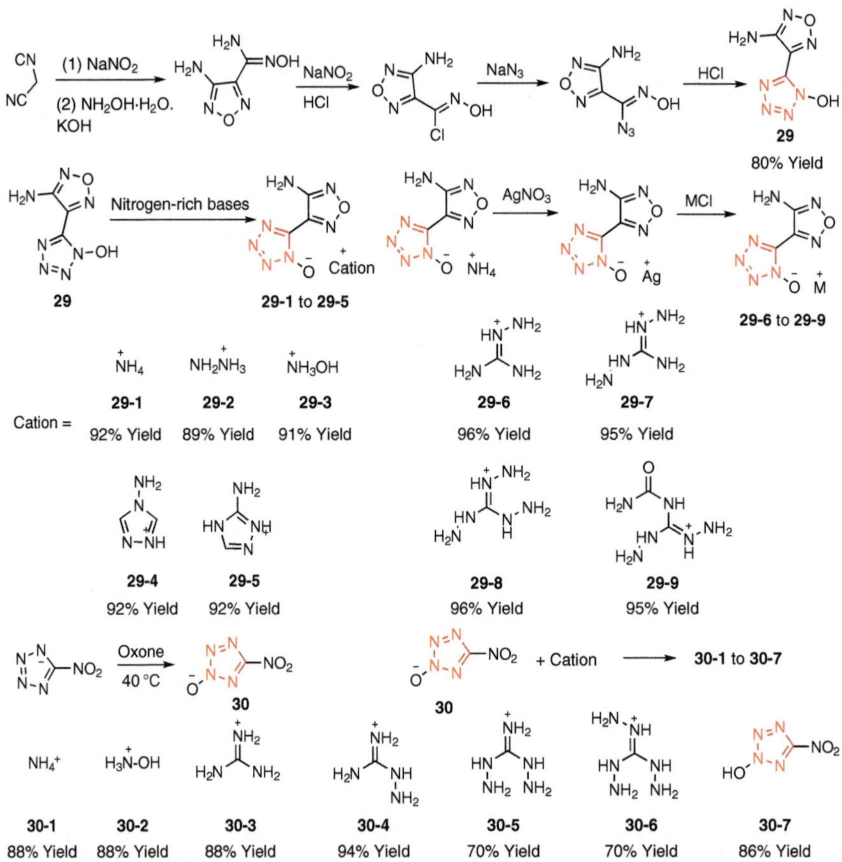

Scheme 7.12 Synthesis of tetrazole *N*-oxide compounds **29**, **30**, and their various nitrogen-rich energetic salts.

Prepared energetic compound **29** and salts **29-1** to **29-9** thermally decomposed between 197 and 277 °C (Table 7.11). Nitrogen-rich ammonium, aminoguanidinium, and 5-amino-4*H*-1,2,4-triazol-1-ium salts (**29-1**, **29-5**, and **29-6**) have

Table 7.11 Properties of tetrazole *N*-oxide compounds **29-31** and their various nitrogen-rich energetic salts.

Compound	T_d (°C)	ρ (g cm^{-3})	D (m s^{-1})	P (GPa)	IS (J)	FS (N)	OB (%)	ΔH_f (kJ mol^{-1})
29	197	1.79	8601	30.0	37	>360	−47	441.35
29-1	277	1.69	8436	26.6	>50	>360	−38	407.62
29-2	216	1.69	8759	29.0	46	>360	−59	561.18
29-3	213	1.80	9100	33.4	>50	>360	−35	454.79
29-4	234	1.75	8600	28.0	>50	>360	−47	762.07
29-5	242	1.74	8363	26.1	>50	>360	−47	625.35
29-6	258	1.64	8173	23.6	>50	>360	−46	488.79
29-7	231	1.70	8580	26.7	>50	>360	−69	593.86
29-8	216	1.71	8859	29.0	>45	>360	−49	713.08
29-9	225	1.77	8303	25.3	>50	>360	−50	262.06
30-1	173	1.73	8885	32.2	7	120	−10	152.0
30-2	157	1.85	9499	39.0	4	60	0	218.7
33-3	211	1.70	8201	26.6	>40	252	−33	136.7
33-4	185	1.70	8514	28.5	20	112	−35	256.4
33-5	174	1.69	8686	29.2	40	120	−36	361.0
33-6	153	1.64	8768	29.4	25	72	−37	471.5
30-7	120	1.94	9447	40.4	—	—	6	308.6
31-1	246.7	1.69	7577	24.49	>60	—	−46	956
31-2	262.0	1.72	7151	22.03	>50	—	−62	717.64
31-3	282.7	1.67	7050	21.05	>50	—	−47	1548.67
31-4	269	1.53	7142	20.23	>60	—	−78	751.13
31-5	303	1.65	7405	23.05	>50	—	−49	885.0
31-6	273	1.70	8151	28.48	>60	—	−58	1385.29
31-7	314	1.68	6820	19.76	>60	—	−66	752.74
31-8	225.7	1.71	7693	22.45	>50	—	−65	751.13
31-9	243	1.67	8218	28.69	>50	—	−49	885.00
31-10	254.0	1.68	1168.85	25.41	>50	—	−53	120.05
31-11	217.6	1.91	8931	36.1	28	120	−39	425.6

high decomposition temperatures (277, 242, and 258 °C) exceeding those of RDX. The measured densities were in the range of 1.64–1.79 g cm^{-3}. The detonation pressures and velocities were 23.6–33.4 GPa and 8173–9100 m s^{-1}, respectively. The hydroxylammonium salt **29-3** has higher detonation pressure and velocity (33.4 GPa and 9100 m s^{-1}) higher than those of TATB and RDX. However, the other prepared compounds have similar detonation pressures and velocities in comparison to standard HEDMs. Moreover, the measured impact and friction sensitivities were lower and acceptable as compared to standard HEDMs.

Furthermore, to achieve high density, HOF, ring strain, and greater oxygen balance, compound **30**, and its nitrogen energetic salts **30-1** to **30-7** were synthesized (Scheme 7.12) [85]. Compound **30** was produced through the oxidation of ammonium 5-nitrotetrazolate hemihydrate in a saturated solution of oxone. In a continued effort the free acid 5-nitrotetrazole-2*N*-oxide**30-7** and nitrogen-rich salts **30-1** to **30-6** were synthesized *via* the metathesis method. The nitrotetrazolate-2*N*-oxide salts **30-1** to **30-6** revealed higher densities from 1.64 to 1.94 g cm^{-3} reaching the level of standard HEDMs (Table 7.11). The free acid 5-nitrotetrazole-2*N*-oxide **30-7** has a high density (1.94 g cm^{-3}) among others comparable to HMX. The prepared salts exhibited positive HOFs in the range of 136.7–308.6 kJ mol^{-1} enhanced in comparison to tetrazole moiety without *N*-oxides. The detonation parameters (pressure and velocity) were 26.6–40.4 GPa and 8201–9499 m s^{-1} respectively. Among others, hydroxylamine salts **30-2** and **30-7** possess higher detonation performance (pressure: 39.9 and 40.4 GPa and velocity: 9499 and 9447 m s^{-1}, respectively) exceeding those of RDX, TATB, and HMX, whereas the impact and friction sensitivities were between the 4 and >40 J and 60 and 252 N, respectively, classifying from sensitive to insensitive energetic compounds. However, the synthesized compounds were found slightly less thermally stable with decomposition temperatures 153–211 °C in assessment to reference HEDMs.

Thermal decomposition temperatures of prepared salts **31-1** to **31-11** were between 217.6 and 314 °C as presented in Table 7.11. All of the energetic salts have higher decomposition values than RDX and salts **31-5** and **31-7** have T_d values 303 and 314 °C greater than TNT. The measured densities were found in the range of 1.53–1.91 g cm^{-3}. Furthermore, HOFs were found positive for all prepared salts ranging from 120.1 to 1385.3 kJ mol^{-1}.The impact sensitivity values were higher than TNT and RDX, i.e. more than 50 and 60 J. However, salt **31-11** has lower impact and friction values which is 28 J and 120 N, respectively. The detonation velocity and pressure were in the range of 6820–8971 m s^{-1} and 19.76–36.71 GPa. Energetic salts **31-1** to **31-11** of 1*H*,1′*H*-5,5′-bitetrazole-1,1′-diolate (BTO)(**31**) were synthesized as shown in Scheme 7.13 [86]. BTO was prepared according to the formerly reported method by using glyoxal as a starting material [83]. The energetic salts **31-1** to **31-9** were prepared normally by treating BTO with nitrogen-rich bases except **35-11**. The salt **35-10** was detached from the mixture comprising barium salt of **31** after the removal of barium sulfate. Moreover, the synthesis of energetic salt **31-11** was successfully achieved by utilizing BTO as starting material as described in Scheme 7.13 [87]. BTO was treated with lithium hydroxide in water and hydrazine dihydrochloride to complete the synthesis of energetic salt **31-11**. The synthesized energetic salt **31-11** has a higher density of 1.91 g cm^{-3} among other salts which is higher than standard HEDMs such as TNT and RDX. The detonation velocity and pressure were also higher than other salts (8931 m s^{-1} and 36.1 GPa, respectively) which was higher than RDX. However, the target compound **31-11** decomposed at a temperature of 217.6 °C which is lower relative to standard HEDMs.

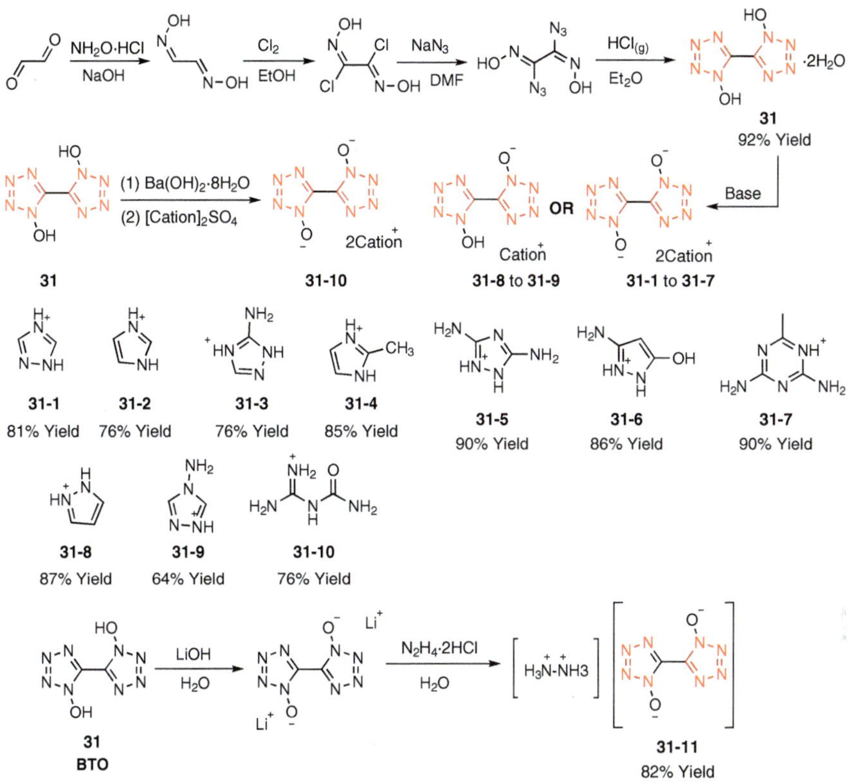

Scheme 7.13 Synthetic route to BTO (**31**) and its nitrogenous salt (**31-1** to **31-11**).

7.2.4 Synthesis of Bistetrazole or Bridged Bistetrazole Energetic Salts

High nitrogen-containing bistetrazole or bridged bistetrazole-based energetic materials and particularly their nitrogen-rich salts are expected to be thermally stable. Therefore, in continuing efforts, Zhao et al. reported the synthesis and characterization of 1,2-bis(3-nitro-1-(1*H*-tetrazol-5-yl)-1*H*-1,2,4-triazol-5-yl)diazene (BNTD)**32** and its different nitrogenous salts illustrated in Scheme 7.14 [60]. Compound **32** was obtained through an oxidation of the starting material 3-nitro-1-(1*H*-tetrazol-5-yl)-1*H*-1,2,4-triazol-5-amine with $KMnO_4$. Corresponding salts **32-1** to **32-8** were prepared by the reductive conversion of an azo substrate (**32**) with hydroxylamine hydrochloride or with other structurally different amino derivatives. The energetic properties of these compounds are shown in Table 7.12. All salts possessed high thermal stability ranged from 212 to 259 °C, competitive to RDX. Most of the salts displayed insensitive behavior toward impact and friction. The calculated densities were similar to RDX for ammonium, hydrazinium, and hydroxylammonium (**32-1** to **32-3**) salts (1.80, 1.82, 1.80 g cm^{-3}, respectively). The salt **32-8** possessed higher HOF, i.e. 1554.7 kJ mol^{-1}, which is higher than those of RDX and HMX. Moreover, the salts **32-5** and **32-6** have higher detonation velocities (8991 and 9038 m s^{-1}) and pressures (33.3 and 32.1 GPa) reachable RDX.

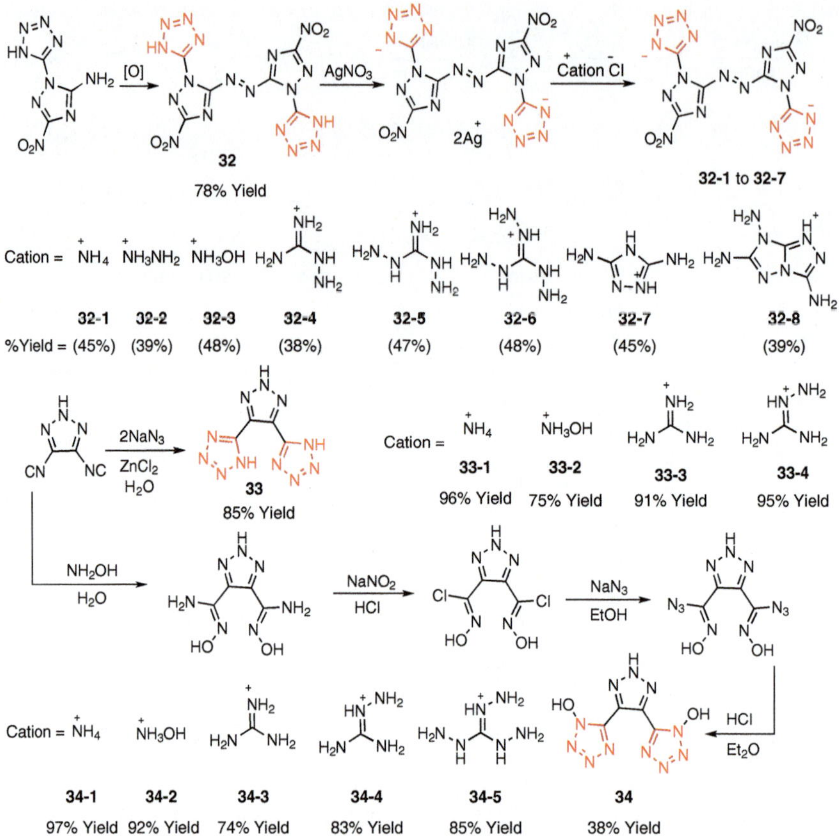

Scheme 7.14 Synthesis of bridged bistetrazole or bistetrazolecompounds **32–34**, and their nitrogen salts.

The synthesis of 4,5-bis(tetrazol-5-yl)-1,2,3-triazole (**33**) and 4,5-bis(1-hydroxy-tetrazol-5-yl)-1,2,3-triazole (**34**) was achieved using same starting raw material, i.e. 4,5-dicyano-2*H*-1,2,3-triazole elaborated in Scheme 7.14 [88]. Compound **33** was prepared by an interaction of 4,5-dicyano-2*H*-1,2,3-triazole and NaN_3. The preparation of compound **34** was carried out in three steps after the reaction of 5-dicyano-2*H*-1,2,3-triazole and hydroxylamine in water. Then the raw product was diazotized with diluted HCl to form chloroxime, the chloro-azide interchange succeeded after reaction with NaN_3 which was precisely further converted to the target compound **34** as a result of the ring closure [89]. Furthermore, nitrogen-rich energetic salts (**33-1** to **33-4** and **34-1** to **34-6**) of the neutral compounds **33** and **34** were also prepared in water. The reaction of energetic compound **34** with triaminoguanidinium chloride produced mono triaminoguanidinium salt **34-6**. The energetic properties of synthesized compounds **33**, **34**, and their salts **33-1** to **33-4** and **34-1** to **34-6** are illustrated in Table 7.12. The decomposition temperatures of thus-prepared compounds were in the range of 160–301 °C. The thermal stability of target compounds was comparable to RDX except hydroxylammonium salts (**33-2**

Table 7.12 Physicochemical and energetic properties of bridged bistetrazole or bistetrazolecompounds **32–34**, and their nitrogenous salts.

Compound	T_d (°C)	ρ (g cm^{-3})	D (m s^{-1})	P (GPa)	IS (J)	FS (N)	OB (%)	ΔH_f (kJ mol^{-1})
32-1	229	1.80	8746	29.9	40	360	−26	411.1
32-2	246	1.82	8991	33.3	15	360	−28	524.7
32-3	212	1.80	9038	32.1	8	360	−17	728.7
32-4	244	1.66	8127	23.9	8	360	−35	629.3
32-5	223	1.67	8346	25.4	6	324	−36	851.9
32-6	228	1.68	8551	26.9	5	288	−37	1082.1
32-7	259	1.73	8105	22.4	40	360	−35	841.6
32-8	252	1.76	8236	25.0	40	360	−36	1554.7
33	277	1.69	8360	24.8	2	240	−74	—
33-1	301	1.62	8587	23.7	>40	>360	−83	—
33-2	160	1.64	8798	27.3	>40	>360	−61	—
33-3	289	1.52	7785	18.4	>40	>360	−91	—
33-4	264	1.66	9022	26.3	>40	>360	−88	—
34	246	1.67	8277	26	1	240	−50	—
34-1	241	1.736 (1.69)	8991	28.8	10	>360	−61	—
34-2	182	1.63	8561	27	7	>360	−44	—
34-3	244	1.56	7659	18.6	>40	>360	−73.3	—
34-4	215	1.63	8836	26.6	>40	>360	−72	—
34-5	212	1.60	8726	26.7	10	>360	−63	—
34-6	240	1.52	7684	18.3	>40	>360	−81	—

and **34-2**). Compounds **33** and **34** showed higher impact and friction sensitivities, while their salts exhibited lower sensitivities meeting the criteria of standard HEDMs. The impact and friction sensitivities of tetrazole comprising compound **33** and salts (**33-1** to **33-4**) were better in comparison to tetrazole oxides bearing compound **34** and salts (**34-1** to **34-6**). The detonation velocities of energetic salts **33-2, 33-4, 34-1,** and **34-4** were 8798, 9022, 8991, and 8836 m s^{-1}, respectively, which are higher than those of TNT, RDX, and TATB. However, the detonation pressures and calculated densities of synthesized compounds were less as compared to standard HEDMs. Dense and thermally stable (H_2BTF) 3,4-bis(1*H*-5-tetrazolyl)furoxan (**35**) and its mono ionic salts (**35-1** to **35-8**) were prepared (Scheme 7.15) and characterized by Zhou and coworkers [90]. The prepared compound **35** and its salts have judicious physicochemical and energetic properties in comparison to reference HEDMs. The thermal stabilities ranged 177–251 °C and high positive HOFs 859.8–1063.5 kJ mol^{-1}, indicating their good energetic properties (Table 7.13). Their detonation velocities and pressures were predicted to be 7740–8790 m s^{-1} and 23.1–32.5 GPa, respectively. The detonation performance was higher than that of

TNT but only reached RDX in the case of the hydrazinium salt **35-1** (8790 m s^{-1} and 32.5 GPa). Impact sensitivities (4–25 J) of **35-2**, **35-5**, and **35-6** salts were between RDX and TNT, whereas **35-3** was less sensitive than TNT. Although hydrazinium salt **35-1** has high density, detonation performance, and positive HOF, the thermal stability, impact, and friction sensitivities values were not measured owing to the impurity of the sample. Synthesis of compound **35** aimed to enhance HOFs, detonation performance, and density due to the connection of two tetrazole rings with furoxan. The synthesis was started from 3,4-dicyanofuroxan and carried out by the literature method, which was promoted to react with NaN_3 catalyzed by ammonium chloride to yield H_2BTF (**35**) [91]. The salts **35-1** to **35-8** were prepared upon neutralization of H_2BTF with certain nitrogen-rich bases in methanol. However, the salt formation protocol was picked cautiously owing to the di-basicity of H_2BTF.

The group of Zhou prepared another series of H_2BTF salts (Scheme 7.15) [92]. The barium salt **35A** was prepared by the reaction of H_2BTF with barium hydroxide. Energetic salts **35A-1** to **35A-8** were easily synthesized by implementing a metathesis

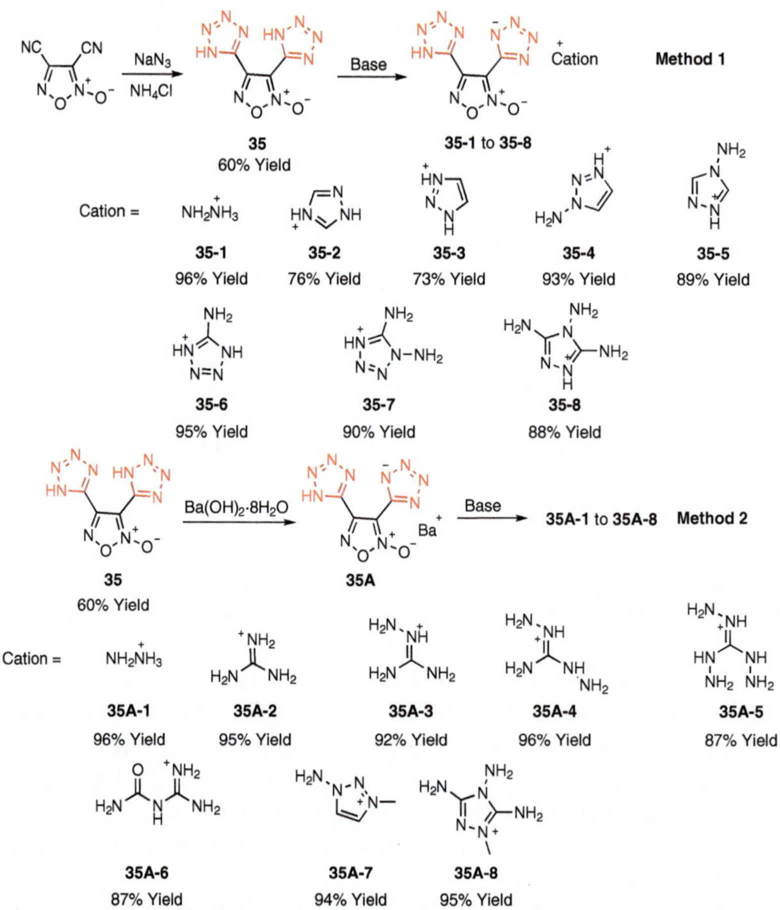

Scheme 7.15 Synthesis of H_2**BTF** (**35**), its barium (**35A**), and nitrogen-rich salts.

Table 7.13 Physicochemical and energetic properties of bridged bistetrazole or bistetrazolecompounds **35** and its nitrogenous salts.

Compound	T_d (°C)	d (g cm^{-3})	D (m s^{-1})	P (GPa)	IS (J)	FS (N)	OB (%)	ΔH_f (kJ mol^{-1})
35	220	1.62	7778	23.9	<2	—	−50	727.8
35-1	—	1.82	8790	32.5	—	—	−56	775.2
35-2	225	1.78	8188	27.3	8	—	−68	859.8
35-3	223	1.63	7740	23.1	25	—	−68	946.9
35-4	177	1.78	8469	29.6	4	—	−67	1063.5
35-5	225	1.76	8278	27.9	8	—	−67	968.4
35-6	218	1.67	8049	19.5	14	—	−54	1014.3
35-7	204	1.68	8153	26.4	9	—	−54	1062.0
35-8	251	1.79	8326	28.1	9	—	−66	883.1
35A-1	233	1.68	8915	32.0	14	—	−39	1163.6
35A-2	264	1.71	8288	26.9	14	—	−47	851.6
35A-3	241	1.64	8278	26.3	28	—	−47	1093.8
35A-4	290	1.68	8641	29.3	12	—	−48	1321.5
35A-5	220	1.62	8597	28.4	7	—	−48	1574.5
35A-6	253	1.85	8495	29.5	29	—	−44	471.6
35A-7	239	1.59	8062	24.8	6	—	−57	1762.0
35A-8	252	1.59	7939	23.4	5	—	−56	1719.8

protocol with barium salt **35A**. Resulting compounds were highly dense in the range of 1.59–1.89 g cm^{-3} and thermally stable in a range of 220–290 °C, impending the criteria of HEDMs (Table 7.13). The values of calculated HOFs were from 471.6 to 1762.0 kJ mol^{-1}. Due to good densities and HOFs, the detonation performance was also looming the constraint of HEDMs. Detonation pressures and velocities were higher than those of TNT in the range of 23.4–32.0 GPa and 8915–9390 m s^{-1}. Impact sensitivities (5–29 J) values listed them in the class of sensitive energetic compounds.

Keeping in view the importance of 1,2,5-oxadiazole ring conjugated with tetrazole, 4,4′-oxybis[3,3′-(1*H*-5-tetrazol)]furazan (H_2BTFOF, **36**) and its salts **36-1** to **36-10** were intended to be synthesized. The synthesis of H_2BTFOF (**36**) came into existence from 3,3′-dicyanodifurazanyl ether as explicated in Scheme 7.16 [93]. 3,3′-Dicyanodifurazanyl ether was reacted with NaN_3 catalyzed by $ZnBr_2$ yielded H_2BTFOF (**36**). The reaction with barium hydroxide produced Ba(BTFOF)$_2$ which was further utilized to generate a series of nitrogen-rich salts **36-1** to **36-10** by the metathesis approach. However, utilization of the dicationic biguanidinium sulfate resulted in a mixture of energetic salts **36-9** and **36-10**, which was detached *via* crystallization technique. The thermal decomposition temperatures were in the range of 212–289 °C (Table 7.14), while the aminoguanidinium (**36-4**) and guanidinium salts (**36-3**) exhibited almost equal decomposition temperatures

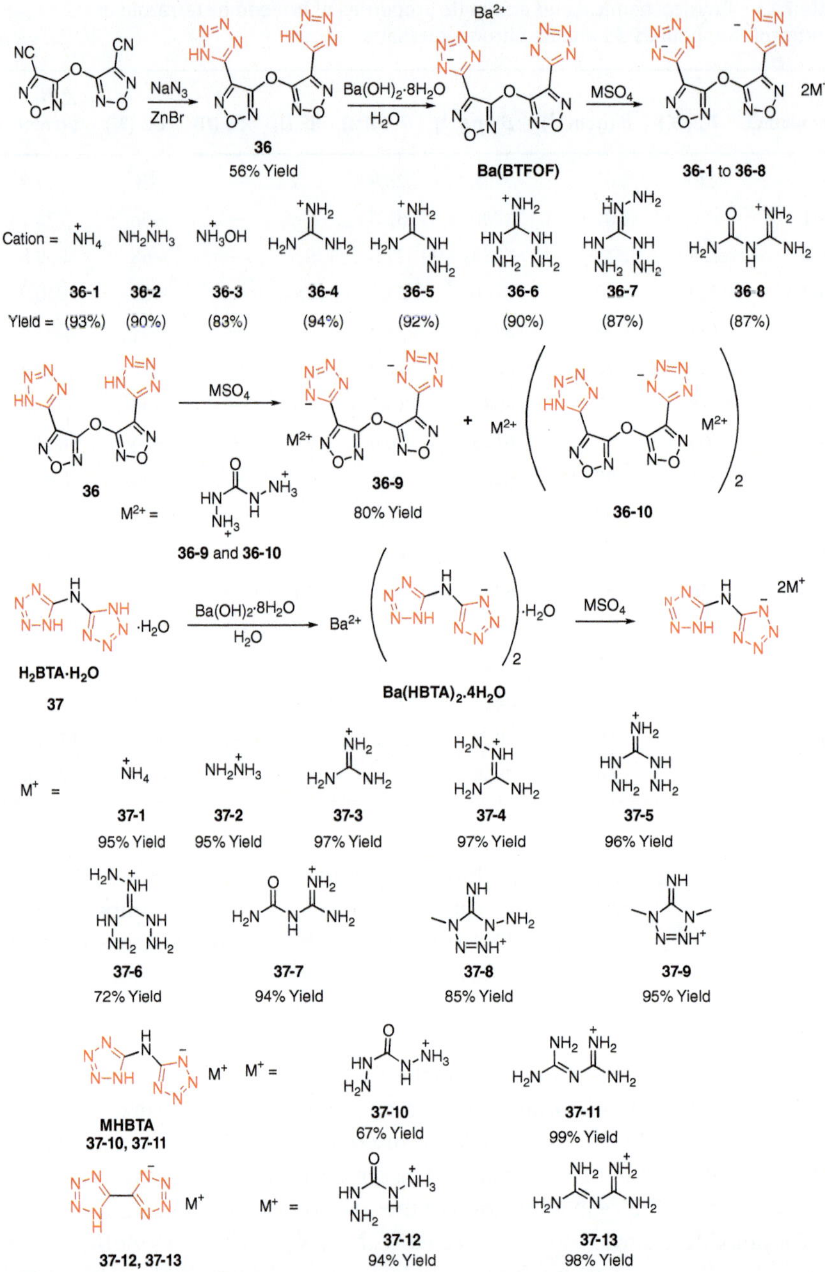

Scheme 7.16 Synthetic route to bistetrazole or bridged bistetrazole compounds **36**, **37**, and their various salts.

Table 7.14 Physicochemical and energetic properties of bistetrazole or bridged bistetrazole compounds and their various nitrogenous salts.

Compound	T_d (°C)	ρ (g cm^{-3})	D (m s^{-1})	P (GPa)	IS (J)	FS (N)	OB (%)	ΔH_f (kJ mol^{-1})
36	240	1.74	7977	26.5	2	—	−22	858.6
36-1	212	1.69	7953	23.4	5	—	−31	693.7
36-2	242	1.74	8517	29.4	18	—	−36	1009.6
36-3	287	1.64	7840	23.3	10	—	−41	722.1
36-4	289	1.64	7804	23.0	12	—	−43	921.4
36-5	213	1.67	8129	25.5	36	—	−44	1152.1
36-6	237	1.75	8624	29.9	25	—	−44	1370.6
36-7	243	1.77	8548	30.8	26	—	−22	802.9
36-8	241	1.65	7266	19.7	5	—	−38	344.3
36-9	247	1.86	8318	28.9	22	—	−26	584.3
37-1	269	1.66	8936	28.1	—	—	−52	594.4
37-2	249	1.66	9257	30.7	—	—	−52	749.8
37-3	231	1.63	7677	19.4	—	—	−60	374.4
37-4	268	1.62	8343	23.2	—	—	−52	565.0
37-5	234	1.64	8696	25.7	—	—	−52	668.8
37-6	213	1.55	8413	24.0	—	—	−47	792.5
37-7	213	1.58	8748	26.3	—	—	−50	900
37-8	201	1.64	8401	24.7	—	—	−42	1002.9
37-9	258	1.58	7860	21.3	—	—	−56	920.3
37-10	189	1.75	9487	33.6	—	—	−42	672.3
37-11	258	1.69	8645	25.1	—	—	−56	634.8
37-12	224	1.68	8749	27.2	—	—	−42	589.2
37-13	251	1.60	7906	20.4	—	—	−56	558.8
38	288.7	1.74	8226	27.4	>40	—	−31	590.5
38A	247.6	1.71	8307	25.3	25	—	−44	646.5
38-1	262.9	1.56	7182	16.4	35	—	−51	148.4
38-2	211.5	1.62	8222	23.0	40	—	−52	485.0
38-3	260.4	1.60	7756	17.8	40	—	−38	294.6
38-4	197.7	1.48	7565	18.6	>40	—	−34	503.2
38-5	278.7	1.58	6896	15.0	>40	—	−70	534.0
38-6	325.1	1.63	7123	16.4	>40	—	−59	516.0
38-7	249.0	1.66	8031	22.2	>40	—	−54	936.1
38A-1	239.9	1.45	7381	17.2	>40	—	−59	560.2
38A-2	212.1	1.68	9050	28.7	35	—	−57	838.2

(Continued)

Table 7.14 (Continued)

Compound	T_d (°C)	ρ (g cm^{-3})	D (m s^{-1})	P (GPa)	IS (J)	FS (N)	OB (%)	ΔH_f (kJ mol^{-1})
38A-3	237.8	1.68	8839	28.8	20	—	−51	642.2
38A-4	191.0	1.66	8346	24.0	>40	—	−56	1299.1
39	145	1.73	8355	28.2	<1	<5	−57	622
40	135	1.57	7593	21.3	9	240	−84	592
39-1	220	1.57	7900	22.9	10	300	−67	562
39-2	136	1.56	8227	25.7	3	40	−51	628
39-3	145	1.57	8120	24.1	15	160	−72	859

(289 and 287 °C, respectively). Prepared compounds were dense in the range of 1.64–1.86 g cm^{-3}, reaching TNT and higher than RDX in the case of **36-9**. Owing to the superior nitrogen content the HOF values were established highly positive from 344.3 to 1370.6 kJ mol^{-1}. However, the detonation pressure and velocity were between 9.7 and 30.8 GPa and 7266 and 8778 m s^{-1} correspondingly impending TNT and in the case of salts **36-6** and **36-7** comparable to TATB.

Monoanionic salts (**37-1** to **37-13**) of bis(1*H*-tetrazol-5-yl)amine H_2BTA (**37**) paired with nitrogen-rich cations anticipated to highly dense and thermally stable compounds (Scheme 7.16) [94]. H_2BTA (**37**) reacted first with barium hydroxide to yield white Ba(HBTA)$_2$·4H_2O salt, which further proceeded to give **37-1** to **37-9** following metathesis with sulfate salts in water. As illustrated in Table 7.14 the ammonium salt **37-1** was the most stable with decomposition temperature 269 °C greater than RDX. The overall decomposition temperatures for all nitrogen-rich salts ranged 189–269 °C. The densities and oxygen balance of prepared salts were 1.55–1.75 g cm^{-3} and −42.1% to −60.1% correspondingly, approaching TNT in the case of some salts but lower than other standard HEDMs. However, the HOF values were highly positive in the range of RDX and HMX. Moreover, the detonation pressures and velocities were 19.4–33.6 GPa and 7677–9487 m s^{-1}, respectively, enhanced than TNT for all salts. However, the salt **37-10** has detonation performance (pressure: 33.6 GPa and velocity: 9487 m s^{-1}) comparable to RDX and TATB. Therefore, the prepared salts were proved painstaking in thermal stability and detonation performance.

Scheme 7.17 shows the synthesis of nitrogen-rich compounds in closing two tetrazole rings, di(1*H*-tetrazol-5-yl)methanone oxime (**38**), and5,5′-(hydrazonomethylene)bis(1*H*-tetrazole) (**38A**) from potassium dicyanonitrosomethanide [95]. The dicyanonitrosomethanide directly reacted with sodium azide in the presence of zinc chloride in water-yielding compound **38**. Numerous salts (**38-1** to **38-7** and **38A-1** to **38A-4**) were obtained after the treatment of **38** and **38A** with nitrogen-rich bases. These compounds decomposed thermally in a range of 197.7–325.1 °C (Table 7.14). Thermal decomposition values suggest their good stability competing with RDX, except for dihydrazinium and dicarbohydrazonic diamide salts (**38-2**

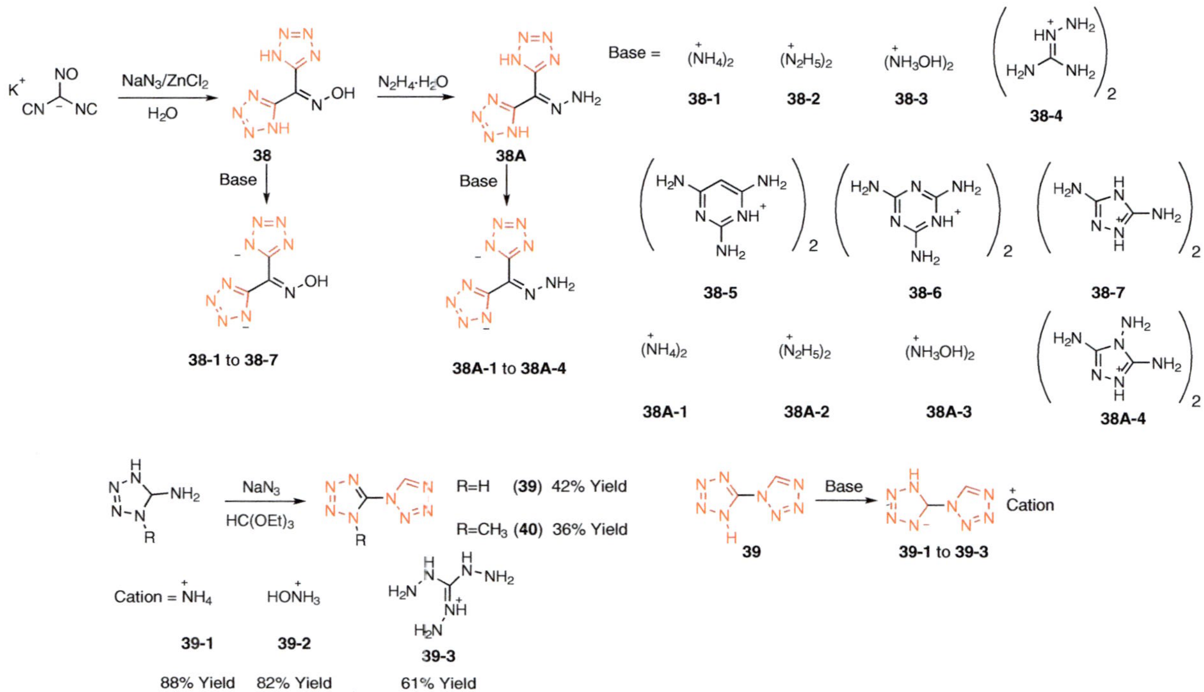

Scheme 7.17 Synthetic route to bridged bistetrazole or bistetrazole compounds **38**, **39**, **40**, and their various salts.

and **38-4**). Although the detonation velocities are reasonable in comparison to standard HEDMs from 7123 to 9050 $m\,s^{-1}$, the detonation pressures were lower (15.0–27.4 GPa). However, friction sensitivities remained unevaluated and the values of densities (1.48–1.74 $g\,cm^{-3}$) were lower than those of standard HEDMs. Target bis-tetrazole compounds **39** and **40** were easily synthesized [96] from a low-cost starting material 5-aminotetrazoles according to the literature method (Scheme 7.17) [97]. Triethyl orthoformate and sodium azide in warm acetic acid were applied for the cyclization of 5-aminotetrazoles in a good yield. Further, energetic salts (**39-1** to **39-3**) were achieved by reaction of nitrogen-rich bases with compound **39** in water. The prepared compounds suffered from shortcomings of low density and thermal stability as shown in Table 7.14. The thermal decomposition temperature is only reasonable in the case of ammonium salt **39-1** which was 220 °C. The density of neutral compound **39** was 1.73 $g\,cm^{-3}$. Due to low density, the detonation performance (pressure and velocity) was found poor than standard HEDMs. Moreover, these compounds were classified as sensitive explosives due to their high sensitivity toward impact and friction.

7.3 Conclusion and Future Trends

It has been seen that intensive investigation in the design and preparation of new HEDMs is an emergent interest of energetic materials researchers, particularly in the field of energetic salts. The key motivation to synthesize these new HEDMs is to substitute traditional energetic materials with an increasing demand for eco-friendly and less toxic for bio-organism materials. Beyond this, it is also attractive and stimulating to attain energetic materials that comprise high energy, density, and improved detonation properties. Furthermore, the exploration of an optimal equilibrium among high thermal stability, good detonation performance, and reasonable sensitivity to different mechanical stimuli is motivating and challenging.

It is the judgment of energetic material researchers that arrangement of nitrogen-rich heterocyclic moieties with different explosophores ($-NO_2$, $-ONO_2$, $NH-NO_2$, N_3, etc.) and further preparation of their nitrogenous salts lead to the formation of new HEDMs with better detonation properties, densities, and environmentally benign nature; however, the thermal stability may suffer. For this purpose, tetrazole is considered exceptionally significant as a backbone to synthesize an extensive range of energetic materials. The important features of the tetrazole ring are high HOF, ring strain, stability, and density which are valuable properties for high-performing explosives. However, most of the tetrazole-based HEDMs suffer from low mechanical sensitivity and thermal stability which can be resolved by salt formation. In this review, we have summarized the tetrazole-based nitrogen-rich salts with a variety of explosophores (nitro, azido, nitramino, amino, hydrazine, tetrazole *N*-oxide, and bistetrazole or bridged bistetrazole) since 2010 and their impacts on energetic properties. The criterion of assessment is that new HEDMs should potentially replace traditionally used energetic compounds, such as TNT, TATB, RDX, HMX, and CL-20. The major energetic requirements which we discussed here include detonation properties (pressure and velocity) greater than 8500 m s^{-1} and 34.0 GPa, respectively (RDX: D = 8795 m s^{-1} and P = 34.0 GPa), thermal stability (T_d > 200), and high density (RDX: d = 1.80 g cm^{-3}). Based on these requirements we summarize the role of different explosophores with a combination of tetrazole rings in the following four points.

(1) Many azido, nitro, and nitramino tetrazole nitrogenous salts which were discussed have energetic properties competitive to TNT and TATB and some have closer or even improved detonation performance to RDX (**2-2**, **2-3**, **4-4**, **5A-2**, **8-1**, **8-3**, **8-5**, **10-1**, **10-3**, **10-6**, **10-7**, **10-8**, **11-2**, **13-1**, **13-4**, **16-3**, **17A-1** to **17A-3**, **19-1** to **19-3**, **20-2**, **21-5**, **24-8**, **24-5**, **26-1** and **26-5**), various are comparable to HMX (**2-3**, **5B-5**, **7-2**, **10-2**, **23**, and **26-2**) and some are competitive to CL-20 (**19-2**, and **19-20**). Compounds **1-2**, **4-4**, **7-2**, **7-3**, **9-1**, **9-2**, **9-4**, **9B-4**, **10-1**, **10-3**, **11-1**, **11-2**, **11-3**, **11-4**, **11-5**, **11-7**, **11-8**, **14-2**, **16-2**, **15-2**, **20-2**, **20-3**, **20-4**, and **26-1** to **26-5** exhibit better thermal stabilities than RDX. This is worth noting that the polynitro or polynitramino tetrazole compounds and their salts show greater detonation performance and density as compared to the less heteroatom-enriched explosophores. Although these properties are usually reflected with low thermal stability, a salt formation strategy tested to be valuable to encourage thermal stability. Therefore, polynitro or polynitramino tetrazole compounds and their salts evidenced for high densities and detonation performance, whereas the less heteroatom-enriched molecules favor high decomposition temperatures and this can be beneficial in the design of future HEDMs.

(2) From the amino/hydrazine-tetrazole nitrogenous salts, **28-3** to **28-5** exhibited better detonation performance than RDX and **28-2** is competitive with CL-20. However, **27-2** to **27-3**, and **27A-1** to **27A-3** were thermally more stable than RDX, while nitrogenous salt **28-3** has a density closer to RDX. Hence, this trend of designing and synthesis of HEDMs favors thc advance detonation performance and thermal stability on the cost of the density of nitrogenous salts.

(3) Tetrazole N-oxide nitrogenous salts **29-3** and **31-11** have detonation performance comparable to RDX, while **30-2** is similar to HMX. The tetrazole N-oxide salt **31-11** also has a better density than RDX. Moreover, most of the prepared salts (**29-1**, **29-5**, **29-6**, **31-1** to **31-11**) were more thermally stable than RDX. As a result, the synthesis of such materials may improve thermal stability or oxygen balance as well as improving density and detonation performance.

(4) From the high nitrogen-containing bistetrazole or bridged bistetrazole-based nitrogen-rich salts which were discussed here, the following HEDMs (**32-1** to **32-8**, **33-1**, **33-3**, **33-4**, **34**, **34-1**, **34-3**, **34-6**, **35-2**, **35-3**, **35-5**, **35-8**, **35A-1** to **35A-8**, **36-2** to **36-4**, **36-6** to **36-9**, **37-1** to **37-5**, **37-9**, **37-11**, **37-12**, **37-13**, **38-1**, **38-3**, **38-5**, **38-6**, **38-7**, **38A-1**, **38A-3**, and **39-1**) exhibited good thermal stability which were higher than that of RDX, while only few of them had detonation performances (**35-1**, **35A-1**, and **37-10**) and densities (**35-1** and **36-9**) similar to RDX. Therefore, this technique to design and synthesis of energetic materials may be adopted in the future to increase thermal stability, nitrogen content, and HOF. However, density and detonation properties will be compromised.

7.4 Cautions

It is to remind the readers that this review is about the current progress of tetrazole-based energetic salts with the combination of various explosophores. Many of the compounds (salts) included in this review are energetic materials and they have the probability to explode under certain conditions. Therefore, the proper protective measures such as the use of leather gloves, fume hoods, face shields, and careful handling are highly recommended throughout the preparation of targeted compounds. Moreover, the original references should be referred for a detailed synthetic method and safety information.

Acknowledgments

We are thankful to the NSAF (U1830134), the project of State Key Laboratory of Explosion Science and Technology (YBKT21-02) for their generous financial support.

References

1 Hu, L., He, C., Zhao, G. et al. (2020). *ACS Appl. Energy Mater.* 3: 5510–5516.
2 Feng, Y., Deng, M., Song, S. et al. (2020). *Engineering* 6: 1006–1012.
3 Ma, J., Zhang, J., Imler, G.H. et al. (2020). *Org. Lett.* 22: 4771–4775.
4 Tang, J., Yang, H., Xiong, H. et al. (2020). *New J. Chem.* 44: 5895–5902.
5 Fershtat, L.L. and Makhova, N.N. (2020). *ChemPlusChem* 85: 13–42.
6 Klapötke, T.M., Krumm, B., Reith, T., and Unger, C.C. (2018). *J. Org. Chem.* 83: 10505–10509.
7 Gao, H., Zhang, Q., and Shreeve, J.M. (2020). *J. Mater. Chem. A* 8: 4193–4216.
8 Fischer, D., Klapötke, T.M., Reymann, M. et al. (2015). *New J. Chem.* 39: 1619–1627.
9 Bolton, O. and Matzger, A.J. (2011). *Angew. Chem. Int. Ed.* 50: 8960–8963.
10 Chavez, D.E., Parrish, D.A., Mitchell, L., and Imler, G.H. (2017). *Angew. Chem. Int. Ed.* 56: 3575–3578.
11 Yin, P., Zhang, Q., and Shreeve, J.M. (2016). *Acc. Chem. Res.* 49: 4–16.
12 Gao, H. and Shreeve, J.M. (2011). *Chem. Rev.* 111: 7377–7436.
13 Qu, Y. and Babailov, S.P. (2018). *J. Mater. Chem. A* 6: 1915–1940.
14 Kommu, N., Balaraju, M., Ghule, V.D., and Sahoo, A.K. (2017). *J. Mater. Chem. A* 5: 7366–7371.
15 Wang, P., Xu, Y., Lin, Q., and Lu, M. (2018). *Chem. Soc. Rev.* 47: 7522–7538.
16 He, P., Zhang, J.-G., Wu, L. et al. (2017). *Inorg. Chim. Acta* 455: 152–157.
17 Yin, X., Wu, J.-T., Jin, X. et al. (2015). *RSC Adv.* 5: 60005–60014.
18 Manzoor, S., Yin, X., and Zhang, J.-G. (2021). *Defence Technol.*
19 Aureggi, V. and Sedelmeier, G. (2007). *Angew. Chem. Int. Ed.* 46: 8440–8444.

20 Klapötke, T.M. (2019). *Chemistry of High-Energy Materials*. Walter de Gruyter GmbH & Co KG.

21 Gökçınar, E., Klapötke, T.M., and Bellamy, A.J. (2010). *J. Mol. Struct. THEOCHEM* 953: 18–23.

22 Kovacevic, L.S., Klapötke, T.M., and Sproll, S.M. (2010). *Z. Anorg. Allg. Chem.* 636: 1079–1084.

23 Garg, S. and Shreeve, J.M. (2011). *J. Mater. Chem.* 21: 4787–4795.

24 S. Bastea, L. Fried, K. Glaesemann, W. Howard, P. Souers and P. Vitello (2011). Cheetah 5.0 User's manual. *Lawrence Livermore National Laboratory*.

25 Bertrand, R., Marco, H., Patric, F., Alexandre, S., Pierre, B., & Richard, B. (2009). UN Recommendations on the Transport of Dangerous Goods, Manual and Tests and Criteria. *5th revise edition United Nations, New York and Geneva, USA*, 297–316.

26 Klapötke, T.M., Stein, M., and Stierstorfer, J. (2008). *Z. Anorg. Allg. Chem.* 634: 1711–1723.

27 Benz, M., Klapötke, T.M., and Stierstorfer, J. (2020). *Z. Anorg. Allg. Chem.*

28 Fischer, D., Klapötke, T.M., and Stierstorfer, J. (2012). *Propellants Explos. Pyrotech.* 37: 156–166.

29 Manzoor, S., Yang, J.Q., Tariq, Q.u.n. et al. (2020). *ChemistrySelect* 5: 5414–5421.

30 Joo, Y.H. and Shreeve, J.M. (2010). *Angew. Chem. Int. Ed.* 49: 7320–7323.

31 Srinivas, D., Ghule, V.D., and Muralidharan, K. (2014). *RSC Adv.* 4: 7041–7051.

32 Klapötke, T.M. and Sabaté, C.M. (2008). *Chem. Mater.* 20: 3629–3637.

33 Damse, R. and Sikder, A. (2009). *J. Hazard. Mater.* 166: 967–971.

34 Zhang, Q., Chen, D., Jing, D. et al. (2019). *Green Chem.* 21: 1947–1955.

35 Nedel'ko, V., Shastin, A., Korsunskii, B. et al. (2005). *Russ. Chem. Bull.* 54: 1710–1714.

36 Karaghiosoff, K., Klapötke, T.M., and Sabaté, C.M. (2009). *Eur. J. Inorg. Chem.* 2009: 238–250.

37 Klapötke, T.M. and Stierstorfer, J. (2009). *Dalton Trans.* 643–653.

38 Klapötke, T.M. and Miró Sabaté, C. (2008). *Eur. J. Inorg. Chem.* 2008: 5350–5366.

39 Klapötke, T.M., Miró Sabaté, C., Penger, A. et al. (2009). *Eur. J. Inorg. Chem.* 2009: 880–896.

40 Stierstorfer, J., Tarantik, K.R., and Klapötke, T.M. (2009). *Chem. Eur. J.* 15: 5775–5792.

41 Tao, G.-H., Guo, Y., Joo, Y.-H. et al. (2008). *J. Mater. Chem.* 18: 5524–5530.

42 Xue, H., Gao, H., Twamley, B., and Shreeve, J.M. (2007). *Chem. Mater.* 19: 1731–1739.

43 Klapötke, T.M., Sabaté, C.M., and Welch, J.M. (2008). *Dalton Trans.* 6372–6380.

44 Klapötke, T.M., Sabaté, C.M., and Rasp, M. (2009). *J. Mater. Chem.* 19: 2240–2252.

45 Klapötke, T.M., Mayer, P., Stierstorfer, J., and Weigand, J.J. (2008). *J. Mater. Chem.* 18: 5248–5258.

46 Wang, Q., Pang, F., Wang, G. et al. (2017). *Chem. Commun.* 53: 2327–2330.

47 Witkowski, T.G., Richardson, P., Gabidullin, B. et al. (2018). *ChemPlusChem* 83: 984–990.

48 Zhu, W., Yan, Q., Li, J. et al. (2012). *J. Comput. Chem.* 33: 1781–1789.
49 Huber, S., Izsák, D., Karaghiosoff, K. et al. (2014). *Propellants Explos. Pyrotech.* 39: 793–801.
50 Badgujar, D., Talawar, M., Asthana, S., and Mahulikar, P. (2008). *J. Hazard. Mater.* 151: 289–305.
51 Ye, C., Gao, H., Boatz, J.A. et al. (2006). *Angew. Chem.* 118: 7420–7423.
52 Dorofeeva, O.V., Ryzhova, O.N., and Suntsova, M.A. (2013). *J. Phys. Chem. A* 117: 6835–6845.
53 Huang, Y., Zhang, Y., and Shreeve, J.M. (2011). *Chem. Eur. J.* 17: 1538–1546.
54 Wang, Q., Shao, Y., and Lu, M. (2019). *Chem. Commun.* 55: 6062–6065.
55 Liu, Y., He, C., Tang, Y. et al. (2019). *Dalton Trans.* 48: 3237–3242.
56 Zhang, J., Parrish, D.A., and Shreeve, J.M. (2014). *Chem. Asian J.* 9: 2953–2960.
57 Xia, H., Zhang, W., Jin, Y. et al. (2019). *ACS Appl. Mater. Interfaces* 11: 45914–45921.
58 Haiges, R. and Christe, K.O. (2013). *Inorg. Chem.* 52: 7249–7260.
59 Yu, Q., Imler, G.H., Parrish, D.A., and Shreeve, J.M. (2019). *Org. Lett.* 21: 4684–4688.
60 Zhao, B., Li, X., Wang, P. et al. (2018). *New J. Chem.* 42: 14087–14090.
61 Tang, Y., Huang, W., Imler, G.H. et al. (2020). *J. Am. Chem. Soc.*
62 Temple, C. Jr., McKee, R.L., and Montgomery, J.A. (1965). *J. Org. Chem.* 30: 829–834.
63 Kumar, D., Imler, G.H., Parrish, D.A., and Shreeve, J.M. (2017). *J. Mater. Chem. A* 5: 16767–16775.
64 Zhang, Y., Parrish, D.A., and Shreeve, J.M. (2012). *Chem. Eur. J.* 18: 987–994.
65 Tang, Y., Ma, J., Imler, G.H. et al. (2019). *Dalton Trans.* 48: 14490–14496.
66 Zhang, M., Gao, H., Li, C. et al. (2017). *J. Mater. Chem. A* 5: 1769–1777.
67 Hervé, G., Roussel, C., and Graindorge, H. (2010). *Angew. Chem. Int. Ed.* 49: 3177–3181.
68 Dalinger, I.L., Vatsadze, I.A., Shkineva, T.K. et al. (2012). *Synthesis* 44: 2058–2064.
69 Bian, C., Zhang, M., Li, C., and Zhou, Z. (2015). *J. Mater. Chem. A* 3: 163–169.
70 Kettner, M.A. and Klapötke, T.M. (2015). *Chem. Eur. J.* 21: 3755–3765.
71 Liang, L., Wang, K., Bian, C. et al. (2013). *Chem. Eur. J.* 19: 14902–14910.
72 Demko, Z.P. and Sharpless, K.B. (2001). *J. Org. Chem.* 66: 7945–7950.
73 Dachs, M., Dippold, A.A., Gaar, J. et al. (2013). *Z. Anorg. Allg. Chem.* 639: 2171–2180.
74 Dippold, A.A. and Klapötke, T.M. (2013). *Chem. Asian J.* 8: 1463–1471.
75 Dippold, A.A., Izsak, D., and Klapötke, T.M. (2013). *Chem. Eur. J.* 19: 12042–12051.
76 Tang, Y., He, C., Imler, G.H. et al. (2016). *J. Mater. Chem. A* 4: 13923–13929.
77 Fischer, D., Klapötke, T.M., and Stierstorfer, J. (2015). *Angew. Chem. Int. Ed.* 54: 10299–10302.
78 Huang, H., Shi, Y., Li, Y. et al. (2016). *RSC Adv.* 6: 64568–64574.
79 Joo, Y.-H., Chung, J.H., Cho, S.G., and Goh, E.M. (2013). *New J. Chem.* 37: 1180–1188.

80 Joo, Y.-H. and Shreeve, J.M. (2010). *J. Am. Chem. Soc.* 132: 15081–15090.

81 Wei, H., Zhang, J., He, C., and Shreeve, J.M. (2015). *Chem. Eur. J.* 21: 8607–8612.

82 Moebius, D.C. and Kingsbury, J.S. (2009). *J. Am. Chem. Soc.* 131: 878–879.

83 Lin, Q.-H., Li, Y.-C., Qi, C. et al. (2013). *J. Mater. Chem. A* 1: 6776–6785.

84 Zhang, J., Mitchell, L.A., Parrish, D.A., and Shreeve, J.M. (2015). *J. Am. Chem. Soc.* 137: 10532–10535.

85 Göbel, M., Karaghiosoff, K., Klapötke, T.M. et al. (2010). *J. Am. Chem. Soc.* 132: 17216–17226.

86 Shang, Y., Jin, B., Peng, R. et al. (2016). *RSC Adv.* 6: 48590–48598.

87 Zhang, Z.-B., Xu, C.-X., Yin, X., and Zhang, J.-G. (2016). *Dalton Trans.* 45: 19045–19052.

88 Dippold, A.A., Izsák, D., Klapötke, T.M., and Pflüger, C. (2016). *Chem. Eur. J.* 22: 1768–1778.

89 Chavez, D.E., Bottaro, J.C., Petrie, M., and Parrish, D.A. (2015). *Angew. Chem. Int. Ed.* 54: 12973–12975.

90 Huang, H., Zhou, Z., Liang, L. et al. (2012). *Chem. Asian J.* 7: 707–714.

91 Zhou, Y., Yao, C., Ni, R., and Yang, G. (2010). *Synth. Commun.* 40: 2624–2632.

92 Huang, H., Zhou, Z., Liang, L. et al. (2012). *Z. Anorg. Allg. Chem.* 638: 392–400.

93 Liang, L., Huang, H., Wang, K. et al. (2012). *J. Mater. Chem.* 22: 21954–21964.

94 Guo, Y., Tao, G.H., Zeng, Z. et al. (2010). *Chem. Eur. J.* 16: 3753–3762.

95 Chand, D., Parrish, D.A., and Shreeve, J.M. (2013). *J. Mater. Chem. A* 1: 15383–15389.

96 Fischer, N., Izsák, D., Klapötke, T.M., and Stierstorfer, J. (2013). *Chem. Eur. J.* 19: 8948–8957.

97 Gaponik, P., Karavai, V., and Grigor'ev, Y.V. (1985). *Chem. Heterocycl. Compd.* 21: 1255–1258.

8

Properties and Application of Nitrogen-Rich Compound BTATz in Low-Signature Propellants

Jianhua Yi, Zhihua Sun, Yi Xu, Zhao Qin, Changjian Wang, Bozhou Wang, Hui Li, Haijian Li, Chao Chen, Xiao Xie, and Fengqi Zhao

Xi'an Modern Chemistry Research Institute, Science and Technology on Combustion and Explosion Laboratory, Zhangba East Road, Xi'an 710065, PR China

8.1 Introduction

The quick responses of the tactical missiles like the kinetic-energy missile and the new individual weapon are required by the modern wars. It is very important for the power source, the propellant, with the nature of high burning rate, high energy, and minimum signature. The composite or the composite-modified double-base (CMDB) propellants containing the ultra-fine ammonium perchlorate (UFAP) have a high burning rate, but it made the propellant show a heavy smoke and high mechanism sensitivity [1, 2]. The CMDB propellants containing nitroamino compound have a minimum smoke, but they have low burning rate and high burning-rate pressure exponent [3].

It is well known that the nitrogen-rich energetic compounds derive most of their energy from the very large positive enthalpy of formation rather than from oxidation of the fuel-like carbon backbone. The materials are particularly suitable for consideration in high-performance propellant applications, because of their large positive enthalpy of formation, insensitivity to impact, friction and electrostatic discharge, and low-molecular-weight reaction products. 3,6-Bis(1*H*-1,2,3,4-tetrazol-5-yl-amino)-1,2,4,5-tetrazine (BTATz) is one of the nitrogen-rich energetic compounds, with nitrogen content of 79.02%, density of $1.76\,g\,cm^{-3}$, enthalpy of formation of $+883\,kJ\,mol^{-1}$, and moderate mechanical sensitivity [4–9]. As a substitute of hexogen (RDX) in the CMDB propellant formulation, BTATz can improve the burning rate and reduce the pressure exponent, it has a prospect using as a primary component in the high burning rate propellant for the booster rocket motor and the kinetic energy ammunition, and it also can be used in the minimum signature propellant for the smokeless ammunition [10–13].

But from the research results, one can see that the energy property parameters of the original control propellant will be decreased with RDX being substituted by BTATz with a low oxygen balance (OB). It is suggested that BTATz should be used as a component of the CMDB propellant matching with the high-energy additives,

Nitrogen-Rich Energetic Materials, First Edition. Edited by Michael Gozin and Leonid L. Fershtat.

like hexanitrohexaazaisowurtzitane (HNIW), if the energy level is wanted. The good matching of HNIW with BTATz can not only show the high-energy nature of HNIW, but also avoid properly the shortage of the high burning-rate pressure exponent brought by HNIW. The work on BTATz–HNIW–CMDB propellant was proceeded, and one has been applied for intellectual property protection [14].

In this chapter, the preparation, characterization of BTATz, formulation design, and preparation of the rocket propellants containing BTATz were described, and their thermal decomposition reaction kinetics, thermal safety, and combustion properties were investigated.

8.2 Synthesis of BTATz

BTATz used in the experiment was synthesized from triaminoguanidine nitrate and 2,4-pentanedione by condensation, oxidation, and substitution. Sodium nitrite and acetic acid were used instead of nitrogen dioxide/N-methylpyrrolidone. The oxidation process was improved, and the synthesis process was simplified with a low cost. The synthesis process of BTATz has been shown in Scheme 8.1 [6].

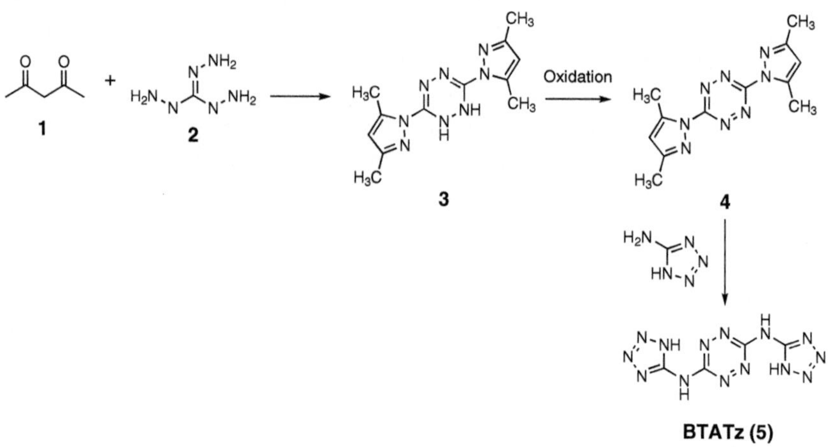

Scheme 8.1 Synthesis of BTATz.

8.2.1 Experimental

Triaminoguanidine nitrate, 5-aminotetrazole, 2,4-pentanedione, sodium nitrite, *N*-methylpyrrolidone, sulfolane, ethanol, dimethylformamide (DMF), and other main reagents were chemically pure.

8.2.2 Synthesis of 3,6-Bis(3,5-Dimethylpyrazole)-1,2-Dihydro-1,2,4,5-Tetrazine (3)

2,4-Pentanedione (40 g, 0.4 mol) was dropped into the aqueous solution of triaminoguanidine nitrate (33.4 g, 0.2 mol) under stirring, then, raised the solution temperature to 70 °C, continued to stir for two hours, and cooled the solution to room temperature. After filtering, washing, and drying, 44 g of solid of orange powder (**3**) was obtained with a yield of 80%. Melting point: 149–151 °C. ^{1}H NMR (CD_3Cl, 500 MHz) δ (ppm): 8.09 (s, 2H, NH), 5.97 (s, 2H, CH), 2.49 (s, 6H, $NCCH_3$), 2.23(s, 6H, $N{=}CCH_3$). ^{13}C NMR (CD_3Cl, 500 MHz) δ (ppm): 150.01, 145.88, 142.39, 109.92, 13.84, 13.52. IR (KBr) ν (cm^{-1}): 3250, 1568, 1414, 967.

8.2.3 Synthesis of 3,6-Bis(3,5-Dimethylpyrazole)-1,2,4,5-Tetrazine (4)

Sixty grams (0.22 mol) of compound (**3**) and 450 ml of acetic acid were mixed, stirred, and heated to 40 °C, 15.5 g of $NaNO_2$ solid was added in batches, after feeding the reaction lasted for two hours, the cool reactant was poured into a large amount of cold water, after filtering, washing, and drying, 52.6 g of red solid (**4**) was obtained with a yield of 88%. Melting point: 225–227 °C. ^{1}H NMR (CD_3Cl, 500 MHz) δ (ppm): 6.19 (s, 2H, CH), 2.76 (s, 6H, $NCCH_3$), 2.43(s, 6H, $N{=}CCH_3$). IR (KBr) ν (cm^{-1}): 1578, 1484, 1425, 970. Analytic calculated for $C_{12}H_{14}N_8$: C 53.33, H 5.19, N 41.48. Found: C 53.43, H 4.85, N 41.18.

Compound **4** is an important intermediate of energetic compounds, which can be used to synthesize a variety of nitrogen-rich tetrazine energetic compounds. In the chapter, the method of preparing compound **4** by the oxidation of compound **3** is improved with a higher yield by using sodium nitrite/acetic acid as oxidant, the ratio of acetic acid as solvent is much cheaper than *N*-methyl pyrrolidone, and the reaction is easy to operate.

8.2.4 Synthesis of BTATz (5)

The mixture of sulfolane (400 ml), compound **4** (40 g, 0.147 mol) and 5-aminotetrazole (32 g, 0.340 mol), was heated slowly to 135 °C, and maintained for 18 hours with protection by nitrogen, then cooled to 50 °C, added 40 ml DMF, after filtering, washing with DMF, and drying at 100 °C for three days, 33 g of red brown solid was obtained, after refining, 14.8 g of orange yellow solid (**5**) was obtained with a yield of 40%. ^{1}H NMR (DMSO-d_6, 500 MHz) δ (ppm): 12.5 (s, 2H, NH). IR (KBr) ν (cm^{-1}): 3428, 3336, 1615, 1491, 1443, 1067. Anal. calcd. for $C_4H_4N_{14}$: C 19.36, H 1.62, N 79.02. Found: C 19.51, H 1.78, N 78.90.

The different products can be produced under different temperature conditions in the reaction of compound **4** with 5-amino-1,2,3,4-tetrazole (5-AT). When the temperature is 120 °C, 1 mol compound **4** and 1 mol 5-AT will take place in a substitution reaction to form the monosubstituted product. When the temperature rises to 135 °C, the disubstituted product can be completed, and the wanted product (**5**) can be formed. In order to reduce the by-products, it is strongly recommended that the reaction temperature be maintained above 135 °C for quite a long time.

8.3 Structure of BTATz

8.3.1 Geometric Configuration

The method of DFT-B3LYP/6-31G(d,p) in the second edition of Gaussian 98 application program was used for quantum chemical calculation. The optimized

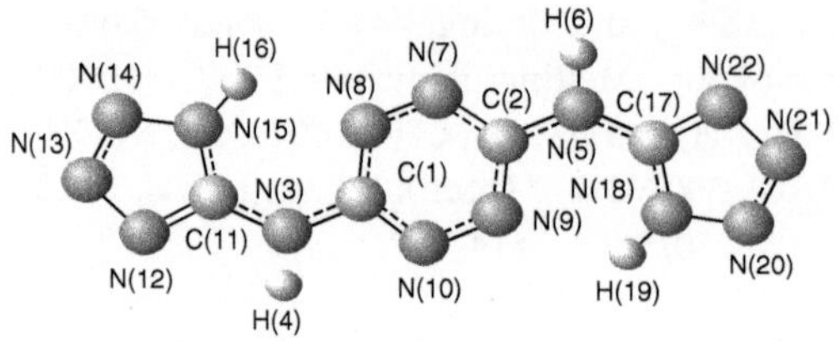

Figure 8.1 The geometric configuration of BTATz after the optimization. Source: Adapted from Wang et al. [7].

Table 8.1 The geometric parameters of BTATz.

Bond	Length (Å)	Bond	Angle (°)	Dihedral	Angle (°)
C(1)—N(3)	1.368	C(1)—N(3)—H(4)	116.850	C(2)—C(1)—N(3)—H(4)	0.023
N(3)—H(4)	1.012	C(2)—N(7)—N(8)	117.419	N(3)—C(1)—C(2)—N(5)	179.967
C(1)—N(8)	1.351	C(1)—N(8)—N(7)	117.713	N(3)—C(1)—C(2)—N(7)	179.976
C(1)—N(10)	1.348	C(1)—N(3)—C(11)	127.435	C(1)—C(2)—N(7)—N(8)	0.000
N(7)—N(8)	1.311	N(8)—C(1)—N(10)	124.869	N(8)—C(1)—C(2)—N(9)	−180.000
N(3)—C(11)	1.378	N(3)—C(11)—N(12)	123.631	N(10)—C(1)—N(3)—C(11)	180.000
C(11)—N(12)	1.320	C(11)—N(12)—N(13)	105.120	C(2)—C(1)—N(3)—C(11)	−179.980
N(12)—N(13)	1.362	N(12)—N(13)—N(14)	111.809	C(1)—N(3)—C(11)—N(12)	−179.997
N(13)—N(14)	1.289	C(11)—N(15)—N(16)	128.441	C(1)—N(3)—C(11)—N(15)	0.000
N(14)—N(15)	1.362	C(11)—N(15)—N(14)	107.757	N(3)—C(11)—N(15)—H(16)	0.000
C(11)—N(15)	1.346	N(15)—N(14)—N(13)	106.096		
N(15)—H(16)	1.012				

configuration and atomic number of BTATz are shown in Figure 8.1, and the data of bond length, bond angle, and dihedral angle are shown in Table 8.1 [7].

All the atoms of BTATz are basically in the same plane. Two five-membered rings and one six-membered ring form a conjugated system, respectively. In a five-membered ring, N atom with H originally provides a pair of electrons, and the other four atoms provide one electron to satisfy $4n + 2$ aromaticity rule. In the six-membered ring, the six atoms provide one electron to satisfy $4n + 2$ aromaticity rule. The two five-membered rings and one six-membered ring provide six electrons, respectively, and the two N atoms connecting the three rings provide two electrons each. The whole molecular system provides 22 electrons in total, which also meets $4n + 2$ aromaticity rule, and the whole molecule forms a large conjugate π bond. Due to the formation of conjugated system, the N—N and C—N bond length (in the range of 1.289–1.378 Å) of the ring is longer than the standard double bond, shorter than the single one, and it tends to average, making the whole molecule have good aromaticity and stability.

Table 8.2 The bond orders of BTATz.

Bond	Bond order	Bond	Bond order	Bond	Bond order
C(1)—N(3)	0.9580	C(1)—N(8)	1.0939	N(9)—N(10)	1.0379
N(3)—H(4)	0.6608	N(3)—C(11)	0.9284	N(12)—C(11)	1.1734
C(1)—N(10)	1.1050	C(11)—N(15)	1.0227	N(12)—N(13)	0.9585
N(13)—N(14)	1.1583	N(14)—N(15)	0.8841	N(15)—H(16)	0.6701

8.3.2 Bond Orders

The magnitude of vibration frequency and strength bond order plays a very important role in judging the strength of bonds in molecules. Table 8.2 shows the bond orders obtained from the analysis of the geometrically optimized natural bond orbit (NBO) of BTATz.

The bond order between the bonding atoms of the tetrazine ring framework is large, which is not easy to break and open the ring; the bond order between N(14)—N(15) and N(18)—N(20) of the tetrazole ring is small, which indicates that the bond is easy to break and makes the tetrazole ring open. In addition, the C—N bond between tetrazole and tetrazine ring is also small, which makes it easy to separate the rings.

8.3.3 Vibration and Infrared Spectrum

The calculation results of vibration frequency and strength of BTATz (correction factor, 0.96) are shown in Table 8.3. BTATz mainly has the following strong absorption peaks. The absorption peaks at 3495 and 3480 cm^{-1} belong to the stretching vibration of N—H bond on tetrazole ring and N—H bond on connected amino group, respectively. The absorption peak at 1587 cm^{-1} belongs to C atom on tetrazole ring and N atom on connected amino group; among the atoms, the stretching vibration of the bond between C(11)—N(3), C(17)—N(5) and the bending vibration of the ring caused by the stretching vibration. The absorption peak at 1503 cm^{-1} belongs to the stretching vibration of the bond between C(11)—N(15), C(17)—N(18) on the tetrazole ring. The absorption peak at 1429 cm^{-1} belongs to the stretching vibration between the C atom on the tetrazine ring and the N atom on the connected amino group C(1)—N(3), C(2)—N(5) bond stretching vibration and the bending vibration of tetrazole ring caused by it. The absorption peak at 1391 cm^{-1} belongs to the stretching vibration of N—N bond on tetrazine ring and the bending vibration of ring caused by it. The absorption peak at 1029 cm^{-1} belongs to the in-plane bending vibration of tetrazine ring. The calculated frequency is in good agreement with the experimental one, but there is a certain deviation, which is mainly due to the solvent effect in the experiment.

Table 8.3 The vibration frequencies and intensities of BTATz.

Vibration frequency (cm^{-1})	Intensity ($kJ\ mol^{-1}$)	Vibration frequency (cm^{-1})	Intensity ($kJ\ mol^{-1}$)
v1	792.47 (0.00)	v17	1255.88 (0.00)
v2	864.82 (13.23)	v18	1319.02 (0.00)
v3	872.95 (0.00)	v19	1320.32 (11.24)
v4	977.89 (0.49)	v20	1385.12 (0.00)
v5	978.24 (0.00)	v21	1391.90 (221.52)
v6	984.79 (174.21)	v22	1429.95 (251.09)
v7	985.87 (0.00)	v23	1459.93 (0.00)
v8	1009.13 (42.67)	v24	1480.90 (0.00)
v9	1029.84 (155.27)	v25	1503.38 (294.53)
v10	1043.13 (0.00)	v26	1520.58 (0.00)
v11	1047.15 (57.46)	v27	1587.50 (1924.96)
v12	1089.87 (0.00)	v28	1609.64 (0.00)
v13	1095.88 (4.51)	v29	3480.52 (236.73)
v14	1211.10 (25.90)	v30	3482.04 (0.10)
v15	1215.77 (0.00)	v31	3495.62 (217.50)
v16	1244.39 (35.47)	v32	3495.94 (2.00)

8.4 Properties of BTATz

8.4.1 Energy Properties of BTATz and Other Energetic Compounds

The four energetic additives, ammonium perchlorate (AP), BTATz, HNIW, RDX, and the two plasticizers, 1,5-diazido-3-nitrazapentane (DIANP) and triethyleneglycol dinitrate (TEGDN), are the commonly used ones in the rocket propellants. Among several energetic compounds, AP has the largest and positive oxygen balance. BTATz has the largest standard enthalpy of formation, the highest nitrogen content, and the smaller oxygen balance. HNIW has the highest density and relatively large oxygen balance. For the calculation purposes, the physical and chemical properties of the compounds are listed in Table 8.4.

BTATz unit propellant has the smallest theoretical specific impulse, characteristic velocity, combustion temperature, and constant volume detonation heat among several unit propellants. HNIW unit propellant has the highest theoretical specific impulse, combustion temperature, and the average relative molecular mass of the combustion products; the constant volume explosion heat is also next to RDX. RDX has the highest constant volume explosion heat and characteristic velocity. For the calculation purposes, the energy characteristic parameters of each energetic compound unit propellant calculated at 10 MPa are listed in Table 8.5.

Table 8.4 Physicochemical properties of some energetic compounds.

Property	AP	BTATz	HNIW	RDX	DIANP	TEGDN
Chemical formula	NH_4ClO_4	$C_4H_4N_{14}$	$C_6H_6N_{12}O_{12}$	$C_3H_6N_6O_6$	$C_4H_8N_8O_2$	$C_6H_{12}N_2O_8$
Relative molecular mass	117.49	248.17	438.19	222.12	200.16	240.17
Oxygen balance (%)	34.0	−64.5	−10.9	−21.6	−80.0	−66.6
Nitrogen content (%)	11.9	79.0	38.4	37.8	56.0	11.7
Density ($g\,cm^{-3}$)	1.95	1.76	2.04	1.80	1.32	1.34
Enthalpy of formation ($kJ\,mol^{-1}$)	−296.19	883	429.42	61.55	592.50	−631.79

Table 8.5 Energy characteristics for monopropellants of several energetic additives.

	Φ	I_{sp} ($N\,s\,kg^{-1}$)	C^* ($m\,s^{-1}$)	T_c (K)	M_c	Q_v ($kJ\,kg^{-1}$)
AP	2.250	2266.21	1394.10	2677.67	27.69	4284.22
BTATz	0	1450.44	849.28	1071.08	20.24	1680.82
HNIW	0.80	2667.02	1602.67	3509.96	29.21	5452.59
RDX	0.667	2632.55	1620.66	3225.35	24.68	5795.64
DIANP	0.167	1961.49	1157.91	1440.048	17.74	2899.64
TEGDN	0.444	2145.37	1273.67	1562.91	18.651	3258.97

Note: Φ is oxygen coefficient, I_{sp} is theoretical specific impulse, C^* is characteristic velocity, T_c is combustion temperature, M_c is average relative molecular mass, and Q_v is constant volume explosion heat, whose value is independent of the calculated pressure.

8.4.2 Physicochemical Properties Detected for BTATz

The basic physicochemical properties of BTATz were measured: purity, 99.90% (measured by high-performance liquid chromatography); density, 1.72 $g\,cm^{-3}$ (measured by liquid statical gravimeter); impact sensitivity (drop height, H_{50}), 22.9 cm (measured by Bruceton staircase method, 5 kg drop weight, 50 mg sample mass); friction sensitivity (probability of explosion, P), 64% (1.5 kg pendulum hammer fixed on 90° tilt angle, 3.92 MPa gauge pressure, 20 mg sample mass); electrostatic spark sensitivity (E_{50}), 31.98 mJ [15].

8.5 Energetic Properties of the Propellants

The effects of BTATz on the energy characteristics of the propellant were investigated, and different amounts of BTATz were used to replace the nitrocellulose (NC) and nitroglycerin (NG) in the propellant which is based on the double-base formulations without aluminum and containing aluminum, respectively. The energy characteristics of the substituted propellant at 10 MPa were calculated. The energetic characteristic parameters of propellant become smaller with the increased content of the nitrogen-rich compound BTATz, no matter whether the propellant contains aluminum powder or not. But when the energetic additive BTATz content is the same, the energy characteristic parameters of the propellant containing aluminum powder are significantly higher than those of the propellant without aluminum powder.

The burning rate of propellant containing BTATz is very high, but the energy is not high. As the low energy of BTATz, when it is used to replace NC and NG in the propellant, the energy characteristic parameters of the original propellant will be reduced to some extent. Therefore, it is necessary to use the high burning-rate characteristics of BTATz and ensure the energy of the propellant; the best way is to introduce a certain amount of energetic additives into the modified propellant containing BTATz. The specific propellant formulations are listed in Table 8.6, and the calculated values of energy characteristic parameters are listed in Table 8.7.

The energetic additives, such as AP, RDX, HNIW, Al, and BTATz, were used together in the propellants, and the energy characteristics were calculated by the energy calculation program. The effects of the mass fraction (W_{HEM}) of the energetic additives on the energy characteristics of the propellants, such as oxygen coefficient, combustion temperature, average relative molecular mass of combustion products, characteristic velocity, theoretical specific impulse, and constant detonation heat, were studied. The basic composition (mass fraction) of the original aluminum-free CMDB propellant was used as following: NC (nitrogen content, 12%), 31.7%; NG, 23.0%; BTATz, 38.0%; other auxiliaries, 7.3%. The basic composition (mass fraction) of aluminized propellant was used as following: NC, 31.7%; NG, 23.0%; BTATz, 33.0%; Al powder, 5.0%; other auxiliaries, 7.3%. The other auxiliaries included DIANP, middle fixative, and Vaseline. The influence of the

Table 8.6 The specific propellant formulations (%).

Number	NC	NG	BTATz	Other	Al
B01	52.7	40.0	0	7.3	0
B02	48.7	36.0	8.0	7.3	0
B03	42.7	32.0	18.0	7.3	0
B04	37.7	27.0	28.0	7.3	0
B05	31.7	23.0	38.0	7.3	0
B06	25.7	19.0	48.0	7.3	0
B07	50.7	37.0	0	7.3	5.0
B08	45.7	34.0	8.0	7.3	5.0
B09	39.7	30.0	18.0	7.3	5.0
B10	34.7	25.0	28.0	7.3	5.0
B11	28.7	21.0	38.0	7.3	5.0
B12	22.7	17.0	48.0	7.3	5.0

Table 8.7 Energy characteristic parameters of propellant.

Number	Φ	T_c (K)	C^* (m s^{-1})	M_c	I_{sp} (N s kg^{-1})	Q_v (kJ kg^{-1})
B01	0.690	2948.46	1509.90	25.702	2461.13	5369.98
B02	0.641	2809.59	1487.81	24.88	2414.46	5069.35
B03	0.581	2639.47	1458.89	23.98	2356.40	4701.00
B04	0.516	2439.03	1421.05	23.09	2281.82	4286.95
B05	0.451	2253.37	1385.51	22.32	2218.54	3901.71
B06	0.384	2066.43	1341.04	21.75	2169.94	3544.92
B07	0.648	3158.75	1538.61	26.26	2538.47	5769.64
B08	0.602	3055.16	1526.94	25.47	2507.93	5523.46
B09	0.542	2905.31	1507.02	24.53	2462.03	5188.98
B10	0.477	2723.22	1477.93	23.60	2400.96	4808.60
B11	0.411	2552.64	1481.33	22.86	2345.55	4457.10
B12	0.343	2388.35	1447.69	22.08	2344.21	4135.61

mass fraction of energetic additives (W_{HEM}) on the energy characteristic parameters of propellant at 10 MPa is shown in Figure 8.2.

The energy characteristic parameters of the two propellants are increased with the increase in BTATz substitution in the propellant. The theoretical specific impulse ($I_{sp,}$ N s kg^{-1}), characteristic velocity (C^*, m s^{-1}), combustion temperature (T_c, K), and constant volume detonation heat (Q_v, kJ kg^{-1}) of the aluminized propellants are significantly higher than those of the nonaluminized propellants. In terms of theoretical specific impulse (I_{sp}, N s kg^{-1}) and combustion temperature (T_c, K), no matter whether the propellant contains aluminum or not, the increment value after replacing BTATz with AP and HNIW is greater than that after replacing with RDX. When the replacement amount of AP and HNIW is small (nonaluminized propellant 20%, aluminized propellant 33%, nonaluminized propellant 29%, aluminized propellant 34%), the two energy characteristic values replaced by AP are slightly higher than those replaced by HNIW, but the two energy characteristic parameters replaced by HNIW are higher than those replaced by AP with the increase in the substitution amount.

In terms of characteristic velocity, no matter whether it contains aluminum or not, HNIW is the best substitute for propellant. When the substitution amount is small (nonaluminized propellant 22%, aluminized propellant 33%), the characteristic velocity replaced by AP is greater than that replaced by RDX. However, when the amount of substitution increases, the effect of substitution with RDX is better than that with AP. In terms of oxygen coefficient (Φ), no matter whether the propellant contains aluminum or not, the substitution of AP is significantly higher than that of HNIW and RDX, and the substitution of RDX is the smallest.

For the constant volume detonation heat of propellant, the substitution of AP is higher than that of the other two nitramines, of which RDX is the smallest. It is possible that the theoretical specific impulse and characteristic velocity properties

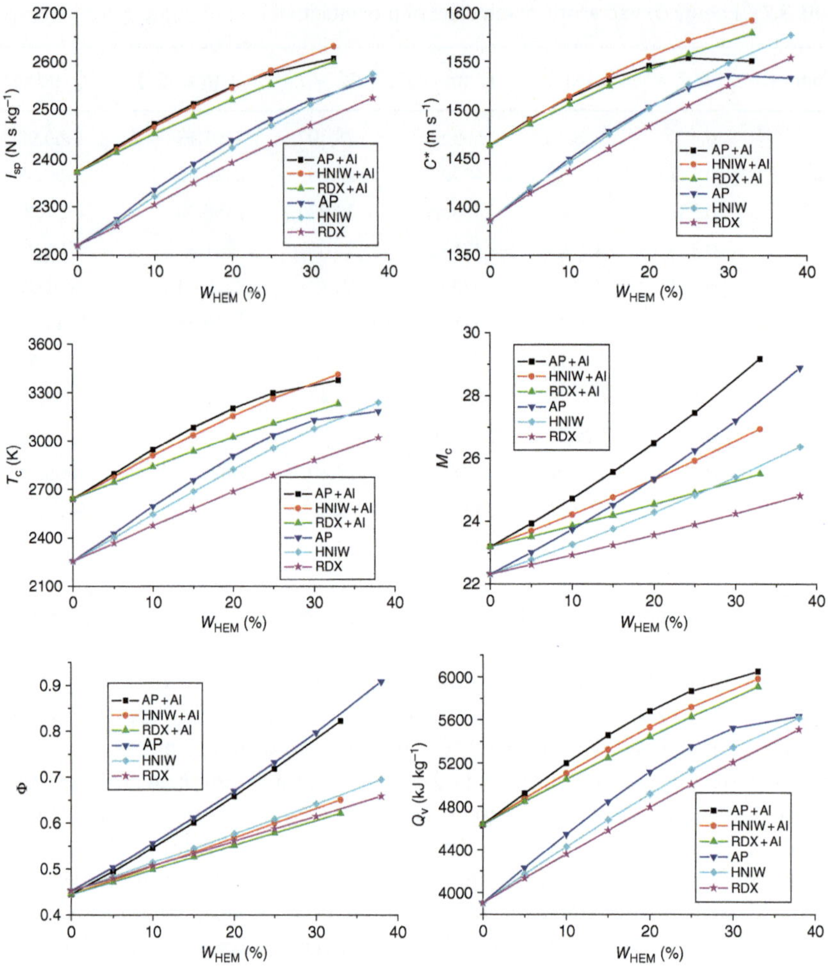

Figure 8.2 Effect of content of energetic additives on energy characteristics of the propellant.

of the propellant are better when HNIW and BTATz in a certain ratio. With the decrease in the ratio of BTATz to HNIW, the energy characteristic parameters of the propellant increase gradually, whether or not the propellant formulation contains aluminum. However, when the ratio of BTATz to HNIW is the same, the energy characteristic parameters of the propellant formulation containing aluminum are significantly higher than those of the one without aluminum. The theoretical specific impulse of the propellant is equivalent to that of RDX–CMDB propellant with 26% RDX at BTATz 8/HNIW 30, which can reach 241.7 seconds (without aluminum) and 251.0 seconds (5% aluminum) at 10 MPa.

On the other hand, in order to make the propellant has an insensitive nature, an effective way is to replace NG with a certain amount of plasticizer with a lower sensitivity. The influence rules of the contents of two insensitive plasticizers, DIANP and TEGDN, on the energy property of the propellant which was obtained as HNIW and BTATz proportioning in a certain amount were calculated, respectively. When NG in the original propellant is replaced by DIANP or TEGDN, the energy characteristic parameters decrease with the increase in the substitution amount of the two auxiliary plasticizers; except for the oxygen coefficient, the other energy characteristic parameters including the aluminum propellant are larger than those without aluminum. The energy of aluminized propellants with NG as plasticizer is the highest if the requirements for insensitive properties of propellants are not high.

8.6 Plume Smoke Signature of the Propellants

The plume smoke signature of BTATz–CMDB propellants in *N*-nitro-dihydroxy-ethylamine-dinitrate (DINA) system, and BTATz–HNIW–CMDB propellants in DIANP system, were calculated using National Aeronautics and Space Administration (NASA) smoke-level calculation program, regardless of the type and content of the combustion catalysts. As the smoke signature of BTATz–CMDB propellant is small, aluminum powder has great influence on the smoke level. When the content of aluminum powder is reduced from 5% to 3%, the characteristic value of primary smoke of propellant can be increased from level C to level B. Therefore, the addition of aluminum powder should not exceed 3%. This is of great significance to the follow-up work. The results are shown in the Tables 8.8 and 8.9.

Table 8.8 The smoke level of the CMDB propellants in DINA system.

	Mass fraction (%)						Characteristic value	
Number	NC	NG	Additive	DINA	Stabilizer	Vaseline	Primary smoke	Secondary smoke
A00	38	28	RDX 26	5	2.5	0.5	0(A)	0.97(A)
A01	38	28	BTATz 26	5	2.5	0.5	0(A)	0.98(A)
A02	38	28	BTATz 26	5	2.5	0.5	0(A)	0.98(A)

Table 8.9 The smoke level of the CMDB propellants in DIANP system.

	Mass fraction (%)								Characteristic value	
Number	NC	NG	BTATz	HNIW	Al	DIANP	Stabilizer	Vaseline	Primary smoke	Secondary smoke
B01	31.7	23	28	10	0	5	1.8	0.5	0(A)	0.98(A)
B02	31.7	23	18	20	0	5	1.8	0.5	0(A)	0.98(A)
B03	31.7	23	18	15	5	5	1.8	0.5	0.91(C)	0.98(A)
B04	31.7	23	18	15	5	5	1.8	0.5	0.91(C)	0.98(A)
B05	31.7	23	18	15	5	5	1.8	0.5	0.91(C)	0.98(A)
B06	31.7	23	18	15	5	5	1.8	0.5	0.91(C)	0.98(A)
B07	31.7	23	18	17	3	5	1.8	0.5	0.76(B)	0.98(A)

8.7 Preparation of the Propellants

8.7.1 BTATz–CMDB Propellants in DINA System

The CMDB propellant samples (numbers RB0601, RB0602) containing BTATz were used in the chapter. The strand sample of the control propellant RB0601 composed of 500 g ingredients without ballistic modifier was prepared by a solventless propellant extrusion technique. The propellant RB0602 composed of 500 g ingredients with the mixture of lead phthalate, copper adipate, and carbon black as a ballistic modifier was also prepared for the comparison with the control propellant RB0601. The formulations of the BTATz–CMDB propellants are listed in Table 8.10, and the formulation of RDX–CMDB propellant (R09) commonly used was also listed in the table for comparison.

8.7.2 BTATz–HNIW–CMDB Propellants in DIANP System

The formulations of the BTATz–HNIW–CMDB propellants are listed in Table 8.11. The strand samples were prepared by the general solventless extrusion technique for the propellant.

Table 8.10 Formulation of RDX/BTATz–CMDB propellants (mass fraction, %).

Sample	NC + NG	Additive	DINA	Other	Ballistic modifier
R09	66.0	RDX 26.0	5.0	3.0	0
RB0601	66.0	BTATz 26.0	5.0	3.0	0
RB0602	63.8	BTATz 25.1	4.8	2.9	A1, 2.4; B1, 0.5; C1, 0.5

Note: A1 is lead phthalate, B1 is copper adipate, and C1 is carbon black commonly used.

Table 8.11 Formulation of BTATz–HNIW–CMDB propellants (mass fraction, %).

Sample	NC + NG	BTATz	HNIW	Aluminum powder	DIANP	Other	Ballistic modifier
BC01	54.7	28.0	10.0	0	5.0	2.3	0
BC02	54.7	18.0	20.0	0	5.0	2.3	0
BC03	52.9	17.4	19.4	0	4.8	2.2	A1, 2.4; B1, 0.5; C1, 0.4
BC04	52.9	17.4	19.4	0	4.8	2.2	A2, 2.4; B2, 0.5; C1, 0.4
BC1202	52.7	17.3	19.3	0	4.8	2.2	A1, 2.4; B3, 0.5; C2, 0.8
BC1205	52.9	17.4	16.5	2.9	4.8	2.2	A1, 2.4; B3, 0.5; C1, 0.4

Note: A1 is lead phthalate, A2 is lead 2,4-dihydroxy benzoic acid, B1 is copper adipate, B2 is copper salicylic acid, B3 is copper 2,4-dihydroxy benzoic acid, C1 is general carbon black commonly used, and C2 is medium-fine carbon black.

8.8 Decomposition Reaction Kinetics and Thermal Safety of the Propellants

8.8.1 BTATz–CMDB Propellants in DINA System

8.8.1.1 Conditions for Thermal Decomposition Reaction

The TG–DTG (thermogravimetry-derivative thermogravimetry) and differential scanning calorimetry (DSC) curves under the condition of flowing nitrogen gas (purity, 99.999%; atmospheric pressure) were obtained by using a TA2950 thermal analyzer (TA Co., USA). DSC and PDSC curves were obtained by using a 204HP high-pressure differential scanning calorimeter (Netzsch Co., Germany). The conditions of TG–DTG were sample mass of approximately 1 mg; N_2 flowing rate, $40\,cm^3\,min^{-1}$; heating rate (β), $10\,K\,min^{-1}$. The conditions of DSC and PDSC analyses were sample mass of approximately 1 mg; N_2 flowing rate, $50\,cm^3\,min^{-1}$; heating rate, 5, 10, 15, 20, 25, and $30\,K\,min^{-1}$; furnace pressure, 0.1, 4, and 7 MPa; reference sample, α-Al_2O_3; type of crucible, aluminum pan with a pierced lid. The specific heat capacity (C_p, $J\,g^{-1}\,K^{-1}$) was determined with continuous C_p mode on a Micro-DSC III microcalorimeter (Setaram Co., France), Heating rate, $0.15\,K\,min^{-1}$; sample mass, approximately 100 mg; atmosphere, N_2; reference sample, calcined α-Al_2O_3. And the specific heat capacity was determined for RB0601: $C_p\ (J\,g^{-1}\,K^{-1}) = -3.90 + 3.10 \times 10^{-2}T - 4.40 \times 10^{-5}T^2$, and for RB0602: $C_p\ (J\,g^{-1}\,K^{-1}) = -3.33 + 2.73 \times 10^{-2}T - 3.81 \times 10^{-5}T^2$.

8.8.1.2 Thermal Behavior

The TG–DTG and DSC curves for BTATz are shown in Figures 8.3 and 8.4. The TG–DTG curves for the BTATz–CMDB propellants, RB0601 and RB0602, are shown in Figures 8.5 and 8.6, the DSC curves at the heating rates of 5, 10, 15, 20, 25, and $30\,K\,min^{-1}$ are shown in Figures 8.7 and 8.8, and the PDSC curves at a heating rate of $10\,K\,min^{-1}$ are shown in Figures 8.9 and 8.10. The basic data for the main exothermic decomposition processes of the propellants are listed in Table 8.12.

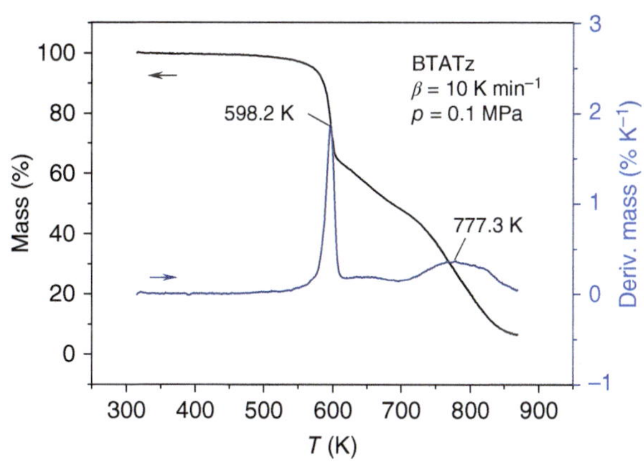

Figure 8.3 TG–DTG curve for pure BTATz.

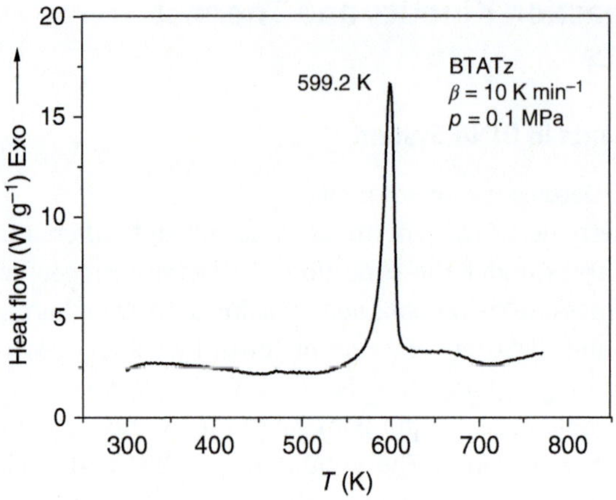

Figure 8.4 DSC curve for pure BTATz.

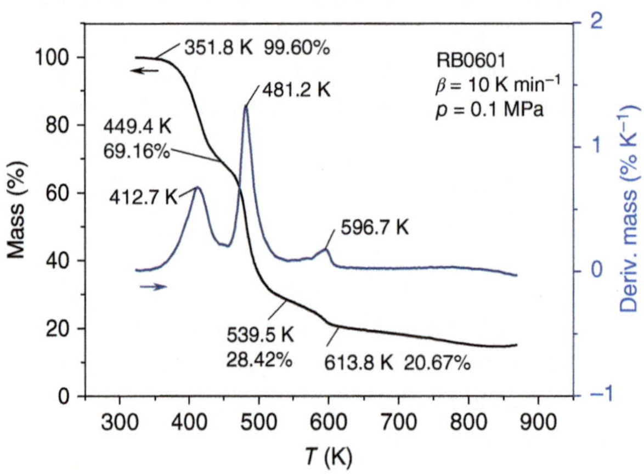

Figure 8.5 TG–DTG curve for propellant RB0601.

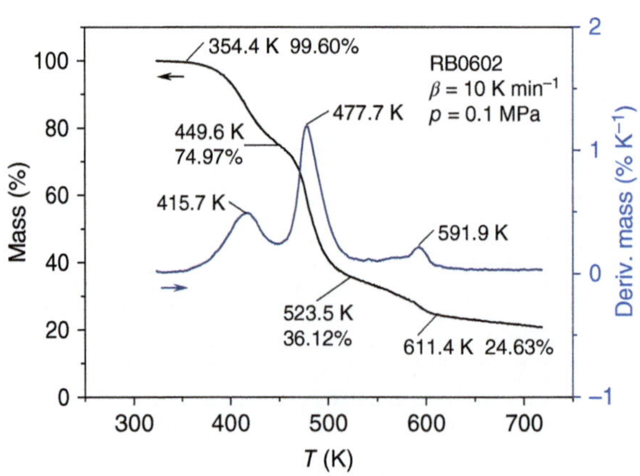

Figure 8.6 TG–DTG curve for propellant RB0602.

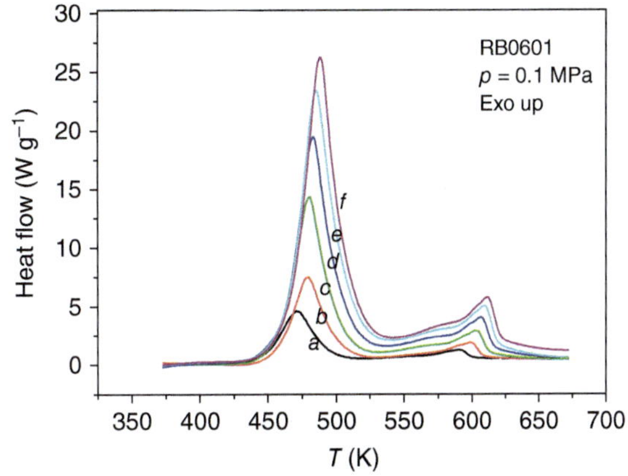

Figure 8.7 DSC curves for propellant RB0601. Heating rate (K min^{-1}): (*a*) 5, (*b*) 10, (*c*) 15, (*d*) 20, (*e*) 25, and (*f*) 30.

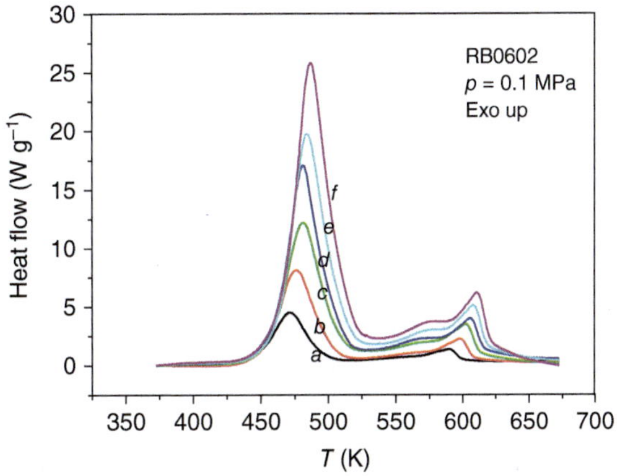

Figure 8.8 DSC curves for propellant RB0602. Heating rate (K min^{-1}): (*a*) 5, (*b*) 10, (*c*) 15, (*d*) 20, (*e*) 25, and (*f*) 30.

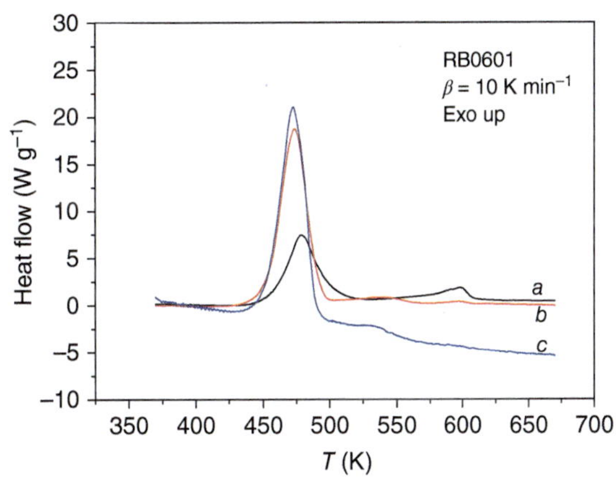

Figure 8.9 PDSC curves for the propellant RB0601. Pressure (MPa): (*a*): 0.1; (*b*): 4; (*c*) 7.

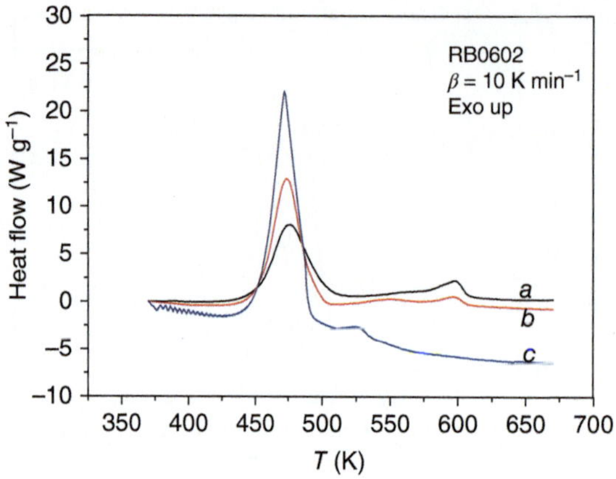

Figure 8.10 PDSC curves for the propellant RB0602. Pressure (MPa): (*a*): 0.1; (*b*): 4; (*c*) 7.

There are three mass loss stages (stages I, II, and III) in TG curve, corresponding to the three peaks in DTG curve as Figures 8.5 and 8.6 show. For TG curve when $\beta = 10\ \mathrm{K\ min^{-1}}$, stage I begins at about 350 K and stops at about 450 K, accompanying 24–30% mass loss, which are close to the mass (28%) of NG, and it likely is attributed to the volatilization and decomposition of NG, inducing the invisible peak in DSC curve of the temperature range of 350–550 K. Stage II begins followed by stage I and stops at about 530 K, with the summit peaks in DTG curves at about 480 K, accompanied by about 40% mass loss, and it is attributed to NG, DINA, NC, and other auxiliaries decomposition in the stage. Stage III stops at about 610 K, with the summit peak in DTG curve at about 590 K, accompanied by about 10% mass loss, and it is attributed to the decomposition of BTATz [16].

There are two exothermic peaks in DSC curve, the main exothermal decomposition peak corresponds to the stage II in TG curve as Figures 8.7 and 8.8 show, and the subordinate one corresponds to the stage III, caused by the decomposition of BTATz as Figures 8.3 and 8.4 show. The reason only one visible exothermic peak appears in the DSC curve during 350–550 K is that the two decomposition stages (stages I and II) occur in succession and the temperature ranges are near, the decomposition heats of the two processes overlap each other in DSC curve, inducing the only one exothermic peak to appear in the temperature range. The decomposition reaction of the propellants occurs with the increase in pressure at one of the heating rates, the peak temperature of the main exothermic process decreases, the initial temperature and the decomposition heat keep a rising trend, and the terminal temperature keeps a decreasing one, the reason may be that high pressure restrains the volatilization of NG and accelerates the decomposition reaction, making the system release more heat. Furthermore, high pressure makes the decomposition peak temperature of BTATz at 7 MPa disappear, and this may be why high pressure and exothermic effect make the tetrazole and tetrazine rings in BTATz molecule break easily at a relatively low temperature [8, 17, 18] as Figures 8.9 and 8.10 show .

Table 8.12 Basic data for the main exothermic decomposition processes of the propellants.

Sample	P (MPa)	β (K min^{-1})	T_0 (K)	T_e (K)	T_p (K)	T_f (K)	ΔT (K)	ΔH_d (10^3 J g^{-1})
RB0601	0.1	5	415.5	449.7	471.4	519.5	104.0	1.55
		10	426.2	456.0	476.9	530.5	104.3	1.45
		15	430.1	458.0	480.5	533.6	103.5	1.57
		20	431.5	460.8	483.6	536.7	105.2	1.63
		25	432.9	463.3	485.6	543.4	110.5	1.59
		30	434.1	465.8	488.6	544.1	110.0	1.46
	4	5	422.8	446.1	467.0	497.2	74.4	2.62
		10	430.8	454.1	473.7	502.2	71.4	2.63
		15	432.3	455.5	478.6	510.2	77.9	2.78
		20	436.7	461.9	481.4	511.3	74.6	2.63
		25	442.1	463.0	485.4	511.4	69.3	2.57
		30	443.1	464.8	487.2	513.4	70.3	2.81
	7	5	423.2	450.2	467.4	494.7	71.5	2.98
		10	428.1	453.4	475.8	497.9	69.8	3.10
		15	431.1	459.8	478.5	504.2	73.1	3.36
		20	439.4	463.2	483.2	508.2	68.8	2.71
		25	443.8	465.5	486.4	511.2	67.4	3.25
		30	450.5	468.2	488.7	509.0	58.5	2.96
RB0602	0.1	5	417.1	450.8	471.5	516.1	99.0	1.60
		10	422.1	451.6	476.1	521.1	99.0	1.49
		15	424.1	456.2	481.4	530.1	106.0	1.45
		20	428.4	458.5	482.5	532.4	104.0	1.45
		25	428.9	461.7	485.3	535.5	106.6	1.39
		30	433.2	463.2	487.1	537.2	104.0	1.51
	4	5	423.2	445.2	465.2	501.9	78.7	2.49
		10	429.5	451.6	474.0	507.3	77.8	2.04
		15	435.8	455.8	478.1	507.6	71.8	2.53
		20	439.1	459.6	481.3	516.3	77.2	2.66
		25	447.3	462.1	484.5	517.5	70.2	2.65
		30	449.3	463.6	486.2	518.3	69.0	2.57
	7	5	430.7	452.9	466.2	489.2	58.5	2.76
		10	432.8	455.1	472.7	495.7	62.9	2.92
		15	435.7	458.3	478.5	502.7	67.0	2.84
		20	444.6	461.6	481.8	502.9	58.3	2.44
		25	447.8	467.4	484.2	508.2	60.4	2.77
		30	453.4	470.6	487.2	510.6	57.2	2.80

Note: T_0 is the initial temperature point at which DSC curve deviates from the baseline, T_e is the extrapolated onset temperature in DSC curve, T_p is the peak temperature, T_f is the finish temperature, and ΔH is the decomposition heat. $\Delta T = T_f - T_0$.

8.8.1.3 Nonisothermal Reaction Kinetics

The PDSC curves at the pressures of 0.1, 4, and 7 MPa, and the heating rates of 5, 10, 15, 20, 25, and 30 K min^{-1} were dealt with mathematic means to explore the reaction mechanisms of the main exothermic decomposition processes of the propellants RB0601 and RB0602 and obtain the corresponding kinetic parameters (apparent activation energy [E_a, kJ mol^{-1}], pre-exponential constant [A, s^{-1}]) and the most probable kinetic model functions, and the temperature data corresponding to the conversion degrees (α) were found. Five integral methods (general integral, MacCallum–Tanner, Šatava–Šesták, Agrawal, Flynn–Wall–Ozawa) and one differential method (Kissinger) were employed [15, 19–21]. The values of E_a were obtained by Ozawa's method from the isoconversional DSC curves, and the $E_a - \alpha$ relations are shown in Figures 8.11, and 8.12. For one of the above curves, the activation energy changes slightly in a proper section of conversion degree, which was selected to calculate the nonisothermal reaction kinetics.

Forty-one types of kinetic model functions and the basic data for one propellant were put into the integral and differential equations for calculation; the values of E_a (kJ mol^{-1}), lg A (s^{-1}), linear correlation coefficient (r), and standard mean square deviation (Q) were calculated on computer with the linear least-squares method,

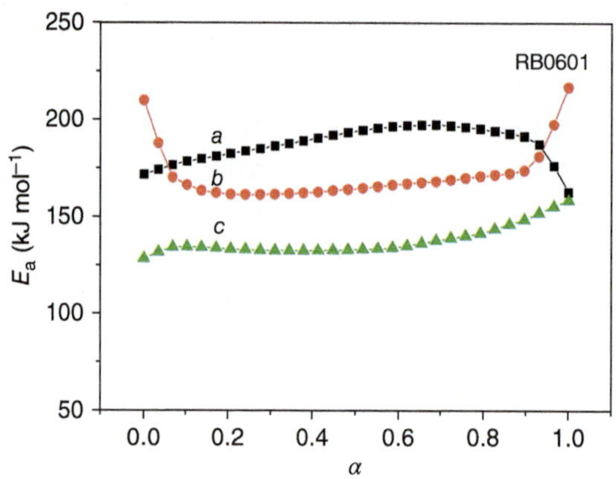

Figure 8.11 $E_a - \alpha$ curves for the main exothermic decomposition process of propellant RB0601. Pressure (MPa): (*a*): 0.1; (*b*): 4; (*c*) 7.

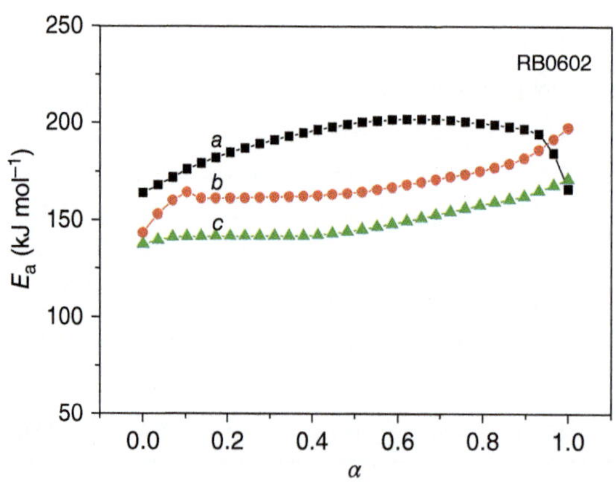

Figure 8.12 $E_a - \alpha$ curves for the main exothermic decomposition process of propellant RB0602. Pressure (MPa): (*a*): 0.1; (*b*): 4; (*c*) 7.

Table 8.13 Kinetic parameters for the main exothermic decomposition process of the propellant RB0601.

		0.1 MPa			4 MPa			7 MPa		
Method	β	E_a	lg A	r	E_a	lg A	r	E_a	lg A	r
General integral	5	188.0	18.70	0.9983	151.0	14.64	0.9956	155.8	15.30	0.9966
	10	190.6	18.99	0.9970	161.6	15.91	0.9973	154.6	15.17	0.9986
	15	201.0	20.19	0.9965	147.1	14.21	0.9914	144.9	14.05	0.9998
	20	196.6	19.69	0.9969	161.3	15.86	0.9973	149.4	14.52	0.9990
	25	187.9	18.70	0.9950	163.0	16.03	0.9993	157.7	15.41	0.9989
	30	193.8	19.32	0.9955	163.4	16.09	0.9990	161.7	15.84	0.9988
Mac Callum–Tanner	5	188.8	18.74	0.9984	151.3	14.61	0.9961	155.9	15.25	0.9970
	10	191.4	19.05	0.9973	162.0	15.89	0.9976	154.9	15.14	0.9987
	15	201.9	20.27	0.9968	147.4	14.20	0.9923	145.2	14.01	0.9998
	20	197.8	19.77	0.9972	161.8	15.87	0.9976	149.8	14.51	0.9991
	25	188.9	18.77	0.9955	163.6	16.04	0.9993	158.2	15.41	0.9991
	30	194.8	19.41	0.9960	164.0	16.11	0.9991	162.3	15.84	0.9989
Šatava–Šesták	5	186.4	18.52	0.9984	151.0	14.62	0.9961	155.4	15.24	0.9970
	10	188.9	18.82	0.9973	161.1	15.84	0.9976	154.4	15.12	0.9987
	15	198.8	19.97	0.9968	147.4	14.24	0.9923	145.2	14.06	0.9998
	20	194.7	19.50	0.9972	161.0	15.81	0.9976	149.6	14.53	0.9991
	25	186.6	18.56	0.9955	162.6	15.98	0.9993	157.5	15.38	0.9991
	30	192.1	19.16	0.9959	163.0	16.04	0.9991	161.4	15.78	0.9989
Agrawal	5	188.0	18.70	0.9983	151.0	14.64	0.9956	155.8	15.30	0.9966
	10	190.6	18.99	0.9970	161.6	15.91	0.9973	154.6	15.17	0.9986
	15	201.0	20.19	0.9965	147.1	14.21	0.9914	144.9	14.05	0.9998
	20	196.6	19.69	0.9969	161.3	15.86	0.9973	149.4	14.52	0.9990
	25	187.9	18.70	0.9950	163.0	16.03	0.9993	157.7	15.41	0.9989
	30	193.8	19.32	0.9955	163.4	16.09	0.9990	161.7	15.84	0.9988
Mean		192.8	19.24		158.0	15.45		154.1	15.04	
Ozawa		192.5		0.9959	157.2		0.9973	152.5		0.9955
Kissinger		194.4	19.51	0.9955	157.4	15.49	0.9970	152.4	14.88	0.9950

Note: The units of β, E_a, and lg A are K min^{-1}, kJ mol^{-1}, and s^{-1}, respectively.

and the most probable mechanism function was selected by the better values of r and Q [19–21]. The results of satisfying the conditions at the same time are the final results as listed in Tables 8.13 and 8.14, and the relevant function was the reaction mechanism function of the decomposition process. The values of E_a (kJ mol^{-1}) and lg$A$ (s^{-1}) obtained from the nonisothermal DSC curves are in approximately good agreement with the values calculated by Kissinger's method and Ozawa's method,

Table 8.14 Kinetic parameters for the main exothermic decomposition process of the propellant RB0602.

		0.1 MPa			4 MPa			7 MPa		
Method	β	E_a	lg A	r	E_a	lg A	r	E_a	lg A	r
General integral	5	197.0	19.76	0.9997	145.2	14.01	0.9978	152.6	14.57	0.9964
	10	207.4	20.98	0.9999	152.8	14.88	0.9982	151.3	14.78	0.9986
	15	196.3	19.67	0.9991	157.1	15.40	0.9985	149.0	14.51	0.9986
	20	203.4	20.56	0.9988	150.1	14.64	0.9993	150.3	14.67	0.9989
	25	202.3	20.36	0.9987	158.2	15.52	0.9994	159.3	15.61	0.9993
	30	206.4	20.79	0.9991	154.8	15.15	0.9995	165.6	16.28	0.9991
Mac Callum–Tanner	5	197.7	19.81	0.9998	145.3	13.97	0.9980	152.7	14.83	0.9967
	10	208.4	21.06	0.9999	153.1	14.87	0.9984	151.5	14.75	0.9988
	15	197.3	19.74	0.9992	157.5	15.39	0.9986	149.2	14.49	0.9988
	20	204.5	20.64	0.9989	150.5	14.63	0.9993	150.6	14.65	0.9991
	25	203.3	20.45	0.9988	158.7	15.53	0.9995	159.8	15.61	0.9994
	30	207.6	20.89	0.9992	155.3	15.16	0.9994	166.2	16.30	0.9992
Šatava–Šesták	5	194.9	19.54	0.9998	145.4	14.02	0.9980	152.4	14.83	0.9967
	10	204.9	20.73	0.9999	152.8	14.86	0.9984	151.2	14.76	0.9988
	15	194.4	19.48	0.9992	156.9	15.36	0.9986	149.1	14.51	0.9988
	20	201.2	20.34	0.9989	150.3	14.65	0.9993	150.4	14.66	0.9991
	25	200.2	20.15	0.9988	158.0	15.49	0.9995	159.0	15.57	0.9994
	30	204.1	20.57	0.9992	154.8	15.15	0.9994	165.1	16.21	0.9992
Agrawal	5	197.0	19.76	0.9997	145.2	14.01	0.9978	152.6	14.57	0.9964
	10	207.4	20.98	0.9999	152.8	14.88	0.9982	151.3	14.78	0.9986
	15	196.3	19.67	0.9991	157.1	15.40	0.9985	149.0	14.51	0.9986
	20	203.4	20.56	0.9988	150.1	14.64	0.9993	150.3	14.67	0.9989
	25	202.3	20.36	0.9987	158.2	15.52	0.9994	159.3	15.61	0.9993
	30	206.4	20.79	0.9991	154.8	15.15	0.9995	165.6	16.28	0.9991
Mean		201.8	20.32		153.1	14.93		154.7	15.08	
Ozawa		204.0		0.9934	152.8		0.9989	152.2		0.9976
Kissinger		206.5	20.88	0.9929	152.8	15.00	0.9988	152.1	14.91	0.9974

and the mechanism function numbers were determined. The decomposition reaction mechanism functions of the propellants are listed in Table 8.15. Substituting $f(\alpha)$ expression, and the values of E_a (kJ mol^{-1}) and A (s^{-1}) into Eq. (8.1), the corresponding kinetic equations (Eqs. (8.2)–(8.7)) of the decomposition reaction of the propellants are obtained and shown in Table 8.15.

$$\frac{d\alpha}{dt} = Af(\alpha)e^{-E/RT} \tag{8.1}$$

Table 8.15 Mechanism functions, apparent activation energies, and kinetic equations for the propellants.

Sample	P (MPa)	Section selected	Mechanism and mechanism function	Mechanism function number	E_a (kJ mol^{-1})	Kinetic equation
RB0601	0.1	$0.08 \le \alpha \le 0.98$	Reaction order model, $G(\alpha)=(1-\alpha)^{-1}-1, f(\alpha)=(1-\alpha)^2$	37	192.8	$\frac{d\alpha}{dt} = 10^{19.24}(1-\alpha)^2 e^{-2.32\times 10^4/T}$ (8.2)
	4	$0.08 \le \alpha \le 0.90$	Avrami–Erofeev equation, assumes random nucleation and its subsequent growth, $n=3/4$, $G(\alpha)=[-\ln(1-\alpha)]^{3/4}$, $f(\alpha)=\left(\frac{4}{3}\right)(1-\alpha)[1-(1-\alpha)]^{1/4}$	15	158.0	$\frac{d\alpha}{dt} = 10^{15.57}(1-\alpha)[1-(1-\alpha)]^{1/4} e^{-1.90\times 10^4/T}$ (8.3)
	7	$0.02 \le \alpha \le 0.80$	Avrami–Erofeev equation, assumes random nucleation and its subsequent growth, $n=2/3$, $f(\alpha)=\left(\frac{3}{2}\right)(1-\alpha)[1-(1-\alpha)]^{1/3}$	14	154.1	$\frac{d\alpha}{dt} = 10^{15.21}(1-\alpha)[1-(1-\alpha)]^{1/3} e^{-1.85\times 10^4/T}$ (8.4)
RB0602	0.1	$0.20 \le \alpha \le 0.96$	Reaction order model, $G(\alpha)=(1-\alpha)^{-1}-1, f(\alpha)=(1-\alpha)^2$	37	201.8	$\frac{d\alpha}{dt} = 10^{20.32}(1-\alpha)^2 e^{-2.43\times 10^4/T}$ (8.5)
	4	$0.10 \le \alpha \le 0.80$	Avrami–Erofeev equation, assumes random nucleation and its subsequent growth, $n=3/4$, $G(\alpha)=[-\ln(1-\alpha)]^{3/4}$, $f(\alpha)=(4/3)(1-\alpha)[1-(1-\alpha)]^{1/4}$	15	153.1	$\frac{d\alpha}{dt} = 10^{15.05}(1-\alpha)[1-(1-\alpha)]^{1/4} e^{-1.84\times 10^4/T}$ (8.6)
	7	$0.02 \le \alpha \le 0.80$	Avrami–Erofeev equation, assumes random nucleation and its subsequent growth, $n=2/3$, $G(\alpha)=[-\ln(1-\alpha)]^{2/3}$, $f(\alpha)=\left(\frac{3}{2}\right)(1-\alpha)[1-(1-\alpha)]^{1/3}$	14	154.7	$\frac{d\alpha}{dt} = 10^{15.25}(1-\alpha)[1-(1-\alpha)]^{1/3} e^{-1.86\times 10^4/T}$ (8.7)

Note: The mechanism function numbers are cited from Refs. [20, 21].

The high pressure can change the decomposition reaction mechanism function and kinetics of the propellants RB0601 and RB0602, and the effect on them was identical approximately with the increase in pressure. The control process of the main decomposition reaction obeys the rule of Avrami–Erofeev equation at high pressure instead of chemical reaction order model at normal pressure. High pressure makes the apparent activation energy decrease fast and the decomposition reaction occurs easily. The mechanism functions are not changed by the ballistic modifier, the apparent activation energies are merely little changed, and the kinetic equations have a little difference between the two propellants.

8.8.1.4 Thermal Safety

8.8.1.4.1 Self-Accelerating Decomposition Temperature (T_{SADT})

The values (T_{00}, T_{e0} and T_{p0}) of the initial temperature point at which DSC curve deviates from the base line (T_0), onset temperature (T_e) and peak temperature (T_p), corresponding to $\beta \to 0$ were obtained by Eq. (8.8), and the self-accelerating decomposition temperature (T_{SADT}) was obtained by Eq. (8.9) [19–21]. The values of the propellants are listed in Table 8.16 [16].

$$T_{0\,(\mathrm{or\,e\,or\,p})i} = T_{00\,(\mathrm{or\,e0\,or\,p0})} + b\beta + c\beta_i^2 + d\beta_i^3 + e\beta_i^4 \tag{8.8}$$

where *b*, *c*, *d*, and *e* are coefficients.

$$T_{\mathrm{SADT}} = T_{\mathrm{e0}} \tag{8.9}$$

8.8.1.4.2 Thermal Ignition Temperature (T_{TIT}) and Critical Temperatures of Thermal Explosion (T_b)

The thermal ignition temperature (T_{be0} or T_{TIT}) was obtained by substituting E_{eO} and T_{e0} into Zhang–Hu–Xie–Li equation (Eq. (8.10)) [22], and the critical temperature of thermal explosion (T_{bp0} or T_{b}) was obtained by substituting E_{pO}

Table 8.16 The derivative parameters for the propellants.

	RB0601	RB0602
T_{00} (K)	391.2	413.3
T_{SADT} (K)	434.1	456.6
T_{p0} (K)	464.5	467.1
T_{TIT} (K)	442.7	465.5
T_{b} (K)	474.2	476.4
t_{TIAD} (s)	51.5	3.4
$\Delta S^{\neq}$ (J mol^{-1} K^{-1})	116.58	142.76
$\Delta H^{\neq}$ (kJ mol^{-1})	190.54	202.62
$\Delta G^{\neq}$ (kJ mol^{-1})	136.38	135.93
$T_{\mathrm{cr,\,hot\text{-}spot}}$ (K)	642.6	633.9

and T_{p0} into the equation. The values are listed in Table 8.16. The high values of T_b for the BTATz–CMDB propellants show that the transition from thermal decomposition to thermal explosion is not easy to take place.

$$T_{be0\,(or\,bp0)} = \frac{E_O - \sqrt{E_O^2 - 4E_O RT_{e0\,(or\,p0)}}}{2R} \tag{8.10}$$

8.8.1.4.3 Adiabatic Time-to-Explosion (t_{TIAD})

The adiabatic time-to-explosion (t_{TIAD}) of energetic materials (EMs) is the time of decomposition transiting to explosion under the adiabatic conditions and is an important parameter for assessing their thermal stability and the safety.

$$t_{TIAD} = \int_0^t dt = \frac{1}{Q_d A}\int_{T_{e0}}^{T_{bp0}} \frac{C_p \exp(E/RT)}{f(\alpha)} dT \tag{8.11}$$

$$\alpha = \int_{T_{e0}}^{T_{bp0}} \frac{C_p}{Q_d} dT \tag{8.12}$$

where C_p is the specific heat capacity measured by micro-calorimeter in J g^{-1} K^{-1}, Q_d is the average value of the decomposition heats (ΔH) at the six heating rates in J g^{-1}, A is the pre-exponential constant in s^{-1} and $A = A_K$, E is the apparent activation energy in J mol^{-1} and $E = E_K$, R is the gas constant in J mol^{-1} K^{-1}, $f(\alpha)$ is the decomposition reaction mechanism function.

The values of t_{TIAD} for the propellants were acquired by substituting the corresponding data into Smith equations (8.11) and (8.12) [23, 24], listed in Table 8.16, and it was found that:

(1) $T_{SADT,\,RB0601} < T_{SADT,\,RB0602}$, $T_{TIT,\,RB0601} < T_{TIT,\,RB0602}$, and $T_{b,\,RB0601} < T_{b,\,RB0602}$. The propellant RB0602 has a higher resistance to heat and thermal safety than RB0601 under the same experimental conditions.

(2) $t_{TIAD,\,RB0601} > t_{TIAD,\,RB0602}$. The transition from adiabatic decomposition to explosion for propellant RB0602 is easier than RB0601.

8.8.1.4.4 Thermodynamic Parameters of Activation Reaction

The entropy of activation ($\Delta S^{\#}$), enthalpy of activation ($\Delta H^{\#}$), and free energy of activation ($\Delta G^{\#}$) of the main exothermic decomposition reaction of the propellants, corresponding to $T = T_{p0}$, $E = E_K$, and $A = A_K$, were obtained by Eqs. (8.13)–(8.15) [19, 20], and are listed in Table 8.16. The positive values of $\Delta G^{\#}$ indicate that the exothermic decomposition reaction of the propellants must proceed under the heating condition.

$$A \exp\left(\frac{-E}{RT}\right) = \left(\frac{k_B T}{h}\right) \exp\left(\frac{-\Delta G^{\neq}}{RT}\right) \tag{8.13}$$

$$\Delta H^{\neq} = E - RT \tag{8.14}$$

$$\Delta G^{\neq} = \Delta H^{\neq} - T\Delta S^{\neq} \tag{8.15}$$

8.8.1.4.5 Critical Temperature of Hot-spot Initiation ($T_{cr,\ hot\text{-}spot}$)

In order to obtain the critical temperature of hot-spot initiation ($T_{cr,\ hot\text{-}spot}$) of the propellants, assuming that $T_{cr,\ hot\text{-}spot}$ is a function of the size and duration of the hot-spot and the physical and chemical properties. Bruckman–Guillet first-order decomposition reaction estimation equation (8.16) [25, 26] was used for the calculation of $T_{cr,\ hot\text{-}spot}$.

$$\left(\frac{4}{3}\pi a^3\right)\rho Q_d\left\{1-\exp\left[-(t-t_0)Ae^{-E/RT_{cr,\ hot\text{-}spot}}\right]\right\}$$
$$=\int_a^{\infty}4\pi r^2\rho C_p\left[\frac{a\theta_0}{r}erfc\left[\frac{r-a}{2\sqrt{Bt}}\right]\right]dr$$
$$=\int_a^{\infty}4\pi r^2\rho C_p\left[\frac{a(T_{cr,\ hot\text{-}spot}-T_{room})}{r}erfc\left[\frac{(r-a)}{2\sqrt{\frac{\lambda}{\rho C_p}t}}\right]\right]dr \tag{8.16}$$

where a is the radius of the hot-spot in 10^{-3} cm, ρ is the density in $g\,cm^{-3}$, Q_d is the heat of reaction in $J\,g^{-1}$, $t-t_0$ is the time interval in 10^{-4} s, A is the pre-exponential constant in s^{-1} and $A=A_K$, E is the apparent activation energy in $J\,mol^{-1}$ and $E=E_K$, R is the gas constant in $J\,mol^{-1}\,K^{-1}$, $T_{cr,\ hot\text{-}spot}$ is the critical temperature of hot-spot initiation in K, C_p is the specific heat in $J\,mol^{-1}\,K^{-1}$, T_{room} is the ambient temperature in 293.15 K, and λ is the thermal conductivity in $J\,cm^{-1}\,s^{-1}\,K^{-1}$.

For the propellant RB0601: the thermal conductivity $\lambda=20\times10^{-4}\,J\,cm^{-1}\,s^{-1}\,K^{-1}$, $\rho=1.655\,g\,cm^{-3}$; for the propellant RB0602: $\lambda=30\times10^{-4}\,J\,cm^{-1}\,s^{-1}\,K^{-1}$, $\rho=1.660\,g\,cm^{-3}$. The values of $T_{cr,\ hot\text{-}spot}$ were acquired by substituting the above corresponding data into Eq. (8.16) and listed in Table 8.16.

8.8.1.4.6 Characteristic Drop Height of Impact Sensitivity (H_{50})

To obtain the characteristic drop height of impact sensitivity (H_{50}) of the propellants, substituting the values of the thermal conductivity (λ, 10^{-4} J cm^{-1} s^{-1} K^{-1}), the density (ρ, g cm^{-3}), A (s^{-1}), Q_d (J g^{-1}), and E (J mol^{-1}) of eight explosives with known 50% drop height listed in Table 8.17 into Eq. (8.17) [27–30], the corresponding values of the parameter n of 0.564 623, D_2 of 33.8765, and D_3 of −0.347 174 were obtained. By substituting the values of λ, ρ, A, Q_d, and E of the propellants listed in Table 8.17 and the values of n, D_2, and D_3 into Eq. (8.17), the corresponding the value of H_{50} (cm) of 17.4 for RB0601 and 18.1 for RB0602 were obtained, which shows that the two propellants have an impact sensitivity approaching that of RDX.

$$\frac{1}{2}n\lg H_{50}+\lg\sqrt{\frac{\lambda}{A\rho Q_d}}+D_3+\frac{0.02612E}{T_1+D_2H_{50}^n}=0 \tag{8.17}$$

Table 8.17 Explosive parameters and comparison of experimental and predicted 50% drop height of impact sensitivity (H_{50}).

Number	Sample	λ (10^{-4} J cm^{-1} s^{-1} K^{-1})[(1)]	ρ (g cm^{-3})	lg[A (s^{-1})][(2)]	Q_d (J g^{-1})[(3)]	E (J mol^{-1})[(2)]	H_{50} (cm) (Exp.)[(1)]	H_{50} (cm) (Predicted)	n	D_2	D_3
1	HMX	34.43	1.79	33.80	2764	373 700	32	33.4	0.564 62	33.8765	−0.347 174
2	RDX	10.58	1.66	12.50	2810	140 000	26	20.1			
3	TNT	21.30	1.57	11.10	1506	155 017	59	56.4			
4	PETN	25.10	1.68	10.40	3263	112 300	16	15.60			
5	BTF	20.92	1.81	22.81	2949	255 000	28	30.0			
6	HNS	8.53	1.65	22.63	1389	289 000	54	50.1			
7	Tetryl	18.74	1.67	16.90	1904	172 500	17[(4)]	17.6			
8	NG	12.55	1.60	16.09	2092	150 122	7[(4)]	9.4			
9	RB0601	20.0	1.655	19.51	1540	194 400		17.4			
10	RB0602	30.0	1.660	20.88	1480	206 530		18.1			

Note: (1) cited from Dong and Zhou [27], (2) cited from Dong et al. [28], (3) cited from Hu et al. [29], and (4) cited from Friedman [30]. $T_{room} = 350.15$ K.

8.8.1.4.7 Critical Thermal Explosion Ambient Temperature (T_{acr}), Thermal Sensitivity Probability Density Function ($S(T)$), Safety Degree (S_d), and Thermal Explosion Probability (P_{TE})

In order to explore the heat-resistance ability of the propellants, the values of the critical thermal explosion ambient temperature (T_{acr}), the thermal sensitivity probability density function ($S(T)$), the safety degree (S_d), and the thermal explosion probability (P_{TE}) were calculated by Frank–Kamenetskii formula (Eq. (8.18)) [31], Wang–Du formula (Eqs. (8.19)–(8.23)), and Eq. (8.24) [32–34]. In these formulas, A is the pre-exponential constant in s^{-1} and $A = A_K$, E is the apparent activation energy in $J\,mol^{-1}$ and $E = E_K$, R is the gas constant in $J\,K^{-1}\,mol^{-1}$, λ is the thermal conductivity in $W\,m^{-1}\,K^{-1}$, δ is the Frank–Kamenetskii (FK) parameter, δ_{cr} is the criticality of thermal explosion of exothermic system, r is characteristic measurement of reactant in 1.0 m, Q_d is decomposition heat in $J\,g^{-1}$, ρ is density in $g\,cm^{-3}$, μ_T is the average value of temperature in K, σ_δ is the standard deviation of FK parameter, σ_T is the standard deviation of ambient temperature in 10 K, T is the surrounding temperature in 350.15 K.

The maximum value of $S(T)$ versus T relation curve ($T_{S(T)\max}$), S_d, T_{acr}, and P_{TE} of the propellant are obtained and shown in Table 8.18.

$$T_{acr} = \frac{-E}{2R\text{Lambert}W_{-1}\left(-\frac{1}{2}\sqrt{\frac{\lambda E\delta_{cr}}{r^2 Q_d \rho A R}}\right)} \tag{8.18}$$

$$W = \frac{r^2 Q_d E_K \rho A_K}{\lambda R} \tag{8.19}$$

$$\mu_T = \frac{-E_K}{2R\text{Lambert}W_{-1}\left(-\frac{1}{2}\sqrt{\frac{\lambda E_K\delta_{cr}}{r^2 Q_d \rho A_K R}}\right)} \tag{8.20}$$

$$\sigma_\delta = W\left(\frac{E_K - 2R\mu_T}{R\mu_T^4}\right)\exp\left(-\frac{E_K}{R\mu_T}\right)\sigma_T \tag{8.21}$$

$$S(T) = \frac{W(E_K - 2RT)}{\sqrt{2\pi}\sigma_\delta RT^4}\exp\left\{-\left[W\frac{\exp\left(-\frac{E_K}{RT}\right)}{T^2} - \delta_{cr}\right]^2 \Big/ \left[2\sigma_\delta^2 - \frac{E_K}{RT}\right]\right\} \tag{8.22}$$

$$S_d = \int_0^{+\infty}\int_0^{+\infty}\frac{W(E_K - 2RT)}{2\pi\sigma_\delta\sigma_T RT^4}\exp\left\{-\left[W\frac{\exp\left(-\frac{E_K}{RT}\right)}{T^2} - \mu_\delta\right]^2 \Big/ \left[2\sigma_\delta^2 - \frac{E_K}{RT} - \frac{(Y - T + \mu_T)^2}{2\sigma_T^2}\right]\right\} dT dy \tag{8.23}$$

$$P_{TE} = 1 - S_d \tag{8.24}$$

Table 8.18 The calculated values of $T_{S(T)max}$, S_d, T_{acr}, and P_{TE} for propellants.

	Infinite plate				Infinite cylinder				Sphere sample			
	$T_{S(T)max}$ (K)	S_d (%)	T_{acr} (K)	P_{TE} (%)	$T_{S(T)max}$ (K)	S_d (%)	T_{acr} (K)	P_{TE} (%)	$T_{S(T)max}$ (K)	S_d (%)	T_{acr} (K)	P_{TE} (%)
RB0601	355.5	41.46	350.7	58.54	360.0	51.82	355.2	48.18	362.8	57.24	358.0	42.76
RB0602	362.7	56.42	357.9	43.58	367.1	62.79	362.3	37.21	369.9	65.63	365.1	34.37

The calculated values of $T_{S(T)max}$, S_d, T_{acr}, and P_{TE} for the propellants RB0601 and RB0602 showed that:

(1) $T_{S(T)max,\ RB0601} < T_{S(T)max,\ RB0602}$, $S_{d,\ RB0601} < S_{d,\ RB0602}$, and $T_{acr,\ RB0601} < T_{acr,\ RB0602}$. The propellant RB0602 has a higher resistance to heat and thermal safety than RB0601 under the same experimental conditions.

(2) $T_{S(T)max,\ infinite\ plate} < T_{S(T)max,\ infinite\ cylinder} < T_{S(T)max,\ sphere\ sample}$, $S_{d,\ infinite\ plate} < S_{d,\ infinite\ cylinder} < S_{d,\ sphere\ sample}$, and $T_{acr,\ infinite\ plate} < T_{acr,\ infinite\ cylinder} < T_{acr,\ sphere\ sample}$. The sphere charge is much safer than another with the same characteristic measurement under the same experimental conditions.

8.8.2 BTATz–HNIW–CMDB Propellants in DIANP System

8.8.2.1 Conditions for Thermal Decomposition Reaction

The DSC curves under the condition of flowing nitrogen gas (purity, 99.999%) were obtained by using a 204HP differential scanning calorimeter (Netzsch Co., Germany). Sample mass, approximately 1 mg; N_2 flowing rate, 50 cm^3 min^{-1}; heating rate, 5, 10, 15, 20, 25, and 30 K min^{-1}; furnace pressures, 0.1 MPa; reference sample, α-Al_2O_3; type of crucible, aluminum pan with a pierced lid. The specific heat capacity (C_p, J g^{-1} K^{-1}) was determined with continuous C_p mode on a Micro-DSC III micro-calorimeter (Setaram Co., France). Heating rate, 0.15 K min^{-1}; temperature range, 293.15–303.15 K; sample mass, approximately 100 mg; atmosphere, N_2; reference sample, calcined α-Al_2O_3. The coefficient of heat conductivity (λ, W m^{-1} K^{-1}) was determined on a KT-6 coefficient of heat conductivity instrument. Sample shape, disc; temperature, room temperature; humidity, 55%. The density (ρ, g cm^{-3}) was determined by liquid statical gravimeter with China Military Standard.

8.8.2.2 Thermal Behavior

The TG–DTG curves and DSC curves for BTATz–HNIW–CMDB propellants are shown in Figures 8.13–8.16, and the basic data of the major exothermic decomposition process are listed in Table 8.19. There are four mass loss stages (stages I–IV) in one of the TG curves, corresponding to the four peaks in DTG curve. For TG curve when $\beta = 10$ K min^{-1}, stage I begins at approximately 350 K and stops at approximately 440 K, accompanying approximately 25% mass loss, which are close to the mass (23%) of NG, and it likely is attributed to the volatilization and

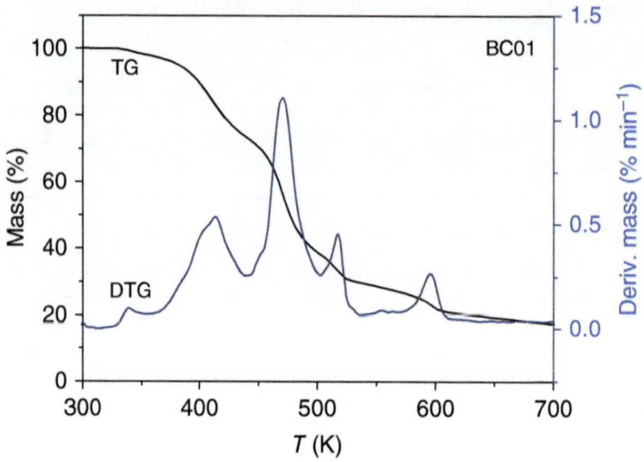

Figure 8.13 TG–DTG curve for the propellant BC01. Heating rate (K min^{-1}): 10; Pressure (MPa): 0.1.

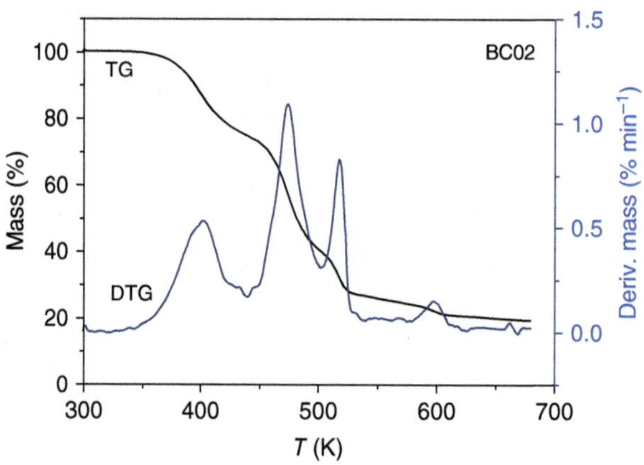

Figure 8.14 TG–DTG curve for the propellant BC02. Heating rate (K min^{-1}): 10; Pressure (MPa): 0.1.

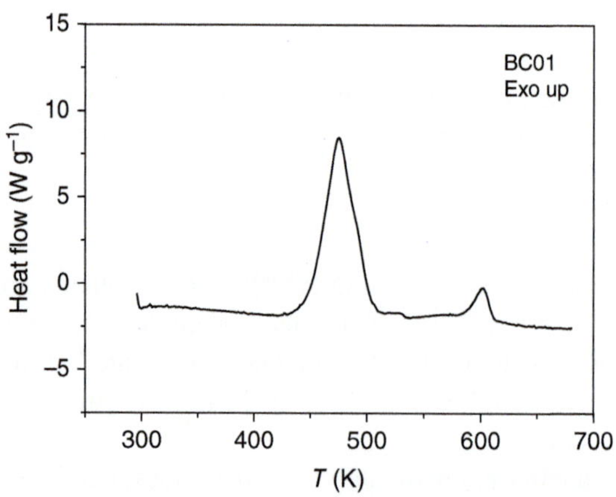

Figure 8.15 DSC curves for the propellant BC01. Heating rate (K min^{-1}): 10; Pressure (MPa): 0.1.

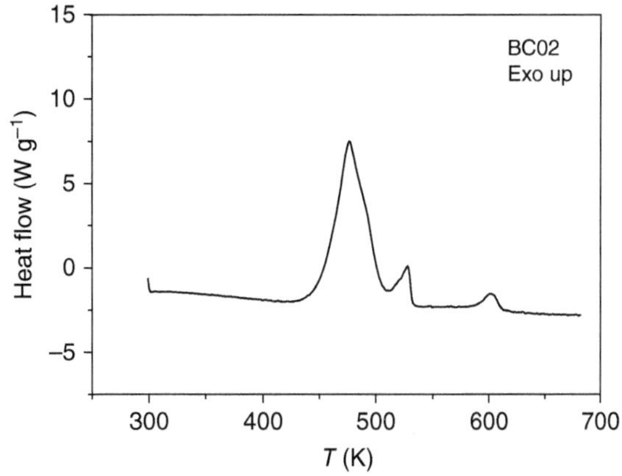

Figure 8.16 DSC curve for the propellant BC02. Heating rate (K min^{-1}): 10; Pressure (MPa): 0.1.

decomposition of NG, inducing the invisible peak in DSC curve of the temperature range of 350–440 K. Stage II begins followed by stage I and stops at approximately 503 K, with a summit peak in DTG curve at approximately 471 K, accompanied by about 35% mass loss, and it is attributed to NC and other auxiliary decomposition in the stage. Stage III stops at approximately 537 K, with a summit peak in DTG curve at approximately 515 K, and it is attributed to the decomposition of HNIW (T_p = 513.2 K). Stage IV begins at approximately 572 K and stops at approximately 614 K, with the summit peaks in DTG curves at approximately 595 K, and it is attributed to the decomposition of BTATz (T_p = 599.2 K).

8.8.2.3 Nonisothermal Reaction Kinetics

There are lots of the decomposition reaction kinetics study methods, such as integral method and differential one. The equations, general-integral (i), Callum–Tanner (ii), Šatava–Šesták (iii), Ozawa (iv), all belong to the integral methods, Kissinger (v) belongs to the differential one. Equations (8.1)–(8.3) are usually employed to calculate the nonisothermal reaction kinetics, but Eq. (8.4) is quite suitable for iso-conversion degree method. When $T = T_p$, Eq. (8.5) is suitable for use. To explore the reaction mechanisms of the major exothermic decomposition processes of the propellant and obtain the corresponding kinetic parameters and the most probable kinetic model functions, the DSC curves at the heating rate of 5, 10, 15, 20, 25, and 30 K min^{-1} were dealt with mathematic means, and T–α relationship was found. The five integral methods (Eqs. (8.1)–(8.4)) and one differential method (Eq. (8.5)) were employed. The values of E_a were obtained by Ozawa's method from the iso-conversional DSC curves at the six heating rates; the E_a–α relation and the calculation section were obtained. The calculation results are listed in Table 8.20, the kinetic parameters, the most probable mechanism function, and the kinetic equation are listed in Table 8.21. The content of HNIW has a great effect on the decomposition reaction mechanism of BTATz–HNIW–CMDB propellant. When the content of HNIW is 10%, the decomposition reaction is controlled by the random nucleation and subsequent growth ($n = 1$), and the reaction mechanism obeys Mampel law. But when the content of HNIW is 20%, the reaction is controlled by

Table 8.19 Basic data of the major exothermic decomposition process for the propellants (0.1 MPa).

Sample	β	T_0	T_e	T_p	T_f	ΔT	Q_d	Q_d (mean)
BC01	5	423.4	442.6	468.4	502.9	79.5	1.81×10^3	1.71×10^3
	10	427.5	450.4	475.2	510.3	82.8	1.84×10^3	
	15	430.1	454.0	479.7	515.1	85.0	1.70×10^3	
	20	432.9	458.1	483.9	520.6	87.7	1.81×10^3	
	25	436.3	460.0	486.9	525.6	89.3	1.54×10^3	
	30	439.8	461.7	489.2	529.8	90.0	1.58×10^3	
BC02	5	420.9	445.9	468.2	500.0	79.1	1.71×10^3	1.58×10^3
	10	427.5	454.0	476.4	510.3	82.8	1.64×10^3	
	15	434.1	458.2	482.0	514.0	79.9	1.52×10^3	
	20	437.5	462.3	486.2	520.0	82.5	1.50×10^3	
	25	438.1	465.5	489.1	523.1	85.0	1.57×10^3	
	30	439.8	468.6	492.1	526.8	87.0	1.52×10^3	
BC03	5	424.0	446.0	468.3	498.5	74.5	1.48×10^3	1.45×10^3
	10	427.1	451.6	475.1	510.2	83.1	1.59×10^3	
	15	430.7	455.6	480.2	513.2	82.5	1.33×10^3	
	20	434.6	460.6	484.4	520.6	86.0	1.51×10^3	
	25	436.7	464.3	488.2	521.2	84.5	1.50×10^3	
	30	444.8	465.4	491.0	524.9	80.1	1.28×10^3	
BC04	5	423.6	445.0	467.2	498.5	74.9	1.63×10^3	1.52×10^3
	10	429.0	451.8	474.8	508.3	79.3	1.50×10^3	
	15	435.6	456.5	480.2	515.1	79.5	1.63×10^3	
	20	436.4	461.6	484.3	518.8	82.4	1.32×10^3	
	25	434.7	462.7	487.5	523.1	88.4	1.47×10^3	
	30	438.7	464.9	489.9	526.2	87.5	1.59×10^3	

the chemical reaction ($n = 1/4$). The mechanism can't be changed by the catalysts, who just make the apparent activation energy change slightly [35].

8.8.2.4 Thermal Safety

The thermal safeties of the propellants were calculated by the methods mentioned above; the basic parameters and the calculation results are listed in Tables 8.22–8.24. The eight important thermal safety parameters were obtained and compared. For the sample, from BC01 to BC04, the values of T_{SADT} and T_{TIT} make an upward tendency, the resistivity to heat: BC04 > BC03 > BC02 > BC01. The values of T_{acr} and S_d for BC01 are the maximum and for BC02 are the minimum, the heat sensitivity: BC01 > BC03 > BC04 > BC02. For the same radius, the thermal safety of the sphere sample is greater than that of the infinite cylinder one.

Table 8.20 Kinetic calculation results for BTATz–HNIW–CMDB propellants (0.1 MPa).

		BC01			BC02			BC03			BC04		
Method	β	E_a	lg A	r	E_a	lg A	r	E_a	lg A	r	E_a	lg A	r
(1)	5	155.0	14.98	0.9930	141.3	12.73	0.9905	139.9	12.59	0.9911	147.4	13.48	0.9945
	10	155.9	15.08	0.9942	142.2	12.84	0.9896	130.2	11.54	0.9905	149.1	13.69	0.9951
	15	152.8	14.75	0.9952	142.7	12.89	0.9925	133.6	11.93	0.9904	143.0	12.99	0.9954
	20	156.1	15.09	0.9948	141.1	12.71	0.9927	131.1	11.66	0.9902	154.8	14.28	0.9952
	25	154.5	14.92	0.9953	143.4	12.95	0.9921	136.7	12.26	0.9919	142.8	12.95	0.9958
	30	153.8	14.84	0.9962	144.7	13.09	0.9934	138.3	12.42	0.9904	147.3	13.44	0.9959
(2)	5	155.3	14.96	0.9937	141.5	12.69	0.9915	140.1	12.56	0.9921	147.7	13.45	0.9951
	10	156.4	15.08	0.9948	142.6	12.82	0.9908	130.5	11.52	0.9916	149.4	13.67	0.9956
	15	153.3	14.75	0.9957	143.1	12.89	0.9933	134.1	11.92	0.9915	143.4	12.98	0.9959
	20	156.7	15.10	0.9954	141.6	12.71	0.9935	131.6	11.65	0.9914	155.4	14.29	0.9957
	25	155.2	14.93	0.9958	144.0	12.96	0.9930	137.3	12.27	0.9928	143.3	12.95	0.9963
	30	154.5	14.86	0.9966	145.3	13.10	0.9942	138.9	12.43	0.9916	147.9	13.46	0.9964

(Continued)

Table 8.20 (Continued)

Method	β	BC01			BC02			BC03			BC04		
		E_a	lg A	r	E_a	lg A	r	E_a	lg A	r	E_a	lg A	r
(3)	5	154.8	14.95	0.9937	141.8	12.77	0.9915	140.5	12.65	0.9921	147.6	13.48	0.9951
	10	155.8	15.06	0.9948	142.8	12.89	0.9908	131.4	11.67	0.9916	149.3	13.70	0.9956
	15	152.9	14.75	0.9957	143.3	12.95	0.9933	134.8	12.05	0.9915	143.6	13.05	0.9959
	20	156.1	15.08	0.9954	141.9	12.79	0.9935	132.4	11.79	0.9914	154.9	14.27	0.9957
	25	154.7	14.92	0.9958	144.1	13.01	0.9930	137.8	12.37	0.9929	143.5	13.01	0.9963
	30	154.0	14.85	0.9966	145.4	13.15	0.9942	139.3	12.53	0.9916	147.9	13.49	0.9964
Mean		154.8	14.94		142.8	12.88		135.3	12.09		147.6	13.48	
(4)		150.1 (E_{eO})		0.9986	132.0 (E_{eO})		0.9981	141.9 (E_{eO})		0.9894	142.5 (E_{eO})		0.9960
		153.8 (E_{pO})		0.9976	136.4 (E_{pO})		0.9994	141.4 (E_{pO})		0.9952	141.0 (E_{pO})		0.9989
(5)		153.8	15.02	0.9973	135.4	12.91	0.9993	140.7	13.54	0.9946	140.3	13.51	0.9988

Note: E with the subscript of eO, and pO is the apparent activation energy obtained from the onset temperature (T_e) and the peak temperature (T_p) by Ozawa's method.

Table 8.21 Kinetic parameters, the most probable mechanism functions, and the kinetic equations for BTATz–HNIW–CMDB propellants (0.1 MPa).

	BC01 ($\alpha = 0.10 - 0.98$)	**BC02 ($\alpha = 0.10 - 0.98$)**	**BC03 ($\alpha = 0.2 - 0.98$)**	**BC04 ($\alpha = 0.20 - 0.84$)**
Mechanism function	$f(\alpha) = 1 - \alpha$, Mampel law, random nucleation and subsequent growth, $n = 1$	$f(\alpha) = 4(1-\alpha)^{3/4}$, Reaction order, $n = 1/4$	$f(\alpha) = 4(1-\alpha)^{3/4}$, Reaction order, $n = 1/4$	$f(\alpha) = 4(1-\alpha)^{3/4}$, Reaction order, $n = 1/4$
Mech function number [19, 20]	16	28	28	28
E_a (kJ mol^{-1})	154.8	142.8	135.3	147.6
A (s^{-1})	$10^{14.94}$	$10^{12.88}$	$10^{12.09}$	$10^{13.48}$
Kinetic equation	$\frac{d\alpha}{dt} = 10^{14.94}(1-\alpha)e^{-1.86\times10^4/T}$ (8.25)	$\frac{d\alpha}{dt} = 10^{13.48}(1-\alpha)^{3/4}e^{-1.72\times10^4/T}$ (8.26)	$\frac{d\alpha}{dt} = 10^{12.69}(1-\alpha)^{3/4}e^{-1.63\times10^4/T}$ (8.27)	$\frac{d\alpha}{dt} = 10^{14.08}(1-\alpha)^{3/4}e^{-1.78\times10^4/T}$ (8.28)

Table 8.22 Parameters for BTATz–HNIW–CMDB propellants.

	BC01	BC02	BC03	BC04
C_p (J g^{-1} K^{-1})	$-0.755+1.10\times10^{-2}T-1.28\times10^{-5}T^2$	$-2.28+2.01\times10^{-2}T-2.64\times10^{-5}T^2$	$-1.21+1.34\times10^{-2}T-1.62\times10^{-5}T^2$	$-0.40+0.83\times10^{-2}T-0.83\times10^{-5}T^2$
λ (J cm^{-1}s K^{-1})	20×10^{-4}	20×10^{-4}	30×10^{-4}	30×10^{-4}
ρ (g cm^{-3})	1.64	1.65	1.66	1.66
a (cm)	10^{-3}	10^{-3}	10^{-3}	10^{-3}
$t-t_0$ (s)	10^{-4}	10^{-4}	10^{-4}	10^{-4}
T_{room} (for $T_{cr\text{-}hot\text{-}spot}$ calculation)	293.15	293.15	293.15	293.15
r (m)	0.5	0.5	0.5	0.5
T (K) (for T_{acr} calculation)	330	330	330	330
σ_T (K)	10	10	10	10

Table 8.23 Thermal safety values for BTATz–HNIW–CMDB propellants (1).

	BC01	BC02	BC03	BC04
T_{00} (K)	415.1	421.0	429.5	432.2
T_{e0} (=T_{SADT}) (K)	428.6	430.7	436.2	440.4
T_{p0} (K)	458.0	457.1	458.0	456.2
T_{TIT} (K)	439.3	443.1	448.0	452.3
T_b (K)	469.9	470.6	471.0	469.2
$T_{cr,\ hot\text{-}spot}$ (K)	712.8	783.8	781.1	777.1
H_{50} (cm)	17.1	17.6	18.6	18.4

Table 8.24 Thermal safety values for BTATz–HNIW–CMDB propellants (2).

	Infinite plate				Infinite cylinder				Sphere sample			
	$T_{S(T)max}$ (K)	T_{acr} (K)	S_d (%)	P_{TE} (%)	$T_{S(T)max}$ (K)	T_{acr} (K)	S_d (%)	P_{TE} (%)	$T_{S(T)max}$ (K)	T_{acr} (K)	S_d (%)	P_{TE} (%)
BC01	341.9	337.1	56.96	43.04	347.3	342.4	65.30	34.70	350.7	345.8	68.90	31.10
BC02	331.4	326.5	32.66	67.34	337.1	332.2	47.32	52.68	340.7	335.8	55.60	44.40
BC03	337.7	332.8	48.55	51.45	343.5	338.5	60.46	39.54	347.0	342.1	66.05	33.95
BC04	336.9	332.0	46.61	53.39	342.6	337.7	58.99	41.01	346.2	341.3	64.93	35.07

8.9 Combustion Properties of the Propellants

8.9.1 BTATz-CMDB Propellants in DINA System

8.9.1.1 Burning Rate Properties

For the sake of the possible application of BTATz in CMDB propellant, the burning rates of the RDX-CMDB propellant (number R09. Mass fraction of NC/NG/RDX/auxiliary, 38/28/26/8) and the two BTATz-CMDB propellants (numbers RB0601 and RB0602) were measured under different pressures. The results are listed in Table 8.25 [36].

In order to evaluate the effects of the ballistic modifier on the burning rates of the BTATz-CMDB propellant, the pressure exponent (n) of the burning rate (u) was calculated, and the mean value of the catalysis efficiency ($\overline{Z}$) was compared before and after ballistic modifier was added into the propellant formulation. The values of n and $\overline{Z}$ were obtained by Eqs. (8.29) and (8.30) [37, 38].

$$u_i = aP_i^n, \quad i = 1 - 11 \tag{8.29}$$

$$\overline{Z} = \sum_{i=1}^{k} \frac{(u_{\mathrm{III},i}/u_{\mathrm{II},i})}{k} \tag{8.30}$$

where a is a factor, mm s^{-1} MPa^{-1}.

For RB0601:

at 8–22 MPa, $u = 3.25P^{0.723}$, $r = 0.999$, $\overline{Z} = 1$;
at 10–22 MPa, $u = 3.33P^{0.714}$, $r = 0.999$, $\overline{Z} = 1$;
at 14–20 MPa, $u = 3.28P^{0.719}$, $r = 0.995$, $\overline{Z} = 1$.

Table 8.25 Burning rates of BTATz-CMDB propellants.

	Burning rate (u, mm s^{-1})		
***P* (MPa)**	**R09**	**RB0601**	**RB0602**
2	3.1	4.7	6.2
4	5.3	8.4	9.9
6	7.4	11.5	14.0
8	9.9	14.5	18.6
10	11.9	17.4	22.0
12	14.0	19.5	25.0
14	15.8	21.9	27.1
16	17.5	24.0	28.5
18	19.2	26.5	29.6
20	20.9	28.1	30.8
22	21.9	30.3	32.2

For RB0602:

at 8–22 MPa, $u = 6.58P^{0.522}$, $r = 0.973$, $\overline{Z} = 1.19$;
at 10–22 MPa, $u = 13.57P^{0.459}$, $r = 0.978$, $\overline{Z} = 1.17$;
at 14–20 MPa, $u = 10.69P^{0.353}$, $r = 0.999$, $\overline{Z} = 1.16$.

The burning rate of CMDB propellant containing BTATz is significantly increased, especially when it is used with high-burning-rate ballistic modifier. This is consistent with the high-burning-rate property of BTATz reported in the literature. As a ballistic modifier, the mixture of lead phthalate, copper adipate, and carbon black can be effective to accelerate the burning rate and reduce the pressure exponent of the BTATz–CMDB propellant. With the substitution of 26% RDX by BTATz in CMDB propellant formulation, the burning rate can be improved by 47.3% at 8 MPa and 38.6% at 22 MPa, and even higher with the help of the ballistic modifier, it can be improved by 89.0% at 8 MPa and 47.1% at 22 MPa, and the pressure exponent can be reduced to 0.353 at 14–20 MPa.

8.9.1.2 Structures of Combustion Flames and Quenched Surfaces

The structures of combustion flames and quenched surfaces for RDX-CMDB propellant (R09), and the two BTATz–CMDB propellants (RB0601 and RB0602) at 6 MPa are shown in Figure 8.17.

The flame structure of the CMDB propellant containing BTATz is similar to that of the conventional RDX–CMDB propellant, which consists of a continuous reaction zone: surface and subsurface reaction zone, hissing zone, dark zone, and luminous flame zone. With the increase in pressure, the thickness of the hissing zone and the dark zone becomes thinner, the luminous flame zone approaches the combustion surface, and the flame zone is much bright. This is consistent with the general characteristics of combustion flame of double-base propellant. However, there are some differences from BTATz–CMDB propellant sample RB0601 without ballistic modifier and RDX–CMDB propellant (R09), although there is a dark zone in the flame structure of RB0601, which is different from that of the propellant R09. The combustion surface of RB0601 is not completely separated from the flame of the propellant R09, but shows a unique dark zone morphology, that is, in the dark region, there are some divergent flame beams from some points on the burning surface of propellant. It is mainly due to the fact that BTATz itself does not have a melting process like RDX.

A large number of lava-like bright particles appear on and near the combustion surface of the sample containing the ballistic modifier. It is these carbonaceous materials that cause the propellant to produce over speed combustion or platform combustion. This also shows that the ballistic modifier can effectively promote the chemical reactions in the condensation phase and/or gas phase region of the combustion surface and near the combustion surface of the propellant. In addition, the dark region with lower chemical reaction rate becomes thinner, and the high-temperature flame zone is closer to the combustion surface, which accelerates the heat transfer to the hissing zone and combustion surface and improves the burning rate of the propellant.

Figure 8.17 Structures of combustion flames and quenched surfaces for the propellants. Source: Adopted from Yi et al. [39], China Ordnance Industry Corporation. (a) R09 (quenched surface ×2000), (b) RB0601 (quenched surface ×2000), (c) RB0602 (quenched surface ×2000).

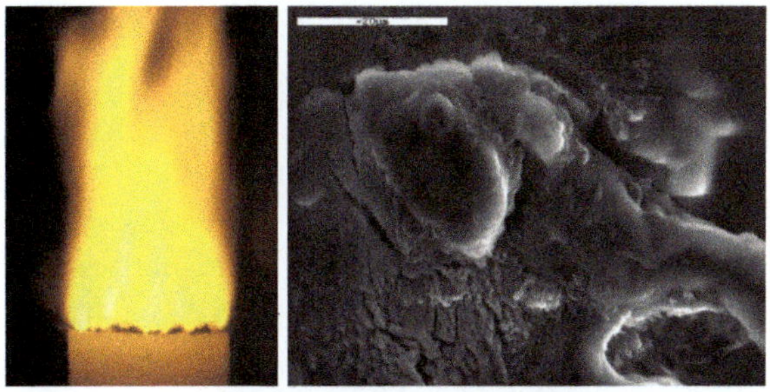

(a) R09 (quenched surface ×2000)

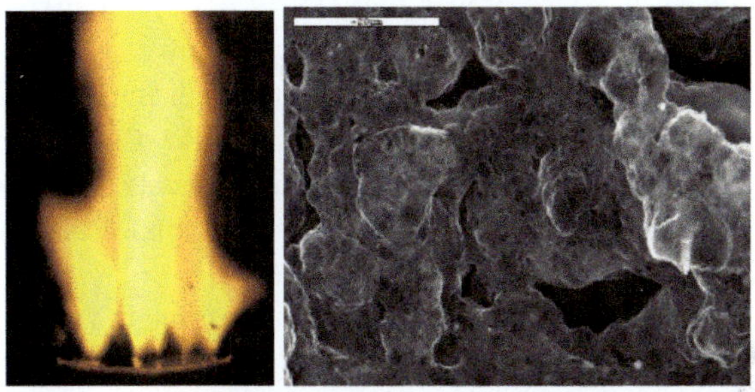

(b) RB0601 (quenched surface ×2000)

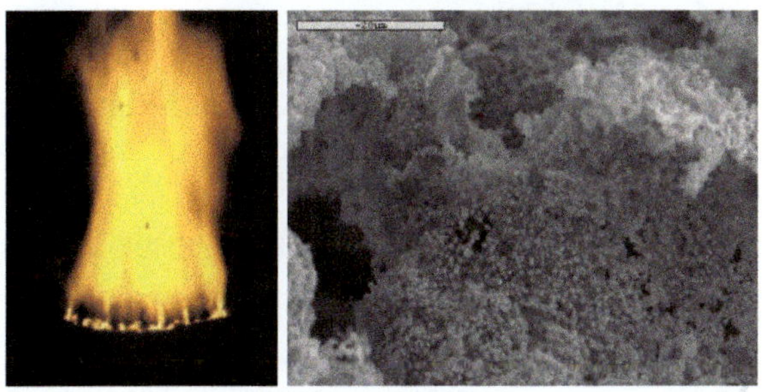

(c) RB0602 (quenched surface ×2000)

For BTATz–CMDB propellant samples, there is a great difference between RB0602 containing ballistic modifier and RB0601 without ballistic modifier: there are more molten substances on the quenched surface of RB0601, and the chain-like nitrogen carbon skeleton is larger. However, the morphology of extinguished surface of RB0602 is greatly changed, which is loose coral-like, and there are many spherical materials with small particle size, which is determined to be metal oxide. There are a lot of micropores and pits on the coral-like structure, especially under high pressure. According to the results of energy spectrum analysis, the metal content on the quenched surface of RB0602 is much higher than its actual addition, which indicates that the activated catalyst is enriched on the combustion surface of the propellant during the combustion process, which catalyzes the decomposition and combustion process of the propellant. This also explains the reason why the ballistic modifier can improve the burning rate of propellant [39].

8.9.2 BTATz–HNIW–CMDB Propellants in DIANP System

8.9.2.1 Burning Rate Properties

The burning rate determination results of BTATz–HNIW–CMDB propellants in DIANP system are listed in Table 8.26 [39].

The burning rate of CMDB propellant is not affected by the content of BTATz from 18.0% to 28.0%, which is related to the higher burning rate of BTATz and HNIW, and the effect of the ratio of BTATz and HNIW on the burning rate of CMDB propellant is not obvious. Due to the presence of HNIW, the pressure exponents of the burning rates of the six samples are higher; in particular, there is no ballistic modifier added. It is also proved that HNIW with high energy density has the potential to improve the burning rate, but the pressure exponent is slightly higher. The increase in burning rate has little to do with a small amount of aluminum powder and HNIW content. In other words, in BTATz–HNIW–CMDB propellant, the effect of the ballistic modifier, which contains lead phthalate (A1), lead 2,4-dihydroxy benzoic acid (A2), carbon black with medium grain size (C2), plays a dominant role in the burning rate, rather than aluminum powder.

Carbon black plays a very important role in the ballistic modifier system, but its role has a lot to do with its particle size distribution, specific surface area, preparation methods, raw material sources, and so on. The particle size of the medium–fine carbon black (C2) is mainly concentrated in 1 and 0.1 μm (as Figure 8.18 shown), and its catalytic effect is better than that of ultra-fine carbon black. It is speculated that one of the reasons is that the medium–fine carbon black blends with large and small particles is more dispersed than the ultra-fine carbon black.

Table 8.26 Burning rates of BTATz–HNIW–CMDB propellants.

	Burning rate (u, mm s^{-1})					
P (MPa)	BC01	BC02	BC03	BC04	BC1202	BC1205
2	6.4	5.8	8.6	5.8	8.1	7.59
4	10.4	9.3	12.7	11.4	13.8	12.9
6	13.7	12.3	16.4	14.9	17.8	17.4
8	16.9	15.0	19.4	18.5	21.1	20.4
10	19.2	17.7	22.0	22.0	24.0	22.9
12	22.4	20.5	24.6	25.0	26.5	24.8
14	24.5	23.1	27.5	28.0	28.6	26.8
16	27.1	26.4	29.4	30.0	30.4	28.4
18	29.6	28.8	32.1	32.4	32.6	30.2
20	32.4	31.3	33.9	34.2	34.5	32.5
22	34.6	34.3	36.0	36.5	36.1	34.7
n (8–22 MPa)	0.72	0.82	0.62	0.66	0.53	0.51
n (14–22 MPa)	0.77	0.85	0.60	0.59	0.52	0.57

Note: n is the pressure exponent of the burning rates within a specific pressure range.

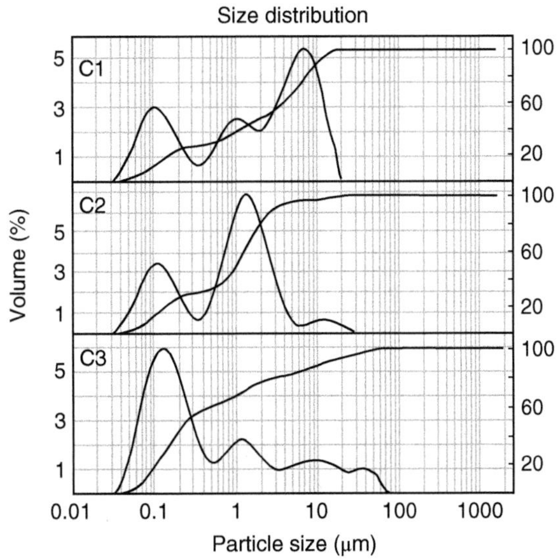

Figure 8.18 Particle size distribution of general carbon black (C1), medium-fine carbon black (C2), and ultra-fine carbon black (C3).

8.9.2.2 Structures of Combustion Flames and Quenched Surfaces

The structures of combustion flames and quenched surfaces for the three BTATz–HNIW–CMDB propellants (BC01, BC02, and BC1202) at 6 MPa are shown in Figure 8.19.

The noncatalytic BTATz–HNIW–CMDB propellant is similar to BTATz–CMDB propellant in that there are dark areas and some divergent flame beams from the combustion surface. When the mass percentage of HNIW increases from 10% to 20%, the brightness of the combustion flame and the active points of the divergent flame beam on the combustion surface will increase. The flame of the sample containing ballistic modifier is brighter than that of the sample without ballistic modifier, and the combustion is obvious at low pressure. This is consistent with the burning rate performance obtained before, that is, the ballistic modifier containing lead phthalate, copper 2,4-dihydroxybenzoate, and carbon black have a strong burning rate promotion ability: at 4 MPa, the burning rate of BC02 is 9.3 mm s^{-1}, while that of BC1202 can reach 13.8 mm s^{-1}, which is the highest value under this pressure. Similar to the previous results, there are bright and hot carbon particles on the combustion surface of the sample containing ballistic modifier, and the catalytic effect mainly occurs near the condensed phase and combustion surface.

Similar to the analysis results of the previous two kinds of propellant samples, BC01 and BC02 without ballistic modifier show chain-like skeleton and agglomerate in some places, but this is different from the melting state of conventional RDX–CMDB propellant. However, the samples with ballistic modifier can change this situation obviously, showing loose coral shape. It is also indicated that the combustion surface structure of these propellants is changed by the ballistic modifier containing lead phthalate, copper 2,4-dihydroxybenzoate, and carbon black. The content of Pb and Cu on the quenched surface of the sample containing ballistic modifier is much higher than that in the formulation, and both of them accumulate on the combustion surface and promote the increase in burning rate [39].

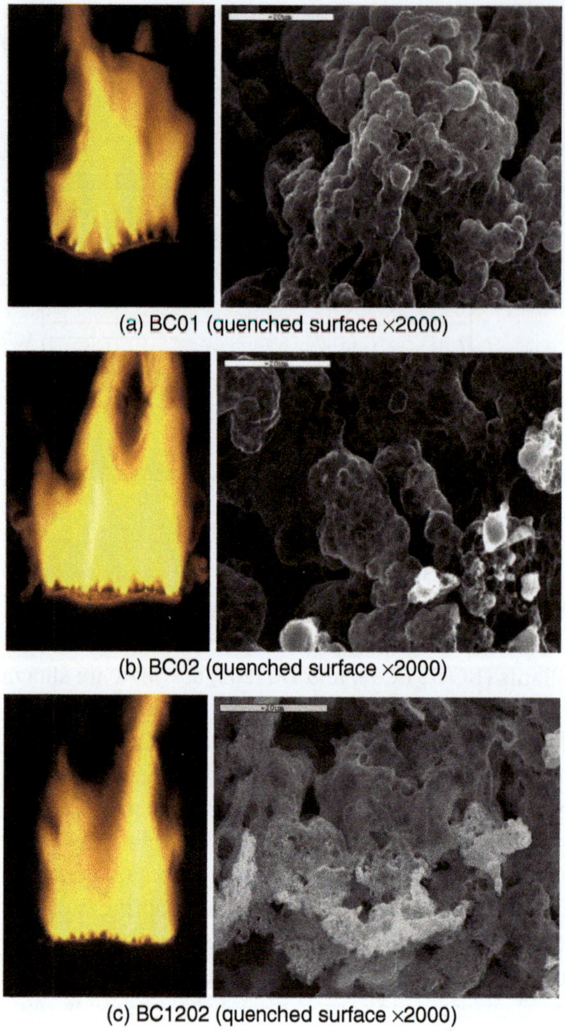

Figure 8.19 Structures of combustion flames and quenched surfaces for the propellants. Source: Adopted from Yi et al. [39], China Ordnance Industry Corporation. (a) BC01 (quenched surface ×2000), (b) BC02 (quenched surface ×2000), (c) BC1202 (quenched surface ×2000).

8.10 Correlation Between PDSC Characteristic Values and Burning Rates

On the combustion models for solid propellant, Miller et al. [40] thought that the burning rate (u) keeps a direct proportion to the decomposition rate on the surface of the solid propellant, and Brill et al. [41] obtained a formula composed of u^2, Arrhenius parameters for the condensed phase decomposition reaction, the decomposition heat, and the temperature on the burning surface. From the research results, one can know that the burning rate keeps a direct proportion to the square root of the reaction rate constant $k = A\,\exp(-E/RT)$ and that of the work pressure. On the basis of the preceding works, Liu et al. [42] correlated the characteristic values of PDSC and the burning rates by a linear empirical formula:

$$u = k_u \left(\frac{p\Delta H_d}{\Delta T} \right)^{1/2} + C \tag{8.31}$$

where u is the burning rate of the propellant in mm s^{-1}; k_u is the correlation factor of the burning rate with PDSC characteristic value in mm (g K)$^{1/2}$ s^{-1} (MPa J)$^{-1/2}$; p is the work pressure in MPa; ΔH_d is the decomposition heat in J g^{-1}; $\Delta T = T_f - T_0$; C is a constant.

As the heating rate of PDSC is controlled by computer program and keeping 10 K min^{-1} during the whole experiment, ΔT corresponds to the decomposition reaction time (Δt), and $\Delta H_d/\Delta T$ can be thought as the rate of the heat release or decomposition rate on condition that the decomposition heat keeps a direct proportion to the conversion degree (α). Moreover, the pressure (p) is an important factor affecting the burning rate and decomposition reaction, and can be introduced into the empirical equation (8.31). As a matter of fact, the formula integrates the two important factors affecting the burning rate: the decomposition rate and the pressure.

The PDSC characteristic values and burning rates for the propellants RB0601 and RB0602 are shown in Table 8.27. The u versus $(p\Delta H_d/\Delta T)^{1/2}$ curves were drawn and the slope coefficients (k_u), the constants (C), and the linear correlation coefficient (r) were obtained by linear regression and shown in Eqs. (8.32) and (8.33).

$$\text{RB0601:} \quad u = 0.833(p\Delta H_d/\Delta T)^{1/2} - 1.65 \quad r = 1 \tag{8.32}$$

$$\text{RB0602:} \quad u = 0.924(p\Delta H_d/\Delta T)^{1/2} + 0.06 \quad r = 0.9902 \tag{8.33}$$

Table 8.27 The PDSC characteristic values and burning rates for the propellants RB0601 and RB0602.

Sample	P (MPa)	T_0 (K)	T_p (K)	T_f (K)	ΔT (K)	ΔH_d (J g^{-1})	$(p\Delta H_d/\Delta T)^{1/2}$ (MPa J g^{-1} K^{-1})$^{1/2}$	u (mm s^{-1})
RB0601	0.1	426.2	476.9	530.5	104.3	1.45×10^3	1.18	—
	2	430.3	473.8	505.3	74.9	2.14×10^3	7.56	4.7
	4	430.8	473.7	502.2	71.4	2.63×10^3	12.1	8.41
	6	435.3	473.4	499.8	64.5	2.70×10^3	15.8	11.5
	7	428.1	475.8	497.9	69.8	3.10×10^3	17.6	13.0
RB0602	0.1	422.1	476.1	521.1	99.0	1.49×10^3	1.23	—
	2	427.0	473.4	505.8	78.8	2.00×10^3	7.12	6.2
	4	429.5	474.0	507.3	77.8	2.04×10^3	10.2	9.9
	6	437.4	473.6	502.0	64.6	2.34×10^3	14.7	14.0
	7	432.8	472.7	495.7	62.9	2.92×10^3	18.0	16.3

The results show the good linear relation of u and $(p\Delta H_d/\Delta T)^{1/2}$ and the good fittings of the formula for the propellants in pressure range (2–7 MPa) of PDSC. k_u keeps a similar changing trend with u, and can be used to study the effect of the ballistic modifier or the other component on the burning rates.

8.11 Conclusions

(1) BTATz was synthesized from triaminoguanidine nitrate and acetylacetone in three steps, and the structure was identified by elemental analysis, IR, NMR, and DSC. The preparation methods of compound **4** with different oxidants were explored. Replacing nitrogen dioxide/*N*-methylpyrrolidone with sodium nitrite/acetic acid can effectively reduce the cost and simplify the operation. All the atoms in BTATz molecular structure are in the same plane. Tetrazine ring and tetrazole ring are connected by amino group and form a large conjugated π bond. The bond order of BTATz is calculated. The bond order between the bonding atoms of tetrazine ring skeleton is large, which is not easy to break open the ring. The bond order between tetrazole ring N(14)—N(15), N(18)—N(20) is small and easy to break, which makes the tetrazole ring open. In addition, the C—N bond order between tetrazole ring and tetrazine ring is also small, which makes it easy to separate each ring. The vibration and infrared spectra of BTATz are obtained. The calculated frequency and intensity are in good agreement with the experimental results.

(2) The application of BTATz in CMDB propellants can improve the burning rate without affecting the smoke signal level of the original propellant, as the energy is not high, so it is necessary to use the high burning rate characteristics of BTATz and ensure the energy of the original propellant. The most desirable method is to introduce a certain amount of energetic additives with a higher energy into the formulation of double-base propellant containing BTATz.

(3) The BTATz–CMDB propellants (numbers RB0601 and RB0602) without and with the ballistic modifier were prepared. There are three mass loss stages in TG curve for the BTATz–CMDB propellant, and they are likely attributes to the volatilization of NG, the decomposition of NG, NC, DINA, and other auxiliary, and the decomposition of BTATz. The reaction mechanisms of the main exothermal decomposition processes of RB0601 and RB0602 are classified as chemical reaction, and $f(\alpha) = (1 - \alpha)^2$. High pressure restrains the volatilization of NG, accelerates the decomposition reaction, and makes the reaction occur easily. High pressure can change the decomposition reaction mechanism function and kinetics, and the control process of the main exothermal decomposition reaction obeys the rule of Avrami–Erofeev equation at high pressure and chemical reaction at normal pressure. The decomposition reaction mechanism function can't be changed by the ballistic modifier. The evaluation for the thermal safety of the propellants was obtained. The propellant RB0602 has a higher resistance to heat and thermal safety than RB0601. The transition from adiabatic decomposition to explosion for propellant RB0602 is easier than

RB0601. The sphere charge is safer than another with the same characteristic measurement. With the substitution of 26% RDX by BTATz and with the help of the ballistic modifier in the CMDB propellant formulation, the burning rate can be improved by 89.0% at 8 MPa and 47.1% at 22 MPa, the pressure exponent can be reduced to 0.353 at 14–20 MPa.

(4) The BTATz–HNIW–CMDB propellants (numbers BC01, BC02, BC03, BC04, etc.) without and with the ballistic modifier were prepared. The thermal decomposition property, nonisothermal reaction kinetics, and thermal safety were carried out, and kinetic function and the eight important thermal safety parameters were calculated and obtained. The content of HNIW has a great effect on the decomposition reaction mechanism of BTATz–HNIW–CMDB propellant: when the content of HNIW is 10%, the decomposition reaction is controlled by the random nucleation and subsequent growth ($n = 1$), and the reaction mechanism obeys Mampel law. But when the content of HNIW is 20%, the reaction is controlled by the chemical reaction ($n = 1/4$). The mechanism can't be changed by the catalysts, who just make the apparent activation energy change slightly. For the samples, from BC01 to BC04, the values of T_{SADT} and T_{TIT} make an upward tendency, the resistivity to heat: BC04 > BC03 > BC02 > BC01. The values of T_{acr} and S_{d} for BC01 are the maximum and for BC02 are the minimum, the heat sensitivity: BC01 > BC03 > BC04 > BC02.

(5) The correlation between PDSC characteristic values and burning rates was carried out and it was found that u and $(p\Delta H_{\mathrm{d}}/\Delta T)^{1/2}$ keep a good linear relation, k_u keeps a similar changing trend with u, and it can be used to study the effect of the ballistic modifier or the other component on the burning rates.

References

1 Sha, H., Li, F.S., Song, H.C., and Chen, S.L. (1995). Surface modification of ultra-fine AP and its influence on the high burning rate propellant. *Chin. J. Energetic Mater.* 3: 26–30.

2 Jia, X.F., Su, C.Y., and Lv, G.L. (2002). Study on combustion properties of HTPB propellant containing nitramine under high and low pressure. *Chin. J. Solid Rocket Tech.* 25: 38–40.

3 Zhao, F.Q., Li, L., Li, S.W. et al. (1999). Research on combustion mechanism of RDX–CMDB propellants with or without catalysts. *Chin. J. Solid Rocket Tech.* 22: 50–53.

4 Hiskey, M.A., Chavez, D.E., and Naud, D. (2002). Preparation of 3,3′-azobis (6-amino-1,2,4,5-tetrazine). US Patent 6342589.

5 Hiskey, M.A., Chavez, D.E., and Naud, D. (2003). 3,6-Bis(1*H*-1,2,3,4-tetrazol-5-yl-amino)-1,2,4,5-tetrazine or salt thereof. US Patent 6657059.

6 Yue, S.T. and Yang, S.Q. (2004). Synthesis and properties of 3,6-bis(1*H*-1,2,3,4-tetrazol-5-yl-amino)-1,2,4,5-tetrazine. *Chin. J. Energetic Mater.* 12: 155–157.

7 Wang, B.Z., Lai, W.P., Liu, Q. et al. (2008). Synthesis, characterization and quantum chemistry study on 3,6-bis(1*H*-1,2,3,4-tetrazol-5-yl-amino)-1,2,4,5-tetrazine. *Chin. J. Org. Chem.* 28: 422–427.

8 Saikia, A., Sivabalan, R., Polke, B.G. et al. (2009). Synthesis and characterization of 3,6-bis(1*H*-1,2,3,4-tetrazol-5-ylamino)-1,2,4,5-tetrazine (BTATz): novel high-nitrogen content insensitive high energy material. *J. Hazard. Mater.* 170: 306–313.

9 Zhang, X.G., Zhu, H., Yang, S.Q. et al. (2007). Study on thermal decomposition kinetics and mechanism of nitrogen-rich compound BTATz. *Chin. J. Propul. Technol.* 28 (3): 322–326.

10 Son, S.F., Berghout, H.L., Bolme, C.A. et al. (2000). Burn rate measurements of HMX, TATB, DHT, DAAF, and BTATz. *Proc. Combust. Inst.* 28: 919–924.

11 Hiskey, M.A., Chavez, D.E., and Naud, D. (2001). Low-smoke pyrotechnic compositions. US Patent 6312537.

12 Li, S.W., Zhao, F.Q., Yuan, C. et al. (2002). Tendency of research and development for overseas solid propellants. *Chin. J. Solid Rocket Tech.* 25: 36–42.

13 Hiskey, M.A., Chavez, D.E., and Naud, D. (2002). Propellant containing 3,6-bis(1*H*-1,2,3,4-tetrazol-5-yl-amino)-1,2,4,5 -tetrazine or salts thereof. US Patent 6458227.

14 Zhao, F.Q., Yi, J.H., and Xu, S.Y. (2014). High-burning-rate composite modified double base propellant. CN 201110012795.4.

15 Yi, J.H., Zhao, F.Q., Gao, H.X. et al. (2008). Preparation, characterization, non-isothermal reaction kinetics, thermodynamic properties, and safety performances of high nitrogen compound: hydrazine 3-nitro-1,2,4-triazol-5-one complex. *J. Hazard. Mater.* 153: 261–268.

16 Yi, J.H., Zhao, F.Q., Wang, B.Z. et al. (2010). Thermal behaviors, nonisothermal decomposition reaction kinetics, thermal safety and burning rates of BTATz–CMDB propellant. *J. Hazard. Mater.* 181: 432–439.

17 Yi, J.H., Zhao, F.Q., Ren, Y.H. et al. (2011). BTATz–CMDB propellants: high-pressure thermal properties and their correlation with burning rates. *J. Therm. Anal. Calorim.* 104: 1029–1036.

18 Singh, G., Shrimal, A.K., Kapoor, I.P.S. et al. (2011). Kinetics of thermolysis of some transition metal perchlorate complexes with 1,6-diaminohexane ligand. *J. Therm. Anal. Calorim.* 103: 149–155.

19 Hu, R.Z., Gao, S.L., Zhao, F.Q. et al. (2008). *Thermal Analysis Kinetics*, 2e. Beijing: Science Press.

20 Zhao, F.Q., Hu, R.Z., Gao, H.X., and Ma, H.X. (2006). Thermochemical properties, nonisothermal decomposition reaction kinetics and quantum chemical investigation of 2,6-diamino-3,5-dinitropyrazine-1-oxide (LLM–105), Chapter 4. In: *New Developments in Hazardous Materials Research* (ed. O.E. Bronna), 93–126. New York: Nova Science Publishers Inc.

21 Ma, H.X., Song, J.R., Zhao, F.Q. et al. (2007). Nonisothermal reaction kinetics and computational studies on the properties of 2,4,6,8-tetranitro-2,4,6,8-tetraazabicyclo [3,3,1] onan-3,7-dione (TNPDU). *J. Phys. Chem. A* 111: 8642–8649.

22 Zhang, T.L., Hu, R.Z., Xie, Y., and Li, F.P. (1994). The estimation of the critical temperature of thermal explosion for energetic materials using non-isothermal DSC. *Thermochim. Acta* 244: 171–176.

23 Smith, L.C. (1975). An approximate solution of the adiabatic explosion problem. *Thermochim. Acta* 13: 1–6.

24 Hu, R.Z., Zhang, H., Xia, Z.M. et al. (2003). Estimation formulae of the critical rate of temperature rise for thermal explosion of exothermic decomposition reaction system of energetic materials. *Chin. J. Energetic Mater.* 11: 130–133, 137.

25 Bruckman, H.J. and Guillet, J.E. (1968). Theoretical calculations of hot-spot initiation in explosives. *Can. J. Chem.* 41: 3221–3228l.

26 Hu, R.Z., Gao, H.X., Zhao, F.Q. et al. (2009). Estimation of critical temperatures of hot-spot initiation in energetic materials. *Chin. J. Energetic Mater.* 17: 127–130.

27 Dong, H.S. and Zhou, F.F. (1989). *Performances of High Explosive and Its Related Materials*. Beijing: Science Press.

28 Dong, H.S., Hu, R.Z., Yao, P., and Zhang, X.X. (2001). *Thermograms of Energetic Materials*. Beijing: National Defence Industry Press.

29 Hu, R.Z., Zhao, F.Q., Gao, H.X. et al. (2009). The estimation of characteristic drop heights of impact sensitivity for polymer bonded explosives JH-94 and JO-96. *Chin. J. Energetic Mater.* 17: 251–254.

30 Friedman, M.H. (1963). A correlation of impact sensitivities by means of the hot spot model. In: *9th (international) Symposium on Combustion*, 294–302. New York: Academic Press Inc.

31 Frank-Kamenetskii, D.A. (1955). *Diffusion and Heat Exchange in Chemical Kinetics*. Princeton: Princeton University Press.

32 Wang, P. and Du, Z.M. (2007). Probability distribution of thermal sensitivity of energetic materials. *Chin. J. Energetic Mater.* 15: 633–636.

33 Wang, P. (2008). Study on thermal safety and ignition reliability of exothermic system. PhD dissertation. Beijing Institute of Technology.

34 Hu, R.Z., Gao, H.X., and Zhao, F.Q. (2008). Thermal safety of plastic bonded explosives JH-94 and JO-96. *Chin. J. Explos. Propellants* 31: 28–31.

35 Yi, J.H., Zhao, F.Q., Wang, B.Z. et al. (2014). BTATz–HNIW–CMDB propellant: decomposition reaction kinetics and thermal safety. *J. Therm. Anal. Calorim.* 115: 1227–1234.

36 Yi, J.H., Zhao, F.Q., Wang, B.Z. et al. (2012). Properties of CMDB propellant containing high-nitrogen compound BTATz. *J. Propul. Tech.* 33: 609–615.

37 Chen, P., Zhao, F.Q., Luo, Y. et al. (2004). Thermal decomposition behavior and non-isothermal decomposition reaction of copper(II) salt of 4-hydroxy-3,5-dinitropyridine oxide and its application in solid rocket propellant. *Chin. J. Chem.* 22: 1056–1063.

38 Yi, J.H., Zhao, F.Q., Hong, W.L. et al. (2010). Effects of Bi–NTO complex on thermal behaviors, nonisothermal reaction kinetics and burning rates of NG/TEGDN/NC propellant. *J. Hazard. Mater.* 176: 257–261.

39 Yi, J.H., Xuan, C.L., Zhao, F.Q. et al. (2017). Combustion properties of CMDB propellant containing high-nitrogen compounds. *Chin. J. Explos. Propellants* 40: 87–94, 107.

40 Miller, M.S. and Anderson, W.R. (2000). Energetic material combustion modeling with elementary gas–phase reactions: a practical approach, Chapter 2.12. In: *Solid Propellant Chemistry, Combustion, and Motor Interior Ballistics*, vol. 185 (ed. V. Yang, T. Brill and W. Ren), 501–531. Reston, VA: American Institute of Aeronautics and Astronautics, Inc.

41 Brill, T.B., Arisawa, H., Brush, P.J. et al. (1995). Surface chemistry of burning explosives and propellants. *J. Phys. Chem.* 99: 1384–1392.

42 Liu, Z.R., Liu, Y., Zhao, F.Q. et al. (2008). Correlation between PDSC characteristic value and burning rate for the solid propellants. *Chin. J. Explos. Propellants* 31: 9–17.

9

Nitro-substituted Oxadiazoles: Important Building Blocks in the Synthesis of Energetic Compounds

Philip Pagoria[1,2]

[1]*M. H. Chew and Associates Inc., 7633 Southfront Road, Suite 170, Livermore, CA 94551, USA*
[2]*Lawrence Livermore National Laboratory, 7000 East Avenue, L-281, Livermore, CA 94550, USA*

9.1 Introduction

The pursuit of novel energetic materials has been a robust area of research for many years. The majority of the early work in the synthesis of novel energetic compound focused on the synthesis of cyclic nitramines, carbocyclic nitroaromatic compounds, nitrate esters, linear aliphatic nitramines, and nitroaliphatics [1–3]. More recently, five- and six-membered nitrogen-rich heterocycles have received a significant amount of interest as substrates for the synthesis of energetic compounds [4, 5]. These have included pyridines [6–8], pyrazines [9], pyrazoles [10, 11], tetrazines [12], 1,2,4-triazoles [13], 1,2,3-triazoles [14, 15], tetrazoles [16], and imidazoles [1]. They have gained interest because, in relation to their carbocyclic analogs, they have higher nitrogen content, a more positive enthalpy of formation and increased oxygen balance because of inherently lower carbon content, all important contributors to the explosive performance of the molecule [17–19].

Recently, oxadiazoles, which possess an oxygen atom in the ring, specifically 1,2,5-oxadiazoles (furazans) [20], 1,2,5-oxadiazole-2-oxides (furoxans) [21–23], 1,2,4-oxadiazoles, and 1,3,4-oxadiazoles, have received growing interest as components of new energetic compounds because of their high enthalpy of formations, good thermal and chemical stability, and synthetic availability [19, 20]. Recent advances in the synthesis and characterization of novel energetic compounds bearing 1,2,5-, 1,2,4- or 1,3,4-oxadiazoles will be described.

9.2 Enthalpy of Formation of Oxadiazoles

The calculated heats of formation for the various oxadiazoles using both the G2 atomization method and the atomization method based on CBS-4M electronic enthalpies have been reported [24–26]. The values for the oxadiazoles are shown in comparison to the values of common azoles, which have higher heats of formation

Nitrogen-Rich Energetic Materials, First Edition. Edited by Michael Gozin and Leonid L. Fershtat.

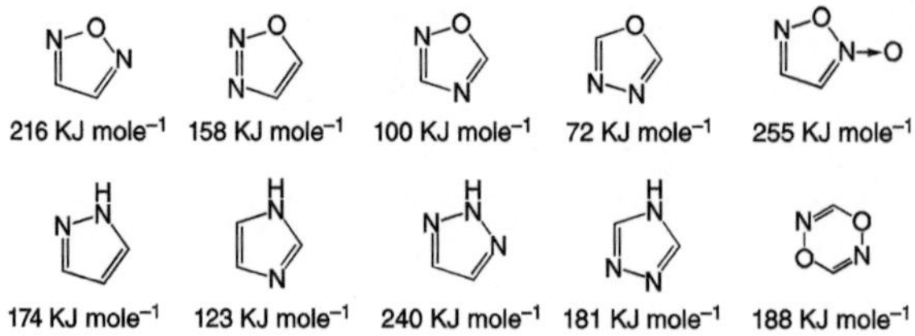

Figure 9.1 Calculated gas-phase heats of formation using the atomization method based on CBS-4M electronic enthalpies. Source: Adapted from Refs. [26, 27].

than the oxadiazoles because of their higher content of nitrogen. The furazan and furoxan have the highest heats of formation, while the 1,3,4-oxadiazole has the lowest. The values shown were calculated using the CBS-4M method (Figure 9.1).

In 1995, Pivina et al. [28] performed a computational, theoretical comparison of energy content, specifically the enthalpy of formation, for a series of nitro-substituted furazans. The encouraging results from these calculations indicated that many of the furazans represented in the study were desirable energetic target compounds. Many of the compounds listed in the study have since been synthesized and their physical properties have been measured. Sheremetev investigated the chemistry of furazans [29] and furoxans [30] fused to six- and seven-membered heterocycles and noted that the furazan ring system when fused to an aromatic ring has significant electron-withdrawing character, compared to the $-CH{=}C(CN)_2$ and $-CF_3$ moieties. Furoxan has a weaker electron-withdrawing effect because the *N*-oxide fragment is electro-donating. The electron-donating effect of an *N*-oxide fragment may also be seen in pyridines and pyrazines where the *N*-oxide increases the aromaticity of the ring system and increases the reactivity of the *N*-oxides to electrophilic substitution [31].

9.3 1,2,4-Oxadiazoles

Until recently, very few energetic compounds incorporating 1,2,4-oxadiazole ring systems were reported. One of the issues that has hindered the development of nitro-substituted 1,2,4-oxadiazoles is the lack of synthetic methodology to convert 3-amino- or 5-amino-1,2,4-oxadiazoles to their corresponding 3-nitro and 5-nitro derivatives. Thus far there are very few reported nitro-substituted 1,2,4-oxadiazoles in which the nitro-group is attached directly to the 1,2,4-oxadiazole ring.

In 1995, Makhova and coworkers [32] reported the conversion of 3,4-dinitrofuroxan (**DNFO**, **1**) to 3-nitro-5-R-1,2,4-oxadiazoles (**2**) and methyl 3-nitro-1,2,4-carboxyl-5-ate (**3**) by treatment of DNFO with excess trichloroacetonitrile or methylcyanoformate in $CHCl_3$ at 20 °C for seven days, respectively (Scheme 9.1). DNFO reportedly undergoes a [3*n*+2] cyclo-reversion reaction at room temperature to give 2 mol of nitroformonitrile oxide (**NFNO**, **4**) which may undergo 1,3-dipolar cycloaddition reactions with various substrates. They subsequently reported [33] the synthesis of various 3-nitroheterocycles by the [3*n*+2] cyclo-addition of NFNO to dipolarophiles including methyl acrylate, acetylenes, dienes, hexafluoroacetone, and 1,2-diphenylcyclopropen-3-one. The syntheses were carried out in CCl_4 with the ionic liquid catalyst, 1,3-dimethyl-imidazolium tetrafluoroborate, although the yields were generally low (<30%).

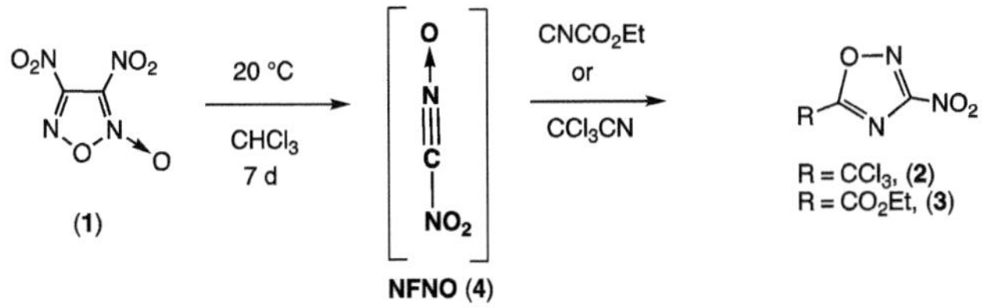

Scheme 9.1 Conversion of **DNFO** (**1**) to 3-nitro-5-*R*-1,2,4-oxadiazoles.

In 2012, Chen and coworkers [34] synthesized the previously unknown, 5-amino-3-nitro-1,2,4-oxadiazole (**NOA**, **5**), by treatment of diaminoglycouril (**6**) with dimethyldioxirane (DMDO) in aqueous CH_3CN to yield the intermediate 5-guanidino-3-nitro-1,2,4-oxadiazole (**NOG**, **7**) which was subsequently treated with mixed acid at 0 °C-rt to give **NOA** (Scheme 9.2). NOG was treated with a mixture of 100% HNO_3 and P_2O_5 to yield 5-nitramino-3-nitro-1,2,4-oxadiazole (**8**) presumably *via* the intermediate, **NOA 5**. NOA **5** has a density of 1.723 g cm^{-3} and a thermal decomposition temperature (DSC peak) (T_d) of 265 °C, while **8** has a density of 1.883 g cm^{-3} and a T_d of 219 °C.

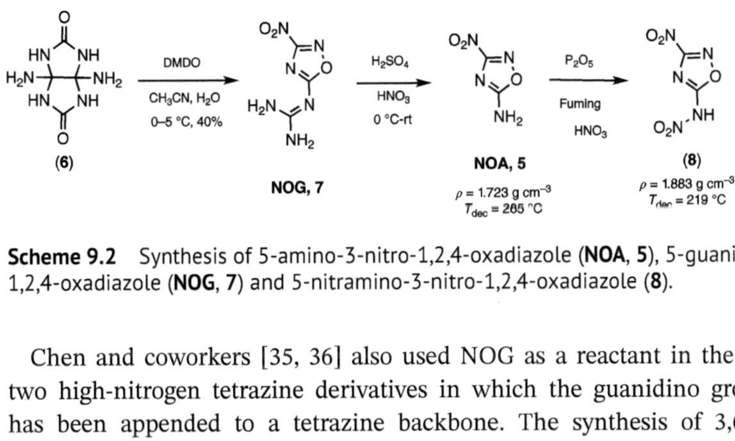

Scheme 9.2 Synthesis of 5-amino-3-nitro-1,2,4-oxadiazole (**NOA**, **5**), 5-guanidino-3-nitro-1,2,4-oxadiazole (**NOG**, **7**) and 5-nitramino-3-nitro-1,2,4-oxadiazole (**8**).

Chen and coworkers [35, 36] also used NOG as a reactant in the synthesis of two high-nitrogen tetrazine derivatives in which the guanidino group of NOG has been appended to a tetrazine backbone. The synthesis of 3,6-bis(3-nitro-1,2,4-oxadiazole-5-guanyl)-1,2,4,5-tetrazine (**9**) involved allowing **NOG** to react with 3,6-bis(3,5-dimethylpyrazol-1-yl)-1,2,4,5-tetrazine (**10**) and NaH in DMF (Scheme 9.3). Compound **9** could not be crystallized in pure form but rather as a bis-DMSO solvate, from which the structure was confirmed by X-ray crystallographic analysis. Because the authors could not obtain a crystal density for **9**, they calculated a density of 1.812 g cm^{-3}. In a similar manner, 3,6-bis(3-azido-1,2,4-oxadiazole-5-guanyl)-1,2,4,5-tetrazine (**11**) was synthesized by allowing **10** to react with 3-azido-5-guanidino-1,2,4-oxadiazole (**AOG**, **12**) with NaH in DMF at 0 °C-RT. **AOG** was synthesized in 36% yield by allowing NaN_3 to react with **NOG** in aqueous DMSO. The calculated crystal densities, using density quantum chemistry

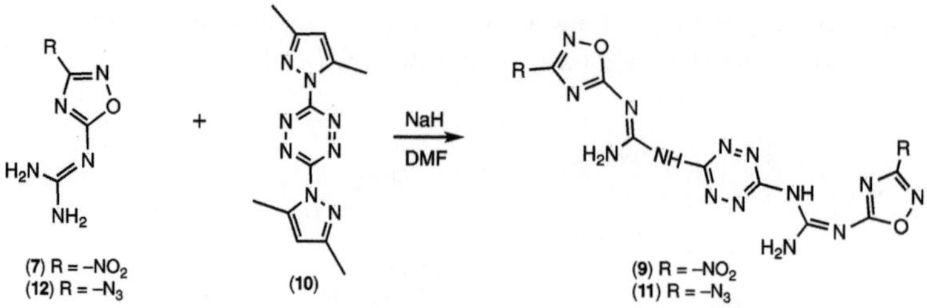

Scheme 9.3 Synthesis of tetrazine derivatives of **NOG**, **7** and **AOG**, **12**.

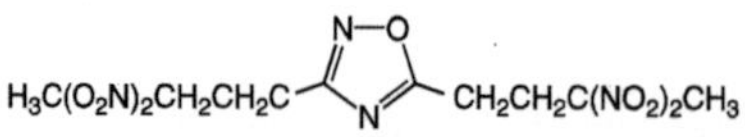

Figure 9.2 3,5-Bis(3,3-dinitrobutyl)-1,2,4-oxadiazole (**13**).

functional method DFT-B3LYP/6-31G*, for **11** and **12** were 1.694 and 1.734 g cm^{-3}, respectively.

In early work, Coburn and Bjorklund [37] reported the synthesis of 3,5-bis(3,3-dinitrobutyl)-1,2,4-oxadiazole (**13**), an oxadiazole-based energetic plasticizer (Figure 9.2). The synthesis involved the condensation of 4,4-dinitropentanenitrile with hydroxylamine in MeOH to give 4,4-dinitropentanamidoxime (**14**) in 63% yield which was condensed with 4,4-dinitropentanoyl chloride to yield *O*-(4,4-dinitropentanoyl)-4,4-dinitro-pent-anamidoxime (**15**). This was cyclized with refluxing $POCl_3$ to give **13** as a viscous liquid.

Recently, there has been a growing interest in the construction of coupled heterocyclic frameworks [18] in which selected heterocycles are linked together, forming a multicyclic array, which appended with various explosophoric functional groups. It has become evident these coupled heterocyclic frameworks offer great potential in the design and synthesis of new energetic compounds in which the component heterocycles may be selected to tailor various attractive physical properties including high thermal stability, high density, less sensitivity, and high detonation performance. This approach has led to the development of new energetic compounds in which oxadiazoles are linked to other heterocycles such as furazans, furoxans, and tetrazoles [18, 38–41]. As an example of this approach, Pagoria et al. [39] and Tsyshevsky et al. [40] reported the computational design and synthesis of a series of multicyclic energetic compounds in which the 1,2,4-oxadiazole moiety served as the backbone for multicyclic target compounds. The work was motivated by the many reports of the synthesis and characterization of 3,4-bis(4-nitrofurazan-3-yl)-furoxan (**BNFF, 16**) [42–47] and the structurally similar 3,4-bis(4-nitrofurazan-3-yl)-furazan (**LLM-172, BNFF-1, BNTF 17**) [48], both highly energetic compounds with good thermal stability that possess the attractive property of having relatively low melting points of 109 and 84 °C, respectively. These attractive properties encouraged the investigation of analogs of BNFF and LLM-172 in which the central core is a 1,2,4-oxadiazole ring instead of a furoxan, leading to the synthesis of 3,5-bis(4-nitrofurazan-3-yl)-1,2,4-oxadiazole (**LLM-191, 18**), 3-(4-aminofurazan-3-yl)-4-(4-nitrofurazan-3-yl)-1,2,4-oxadiazole (**LLM-192, 19**),

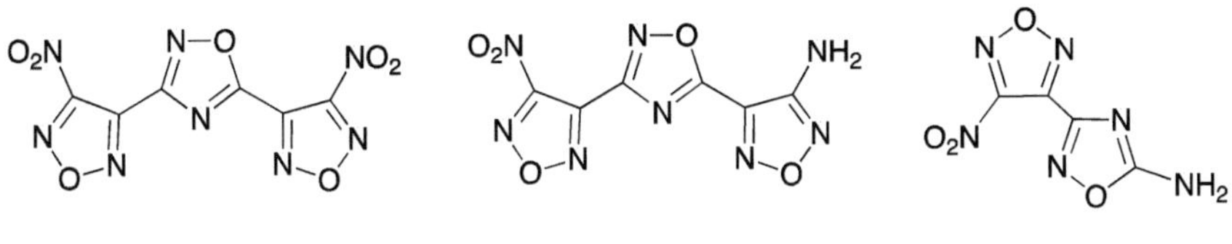

LLM-191 (18) LLM-192 (19) LLM-201 (20)

Figure 9.3 Recent examples of energetic 1,2,4-oxadiazoles.

and 3-(4-nitrofurazan-3-yl)-1,2,4-oxadiazole-5-amine (**LLM-201**, **20**) (Figure 9.3). It was postulated that the incorporation of a 1,2,4-oxadiazole ring system with a 3,5-substitution pattern would allow better crystal packing than the furoxan-based ring systems because of less steric crowding and the possibility of forming a co-planar arrangement, leading to higher density and possibly lower melting points. Indeed, **LLM-191** has a high density of 1.91 g cm^{-3} and a melting point of 61 °C. Tsyshevsky et al. [41] followed up this work with a computational modeling study that predicts the use of the 1,3,4-oxadiazole ring system as the backbone of structurally similar compounds would also lead to target compounds with attractive densities and performance characteristics. LLM-201 and LLM-192, both insensitive energetic compounds, have crystal densities of 1.736 and 1.714 g cm^{-3}, respectively. LLM-201 has a melting point of 97 °C.

The synthesis of LLM-191 involved the condensation of 4-aminofurazan-3-carboxamidoxime (**AFCAO**, **21**) [49] with 4-aminofurazan-3-carbonitrile (**22**). This was condensed at 150 °C with **21** in a mixture of toluene, anisole, and the thermally stable hindered base catalyst, collidine, to give 3,5-bis(4-aminofurazan-3-yl)-1,2,4-oxadiazole (**23**) in 76% yield. The carboxamidoxime **21** is an important intermediate for the synthesis of many target multi-cyclic furazans, 1,2,4-oxadiazoles, and furoxans. The synthesis of **22** involved the oxidation of **AFCAO** with $Pb(OAc)_4$ in a mixture of CH_3CN and AcOH at 15–25 °C. Compound **23** was oxidized with a mixture of trifluoroacetic acid (TFA) and 70% H_2O_2 to give LLM-191. By reducing the amount of 70% H_2O_2, a mixture of LLM-191 and LLM-192 is formed, from which LLM-192 may be isolated by column chromatography (Scheme 9.4).

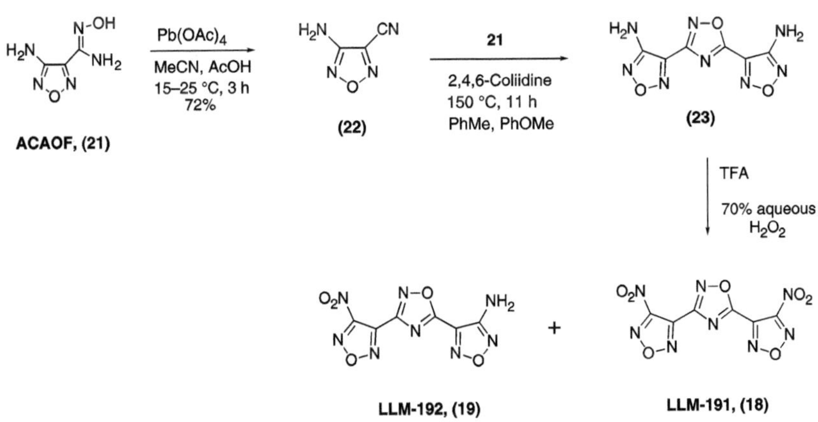

Scheme 9.4 Synthesis of **LLM-191, 18** and **LLM-192, 19**.

LLM-201 was synthesized by the condensation of **AFCAO, 21** with cyanogen bromide (BrCN) in aqueous EtOH, in the presence of $NaHCO_3$, to give 3-(4-aminofurazan-3-yl)-1,2,4-oxadiazole-5-amine [50] (**24**). Oxidation of **24** with trifluoroacetic peracid gave **20** in 63% yield (Scheme 9.5). The authors reported the amino-group on the 1,2,4-oxadiazole shows remarkable resistance to oxidation to the nitro-group under all oxidative conditions attempted thus far.

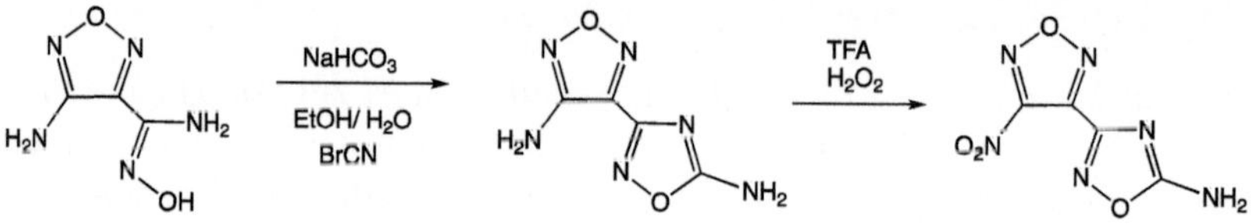

AFCAO, 21 (24) LLM-201, 19

Scheme 9.5 Synthesis of **LLM-201, 20.**

Wei et al. [25] reported the synthesis new energetic, high-nitrogen, nitramines, and their nitrogenous salts using the same intermediates used in the synthesis of LLM-191 and LLM-201. They describe the synthesis of 3,5-bis(4-nitraminofurazan-3-yl)-1,2,4-oxadiazole (**25**) and 3-(4-nitraminofurazan-3-yl)-1,2,4-oxadiazole-5-nitramine (**26**) by the nitration with fuming nitric acid at 5 °C of the amine precursors, **23** and **24**, respectively (Figure 9.4). The various nitrogenous salts of **25** and **26** that were reported all had good thermal stability (T_d = 172–296 °C) and relatively high densities (1.67–1.85 g cm^{-3}), with bis-hydroxylammonium 3,5-bis (4-nitraminofurazan-3-ylate)-1,2,4-oxadiazole (**27**) having the highest density and predicted detonation pressure (37.4 GPa), higher than RDX (35.0 GPa).

Pang et al. [51] investigated the synthesis of some interesting ionic salts derived from 3-nitro-5-amino-1,2,4-oxadiazole (**NOA**). Treatment of **NOA** with sodium dichloroisocyanurate (SDCI) gave a mixture of sodium bis(3-nitro-1,2,4-oxadiazole-5-yl)amine (**28**) and sodium bis(3-nitro-1,2,4-oxadiazole-5-yl)triaz-2-ene-1-ide (**29**) in 48% and 44% yields, respectively, which were converted to various high-nitrogen ionic salts (Scheme 9.6). This was an interesting approach that introduced the triazene functionality as an ionic species that bridges two nitrooxadiazole rings. Many of the reported compounds could not be obtained in pure crystalline form, but instead as complexes with the solvent of crystallization. Also, the triazene

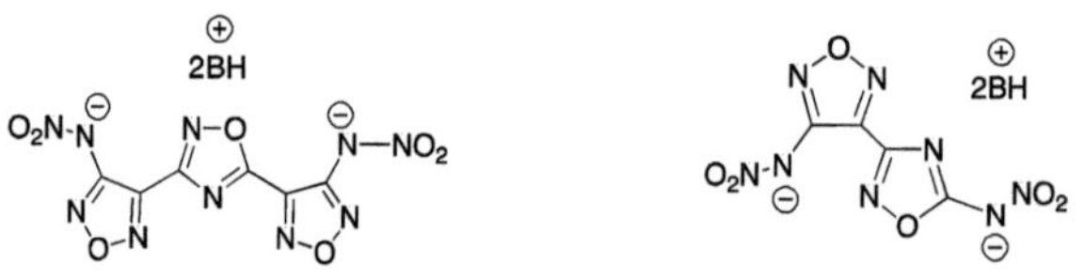

(25) (26)

B = NH_2NH_2, NH_3

B = NH_2OH (**27**), NH_2–(C=NH)–NH_2, NH_2–(C=NH)–$NHNH_2$, NH_2NH–(C=NH)-$NHNH_2$, NH_2NH–(C=NNH_2)–$NHNH_2$

Figure 9.4 High-nitrogen salts of 3,5-bis(4-nitraminofurazan-3-yl)-1,2,4-oxadiazole (**26**) and 3-(4-nitro-furazan-3-yl)-1,2,4-oxadiazole-5-nitramine (**27**).

derivatives had a much lower density than the amine-linked counterparts. Guanidinium bis(3-nitro-1,2,4-oxadiazole-5-yl)-amine (**30**) had the reasonably high density of 1.799 g cm^{-3}, as measured by gas pycnometry.

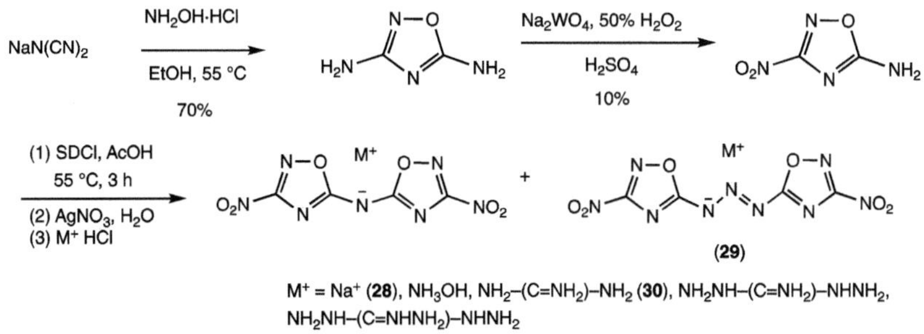

Scheme 9.6 High-nitrogen salts derived from sodium bis(3-nitro-1,2,4-oxadiazole-5-yl)amine (**28**) and sodium 1,3-bis(3-nitro-1,2,4-oxadiazole-5-yl)triaz-2-ene (**29**).

The authors recognized that the addition of acidic groups to an oxadiazole-based precursor compound by the conversion of conventional explosophoric groups or non-energetic precursors to acidic groups such as hydroxyls, dinitromethanide, nitramines, or triazenes allows the selection of various combinations of cations and anions to modify and optimize the detonation and safety properties and greatly expand the number of available energetic compounds. This approach was used extensively in the synthesis of azole-based energetic salts and has become a common method to generate numerous new energetic compounds from common intermediates [5]. Generally, the common nitrogenous salts or nitrogenous bases, NH_4Cl (or gaseous NH_3), hydrazine, guanidinium bicarbonate (Gu), aminoguanidinium bicarbonate (AG), aminonitroguanidine (ANG), diamino-urea (DAU), and 1,5-diaminotetrazole (DAT), and the HCl salts of hydroxylamine, diaminoguanidine (DAG) or triaminoguanidine (TAG) are used in metathesis reactions to synthesize the target energetic salts using this approach.

Cheng and coworkers [18] synthesized a series of energetic compounds derived from 1,2,4-oxadiazole-3-chloroxime (**31**). Compound **31** was treated with trifluoroacetic anhydride (TFAA) and 100% HNO_3 in $CHCl_3$ to give 3-(chlorodinitromethyl)-1,2,4-oxadiazole (**32**). This was reduced with KI to give potassium 1,2,4-oxadiazole-3-dinitromethylide (**33**) in 33% yield, then acidified to give 3-(dinitromethyl)-1,2,4-oxadiazole (**34**). Compound **31** was converted to 5-(1,2,4-oxadiazole-3-yl)-1-hydroxytetrazole (**35**) in a one-pot synthesis by allowing it to react with NaN_3 in aqueous EtOH to give the intermediate 1,2,4-oxadiazole-3-azidoxime (**36**), which was subsequently cyclized with aqueous HCl to give **35** in 95% yield (Scheme 9.7). Both of these compounds were converted to a series of high-nitrogen salts ($^+NH_4$, $^+NH_3NH_2$, $^+NH_3OH$) with good densities (1.69–1.85 g cm^{-3}) and T_ds from 140 to 208 °C. The hydroxylammonium salts of **33** and **35** had the highest densities of 1.84 and 1.79 g cm^{-3}, respectively. Treatment of

31 with aqueous K_2CO_3 in a biphasic mixture with ether generated the intermediate cyanogen-*N*-oxide which dimerized to give 3,4-bis(1,2,4-oxa-diazol-3-yl)furoxan (**37**) in 44% yield.

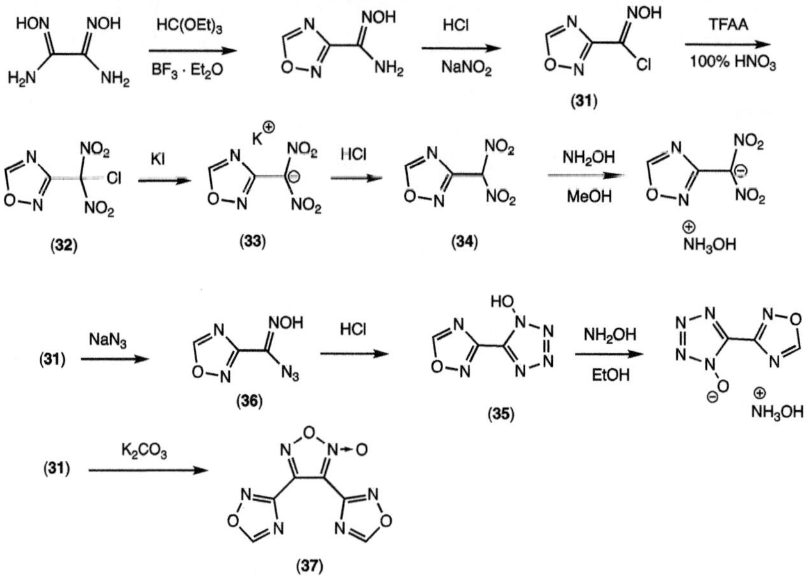

Scheme 9.7 Synthesis of new energetic compounds derived from 1,2,4-oxadiazole-3-chloroxime (**31**).

Klapotke and coworker [52, 53] synthesized several energetic materials derived from the bi-1,2,4-oxadiazole ring system. This work involved the nitration of diethyl 3,3′-bis(1,2,4-oxadiazolyl)-5,5′-diacetate (**38**) with an excess of 100% HNO_3 in concentrated H_2SO_4 to give diethyl 3,3′-bis(1,2,4-oxadiazolyl)-5,5′-bis(2,2-dinitroacetate) (**39**) [54] (Scheme 9.8). Treatment of **39** with various nitrogenous bases, such as aqueous NH_3, guanidinium bicarbonate (Gu), AG, ANG, DAU, and DAT, resulted in the decarboxylation of **39** and the formation of the bis(nitrogenous) salt of the dianion of 5,5′-dinitromethyl-3,3′-bis(1,2,4-oxadiazole) (**40**). The diammonium salt of 5,5′-dinitro-methyl-3,3′-bis(1,2,4-oxadiazole) (**41**) was converted to the bis-silver salt by treatment with $AgNO_3$, which was then treated with the HCl salts of hydroxylamine, diaminoguanidine, or triaminoguanidine, to give the bis-hydroxylammonium (HA) (**42**), bis(diaminoguanidinium) (DAG) (**43**), and bis(triaminoguanidinium) (TAG) (**44**) salts of **40**. The bis(hydroxylammonium) salt (**42**) was reported to have a density of 1.946 g cm^{-3}, a $\Delta H_f = 41.1$ KJ mol^{-1}, and a $T_d = 156\,°C$.

Compound **38** was also converted to 5,5′-bis(fluorodinitromethyl)-3,3′-bi-1,2,4-oxadiazole (**45**) and 5,5′-bis(trinitromethyl)-3,3′-bi-1,2,4-oxadiazole (**46**) as shown in Scheme 9.8. Compounds **45** and **46** are zero-hydrogen compounds with very high crystal densities – 1.960 and 1.936 g cm^{-3}, respectively.

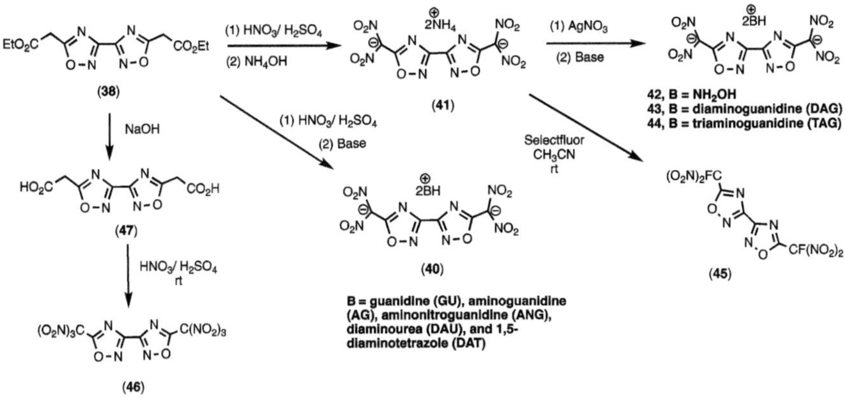

Scheme 9.8 Synthesis of bis-hydroxylammonium salt of 5,5′-dinitromethyl-3,3′-bis(1,2,4-oxadiazole) (**42**), 5,5′-bis(fluorodinitromethyl)-3,3′-bi-1,2,4-oxadiazole (**45**) and 5,5′-bis(trinitromethyl)-3,3′-bi-1,2,4-oxa-diazole (**46**).

9.4 1,3,4-Oxadiazoles

The reports of energetic compounds incorporating the 1,3,4-oxadiazole ring system are relatively rare compared to the other oxadiazoles. Currently, there is only one report of a nitro-substituted 1,3,4-oxadiazole in which the nitro-group is attached directly to the ring. Similar to the 1,2,4-oxadiazole ring system, this may be due to the fact that there is only one reported synthetic methodology for the oxidation of 2-amino- or 5-amino-1,3,4-oxadiazoles to their corresponding nitro compound. Bagal and coworkers [55] reported the conversion of 2-methyl-5-amino-1,3,4-oxadiazole (**47**) to 2-methyl-5-nitro-1,3,4-oxadiazole (**48**) by suspending **47** in 20% aqueous $NaNO_2$, cooling to −5 °C, and slowly adding 7% aqueous H_2SO_4 over a three hour period (Scheme 9.9). They reported that derivatives of 1,3,4-oxadiazoles are unstable in acidic media. Therefore, their synthetic procedure was designed to reduce prolonged contact with acid.

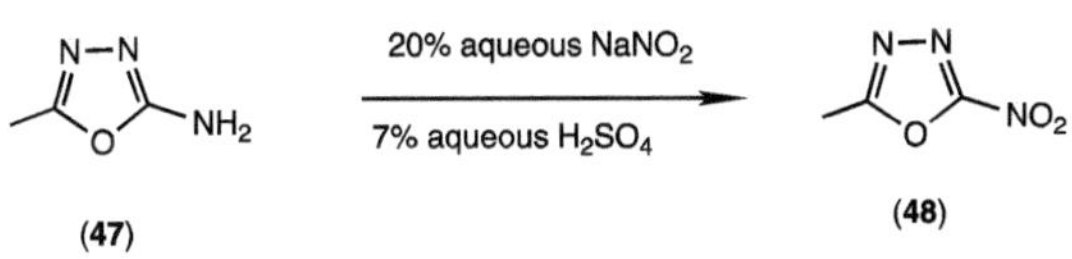

Scheme 9.9 Conversion of 2-methyl-5-amino-1,3,4-oxadiazole (**47**) to 2-methyl-5-nitro-1,3,4-oxadiazole (**48**).

Much of the early work in the study of energetic 1,3,4-oxadiazoles involved the use of the 1,3,4-oxadiazole ring system as a substrate to which the 2- and 5-positions are substituted with stable explosophoric groups. In 1977, Sitzmann and Dacons [56, 57] reported the synthesis of picryl-substituted 1,3,4-oxadiazoles as thermally stable energetic compounds. The synthesis of 2,5-dipicryl-1,3,4-oxadiazole (**DPO, 50**) in 33% yield involved the treatment of *N,N′*-bis(2,4,6-trinitrobenzoyl)hydrazine (**51**) with P_2O_5 in nitrobenzene at 115–120 °C for one hour. DPO has a crystal density of 1.77 g cm^{-3} and a melting point of 335 °C [57] (Scheme 9.10). 2,5-Bis(3-amino-2,4,6-dinitrophenyl)-1,3,4-oxadiazole (**BADPO, 52**) (melting with decomposition at 317 °C) was synthesized in a similar manner. Xiao and coworkers [58] performed a theoretical study of the vibrational and thermodynamic properties of a series of 2,5-dipicryl-1,3,4-oxadiazole derivatives containing various explosophoric groups.

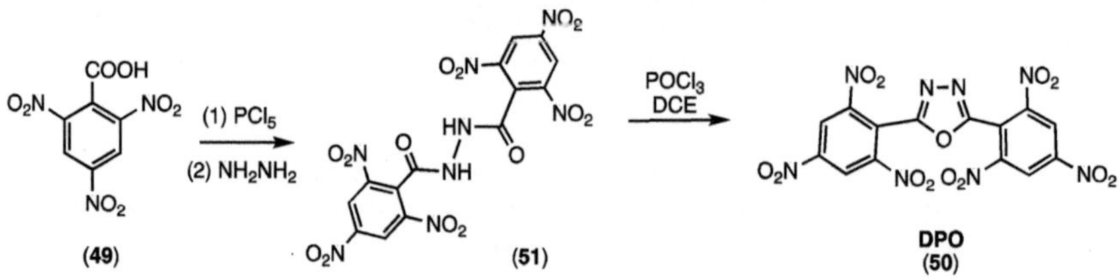

Scheme 9.10 Synthesis of 2,5-dipicryl-1,3,4-oxadiazole (**DPO, 50**).

In 2016, Klapotke and Witkowski [59] reported the synthesis of an analog of DPO, 5,5′-bis(2,4,6-trinitro-phenyl)-2,2′-bi(1,3,4-oxadiazole) (**TKX-55, 53**), a thermally stable energetic compound with attractive physical properties. The synthesis of TKX-55 involved the condensation of 2,4,6-trinitrobenzoyl chloride with oxalyldihydrazide in THF to yield (2,4,6-trinitrobenzoyl)oxalohydrazide (**54**) (Scheme 9.11). This was converted to TKX-55 in 93% yield, by treatment with 20% oleum at 25 °C. TKX has a density of 1.837 g cm^{-3} and showed melting with decomposition at 335 °C.

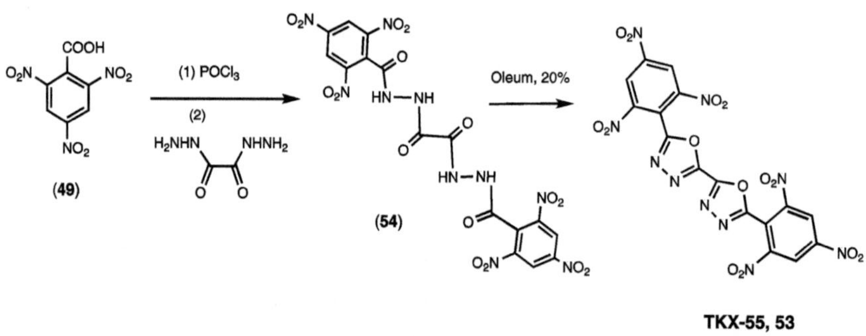

Scheme 9.11 Synthesis of 5,5′-bis(2,4,6-trinitrophenyl)-2,2′-bi(1,3,4-oxadiazole) (**TKX-55, 53**).

In recent years, most of the work on energetic 1,3,4-oxadiazoles followed the same approach as both 1,2,4-oxadiazoles and many 1,2,5-oxadiazoles. The work centered around synthesizing 1,3,4-oxadiazoles with appended acidic explosophoric groups, derived from synthetically available energetic or non-energetic precursors, followed by the formation of a series of energetic salts of the acidic explosophoric groups.

In 2015, Shreeve and coworkers [60] reported the synthesis of a series of energetic salts of 2,5-bis(dinitromethyl)-1,3,4-oxadiazole (**55**) and 2,5-bis(trinitromethyl)-1,3,4-oxadiazole (**56**). The synthetic route followed an established method to synthesize dinitromethyl groups in which diethyl 1,3,4-oxadiazole-2,5-diacetate (**57**) is treated with 100% HNO_3 and concentrated H_2SO_4 to yield diethyl 1,3,4-oxadiazole-2,5-dinitroacetate (**58**) (Scheme 9.12). Treatment of **58** with aqueous NH_3 yielded the bisammonium salt of 2,5-bis-(dinitromethyl)-1,3,4-oxadiazole (**59**). This could be converted to other nitrogenous salts by treatment with $AgNO_3$ followed by treatment with a HCl or H_2CO_3 salt of a nitrogenous base (Scheme 9.12). The bis(hydroxyl-ammonium), bis(hydrazinium), bis(guanidinium), bis(aminoguanidinium), bis(triaminoguanidinium), and 1,5-diaminotetrazolate salts were synthesized in the similar manner. These salts had T_ds between 142 and 235 °C, with the bis(guanidinium) salt showing the highest T_d. The densities of the salts ranged from 1.65 to 1.92 g cm^{-3} with the bis(hydroxylammonium) salt (**60**) having the highest density of the salts (1.89 g cm^{-3}). The 2,5-bis(trinitro-methyl) derivative (**56**) was also synthesized by nitration of **55** with 100% HNO_3 and concentrated H_2SO_4.

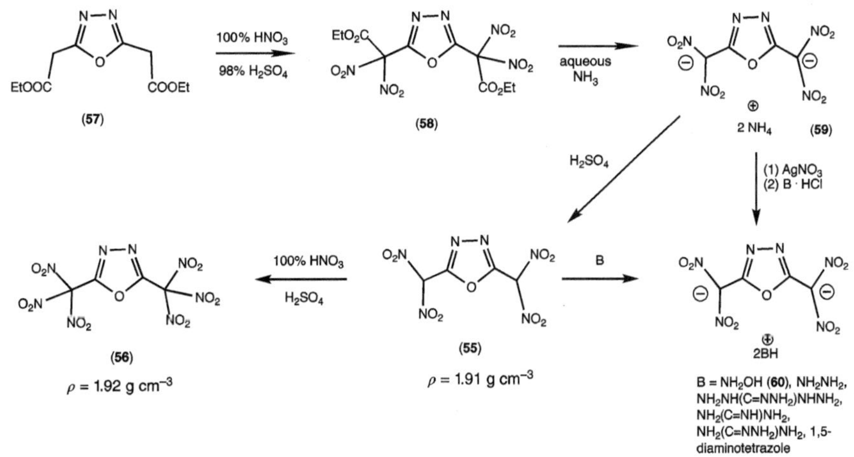

Scheme 9.12 Synthesis of 2,5-bis(trinitromethyl)-1,3,4-oxadiazole (**56**) 2,5-bis(dinitromethyl)-1,3,4-oxa-diazole (**55**) and its energetic salts.

Shreeve and coworkers [61] followed up this work with the synthesis of a series of anionic dinitromethanide energetic salts of a nitromethane-bridged 1,3,4-oxadiazole using a similar synthetic approach as described in the previous work. The synthesis involved the nitration of diethyl 2,2′-methylenebis(1,3,4-oxadiazole-5,2-diyl)-diacetate (**61**) with 100% HNO_3 and concentrated H_2SO_4 to yield diethyl 2,2′-((dinitromethylene)bis-(1,3,4-oxadiazol-5,2-diyl))-bis(2,2-dinitroacetate) (**62**) in 88% yield. Compound **62** was isolated from the reaction mixture as a solid and the reaction mixture was not quenched into water, presumably to avoid the possible hydrolysis of the bridged dinitromethane moiety to the corresponding ketone. Compound **62** was treated with aqueous ammonia in EtOH to yield triammonium

2,2′-((nitromethaneidyl)bis(1,3,4-oxadiazol-5,2-diyl))-bis(2,2-dinitromethanide)(**63**), which was converted to the trisilver salt by treatment with $AgNO_3$ (Scheme 9.13). The trisilver salt was converted to a series of nitrogenous salts by treatment with the corresponding nitrogenous salt hydrochloride, such as tris (hydroxylammonium) (**64a**), tri(hydrazinium) (**64b**), and tris(triaminoguanidine)) (**64c**). The tris(hydroxylammonium) salt (**64a**) had a density of 1.87 g cm^{-3} (gas pycnometry) and a T_d of 180 °C.

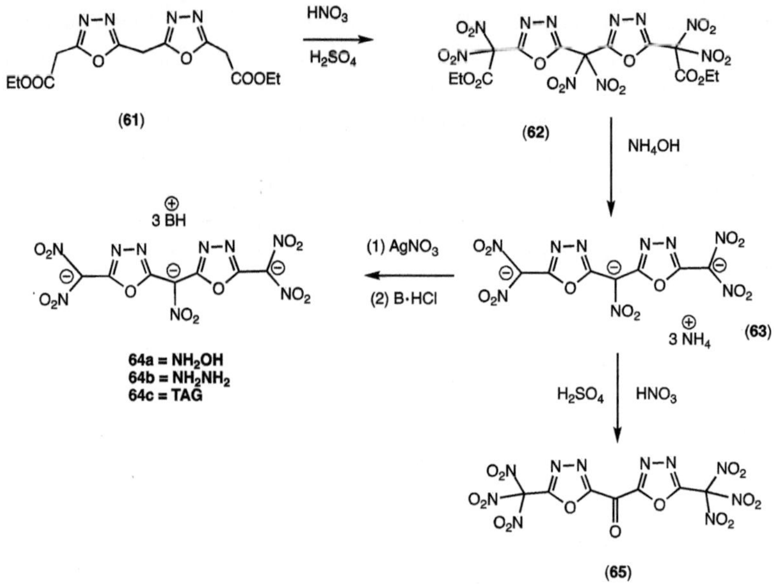

Scheme 9.13 Synthesis of nitromethane-bridged bis(1,3,4-oxadiazoles).

Compound **63** was also converted to bis(trinitromethyl)-1,3,4-oxadiazol-2-yl)methanone (**65**) by nitration with 100% HNO_3 and concentrated H_2SO_4 in CH_2Cl_2, yielding the intermediate dinitro-bis(5-(trinitro-methyl)-1,3,4-oxadiazol-2-yl)methane, which was hydrolyzed to **65** under the reaction conditions.

Shreeve and coworkers [62] synthesized a series of salts of 3-nitramino-4-(5-amino-1,3,4-oxadiazol-2-yl)-furazan (**67**) with good thermal stability but relatively low oxygen balance. The synthesis involved allowing 4-aminofurazan-3-carbohydrazide to react with cyanogen bromide to yield the parent 3-amino-4-(5-amino-1,3,4-oxadiazol-2-yl)furazan (**68**), which was nitrated with 100% HNO_3 to give **67** (Scheme 9.14). A series of nitrogenous bases (**69**), including ammonium, hydrazinium, hydroxylammonium, and 3,5-diamino-1,2,4-triazolate of **67**, were synthesized by simple metathesis reactions. The densities ranged from 1.59 to 1.81 g cm^{-3} with the 4-amino-1,2,4-triazole having the highest density. The hydroxylammonium and hydrazinium salts were isolated as mono-hydrates.

Lu and coworkers [63] synthesized azo-1,3,4-oxadiazoles appended to nitro- and nitramino-furazans to yield energetic compounds with good densities and thermal stability. The synthesis involved the oxidation of **68** with a mixture of 50% H_2O_2,

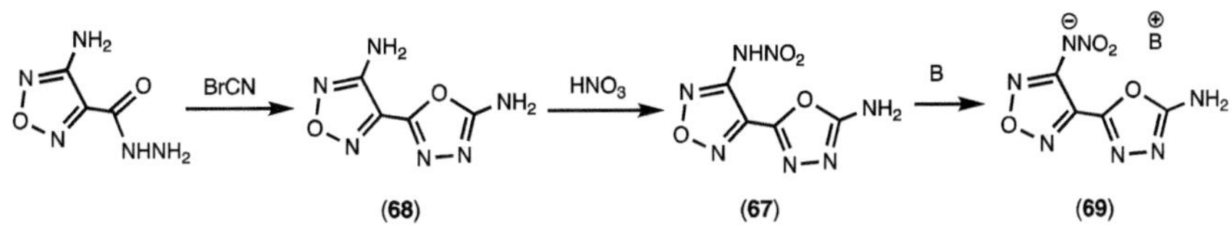

Scheme 9.14 Synthesis of a series of salts of 3-nitramino-4-(5-amino-1,3,4-oxadiazol-2-yl) furazan (**67**).

H_2SO_4, and Na_2WO_4 to yield 2-(4-nitro-1,2,5-oxadiazol-3-yl)-5-amino-1,3,4-oxadiazole (**70**), followed by oxidation with $KMnO_4$, to yield azo-2,2′-bis-1,3,4-oxadiazole-5,5′-(4-nitro-1,2,5-oxadiazol-3-yl) (**71**). Compound **67** was converted to azo-2,2′-bis-1,3,4-oxadiazole-5,5′-(4-nitramino-1,2,5-oxadiazol-3-yl) (**72**) by oxidation with $KMnO_4$ in aqueous HCl (Scheme 9.15). They have densities of 1.767 and 1.825 g cm^{-3}, respectively. They also have better thermal stability than their starting energetic compounds, **67** and **70**, possibly because of the increase in molecular weight and better crystal packing.

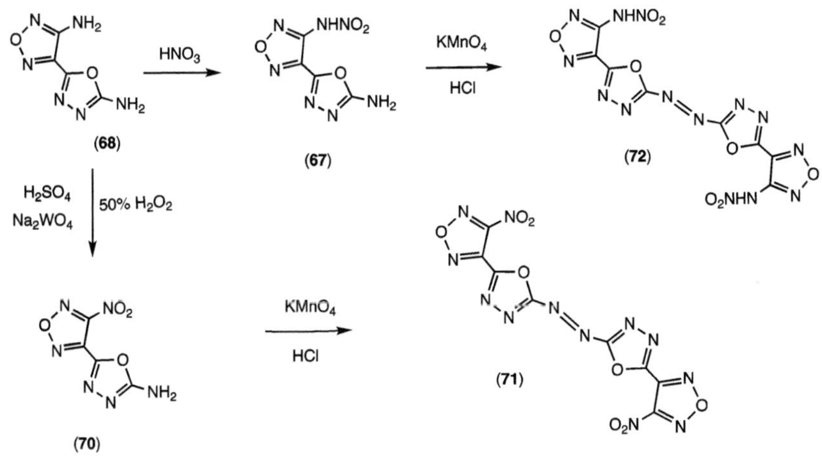

Scheme 9.15 Synthesis of azo-1,3,4-oxadiazoles attached to energetic furazans.

Klapotke and coworkers [64] reported the synthesis of energetic salts of 2,2′-dinitramino-5,5′-bi(1-oxa-3,4-diazole) (**73**), materials with high densities and good thermal stability. The synthesis involved the condensation of oxalylhydrazide with BrCN to yield 2,2′-diamino-5,5′-bi(1,3,4-oxadiazole) (**74**), which was nitrated with 100% HNO_3 to yield 2,2′ dinitramino-5,5′-bi(1-oxa-3,4-diazole) (**73**) (Scheme 9.16). Compound **73** was treated with a series of nitrogenous bases to yield the corresponding dibasic salt, including the diammonium, dihydroxylammonium, and diguandino salts. The diammonium (**75**) and bis(hydroxylammonium) (**76**) salts crystallized as dihydrates but still had impressive densities of 1.95 and 1.92 g cm^{-3}.

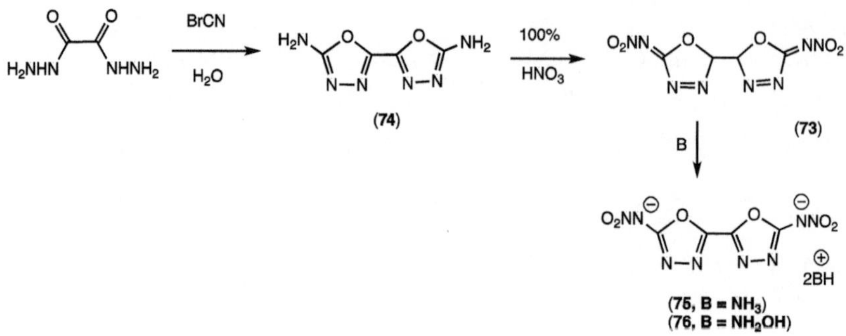

Scheme 9.16 Synthesis of energetic salts of 2,2′-dinitramino-5,5′-bi(1-oxa-3,4-diazole) (**73**).

9.5 Furazans (1,2,5-Oxadiazole) and Furoxans (1,2,5-Oxadiazole-2-Oxides)

The furazan [65] and furoxan ring systems have received the most attention of the various oxadiazoles for the synthesis of energetic compounds in the last 40 years [19–22, 30, 66]. Sheremetev [20] reviewed the early work on the synthesis and reactivity of 3,4-diaminofurazan (**DAF**, **77**), an important and readily accessible starting material for many furazan-based energetic compounds. The early investigations explored the oxidation of the amino-groups to nitro-groups and azo- and azoxy-dimers that were subsequently transformed into energetic target compounds. In addition, the reactivity of 3,4-dinitrofurazan to nucleophiles with subsequent displacement of a nitro-group was reported. A short summary with a description of some of the significant early work will be presented, but generally this chapter will concentrate on recent work which incorporated furazans and furoxans into energetic multicyclic compounds.

Coburn in 1968 [67] pioneered the early work on furazans when he reported the first synthesis of DAF. There have since been many reported syntheses of DAF, all of which proceed through a common intermediate, diaminoglyoxime (**DAGO**, **78**) [68–71]. This work also introduced the use of mixtures of TFAA (or TFA) and H_2O_2 to produce trifluoroacetic peracid (CF_3CO_3H) [72] as an effective oxidant to convert aminofurazans to nitrofurazans in generally good yields and purity. Coburn reported the synthesis of 4,4′-dinitro-3,3′-bifurazan (**DNBF**, **79**) as a zero-hydrogen energetic compound. The synthesis of DNBF involved the synthesis of dicyanoglyoxime (**80**) by allowing aqueous KCN to react with dichlorogly-oxime (**81**) followed by allowing **80** to react with aqueous hydroxylamine (NH_2OH) at 60 °C to give 1,4-diamino-1,2,3,4-tetraoximobutane (**82**). Compound **82** was cyclized in aqueous NaOH at 135 °C to yield 4,4′-diamino-3,3′-bifurazan (**DABF**, **83**). **DABF** was oxidized with 90% H_2O_2 and TFAA in CH_2Cl_2 to give **DNBF** in 35% yield (Scheme 9.17). **DNBF** has a melting point of 85 °C and a density of 1.85 g cm^{-3}.

Coburn also introduced the concept of picrylamino-substituted oxadiazoles, as potential high-thermal stability energetic materials. He reported the synthesis

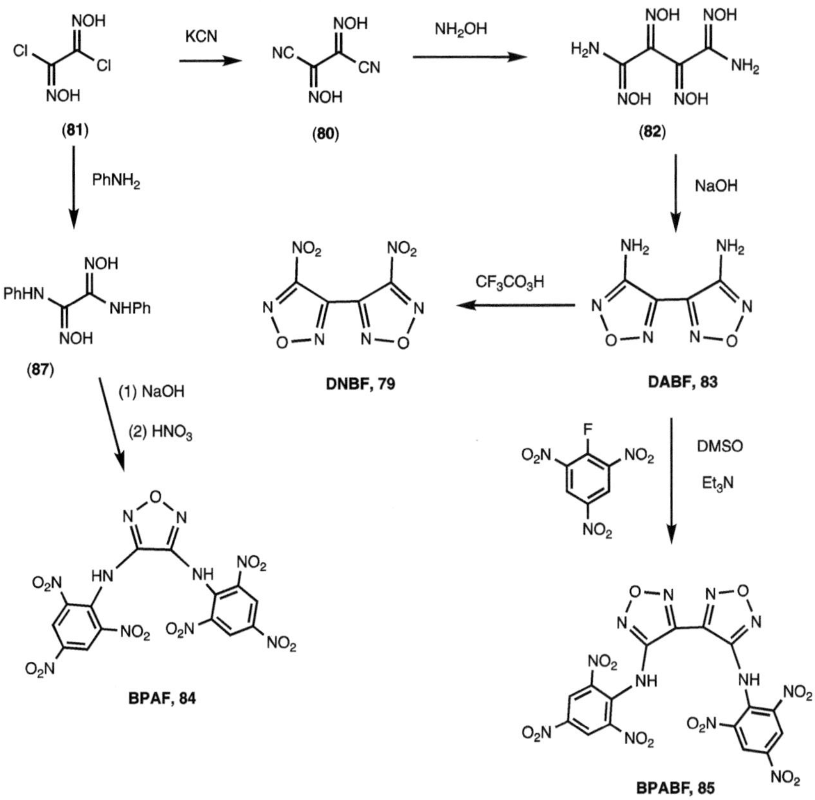

Scheme 9.17 Synthesis of **DNBF** (**79**) and picrylamino-substituted-1,2,4-oxadiazoles, **BPAF** (**84**), **BPAPF** (**85**).

of 3,4-bis(picrylamino)furazan (**BPAF, 84**) (with a melting point of 250 °C), and 3,4-bis(picrylamino)bifurazan (**BPABF, 85**) (with a melting point of 315 °C), highly thermally stable energetic compounds with good densities. **BPAF** was synthesized by allowing dichloro-oglyoxime (**81**) to react with aniline to yield dianilinoglyoxime (**87**), which was cyclized by heating with NaOH in ethylene glycol to yield 3,4-bis(anilino)furazan (**88**). Nitration with concentrated HNO_3 at 25 °C gave **BPAF** in 86% yield. **BPABF** was synthesized by treatment of **83** with picryl fluoride (**89**) and Et_3N in DMSO. The treatment of **DAF** with 1 equiv of picryl fluoride yielded 3-amino-4-(picrylamino)furazan (**90**), which was oxidized with TFA/90% H_2O_2 to yield 3-nitro-4-(picrylamino)furazan (**91**).

In 1996, Sheremetev and Mantseva [73] reported an improved synthesis of **DABF** (14% overall yield starting from CH_3NO_2), by allowing 3,4-bis(hydroxyiminomethyl)furoxan (**92**) to react with hydroxylamine. Epishina et al. [74] reported the synthesis of **DABF** and 3-amino-4-(4-aminofurazan-3-yl)furoxan (**93**) through the common intermediate, 4-aminofurazan-3-carboxamidoxime (**AFCAO, 21**) (Scheme 9.18). The stepwise synthesis of **93** involved reaction of **87** with $NaNO_2$ in aqueous HCl, to yield 4-amino-3-(chloroximino)-furazan (**94**), which was allowed to react with KCN to yield the nitrile **95**. This was followed by reaction of **95** with NH_2OH to yield **96**. Treatment of **96** with Br_2 in 37% HCl or $K_3Fe(CN)_6$ gave **93**, while treatment of **96** with refluxing aqueous NaOH gave **DABF** in 56% yield. Heating **93** in refluxing dioxane converted it to the more thermodynamically stable isomer, 4-amino-3-(4-aminofurazan-3-yl)furoxan (**97**).

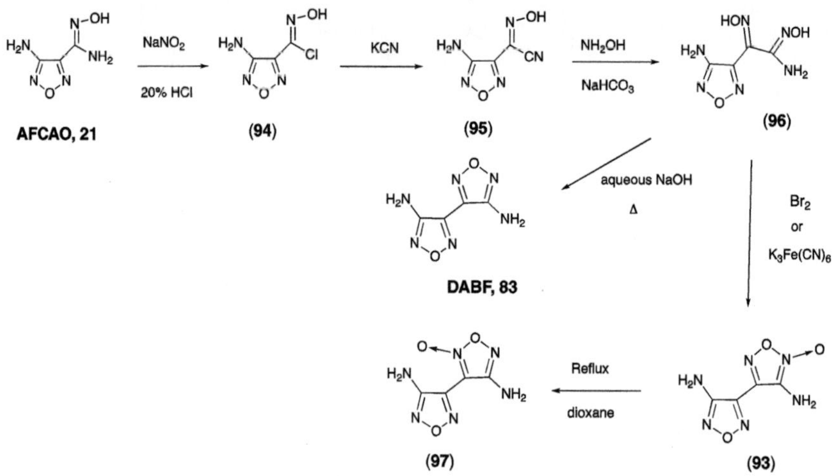

Scheme 9.18 Synthesis of **DABF** and 3-amino-4-(4-aminofurazan-3-yl)furoxan (**93**) from **AFCAO 21**.

In 1981, Solodyuk et al. [75] first reported the synthesis of a series of energetic compounds obtained by the oxidation of **DAF** with H_2O_2 under various reaction conditions, namely 3-amino-4-nitro-furazan (**ANF, 98**), 4,4′-diamino-3,3′-azoxyfurazan (**DAAF, 99**), and 4,4′-diamino-3,3′-azofurazan (**DAAzF, 100**) (Figure 9.5). **DAAF** has a crystal density of 1.747 g cm^{-3}, ΔH_f of 106 kcal mol^{-1}, and a good thermostability. This compound received a considerable interest as an insensitive energetic material. Chavez et al. [76] scaled up the synthesis of **DAAF** and characterized this

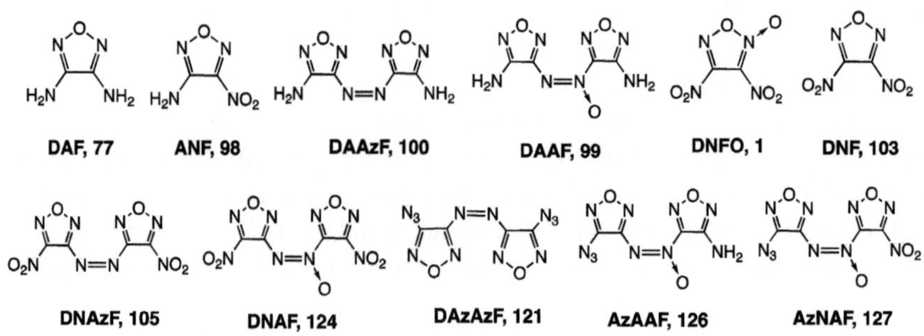

Figure 9.5 Various energetic compounds derived from **DAF**.

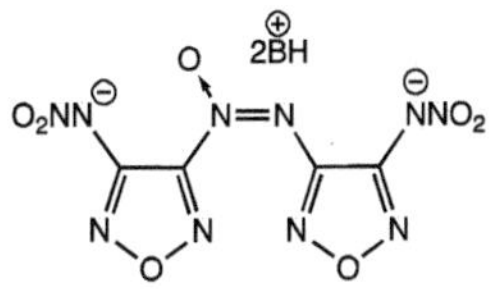

(101)

BH = ammonium, hydrazinium, hydroxylammonium, guanidinium, aminoguanidinium, triaminoguanidinium and guanazyl

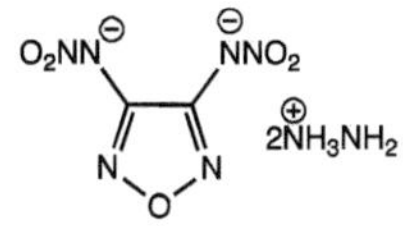

(102)

BH = ammonium, hydrazinium, hydroxylammonium, guanidinium, aminoguanidinium, triaminoguanidinium, 3-amino-1,2,4-triazolate, 4-amino-1,2,4-triazolate

Figure 9.6 Nitramino derivatives of **DAF** and **DAAF**.

material with respect to its explosive and safety properties. In 2009, Chavez and Francois reported an improved synthesis of **DAAF** [77] by the treatment of **DAF** with aqueous Oxone.

Recently, Zhang and Shreeve [78] reported the synthesis of several bis-high-nitrogen salts of 3,3′-dinitramino-4,4′-azoxyfurazan (**101**), including salts of ammonium, hydrazinium, hydroxylammonium, guanidinium, aminoguanidinium, triaminoguanidinium, and guanazyl. Compound **101** was also synthesized by the nitration of **DAAF** with 100% HNO_3 at 0 °C in 86% yield (Figure 9.6). The densities varied in the range of 1.76–1.96 g cm^{-3} and thermal decomposition temperatures (T_ds) (DSC peak) were in the range of 90–246 °C, with the guanidinium salt having the highest T_d. The authors also reported an improved synthesis of **DAF**, **77** that proceeds at lower temperatures (125 °C) to affect the dehydration of diaminoglyoxime (**DAGO 78**) to **DAF**, by introducing urea into the reaction mixture to act as the dehydrating agent. Subsequently, they reported the synthesis of 3,4-di(nitramino)furazan (**102**) and various bis-high-nitrogen salts including ammonium, hydrazinium, hydroxylammonium, guanidinium, aminoguanidinium, triaminoguanidinium, 3-amino-1,2,4-triazolate, 4-amino-1,2,4-triazolate, using very similar synthetic methods as described above [79]. The authors were able to crystallize the diamino- and bis(hydrazinium) salts, as the tri-hydrate and penta-hydrate, respectively, and obtain their crystal structures by X-ray crystallography. The densities ranged 1.645–1.873 g cm^{-3} and T_ds were in the range of 99–282 °C, with the bis(guanidinium) salt having the highest T_d.

In 1994, Novikova et al. [80] found the oxidation of various aminofurazans to the corresponding nitrofurazans could be correlated with the ionization potential (I_0) of the aminofurazans and their protonated forms. They were able to determine the active oxygen content of various oxidizing mixtures and correlate them with the I_0 of each target molecule, determining the best oxidant for each target compound that was proposed. The mixture with the highest oxidizing power was a mixture of 93% H_2O_2, H_2SO_4, and Na_2WO_4 (2 : 1 : 0.05). The methods described in this paper have been used extensively in later work for the synthesis of nitrofurazans and furoxans from their corresponding amino-substituted compounds. The results from this theoretical study guided the authors to the synthesis of the previously unknown 3,4-dinitrofurazan (**DNF, 103**), by the oxidation of **DAF**, along with several other nitrofurazans. **DNF** has a density of 1.62 g cm^{-3}, a melting point of 15 °C, and a boiling point of 168 °C.

In 1998, Zelenin and Trudell [81] reported the synthesis of 4-amino-4′-nitro-3,3′-azoxybisfurazan (**104**) by the oxidation of **DAAzF** in 25% yield, using a mixture 30% H_2O_2, Na_2WO_4, $(NH_4)_2S_2O_8$, and concentrated H_2SO_4 at 50 °C. They also reported the synthesis of 3-amino-4-nitrofurazan (**ANF, 98**) and 4,4′-dinitro-3,3′-azo-furazan (**DNAzF, 105**). **ANF** was synthesized by the oxidation of **DAF** with a mixture 30% H_2O_2, Na_2WO_4, $(NH_4)_2S_2O_8$ and concentrated H_2SO_4 at 30 °C in 35% yield. **ANF** was converted to **DNAzF** in 60% yield by treatment with $KMnO_4$ at 55 °C for two hours. **DNAzF** is a low-melting solid, with a density of 1.742 g cm^{-3}.

Sheremetev and coworkers realized that **DNF** is a useful starting material for the synthesis of various 3-*R*-4-nitrofurazans by the nucleophilic displacement of one of the nitro-groups. They performed a series of studies on the reactivity of **DNF** to nucleophiles, including sulfur nucleophiles [82], amines [83], alcoholates [84], and hydroxide [84–86]. The researchers found that allowing **DNF** (or other nitrofurazans, containing a second electron-withdrawing group) to react with Na_2CO_3 in CH_3CN at 60 °C leads to the formation of symmetrical difurazanyl ethers (Scheme 9.19). Interestingly, it was found that the conversion of **DNF** to 3-nitrofurazanyl ether (**106**) did not involve the intermediate, 3-nitro-4-hydroxyfurazan (**107**), but instead was a result of the intramolecular rearrangement of a nitro-group of **DNF** to the nitrite ester, followed by an intermolecular reaction with another molecule of **DNF** and loss of N_2O_3. 3-Cyano-4-nitrofurazan (**108**) and 3-acetyl-4-nitrofurazan (**109**) were also converted into their symmetrical ethers by treatment under the same reaction conditions. 3-Nitrofurazanyl ether (**106**) has a very attractive density of 1.898 g cm^{-3}. When **DNF** was treated with aqueous base in CH_3CN, 3-hydroxy-4-nitrofurazan (**107**) was found to be the predominant product.

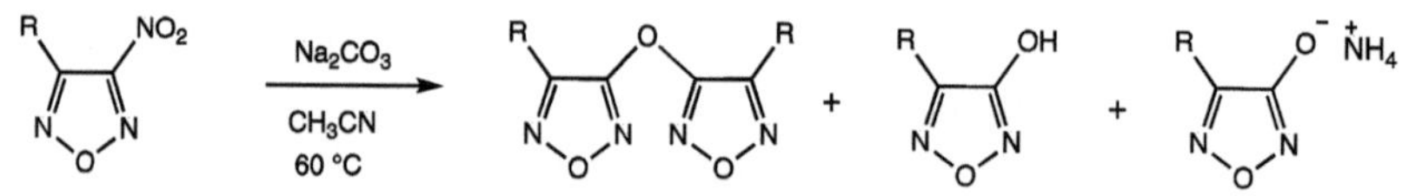

103 R = NO2
108 R = CN
109 R = Ac

106 R = NO2
R = CN
R = Ac

107 R = NO2

110 R = NO2

Scheme 9.19 Synthesis of dinitrofurazanyl ethers from **DNF**.

DNF was also found to react with aqueous ammonia, leading to a mixture of **ANF**, 3-hydroxy-4-nitro-furazan **107** and ammonium 4-nitrofurazan-3-hydroxide (**110**), while treatment with anhydrous NH_3 in $CHCl_3$ or CCl_4 yields exclusively **ANF** in 85–90% yield. When difuranzanyl ether **106** was treated with NH_3 in $CHCl_3$ or CH_2Cl_2, the reaction occurs at the ether linkage and the nitro-group is not displaced, yielding **ANF 85** and compound **110**, as products. The researchers also reported the synthesis of a series of dinitrofurazans linked by various bridging moieties, such as oxy, azo, and azoxy linkages, using precursor molecules derived from **DNF** and **DNAzF** (**105**) [86]. For example, the trifurazan derivative (**111**) was synthesized by allowing **DNF** to react with disodium salt of 3,4-dihydroxyfurazan (**112**) in glyme at 30 °C, forming along with the by-products, compound **106** and a tricyclic product – [1,4]dioxino[2,3-*c*:5,6-*c*′]bis([1,2,5]oxadiazole) (**113**) (Scheme 9.20). In a similar manner, the tricyclic compound, **114**, was synthesized by allowing (*E*)-4-((4-nitro-1,2,5-oxadiazol-3-yl)diazenyl)-1,2,5-oxadiazol-3-ol (**115**), synthesized from **DNAzF**, to react with **DNF** in glyme. The azo compound **114** was oxidized with $(NH_4)_2S_2O_8$ in fuming H_2SO_4 to yield mixtures of the azoxy derivatives **116** and **117**. The trifurazanylazo compound, 3,4-bis((*Z*)-(4-nitro-1,2,5-oxadiazol-3-yl)diazenyl)-1,2,5-oxadiazole (**118**), was synthesized from **ANF** and the mono-amine derivative of **DNAzF**, (*E*)-4-((4-nitro-1,2,5-oxadiazol-3-yl)diazenyl)-1,2,5-oxadiazol-3-amine (**119**), by oxidation with $KMnO_4$ in aqueous HCl (Scheme 9.20).

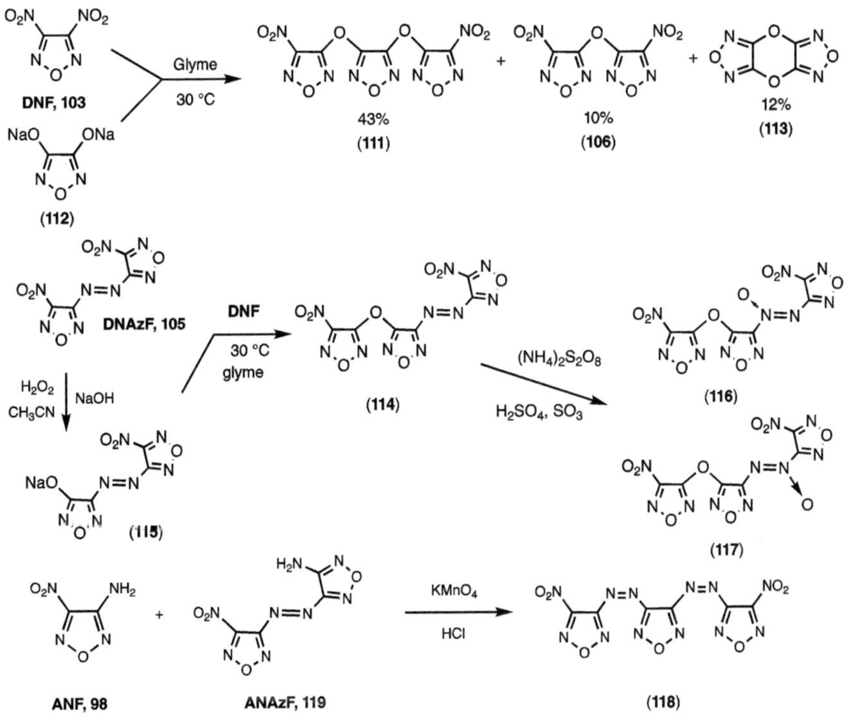

Scheme 9.20 Conversion of **DNF** to multicyclic linear nitrofurazans.

Gunasekaran and Boyer [87] synthesized a highly energetic liquid (with boiling point of 160–165 °C), 5-(4-nitro-(1,2,5)oxadiazolyl]-5*H*-[1,2,3]triazolo[4,5-*c*][1,2,5] oxadiazole (**NOTO**, **120**). The synthesis of **120** involved treatment of **DAAzF** with $NaNO_2$ in concentrated H_2SO_4 and AcOH, followed by NaN_3, to yield the diazide (**121**). Heating the diazide **121** in CH_3CN yielded 5-(4-azido-(1,2,5)oxadiazolyl]-5*H*-[1,2,3]triazolo-[4,5-*c*][1,2,5]oxadiazole (**122**) [88]. The latter compound was reduced to the amine with $SnCl_2$ in MeOH, to give 5-(4-amino-(1,2,5)oxadiazolyl]-5*H*-[1,2,3]triazolo[4,5-*c*][1,2,5]oxadiazole (**123**), which was oxidized with a mixture of 30% H_2O_2, $(NH_4)_2S_2O_8$, and concentrated H_2SO_4 at 35 °C to give **NOTO** (Scheme 9.21). Gunasekaran and Boyer [87] and Khmel'nitskii and coworkers

[88] reported the one-step synthesis of **123** from **DAAF** either by oxidation with iodobenzenediacetate in acetone, $Pb(OAc)_4$ in chlorobenzene at 135–140 °C, or with $SOCl_2$ at reflux, albeit in low yields (25% and 8%, respectively). Gunasekaran and Boyer [87] also reported several energetic compounds derived from **DAAF** and **DAAzF**, including 4,4′-dinitro-3,3′-azoxy-furazan (**DNAF**, **124**), 5,5′-[azobis(1,2,5-oxadiazol-4,3-diyl]-bis[*1H*-[1,2,3]triazolo[4,5-*c*][1,2,5]-oxa-diazolium] (inner salt) (**125**) and 4-aminofurazan-2-*ONN*-azoxyazidofurazan (**126**). Wu et al. [89] described the synthesis of 4-nitrofurazan-2-*ONN*-azoxyazidofurazan (**127**) by the oxidation of **126** with 30% H_2O_2 and Na_2WO_4 in $MeSO_3H$ at 55 °C in 43% yield.

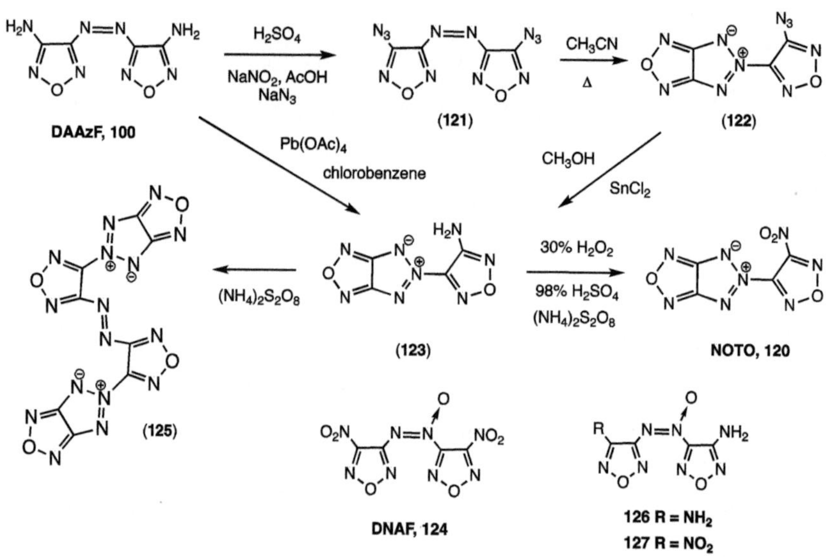

Scheme 9.21 Synthesis of **NOTO, 120** and 5,5′-[azobis(1,2,5-oxadiazol-4,3-diyl]-bis[*1H*-[1,2,3]triazolo[4,5-*c*][1,2,5]-oxadiazolium]bis (inner salt) (**125**). 4,4′-Dinitro-3,3′-azoxyfurazan (**DNAF, 124**), 4-aminofurazan-2-*ONN*-azoxyazidofurazan (**126**) and 4-nitrofurazan-2-*ONN*-azoxyazidofurazan (**127**) also depicted.

In the early 1990s, Khmel'nitskii and coworkers [90, 91] reported the synthesis of 3,4-dinitrofuroxan (**DNFO, 1**), a highly oxidized, fully nitrated heterocycle, by the nitration of glyoxime followed by cyclization of the resulting dinitroglyoxime (**128**) by treatment with N_2O_4. They also reported the synthesis of **DNFO** by three different synthetic methods, all of which proceed through the common intermediate, nitroformonitrile oxide (**NFNO 4**) [92]. The three methods involve dehydration of dinitromethane with oleum, nitration of acetylmethane-nitrolic acid, and the nitration of 1-nitro-2-methyl-prop-1-ene. **DNFO** has proven useful as a synthon for other nitro-substituted furoxans, as it readily reacts with nucleophiles to give the corresponding substituted 3-nitrofuroxans. **DNFO** was found to be more reactive to nucleophiles than **DNF**, but the 4-position is usually the reactive position. They reported the synthesis of 4-amino-, 4-alkylamino-, 4-azido, and 4-methoxy-3-nitro-furoxans from **DFNO** [91].

DNFO underwent an interesting conversion to a series of 2-alkyl-4-alkylamino-5-nitro-1,2,3-triazole-1-oxides (**129**) by treatment of **DNFO** with alkylamines to give the intermediate 3-alkylamino-4-nitrofuroxans (**130**) [82, 93]. This was followed by treatment of **130** with primary aliphatic amines, which added to electropositive nitrogen of the *N*-oxide, resulting in a ring opening of the furoxan moiety followed by a ring closure with loss of water to give **129** (Scheme 9.22).

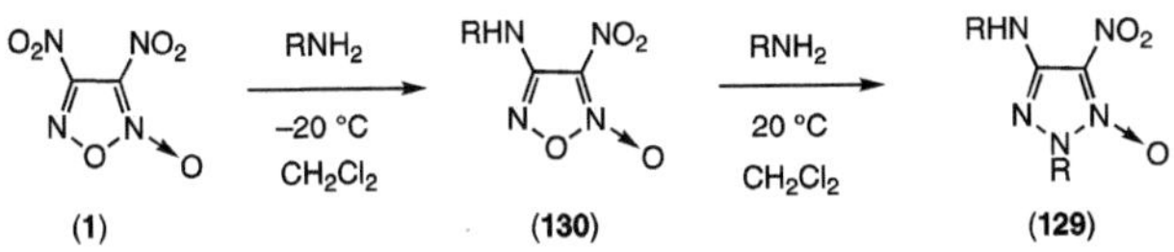

Scheme 9.22 Conversion of **DNFO, 1** to 2-alkyl-4-alkylamino-5-nitro-1,2,3-triazole-1-oxides (**129**).

Klapotke and coworkers [94] reported the synthesis of the previously unknown dinitrobifuroxan structures, 4,4′-diamino-3,3′-bifuroxan (**131**) and 4,4′-dinitro-3,3′-bifuroxan (**DNBFO, 133**), by the treatment of 1,4-diamino-1,2,3,4-tetraoximobutane (**82**) with Br_2 in HCl to yield 3,3′-diamino-4,4′-bisfuroxan (**133**), which was converted to the more thermodynamically stable 4,4′-diamino-3,3′-bisfuroxan (**131**) by heating in dioxane. Oxidation with 50% H_2O_2 and $(NH_4)_2S_2O_8$ in concentrated H_2SO_4 at 0 °C gave **DNBFO** in very low yield. **DNBFO** has a very high density of 1.97 g cm^{-3} and an estimated T_d of 140 °C (Scheme 9.23).

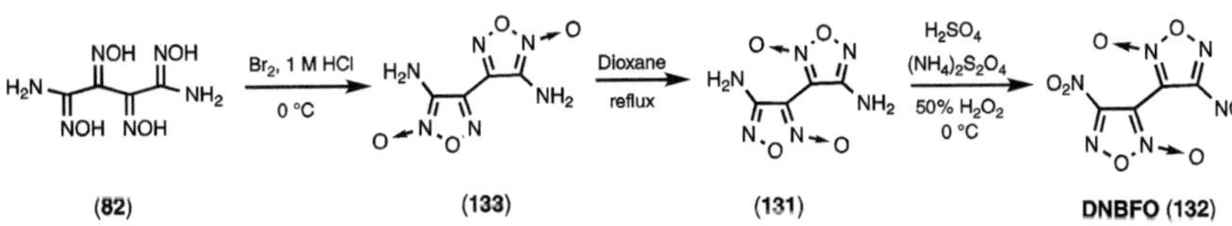

Scheme 9.23 Synthesis of **DNBFO** (**132**) from **82**.

Fershtat et al. [95], building off their considerable experience in the synthesis and reactivity of nitro-furoxans, investigated the synthesis of combinations of nitrofurazans and nitrofuroxans connected *via* C—C bonds, in a similar manner to **DNBFO**. Their synthetic approach also allowed them to explore the possibility of 4-nitrofuroxans undergoing a thermally induced rearrangement to the 3-nitrofuroxan isomer. The synthesis involved the oxidation of 4-amino-3-chloroximinofurazan (**94**) with 85% H_2O_2 in concentrated H_2SO_4 and $CHCl_3$ to 4-nitrofurazancarboximidoyl chloride (**134**) in 48% yield (Scheme 9.24). The latter product was treated with sodium dinitromethylide (**NaDNM**) in DMF at 0 °C, followed by reaction with NaOAc, AcOH, and $NaNO_2$, in a one-pot process to give 3-nitro-4-(4-nitrofurazan-3-yl)-furoxan (**135**) in 25% yield. The mechanism of this reaction involves first, displacement of the chloride group with dinitromethane followed by nitrosation of the dinitromethylide moiety to the nitrosodinitromethane group, which undergoes cyclization to a furoxan and loss of a nitrite ion. The 3-nitro isomer **135** isomerizes into the more thermodynamically stable 4-nitro-3-(4-nitrofurazan-3-yl)-furoxan (**136**) by heating in CCl_4. The structures were confirmed by X-ray crystallography and characterized with ^1H, ^{13}C, and ^{14}N NMR spectroscopies.

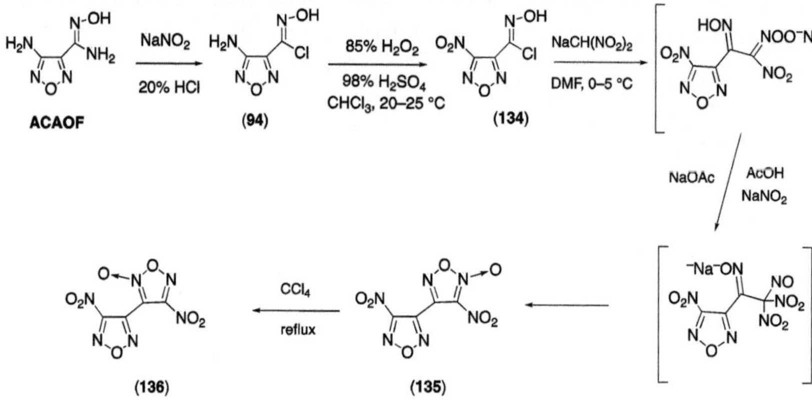

Scheme 9.24 Synthesis of 3-nitro-4-(4-nitrofurazan-3-yl)furoxan (**135**) and 4-nitro-3-(4-nitrofurazan-3-yl)-furoxan (**136**).

Earlier, Makhova and coworkers [74] reported the synthesis of 3-amino-4-(4-aminofurazan-3-yl)furoxan (**137**) and 4,4′-diamino-3,3′-bifurazan (**DABF, 83**) through the common intermediate, 1-amino-2-(4-aminofurazan-3-yl)gloxime (**138**) (Scheme 9.25). The stepwise synthesis of **138** involved allowing **94** to react with KCN to yield the intermediate nitrile (**139**), followed by reaction with NH_2OH to yield **138**. Treatment of **138** with Br_2 in 37% HCl or $K_3Fe(CN)_6$ gave **137**, while treatment of **138** with refluxing aqueous NaOH gave **DABF** in 56% yield. Compound **137** was heated in refluxing dioxane to give the more thermodynamically stable, 4-amino-3-(4-aminofurazan-3-yl)furoxan (**140**). Oxidation of **140** or **DABF** with DBI or $KMnO_4$ yielded the corresponding azo-dimer (**141**) (Scheme 9.25).

In 2017, Fershtat et al. [96] developed a general synthetic method to synthesize previously unknown bifuroxanyl structures comprising the 3-nitrofuroxanyl moiety. The approach was similar to their earlier work shown above. The synthesis involved allowing 4-*R*-(furoxano-2)-3-carboximidoyl chlorides (**142**) to react with **NaDNM** followed by treatment with $NaNO_2$ in an AcOK/AcOH buffer, to yield the corresponding 3-nitro-4-(4-*R*-furoxan-3-yl)furoxans (**143**). The carboximidoyl chlorides were synthesized from the corresponding nitriles by allowing them to react with NH_2OH hydrochloride in aqueous *i*-PrOH to give the carboxamidoxime, which was followed by treatment with $NaNO_2$ in 20% aqueous HCl. The nitriles were synthesized by the dehydration of the corresponding amide with TFAA and pyridine in CH_2Cl_2 [97] (Scheme 9.26).

As mentioned earlier, one of the first tricyclic heterocyclic, zero-hydrogen, energetic compounds reported was 3,4-bis(4-nitrofurazan-3-yl)-furoxan (**DNTF, BNFF, 16**) [42–47, 98]. **BNFF** was synthesized by allowing **AFCAO** to react

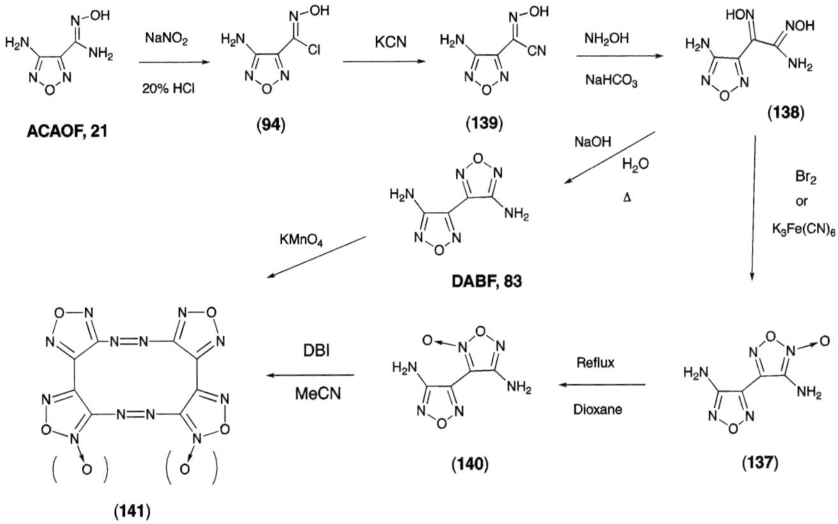

Scheme 9.25 Synthesis of 3-amino-4-(4-aminofurazan-3-yl)furoxan (**137**) and 4,4′-diamino-3,3′-bifurazan (**DABF, 83**) from **138**.

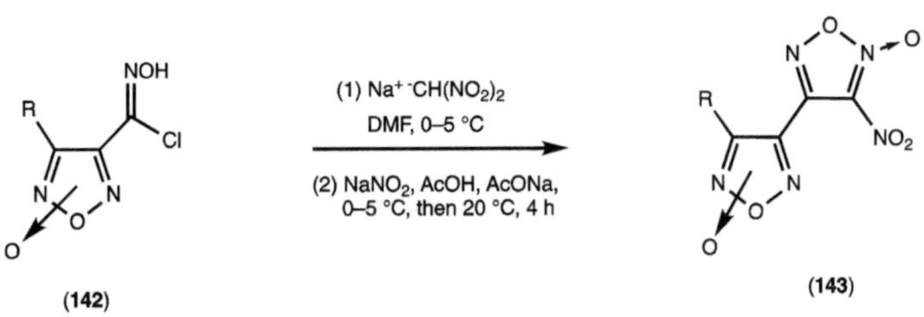

R = –Ph, –Me, –N_3, –N=N–Mes, –SPh, –SBu, –OMe, –NO_2, and –CN

Scheme 9.26 Synthesis of 3-nitro-4-(4-*R*-furoxan-3-yl)furoxans (**143**).

with $NaNO_2$ in aqueous HCl to yield the 4-amino-3-chloroximinofurazan (**94**). The latter product was treated with aqueous Na_2CO_3 in a suspension with ether to yield, the intermediate cyanogen-*N*-oxide, which dimerized to the precursor to **BNFF** – 3,4-bis(4-aminofurazan-3-yl)furoxan (**DATF**, **BAFF**, **144**) [99]. The cyclization of the intermediate cyanogen-*N*-oxide may also proceed in a head-to-tail fashion to yield 3,6-bis(4-amino-1,2,5-oxadiazol-3-yl)-1,4,2,5-dioxadiazine (**145**), depending on the choice of base and solvent [40, 44, 99, 100] (Scheme 9.27). Andrianov and coworkers [99, 100] found the use of soft bases such as NEt_3 in CH_3CN predominantly gave the dioxadiazine (**145**) in 84% yield. Lim et al. [44] performed a study of the base and solvent effect on the product distribution for the dimerization of the cyanogen-*N*-oxide. The use of ether as a solvent and alkali carbonate salts gave predominantly the furoxan product **144**. They also found that K_2CO_3 in solvents, such as THF, MeOH, and MeCN, gave predominantly the dioxadiazine **145** product. Tsyshevsky et al. [40] described the use of Ag_2CO_3 in THF to yield exclusively **BAFF** in 80% yield.

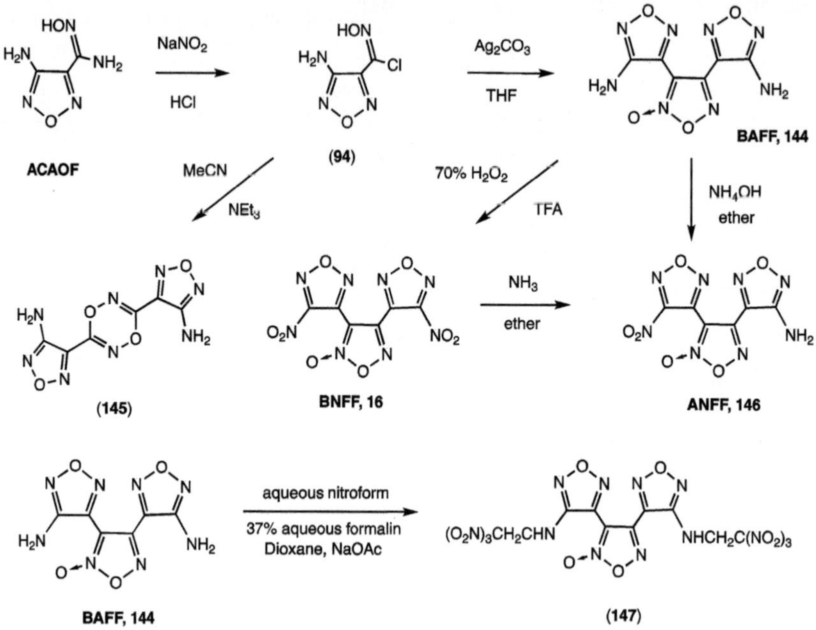

Scheme 9.27 Representative synthesis of **BAFF**, **BNFF**, **ANFF**, and compound **147**.

BAFF was oxidized with a mixture of TFA and 70% H_2O_2 to give **BNFF** in 70% yield. **BNFF** has very attractive energetic properties, including good thermal stability and high density (1.87 g cm^{-3}), and has a melting point of 109 °C, allowing it to be used in a variety of applications [48, 101–106]. Sheremetev et al. [98] reported the synthesis of **BNFF** by treating 3-amino-4-methylfurazan with *n*-BuLi in toluene, to yield the dilithium intermediate, which was silylated with trimethylsilyl chloride to the bistrimethylsilyl intermediate, and then cyclized with N_2O_5 to give **BNFF** in 23% yield. They also reported the oxidation of **BAFF 144** to corresponding **BNFF 16**, using a mixture of TFAA, 96% H_2O_2 in CH_2Cl_2 in 50% yield. Recently Duddu et al. [107] reported an alternative synthesis of **BNFF** in which compound **94** was converted into the product **134** by treatment with 70% H_2O_2 and catalyst $(Bmim)_4W_{10}O_{23}$ in concentrated H_2SO_4 in 67% yield. This was cyclized by treatment with aqueous K_2CO_3 in a suspension with ether to yield **BNFF** in 27% yield. The researchers also reported the synthesis of 3-(4-aminofurazan-3-yl)-4-(4-nitrofurazan-3-yl)furoxan (**ANFF, 146**) by treatment of **BNFF** with aqueous NH_3 in ether and 3,4-bis(4-(2,2,2-trinitroethylaminofurazan-3-yl)furoxan (**147**), by treatment of **BAFF** with aqueous nitroform and 37% formalin in a suspension with dioxane and NaOAc (Scheme 9.27).

Recently, Shreeve and coworkers [108] reported the synthesis of **BNFF** from 4-nitro-3-(dinitromethyl)-furazan (**148**) during an attempted recrystallization of

148 with CH_2Cl_2 (Scheme 9.28). The synthesis of **148** involved the conversion of **134** into 4-nitro-3-(chlorodinitromethyl)furazan (**149**), by oxidation with a mixture of TFAA and 100% HNO_3. This was treated with KI to produce potassium 4-nitrofurazan-3-(dinitromethylide) (**150**), which was acidified with 50% H_2SO_4 to yield **146** (Scheme 9.28). Compound **148** cyclized upon acidification with loss of HNO_2 to form **BNFF**. The hydroxylammonium (**151**), hydrazinium (**152**), and ammonium (**153**) salts of **148** were synthesized by a metathesis reaction, by treatment of the silver salt with the respective amine hydrochloride. The hydrazinium salt had the highest density (1.92 g cm^{-3}) and highest T_d (161 °C).

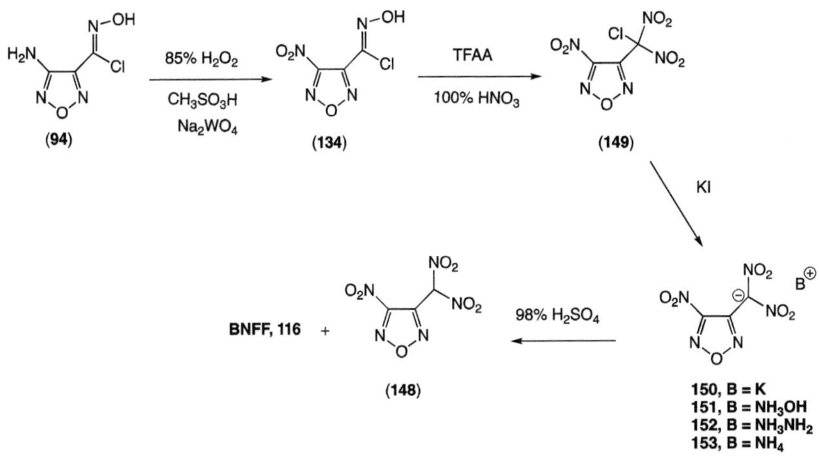

Scheme 9.28 Synthesis of 4-nitro-3-(dinitromethyl)furazan (**148**) and BNFF **116**.

BAFF was converted to the bisnitramino-derivative (**154**) by treatment with 100% HNO_3 at −10 to −5 °C [26]. The bishydroxylammonium, bisguanidinium, bis-aminoguanidinium, and bis(3-amino-1,2,4-triazolyl) salts of **154** were synthesized from the di-metal salt of the nitramine by allowing it to react with the corresponding amine hydrochloride. The bis-aminoguanidinium salt was the only one that crystallized in pure form, the rest with solvents of recrystallization. The researchers also reported the synthesis of 4″-amino-4-(((2,2,2-trinitro-ethyl) amino)-[3,3′:4′,3″-ter(1,2,5-oxadiazole)]-2′-oxide (**155**), 4,4″-bis((2,2,2-trinitroethyl) amino)-[3,3′:4′,3″-ter(1,2,5-oxadiazole)]-2′-oxide (**156**) and 4,4″-bis((*N*-nitro-2,2,2-trinitroethyl)amino)-[3,3′:4′,3″-ter(1,2,5-oxadiazole)]-2′-oxide (**157**). Compound **155** showed a density of 1.825 g cm^{-3} and **156** had a density of 1.844 g cm^{-3} (Figure 9.7).

The dioxadiazine **145** was converted to the bis-nitramino- derivative (**158**) in a similar manner by treatment with 100% HNO_3 at 0 °C. The bishydroxylammonium, diammonium , dihydrazinium, diguanidinium, and bis(aminoguanidinium) salts of **158** were synthesized by treatment with the corresponding base in ether or CH_3CN [109]. The bis(hydroxylammonium) salt (**159**) showed the highest density of 1.846 g cm^{-3}. Yang and coworkers [27] reported the synthesis of 3,6-bis(4-(2,2,2-trinitroethyl)amino)-1,2,5-oxadiazol-3-yl)-1,4,2,5-dioxadiazine

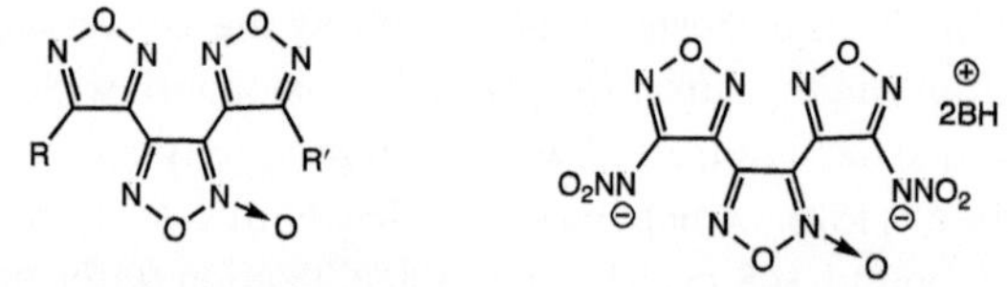

155, R = NH_2, R′ = $NHCH_2C(NO_2)_3$
156, R = R′ = $NHCH_2C(NO_2)_3$
157, R = R′ = $N(NO_2)CH_2C(NO_2)_3$

(154), B = NH_2OH, guanidine, 3-amino-1,2,4-triazolyl, aminoguanidine

Figure 9.7 Energetic compounds derived from **BAFF**.

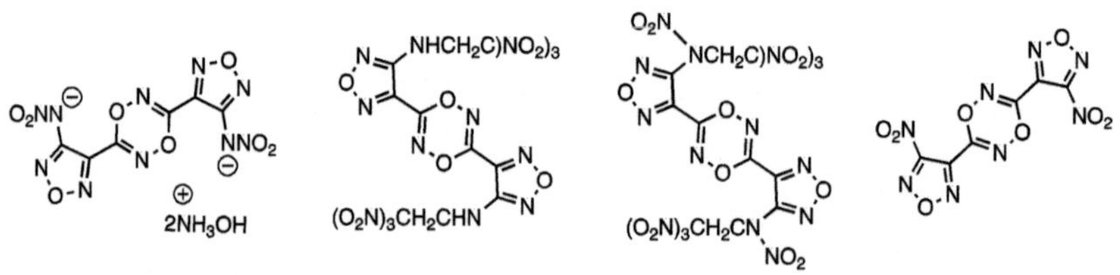

(159) (160) (161) (162)

Figure 9.8 Energetic compounds derived from dioxadiazene, **145**.

(**160**), 3,6-bis((4-*N*-nitro-2,2,2-trinitroethyl)amino)-1,2,5-oxadiazol-3-yl)-1,4,2,5-dioxadiazine (**161**), with densities of 1.94 and 1.95 g cm^{-3}, respectively (Figure 9.8). They also synthesized the bis-nitramino derivative (**158**), and synthesized several energetic salts, the bishydroxylammonium (**159**), diammonium, bis(oxalyldihydrazinium), diguanidinium, bis(aminoguanidinium), bis(1,5-diamino-tetra-zolium), bis(1,2,4-triazolium), bis(3-amino-1,2,4-triazolium), and bis(3,5-diamino-1,2,4-triazolium), all of which were air stable and showed good densities. The bis(hydroxylammonium) salt **158** showed the highest density. The dioxadiazine **145** was converted into 3,6-bis(4-nitro-1,2,5-oxadiazol-3-yl)-1,4,2,5-dioxadiazine (**BNDD, 162**) by oxidation with 90% H_2O_2 and Na_2WO_4 in 98% H_2SO_4 [110, 111] (Figure 9.8). **BNDD** showed a density of 1.87 g cm^{-3} but was sensitive to impact.

Due to the strongly electron-withdrawing character of the furazans of **BNFF**, its nitro-groups are susceptible to nucleophilic displacement. This property was exploited as a method to synthesize a series of energetic compounds derived from **BNFF** (Figure 9.9). Stepanov and coworkers [112, 113] synthesized 3-(4-azido-furazan-3-yl)-4-(4-nitrofurazan-3-yl)furoxan (**163**), with a density of 1.743 g cm^{-3}, by allowing NaN_3 to react with **BNFF** in CH_3CN. **BNFF** was also reacted with an excess of secondary amines to produce a series of bis(4-alkylaminofurazan-3-yl)furoxans (**164**), or to react with primary amines, ammonia or hydrazine in CH_3CN at 20 °C to produce the interesting cyclic compounds, 7-*R*-amino-7*H*-tris[1,2,5]oxa-diazolo[3,4-*b*:3′,4′-*d*:3″,4″-*f*]azepine-1-oxide (**165**), by the intramolecular displacement of both nitro-groups by the nucleophilic base. **BNFF** was also converted to [2,3-*c*;4,5-*c*′:6,7-*c*″]tris[1,2,5-oxadiazole-1-oxide (**166**), a low melting solid (with melting point of 91–92 °C), with good density (1.866 g cm^{-3}), by the intramolecular displacement of both nitro-groups by treatment with hydroxide anion (K_2CO_3 in CH_3CN).

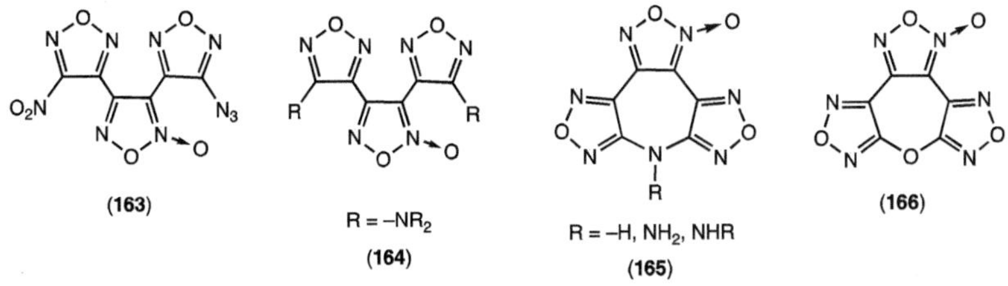

Figure 9.9 Derivatives synthesized from nucleophilic displacement of the nitro-groups of **BNFF**.

BAFF (144) could be converted to 3,4-bis(4-aminofurazan-3-yl)-furazan (**BAFF-1, BATF, 167**) by treatment with $P(OEt)_3$ in refluxing CH_3CN [40] or $SnCl_2$ in a mixture of acetic anhydride, acetic acid, and concentrated HCl [114]. **BATF (167)** could be oxidized with a mixture of 50–70% H_2O_2, Na_2WO_4, and concentrated H_2SO_4 or TFA and 70% H_2O_2, to yield the **BNFF** analog – 3,4-bis(4-nitrofurazan-3-yl)-furazan (**LLM-172, BNFF-1, BNTF, 168**). **BNTF** also has attractive physical properties, including good thermal stability, high density (1.836 g cm^{-3}), and a melting point of 84 °C. **BATF** could be converted into a mixture of **BNTF** and 3-(4-aminofurazan-3-yl)-4-(4-nitrofurazan-3-yl)furazan (**LLM-175, ANFF-1, ANTF, 169**), by treatment with 35% H_2O_2 in concentrated H_2SO_4 in 44% and 45% yield, respectively. The latter two products were separated easily, as the **ANTF** precipitated from the reaction mixture. **ANTF** showed good thermal stability, high density (1.782 g cm^{-3}), and a melting point of 100 °C (Scheme 9.29).

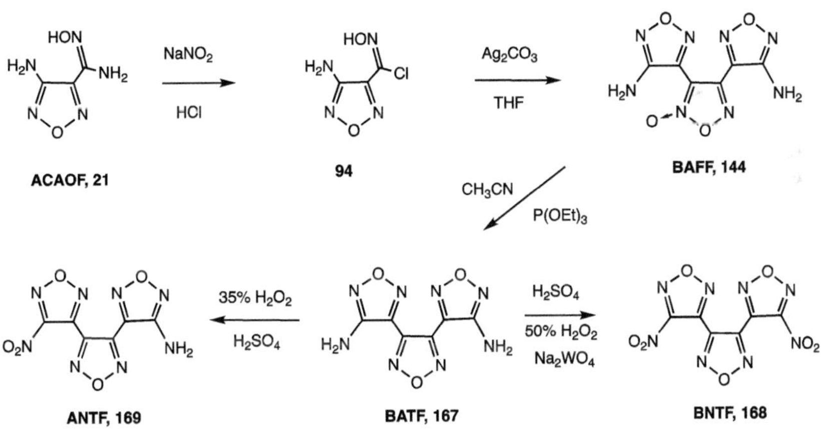

Scheme 9.29 Synthesis of BATF (**167**), BNTF (**168**), and ANTF (**169**).

BATF was converted to the bisnitramino- derivative (**170**) by treatment with 100% HNO_3 at 0 °C [109]. The bishydroxylammonium, diammonium, and dihydrazinium salts of **170** were synthesized by treatment with the corresponding free bases. The bis(hydroxylammonium) salt (**171**) showed the highest density of 1.797 g cm^{-3} (Figure 9.10).

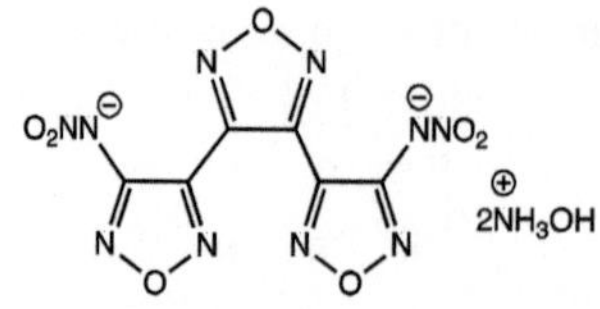

(171)

Figure 9.10 3,4-Bis(4-nitraminofurazan-3-yl)-furazan bis-hydroxylammonium salt (**171**).

Similar to **BNFF**, the nitro-groups of **BNTF** are susceptible to nucleophilic substitution, and a series of energetic compounds derived from **BNTF** were synthesized in a similar manner to **BNFF** [115]. When **BNTF** was allowed to react with excess NaN_3, or with excess of secondary amines, both nitro-groups are substituted to produce 3,4-bis(4-azidofurazan-3-yl)furazan and series of bis(4-alkylaminofurazan-3-yl)furazans (**172**), respectively. When **BNTF** was allowed to react with primary amines, ammonia, or hydrazine in CH_3CN at 20 °C, interesting cyclic compounds – 7-*R*-substituted 7*H*-tris[1,2,5]oxadiazolo[3,4-*b*:3′,4′-*d*:3″,4″-*f*] azepines (**173**) were formed by substitution of both nitro-groups with the amine nucleophile. **BNTF** could be converted to oxepino[2,3-*c*;4,5-*c*′:6,7-*c*″]tris[1,2,5-oxadiazole (**174**), which is a low melting solid (with melting point of 76–77 °C) with good density (1.879 g cm^{-3}), by treatment with K_2CO_3 in dry CH_3CN (Scheme 9.30).

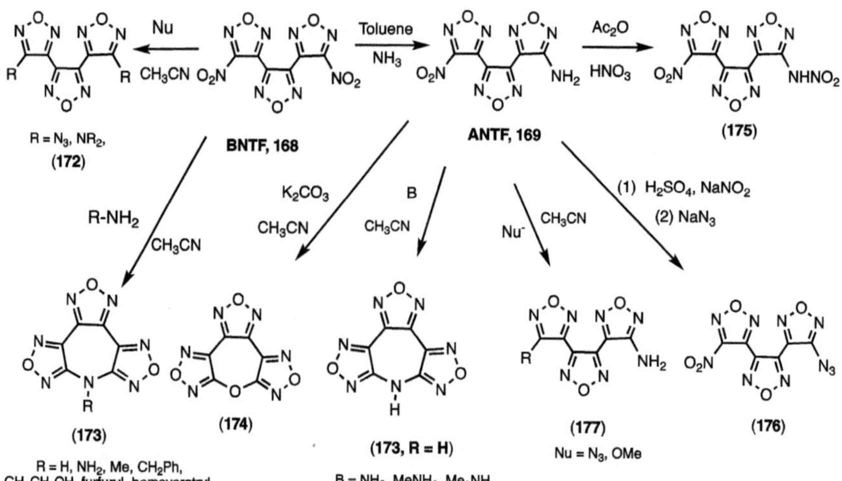

Scheme 9.30 Reactivity of **BNTF** and **ANTF** with various nucleophiles.

The researchers also reported the synthesis of **ANTF** by treatment of **BNTF** with NH_3 in toluene. **ANTF** was converted into 3-(4-nitraminofurazan-3-yl)-4-(4-nitrofurazan-3-yl)furazan (**175**) in 63% yield by treatment with 100% HNO_3 in Ac_2O at 5–10 °C, and 3-(4-azidofurazan-3-yl)-4-(4-nitrofurazan-3-yl)furazan (**176**) by conversion to the diazonium salt, which was followed by treatment with NaN_3. **ANTF** was converted into 7*H*-tris[1,2,5]oxadiazolo[3,4-*b*:3′,4′-*d*:3″,4″-*f*]azepine (**173**, R = H) by treatment with ammonia in MeCN (Scheme 9.30). The authors also found that treatment of **ANTF** with aqueous solutions of stronger bases (e.g. NHMe, MeNH, Me_2NH, K_2CO_3, and KOH) produced the azepine (**173, R = H)**, but treatment of **ANTF** with hard nucleophiles, such as NaN_3, NH_2NH_2, or NaOMe in CH_3CN, resulted in the nucleophilic substitution of the nitro-group, yielding the unsymmetrically substituted open-ring compound (**177**). Upon conversion of **ANTF** into the corresponding –NHAc derivative, the nitro-group could be substituted with various amines, to yield the unsymmetrically substituted products, and does not result in the formation of the azepine (Scheme 9.30).

Tselinsky and coworkers [116] investigated the reactivity of 3,5-bis-(4-amino-1,2,5-oxadiazol-3-yl-1,2,4-triazole (**173**) to alkylation, oxidation, and nitration. Nitration of **178** with a mixture of Ac_2O and 100% HNO_3 at <0 °C yielded 3,5-bis-(4-nitramino-1,2,5-oxadiazol-3-yl-1,2,4-triazole (**179**) as an oily residue which was allowed to react with an alcoholic NaOH solution to yield the trisodium salt of **179**. This was alkylated with MeI in DMF which yielded, upon quenching into water, 1-methyl-3,5-bis-(4-nitramino-1,2,5-oxadiazol-3-yl-1,2,4-triazole (**180**) with alkylation occurring only on the triazole ring. Compound **178** was also converted to 3,5-bis-(4-nitro-1,2,5-oxadiazol-3-yl)-1,2,4-triazole (**181**) by oxidation with 84% H_2O_2 in a mixture of TFA and TFAA in 44% yield and 3,5-bis-(4-azido-1,2,5-oxadiazol-3-yl)-1,2,4-triazole (**182**) by diazotization of **178** with $NaNO_2$ in 95% H_2SO_4 at 0–5 °C followed by treatment with a mixture of NaN_3 and AcOH (Figure 9.11). Shreeve and coworkers [109] reported the conversion of the bis-nitramino-derivative (**179**) to the diammonium (**183**) and dihydrazinium (**184**) salts by treatment with the corresponding free base. The acidic proton of the 1,2,4-triazole moiety was not converted to the conjugate base by treatment with either base. The dihydrazinium salt (**184**) had the highest density of 1.828 g cm^{-3} and a T_d of 229 °C.

Gasco and coworkers [117] reported a series of 4,4″-di-*R*-3,3′:4′3″-terfuroxans as pharmacotherapeutic agents, specifically as nitric oxide release compounds, including 4,4″-bis(nitroxymethyl)-3,3′:4′3″-terfurazan-2′,5,5″-trioxide (**186**) and 4,4″-bis(nitroxymethyl)-3,3′:4′3″-terfurazan-2,2′,2″-trioxide (**186**) (Scheme 9.31). Compound **185** was isolated as an oil, while **186** was isolated as a low melting solid. (mp: 99 °C) The synthesis involved two similar but separate synthetic pathways to

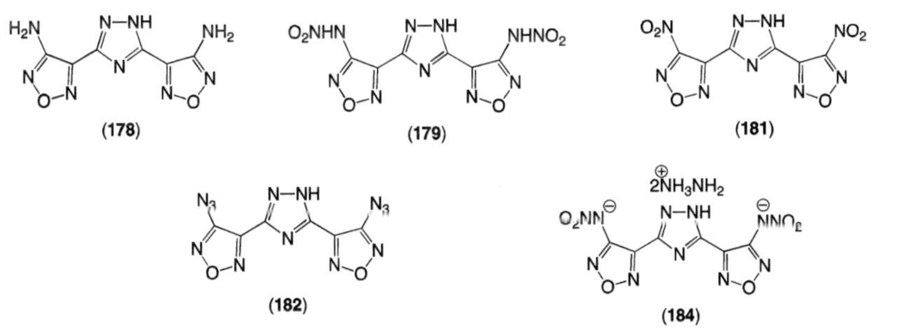

Figure 9.11 Energetic compounds derived from 3,5-bis-(4-amino-1,2,5-oxadiazol-3-yl-1,2,4-triazole (**178**).

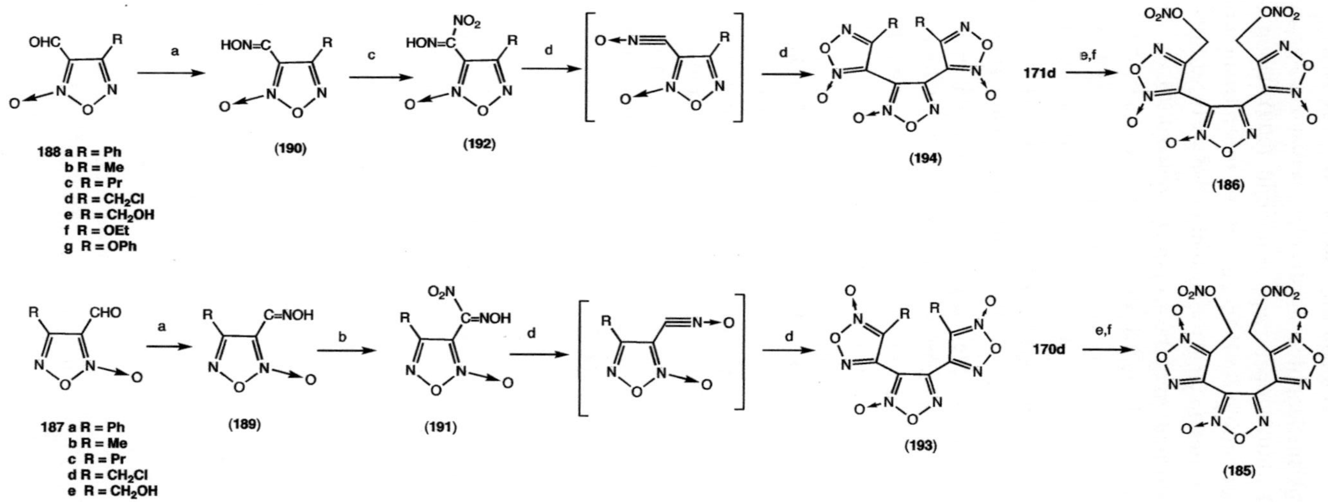

(a) $NH_2OH \cdot HCl$, Pyr, EtOH. (b) N_2O_4, $CHCl_3$, 0 °C. (c) N_2O_4, Et_2O, 0 °C. (d) $CHCl_3$, refux (e) LiBr, $BuP(Ph)_3Br$, $CHCl_3$. (f) $AgNO_3$, CH_3CN.

Scheme 9.31 Synthesis of the isomers, 4,4″-bis(nitroxymethyl)-3,3′:4′3″-terfurazan-2′,5,5″-trioxide (**185**) and 4,4″-bis(nitroxymethyl)-3,3′:4′3″-terfurazan-2,2′,2″-trioxide (**186**).

generate the two different terfuroxan derivatives. The synthesis involved the conversion of 3(4)-*R*-furoxan-(3)4-carbaldehyde (**187** and **188**) with $NH_2OH \cdot HCl$ and pyridine in EtOH to yield 3(4)-*R*-furoxan-(3)4-carboxime (**189** and **190**) followed by treatment with N_2O_4 in $CHCl_3$ or ether to yield 3(4)-*R*-furoxan-(3)4-nitrocarboxime (**191** and **192**). The nitrocarboximes, upon heating in refluxing $CHCl_3$, formed the corresponding cyanogen-*N*-oxide which cyclized to the terfuroxan (**193** and **194**). The nitrate esters, **185** and **186**, were synthesized from the bis(chloromethyl) derivatives by treatment with LiBr, $BuP(Ph)_3Br$ in $CHCl_3$ followed by treatment with $AgNO_3$ in CH_3CN.

Shreeve and coworkers [118, 119] synthesized the structurally similar compound, 3,4-bis(4-nitrofuroxan-3-yl)furoxan (**195**), which has a high oxygen balance, high positive heat of formation, and high density (1.914 g cm^{-3}) along with the furoxan-fused 1,2-diazocine (**196**), from the common intermediate, 3,4-bis(4-aminofuroxan-3-yl)furoxan (**197**) (Scheme 9.32). The formation of **195** required the use of a protecting group strategy on the amino-group of the intermediate 4-aminofuroxan-3-carboxamidoxime (**198**) as it could not be directly converted to the corresponding 4-aminofuroxan-3-carboximidoyl chloride (**199**) using the standard conditions, aqueous HCl and $NaNO_2$. The conversion of **197** to **195** required the use of hypofluorous acid (HOF) in CH_3CN, a very strong oxidizing mixture, to affect the oxidation. The diazocine **196**, with a density of 1.895 g cm^{-3}, was synthesized by allowing **192** to react with trichloroisocyanuric acid (**TCICA**) in CH_3CN at 25 °C (Scheme 9.32).

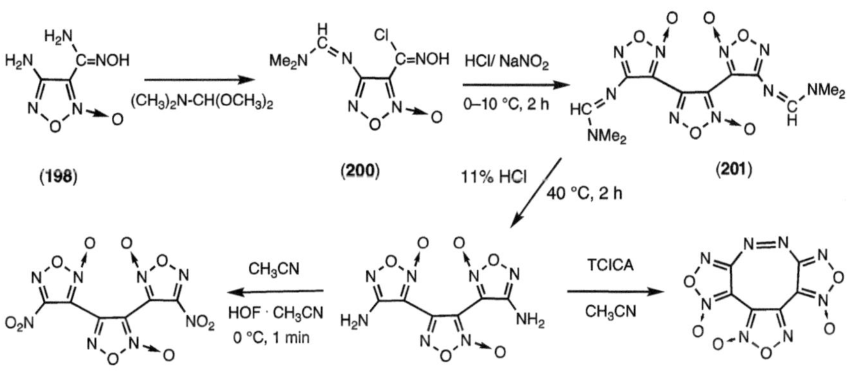

Scheme 9.32 Synthesis of the trisfuroxans, **195** and **196**, with densities of 1.914 and 1.895 g cm^{-3}, respectively.

Fershtat and coworkers [91, 97, 120, 121] investigated the synthesis of 3- and 4-nitrofuroxans from cyclic and acyclic precursors. In general, nitrofuroxans cannot be synthesized by the simple oxidation of the amino-substituted precursor. Thus, the alternative methods had to be developed to access these important energetic compounds. The five general methods for the synthesis of 3- and 4-nitrofuroxans involve: (i) the nitrosation of dipotassium salts of 2-*R*-2-hydroximino-1,1-dinitroethanes to synthesize 3-nitro-4-*R*-furoxans, (ii) the treatment of methacrylic acid and 2-alkylacrylic acids with $NaNO_2$ in a two-phase mixture of dichloroethane (DCE) and 60% H_2SO_4 or in acetic acid, to synthesize 3- and 4-nitro-3(4)-*R*-furoxans, (iii) obtaining 4-nitro-furoxans by cyclization of the *amphi*-from of the corresponding nitroglyoximes, (iv) by oxidation of 4-amino-furoxans; and (v) treatment of β-nitrostyrenes with $NaNO_2$ in acetic acid, with the conversion of acrylic acids to nitrofuroxans, showing the highest yields. 3-Nitro-4-*R*-furoxans could be converted into the more thermodynamically stable 4-nitro-3-*R*-furoxans, by their refluxing in toluene (Scheme 9.33).

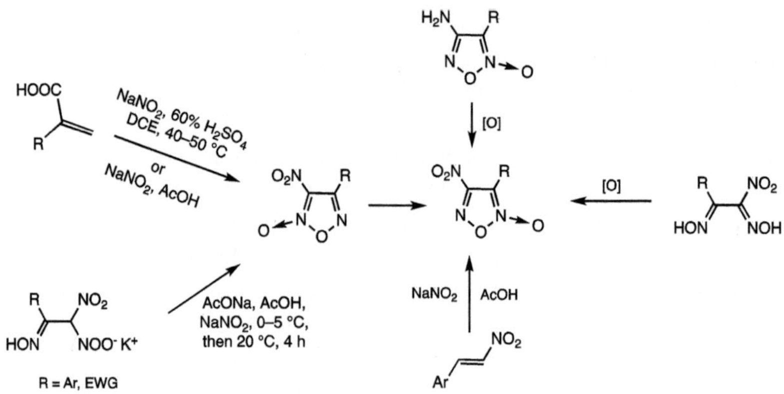

Scheme 9.33 Synthetic methods leading to 3- and 4-nitro-*R*-furoxans.

Recently, Makhova and coworkers [118] reported the synthesis of 4-nitrofuroxanylcarbonitrile oxide (**202**) by the thermolysis of 4-nitrofuroxannitrolic acid (**203**, Scheme 9.34). The authors investigated the synthesis of 3-heterocyclic-4-nitrofuroxans (**204**) by [3*n*+2] cycloaddition reactions of **202** with dienophiles, such as diethyl acetylenedicarboxylate (**204a**), in a similar manner to the [3*n*+2] cycloaddition reaction of **NFNO 4** with dienophiles, to form 3-nitroheterocycles [32, 33].

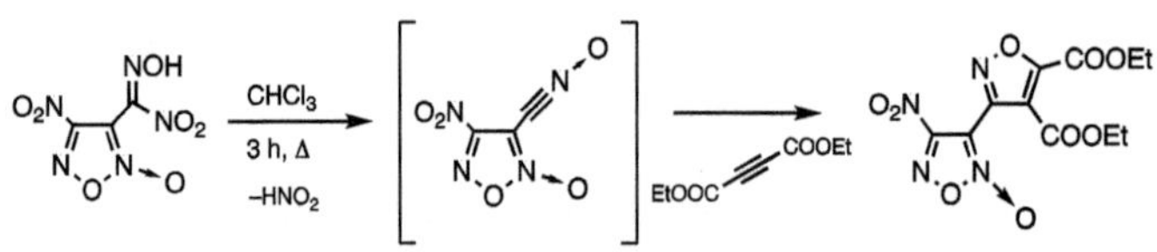

Scheme 9.34 Synthesis of 3-heterocyclic-4-nitrofuroxans (**204**) from 4-nitrofuroxannitrolic acid (**202**).

In 1975, Adolph [122] reported the synthesis of the highly oxidized 4,5-bis(fluorodinitromethyl)furoxan (**205**) while investigating the oxidation reactions of 2,2,2-fluorodinitroethylamine (**206**). The synthesis involved the treatment of **199** with *m*-chloroperbenzoic acid (MCPBA) in CH_2Cl_2, to yield fluorodinitroacetaldoxime (**207**), which was treated with anhydrous N_2O_4, to give product **205** in 10% yield. Although the physical properties of **205** were not mentioned, the latter compound was reported to be unstable in EtOH (Scheme 9.35).

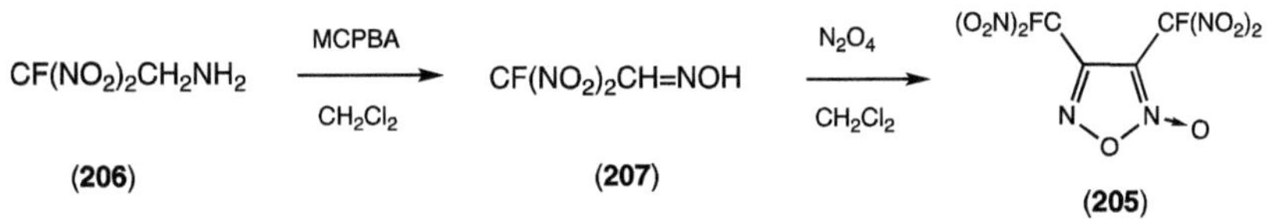

Scheme 9.35 Synthesis of 4,5-bis(fluorodinitromethyl)furoxan (**205**).

Lead azide is a common primary explosive, used in detonators and as a component in formulations for stab initiation mixtures, while lead styphnate is a common component in formulations used as primers and initiating charges [123]. Due to the known acute and chronic toxicity of lead-containing compounds, there has been a growing interest in the development of new "green" primary explosive materials, which are devoid of toxic metals such as lead. Two examples of recently developed green primary explosives that received a considerable attention were Cu(I) 5-nitrotetrazolate (**DBX-1**, **208**) [123–125] and potassium 1,1-dinitramino-5,5′-bistetrazolate (**K_2DNAT**, **209**) [126]. There are many recent reported examples of potassium-containing green primary materials derived from furazans and furoxans, substituted with potassium dinitromethanide or with potassium nitramino groups. Some of the notable examples are dipotassium 3,3′-bis(dinitromethyl)difurazanyl ether [127, 128] (**210**), dipotassium 4,4′-bis (dinitromethyl)-3,3′-azofurazanate [129, 130] (**211**), 4,4′-bis(fluorodinitromethyl)-3,3′-azo-furazan [129] (**212**), dipotassium 3,4-bis(3-dinitromethylfurazan-4-oxy) furazan [131] (**213**), dipotassium dinitraminodifurazano[3,4-*b,e*]pyrazine [132] (**214**), potassium 4,5-Bis(dinitromethyl)furoxanate [133] (**215**), potassium 3-dinitromethyl-4-nitraminofurazan [134] (**216**), and potassium 6-nitro-5-oxidopyrazolo[3,4-c][1,2,5]oxadiazol-4-ide (**217**) (Figure 9.12). These compounds exhibit good density and have good thermal stability.

The preparation of many of these compounds used a similar synthetic method, starting from either bis(carboxamidoxime) or dinitrile, as for example, the synthesis of potassium 4,5-bis(dinitromethyl)-furoxanate [133] (**215**) (Scheme 9.36). The synthesis follows a general approach that has been seen in earlier examples in this review, in which a precursor dinitrile (**219**) was converted to the bis(carboxamidoxime) (**220**) by treatment with hydroxylamine. The latter compound was further converted to the bis(chloroximido) (**221**), by treatment with $NaNO_2$ in aqueous HCl, followed by nitration to give the bis(chlorodinitromethyl) (**222**). Reduction with KI gives the dipotassium bis(dinitromethanide)-substituted target product.

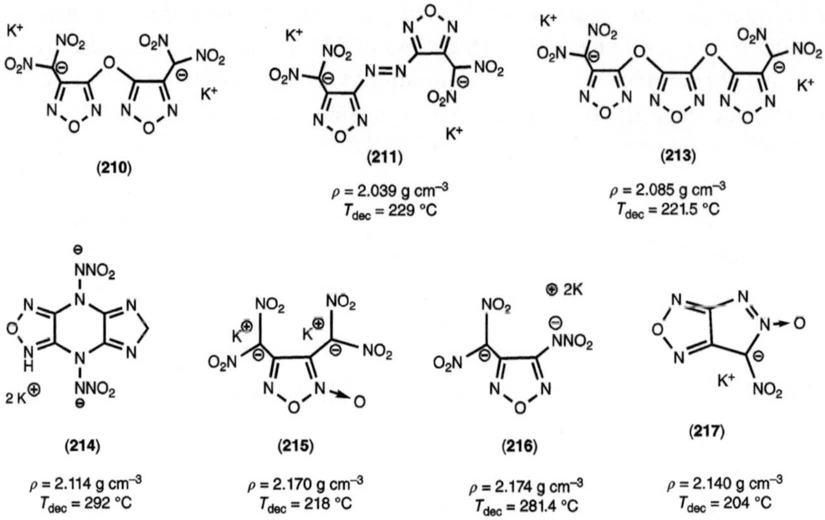

Figure 9.12 New furazan and furoxan-based green primary explosives.

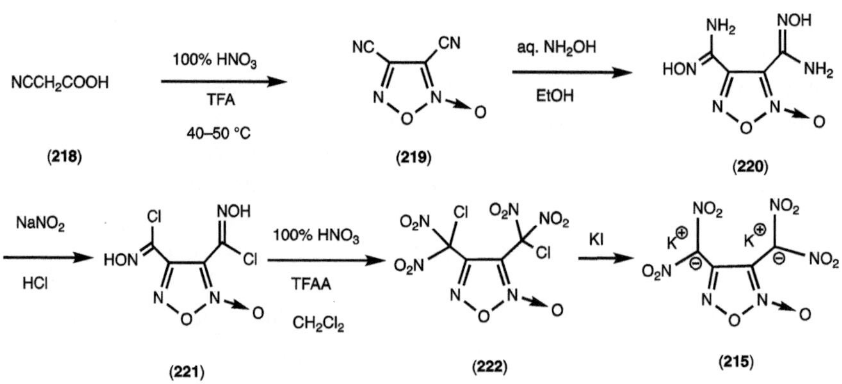

Scheme 9.36 Synthesis of dipotassium 4,5-bis(dinitromethyl)furoxanate [133] (**215**).

The synthesis of the nitramino-compounds listed above, dipotassium 4,8-dinitraminodifurazano[3,4-*b*,*e*]pyrazine (**214**) [133], and potassium 3-dinitromethyl-4-nitraminofurazan (**216**) [134], used a different synthetic approach. Compound **214** was synthesized from the parent 4*H*,8*H*-difurazano[3,4-*b*,*e*]pyrazine (**DFP**, **223**) by amination with hydroxylamine-*O*-sulfonic acid (**HAOSA**), to give 4,8-diamino-difurazano[3,4-*b*,*e*]pyrazine (**224**). Compound **224** was nitrated under carefully controlled conditions with mixed acids to give 4,8-dinitraminodifurazano [3,4-*b*,*e*]pyrazine (**225**), which was further treated with KOH to give compound **214**. The use of 100% HNO_3 or N_2O_5 in the nitration of **224** resulted in the loss of the amino-groups, while use of mixed acids resulted in decomposition, if left for too long time in the reaction mixture (Scheme 9.37).

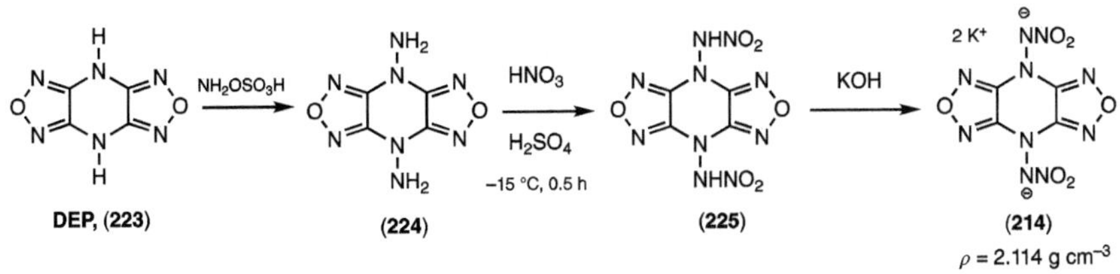

Scheme 9.37 Synthesis of dipotassium 4,8-dinitraminodifurazano[3,4-*b*,*e*]pyrazine (**214**) [132].

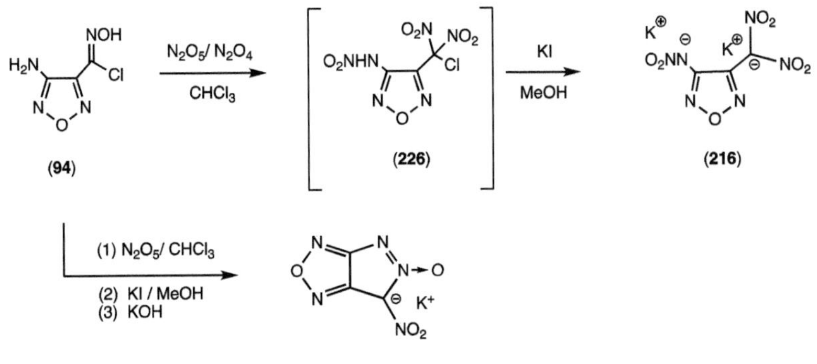

Scheme 9.38 Synthesis of dipotassium 3-dinitromethyl-4-nitraminofurazan (**216**) and potassium 6-nitro-5-oxidopyrazolo[3,4-*c*][1,2,5]oxadiazol-4-ide **(217)**.

The synthesis of potassium 3-dinitromethyl-4-nitraminofurazan (**216**) involved the nitration of 3-amino-4-chloroximinofurazan (**94**) with a N_2O_5/N_2O_4 in $CHCl_3$, to yield 3-amino-4-(chlorodinitromethyl)furazan (**226**), followed by treatment with KI in MeOH, to yield product **216** in 36% yield [135]. Conversely, Yu et al. [136] reported that the nitration of **94** using TFAA/100% HNO_3 at 0 °C followed by KI in MeOH yielded the ring-closed product – potassium 6-nitro-5-oxidopyrazolo[3,4-*c*][1,2,5]oxadiazol-4-ide (**217**), instead of compound **216**. Li et al. [134] studied the nitration of **94** with various nitrating agents, and found that when **94** is treated with N_2O_5, followed by treatment with KI, the product is compound **217**. When **94** is treated with a mixture of N_2O_5 containing ~10% N_2O_4, followed by treatment with KI, instead of **217**, the obtained product is **216**, with a yield of 46.3% (Scheme 9.38).

9.6 Summary

The synthesis of nitro-substituted oxadiazoles has been a robust area of research in the development of novel energetic compounds. Their combination of good oxygen balance, thermal stability, and synthetic availability of the various oxadiazoles, along with the creation of coupled heterocyclic frameworks incorporating the various oxadiazoles, has led to an expansion of the number of energetic oxadiazoles in recent years. The synthesis of energetic nitro-substituted oxadiazoles should continue to be a fertile area of energetic materials research.

This work performed under the auspices of the U.S. Department of Energy by Lawrence Livermore National Laboratory under Contract DE-AC52-07NA27344. Document. ID: LLNL-BOOK-818308-DRAFT.

Abbreviations

AFCAO	4-aminofurazan-3-carboxamidoxime
AG	aminoguanidine
ANF	3-amino-4-nitrofurazan
ANFF	3-(4-aminofurazan-3-yl)-4-(4-nitrofurazan-3-yl)furoxan
ANG	aminonitroguanidine
ANTF	3-(4-aminofurazan-3-yl)-4-(4-nitrofurazan-3-yl)furazan
AO	3-azido-5-guanidino-1,2,4-oxadiazole
BADPO	2,5-bis(3-amino-2,4,6-dinitrophenyl)-1,3,4-oxadiazole
BAFF	3,4-bis(4-aminofurazan-3-yl)-furoxan
BATF	3,4-bis(4-aminofurazan-3-yl)-furazan
BNDD	3,6-bis(4-nitro-1,2,5-oxadiazol-3-yl)-1,4,2,5-dioxadiazine
BNTF	3,4-bis(4-aminofurazan-3-yl)furazan
BNFF	3,4-bis(4-nitrofurazan-3-yl)-furoxan
BPABF	3,4-bis(picrylamino)furazan
BPAF	3,4-bis(picrylamino)furazan
BrCN	cyanogen bromide
DAAF	4,4′-diamino-3,3′-azoxyfurazan
DAAzF	4,4′-diamino-3,3′-azofurazan
DABF	4,4′-diamino-3,3′-bifurazan
DAF	3,4-diaminofurazan
DAG	diaminoguanidine
DAU	diaminourea
DAT	1,5-diaminotetrazole
DAGO	diaminoglyoxime
DBAPF	3,4-bis(picrylamino)furazan
DBX-1	copper(1) 5-nitrotetrazolate
DCE	dichloroethane
DFP	4*H*,8*H*-difurazano[3,4-*b*,*e*]pyrazine
DMDO	dimethyldioxirane
DNAzF	4,4′-dinitro-3,3′-azofurazan
DNBF	4,4′-dinitro-3,3′-bifurazan
DNBFO	4,4′-dinitro-3,3′-bifuroxan
DNE	dichloroethane
DNF	3,4-dinitrofurazan
DNFO	3,4-Ddnitrofuroxan
DPO	2,5-dipicryl-1,3,4-oxadiazole

Gu guanidine
HA hydroxylamine
HAOSA hydroxylamine-*O*-sulfonic acid
HOF hypofluorous acid
K2DNAT potassium 1,1-dinitramino-5,5′-bistetrazolate
LLM-172 3,4-bis(4-nitrofurazan-3-yl)-furazan
LLM-191 3,5-bis(4-nitrofurazan-3-yl)-1,2,4-oxadiazole
LLM-200 3,3′-bis(4-nitro-furazan-3-yl)-5,5′-bi-1,2,4-oxadiazole
LLM-201 3-(4-nitrofurazan-3-yl)-1,2,4-oxadiazole-5-amine
MCPBA *m*-chloroperbenzoic acid
NaDNM sodium dinitromethylide
NFNO nitroformonitrile oxide
NOA 5-amino-3-nitro-1,2,4-oxadiazole
NOG 5-guanidino-3-nitro-1,2,4-oxadiazole
NOTO 5-(4-nitro-(1,2,5)oxadiazolyl]-5*H*-[1,2,3]triazolo[4,5-*c*][1,2,5]oxadiazole
PETN pentaerythritol tetranitrate
SDCI sodium dichloroisocyanurate
TAG triaminoguanidine
TCICA trichloroisocyanuric acid
TFA trifluoroacetic acid
TFAA trifluoroacetic anhydride
TKX-55 5,5′-bis(2,4,6-trinitrophenyl)-2,2′-bi(1,3,4-oxadiazole)

References

1 Spear, R.J. and Wilson, W.S. (1984). Recent approaches to the synthesis of high explosive and energetic materials: a review. *J. Energetic Mater.* 2: 61–149.

2 Agrawal, J.P. (2010). *High Energy Materials: Propellants, Explosives and Pyrotechnics*. Wiley.

3 Agrawal, J.P. and Hodgson, R.D. (2007). *Organic Chemistry of Explosives*. West Sussex: Wiley.

4 Yin, P. and Shreeve, J.M. (2017). Nitrogen-rich azoles as high density energy materials. In: *Heterocyclic Chemistry in the 21st Century – A Tribute to Alan Katritzky*, vol. 121 (ed. E. Scriven and C.A. Ramsden), 89–131.

5 Gao, H. and Shreeve, J.M. (2011). Azole-based energetic salts. *Chem. Rev.* 111 (11): 7377–7436.

6 Licht, H.H. and Ritter, H. (1988). 2,4,6-Trinitropyridine and related compounds, synthesis an characterization. *Propellants Explos. Pyrotech.* 13: 25–29.

7 Hollins, R.A., Merwin, L.H., Nissan, R.A. et al. (1996). Aminonitropyridines and their *N*-oxides. *J. Heterocyclic Chem.* 33: 895–904.

8 Coburn, M.D. (1972). 2,6-Bis(picrylamino-3,5-dinitropyridine and a method for its preparation. US Patent No. 3,678,061.

9 Sheremetev, A.B. and Yudin, I.L. (2003). Advances in the chemistry of furazano[3,4-*b*]pyrazines and their analogues. *Russ. Chem. Rev.* 72 (1): 87–100.

10 Herve, G., Roussel, C., and Graindorge, H. (2010). Selective preparation of 3,4,5-trinitro-1*H*-pyrazole: a stable all-carbon-nitrated arene. *Angew. Chem. Int. Ed.* 49: 3177–3181.

11 Dalinger, I.L., Suponitsky, K.Y., Shkineva, T.K. et al. (2018). Bipyrazole bearing ten nitro groups – novel highly dense oxidizer for forward-looking rocket propulsion. *J. Mater. Chem. A* 6: 14780–14786.

12 Talawar, M.B., Sivabalan, R., Senthilkumar, N. et al. (2004). Synthesis, characterization and thermal studies on furazan- and tetrazine-based high energy materials. *J. Hazard. Mater.* 113 (1–3): 11–25.

13 Lee, K.-Y., Storm, C.B., Hiskey, M.A., and Coburn, M.D. (1991). An improved synthesis of 5-amino-3-nitro-1*H*-1,2,4-triazole (ANTA), a useful intermediate for the preparation of insensitive high explosives. *J. Energetic Mater.* 9: 415–428.

14 Ladyzhnikova, T.D., Manuel, D.V., and Altukhov, K.V. (1993). Reaction of 4-trinitromethyl-1,2,3-triazoles with formaldehyde. *Russ. J. Org. Chem.* 29 (7): 1249–1251.

15 Neuman, P.N. (1970). Picryl derivatives of 4-amino-1*H*-1,2,3-triazole. *J. Heterocyclic Chem.* 7: 1159–1160.

16 Hiskey, M.A., Goldman, N., and Stine, J.R. (1998). High-nitrogen energetic materials derived from azotetrazolate. *J. Energetic Mater.* 16: 119–127.

17 Pagoria, P.F., Zhang, M.-X., Zuckerman, N.B. et al. (2017). Synthesis and characterization of multicyclic oxadiazoles and 1-hydroxytetrazoles as energetic materials. *Chem. Heterocyclic Compd.* 53: 760.

18 Yan, C., Wang, K., Liu, T. et al. (2017). Exploiting the energetic potential of 1,2,4-oxadiazole derivatives: combining the benefits of a 1,2,4-oxadiazole framework with various energetic functionalities. *Dalton Trans.* 46 (41): 14210–14218.

19 Zlotin, S.G., Dalinger, I.L., Makhova, N.N., and Tartakovsky, V.A. (2020). Nitro compounds as the core structures of promising energetic materials and versatile reagents for organic synthesis. *Russ. Chem. Rev.* 89 (1): 1–54.

20 Sheremetev, A.B. (1998). Nitro- and nitraminofurazans. *Mendeleev Chem. J.* 41 (2): 62–80.

21 Makhova, N.N. and Kulikov, A.S. (2013). Advances in the chemistry of monocyclic amino- and nitrofuroxans. *Russ. Chem. Rev.* 82 (11): 1007–1033.

22 Makhova, N.N. and Fershtat, L.L. (2018). Recent advances in the synthesis and functionalization of 1,2,5-oxadiazole 2-oxides. *Tetrahedron Lett.* 59 (24): 2317–2326.

23 Sheremetev, A.B., Makhova, N.N., and Friedrichsen, W. (2001). Monocyclic furazans and furoxans. *Adv. Heterocyclic Chem.* 78: 65.

24 Klapotke, T.M., Mayr, N., Stierstorfer, J., and Weyrauther, M. (2014). Maximum compaction of ionic organic explosives: bis(hydroxylammonium) 5,5′-dinitromethyl-3,3′-bis(1,2,4-oxadiazolate) and its derivatives. *Chem. Eur. J.* 20: 1410–1417.

25 Wei, H., He, C., Zhang, J., and Shreeve, J.M. (2015). Combination of 1,2,4-oxadiazole and 1,2,5-oxadiazole moieties for the generation of high-performance energetic materials. *Angew. Chem. Int. Ed. Engl.* 54 (32): 9367–9371.

26 Yu, Q., Yang, H., Ju, X. et al. (2017). The synthesis and study of compounds based on 3,4-bis(aminofurazano)furoxan. *ChemistrySelect* 2 (2): 688–696.

27 Yu, Q., Cheng, G., Ju, X. et al. (2017). An interesting 1,4,2,5-dioxadiazine-furazan system: structural modification by incorporating versatile functionalities. *Dalton Trans.* 46 (41): 14301–14309.

28 Pivina, T.S., Sukhachev, D.V., Evtushenko, A.V., and Khmel'mitskii, L.I. (1995). Comparative characteristic of energy content calculating methods for the furazan series as an example of energetic materials. *Propellants Explos. Pyrotech.* 20 (1): 5–10.

29 Sheremetev, A.B. (1999). The chemistry of furazans fused to six- and seven-membered heterocycles with one heteroatom. *Russ. Chem. Rev.* 68 (2): 137–148.

30 Sheremetev, A.B. (1995). Chemistry of furoxans fused to five-membered rings. *J. Heterocyclic Chem.* 32: 371–385.

31 Albini, A. and Pietra, S. (1991). *Heterocyclic-N-Oxides*. Boca Raton, FL: CRC Press.

32 Ovchinnikov, I.V., Popov, N.A., Makhova, N.N. et al. (1995). Nitroformonitrile oxide in the reaction of 1,3-dipolar addition. *Mendeleev Commun.* 5 (6): 231–232.

33 Fershtat, L.L., Ovchinnikov, I.V., and Makhova, N.N. (2014). Ionic liquid-promoted [3+2]-cycloaddition reactions of nitroformonitrile oxide generated by the cycloreversion of dinitrofuroxan. *Tetrahedron Lett.* 55 (15): 2398–2400.

34 Fu, Z., Su, R., Wang, Y. et al. (2012). Synthesis and characterization of energetic 3-nitro-1,2,4-oxadiazoles. *Chem. Eur. J.* 18: 1886–1889.

35 Fu, Z., He, C., and Chen, F.-x. (2012). Synthesis and characteristics of a novel, high-nitrogen, heat-resistant, insensitive material (NOG_2Tz). *J. Mater. Chem. A* 22 (1): 60–63.

36 Fu, Z., Wang, Y., Yang, L. et al. (2014). Synthesis and characteristics of novel, high-nitrogen 1,2,4-oxadiazoles. *RSC Adv.* 4 (23): 11859–11861.

37 Coburn, M.D. and Bjorklund, M.D. (1980). 3,3-Dinitrobutyl-1,2,4-oxadiazole. *J. Heterocyclic Chem.* 17: 819–821.

38 Teslenko, F.E., Churakov, A.I., Larin, A.A. et al. (2020). Route to 1,2,4- and 1,2,5-oxadiazole ring assemblies via a one-pot condensation/oxidation protocol. *Tetrahedron Lett.* 61 (13).

39 Pagoria, P.F., Zhang, M.-X., Zuckerman, N.B. et al. (2017). Synthesis and characterization of multicyclic oxadiazoles and 1-hydroxytetrazoles as energetic materials. *Chem. Heterocyclic Compd.* 53 (6/7): 760–778.

40 Tsyshevsky, R., Pagoria, P., Zhang, M. et al. (2017). Comprehensive end-to-end design of novel high energy density materials: I. Synthesis and characterization of oxadiazole based heterocycles. *J. Phys. Chem. C* 121 (43): 23853–23864.

41 Tsyshevsky, R., Pagoria, P.F., Smirnov, A.S., and Kuklja, M.M. (2017). Comprehensive end-to-end design of novel high energy density materials: II. Computational modeling and predictions. *J. Phys. Chem. C* 121: 23865.

42 Zheng, W., Wang, J., Ren, X. et al. (2007). An investigation on thermal decomposition of DNTF-CMDB propellants. *Propellants Explos. Pyrotech.* 32: 520–524.

43 Zheng, W. and Wang, J.-N. (2006). Review on 3,4-bisnitrofurazanfuroxan (DNTF). *Chin. J. Energetic Mater.* 14: 463–466.

44 Lim, C.-H., Kim, T.-K., Kim, K.-H. et al. (2010). Synthesis and characterization of bisnitrofurazanofuroxan. *Bull. Korean Chem. Soc.* 31: 1400–1402.

45 Zhou, Y., Wang, B., Li, J.-K. et al. (2011). Study on synthesis, characterization and properties of 3,4-bis(4′-nitrofurazano-3′-yl)furoxan. *Acta Chim. Sinica* 69 (14): 1673–1680.

46 Sinditskii, V.P., Burzhava, A.V., Sheremetev, A.B., and Aleksandrova, N.S. (2012). Thermal and combustion properties of 3,4-bis(3-nitrofurazan-4-yl) furoxan (DNTF). *Propellants Explos. Pyrotech.* 37: 575.

47 Stepanov, A.I., Dashko, D.V., and Astrat'ev, A.A. (2012). 3,4-Bis(4′-nitrofurazanyl-3′-yl)furoxan: a melt cast powerful explosive and a valuable building block in 1,2,5-oxadiazole chemistry. *Cent. Eur. J. Energetic Mater.* 9 (4): 329–342.

48 Tsyshevsky, R., Pagoria, P., Zhang, M. et al. (2015). Searching for low-sensitivity cast-melt high-energy-density materials: synthesis, characterization, and decomposition kinetics of 3,4-bis(4-nitro-1,2,5-oxadiazol-3-yl)-1,2,5-oxadiazole-2-oxide. *J. Phys. Chem. C* 119 (7): 3509–3521.

49 Ichikawa, T., Kato, T., and Takenishi, T. (1965). A new synthesis of adenine and 4-aminoimidazole-5-carboxamide. *J. Heterocyclic Chem.* 253–255.

50 Shaposhnikov, S.D., Korobov, N.V., Sergievskii, A.V. et al. (2002). Heterocycles with a 3-aminofurazanyl substituent. *Russ. J. Org. Chem.* 38 (9): 1351–1355.

51 Pang, F., Wang, G., Lu, T. et al. (2018). Preparation and characteristics of 1,2,4-oxadiazole-derived energetic ionic salts with nitrogen linkages. *New J. Chem.* 42 (6): 4036–4044.

52 Kettner, M.A. and Klapotke, T.M. (2014). 5,5′-Bis-(trinitromethyl)-3,3′-bi-(1,2,4-oxadiazole): a stable ternary CNO-compound with high density. *Chem. Commun.* 50 (18): 2268–2270.

53 Kettner, M.A., Karaghiosoff, K., Klapotke, T.M. et al. (2014). 3,3′-Bi(1,2,4-oxadiazoles) featuring the fluorodinitromethyl and trinitromethyl groups. *Chemistry* 20 (25): 7622–7631.

54 Kiseleva, V.V., Gakh, A.A., and Fainzel'berg, A.A. (1990). Synthesis of C-azolylacetic acid esters via ethyl carbethoxyacetimidate. *Russ. Chem Bull.* 39: 1888–1895.

55 Bagal, L.I., Pevzner, M.S., Frolov, A.N., and Sheludyakova, N.I. (1970). Heterocyclic nitro compounds. *Chem. Heterocyclic Compd.* 6 (2): 240–244.

56 Dacons, J.C. and Sitzmann, M.E. (1977). Synthesis of 2,4,6-trinitrophenyl derivatives of heterocyclic compounds. *J. Heterocyclic Chem.* 14: 1151–1155.

57 Sitzmann, M.E. (1988). 2,5-Dipicryl-1,3,4-oxadiazole:a shock sensitive explosive with high thermal stability. *J. Energetic Mater.* 6: 129–144.

58 Wang, G., Gong, X., Liu, Y. et al. (2010). A theoretical study on the vibrational spectra and thermodynamic properties for the derivatives of 2,5-dipicryl-1,3,4-oxadiazole with $-NO_2$, $-NH_2$, $-OH$ and $-CH_3$ groups. *Chin. J. Chem.* 28: 1538–1546.

59 Klapotke, T.M. and Witkowski, T.G. (2016). 5,5′-Bis(2,4,6-trinitrophenyl)-2,2′-bi(1,3,4-oxadiazole) (TKX-55): thermally stable explosive with outstanding properties. *ChemPlusChem* 81 (4): 357–360.

60 Yu, Q., Yin, P., Zhang, J. et al. (2017). Pushing the limits of oxygen balance in 1,3,4-oxadiazoles. *J. Am. Chem. Soc.* 139 (26): 8816–8819.

61 Yu, Q., Imler, G.H., Parrish, D.A., and Shreeve, J.M. (2017). Nitromethane bridged bis(1,3,4-oxadiazoles): trianionic energetic salts with low sensitivities. *Chem. Eur. J.* 23: 17682–17686.

62 Tang, Y., He, C., Mitchell, L.A. et al. (2015). Energetic compounds consisting of 1,2,5- and 1,3,4-oxadiazole rings. *J. Mater. Chem. A* 3: 23143–23148.

63 Wang, Q., Shao, Y., and Lu, M. (2019). Azo-1,3,4-oxadiazole as a novel building block to design high-performance energetic materials. *Cryst. Growth Des.* 19 (2): 839–844.

64 Hermann, T.S., Karaghiosoff, K., Klapotke, T.M., and Stierstorfer, J. (2017). Synthesis and characterization of 2,2′-dinitramino-5,5′-bi(1-oxa-3,4-diazole). *Chem. Eur. J.* 23: 12087–12091.

65 Olofson, R.A. and Michelman, J.S. (1965). Furazan. *J. Org. Chem.* 30: 1854–1859.

66 Andrianov, V.G. and Eremeev, A.V. (1984). Aminofurazans (review). *Chem. Heterocyclic Compd.* 20 (9): 937–951.

67 Coburn, M.D. (1968). Picrylamino-substituted heterocycles. II. Furazans. *J. Heterocyclic Chem.* 5 (1): 83–87.

68 Pearse, G.A. and Pflaum, R.T. (1959). Interaction of metal ions with amidoximes. *J. Am. Chem. Soc.* 81: 6505.

69 Ungnade, H.E. and Kissinger, L.W. (1963). Nitration of chloroglyoximes, chlorofuroxans and other nitration products. *Tetrahedron* 19 (Suppl. 1): 143–154.

70 Zelenin, A.K. and Trudell, M.L. (1997). A two-step synthesis of diaminofurazan and synthesis of *N*-monoarylmethyl and *N*,*N*-1-diarylmethyl derivatives. *J. Heterocyclic Chem.* 34 (3): 1057–1060.

71 Gunasekaran, A., Jayachandran, T., Boyer, J.H., and Trudell, M.L. (1995). A convenient synthesis of diaminoglyoxime and diaminofurazan – useful precursors for the synthesis of high density energetic materials. *J. Heterocyclic Chem.* 32 (4): 1405–1407.

72 Emmons, W.D. (1954). Peroxytrifluoroacetic acid. II. The oxidation of aniline to nitrobenzene. *J. Am. Chem. Soc.* 76 (13): 3470–3472.

73 Sheremetev, A.B. and Mantseva, E.V. (1996). One-pot synthesis of 4,4′-diamino-3,3′-bifurazan. *Mendeleev Commun.* 6: 246–247.

74 Epishina, M.A., Kulikov, A.S., and Makhova, N.N. (2008). Synthesis of macrocyclic systems from 4,4′-diamino-3,3′-bi-1,2,5-oxadiazole and 3(4)-amino-4(3)-(4-amino-1,2,5-oxadiazol-3-yl)-1,2,5-oxadiazole 2-oxides. *Russ. Chem. Bull.* 57 (3): 644–651.

75 Solodyuk, G.D., Boldyrev, M.D., Gidaspov, B.V., and Nikolaev, V.D. (1981). Oxidation of 3,4-diaminofurazan by some peroxide reagents. *J. Org. Chem. USSR* 17 (4): 756–759.

76 Chavez, D., Hill, L., Hiskey, M., and Kinkead, S. (2000). Preparation and explosive properties of azo- and azoxy-furazans. *J. Energetic Mater.* 18: 219–236.

77 Chavez, D.E. and Francois, E.G. (2009). Process for preparation of 3,3′-diamino-4,4′-azoxyfurazan or 3,3′-diamino-4,4′-azofurazan from 3,4-diaminofurazan. Los Alamos National Laboratory, Los Alamos, NM, USA. US Patent No. 20090306355.

78 Zhang, J. and Shreeve, J.M. (2014). 3,3′-Dinitroamino-4,4′-azoxyfurazan and its derivatives: an assembly of diverse N–O building blocks for high-performance energetic materials. *J. Am. Chem. Soc.* 136 (11): 4437–4445.

79 Tang, Y., Zhang, J., Mitchell, L.A. et al. (2015). Taming of 3,4-di(nitramino)furazan. *J. Am. Chem. Soc.* 137 (51): 15984–15987.

80 Novikova, T.S., Mel'nikova, T.M., and Kharitonova, T.M. (1994). An effective method for the oxidation of aminofurazans to nitrofurazans. *Mendeleev Commun.* 4: 139–140.

81 Zelenin, A.K. and Trudell, M.L. (1998). Synthesis and structure of dinitroazofurazan. *J. Heterocyclic Chem.* 35: 151–155.

82 Sheremetev, A.B., Mantseva, E.V., and Aleksandrova, N.S. (1995). Reaction of nitrofurazans with sulfur nucleophiles. *Mendeleev Commun.* 1: 25–27.

83 Sheremetev, A.B., Kharitonova, O.V., Mantseva, E.V. et al. (1999). Nucleophilic substitution of DNF. *Russ. J. Org. Chem.* 35: 1555.

84 Sheremetev, A.B., Aleksandrova, N.S., Mel'nikova, T.M. et al. (2000). Synthesis of difurazanyl ethers from 4,4′-dinitroazoxyfurazan. *Heteroat. Chem.* 11 (1): 48–56.

85 Sheremetev, A.B., Kharitonova, O.V., and Mel'nikova, T.M. (1996). Synthesis of symmetrical difurazanyl ethers. *Mendeleev Commun.* 141–143.

86 Sheremetev, A.B., Kukagina, V.O., Aleksandrova, N.S. et al. (1998). Dinitro trifurazans with oxy, azo and azoxy bridges. *Propellants Explos. Pyrotech.* 23: 142–149.

87 Gunasekaran, A., Trudell, M.L., and Boyer, J.H. (1994). Dense energetic compounds of C, H, H, and O atoms. Nitro and azido furazan derivatives. *Heteroat. Chem.* 5: 441–446.

88 Eman, V.E., Sukhanov, M.S., Lebedev, O.V. et al. (1996). 5-[4-Amino(1,2,5)oxadiazolyl]-5*H*-1,2,3-triazolo[4,5-*c*][1,2,5]oxadiazole from 4,4′-diamino-3,3′-azofurazan. *Chem. Heterocyclic Compd.* 32 (2): 227–228.

89 Wu, X., Chu, Y., Zheng, C. et al. (2015). Synthesis and theoretical prediction of detonation properties of 3-azido-4-[(4-nitrofurazan-3-yl)-*ONN*-azoxy]furazan. *Chem. Heterocyclic Compd.* 51 (8): 760–762.

90 Godovikova, T.I., Rakitin, O.A., Golova, S.P. et al. (1993). 3,4-dinitrofuroxan- the first example of a pernitro heterocycle. *Mendeleev Commun.* 3: 209.

91 Godovikova, T.I., Rakitin, O.A., Golova, S.P., Vozchikova, S.A., Povorin, M.V., and Khmel'nitskii, L.I. (1994). Synthesis and nucleophilic substitution reactions of 3,4-dinitrofuroxan. *Chem. Heterocyclic Compd.* 30: 465–469.

92 Ovchinnikov, I.V., Makhova, N.N., and Khmel'mitskii, L.I. (1993). Generation of nitro formonitrile oxide as an intermediate for the preparation of dinitrofuroxan. *Mendeleev Commun.* 3: 210.

93 Godovikova, T.I., Golova, S.P., Vozchikova, S.A. et al. (1995). A new method for the synthesis of 1,2,3-triazole-1-oxides. *Mendeleev Commun.* 3: 194–195.

94 Fischer, D., Klapötke, T.M., and Stierstorfer, J. (2014). Synthesis and characterization of diaminobisfuroxane. *Eur. J. Inorg. Chem.* 2014 (34): 5808–5811.

95 Fershtat, L.I., Ovchinnikov, I.V., Epishina, M.A. et al. (2017). Assembly of nitrofurazan and nitrofuroxan frameworks for high-performance energetic materials. *ChemPlusChem* 82: 1315–1319.

96 Fershtat, L.I., Larin, A.A., Epishina, M.A. et al. (2016). Regioselective synthesis of bifuroxanyl systems with the 3-nitrobifuroxanyl core via a one-pot acylation/nitrosation/cyclization cascade. *Tetrahedron Lett.* 57: 4268.

97 Fershtat, L.L., Epishina, M.A., Kulikov, A.S. et al. (2015). An efficient access to (1*H*-tetrazol-5-yl)furoxan ammonium salts via a two-step dehydration/[3+2]-cycloaddition strategy. *Tetrahedron* 71 (38): 6764–6775.

98 Sheremetev, A.B., Ivanova, E.A., Spiridinova, N.P. et al. (2005). Desilylative nitration of C,N-disilylated 3-amino-4-methylfurazan. *J. Heterocyclic Chem.* 42: 1237–1242.

99 Tselinsky, I.V., Melnikova, S.F., Romanova, T.V. et al. (2001). Dimerization of nitrile oxides of the 1,2,5-oxadiazole series. *Russ. J. Org. Chem.* 37 (9): 1355–1356.

100 Andrianov, B.G., Semenikhina, V.G., and Eremeev, A.V. (1992). 4-Aminofurazan-3-hydroximic halides. *Chem. Heterocyclic Compd.* 28 (5): 581–585.

101 Stepanov, A.I., Dashko, D.V., and Astrat'ev, A.A. (2012). 3,4-Bis(4′-nitrofurazan-3′-yl)furoxan: a melt castable powerful explosive and a valuable building block in 1,2,5-oxadiazole chemistry. *Cent. Eur. J. Chem.* 9 (4): 329–342.

102 Wang, J., Dong, H.-S., Huang, Y.-G., and Zhou, F.-Q. (2007). Properties of 3,4-bis(nitrofurazano)furoxan. In: *International Conference of ICT, 37th (Energetic Materials)*, (ed. ICT), 171/1–171/16. Karlsruhe: ICT.

103 Wang, J., Li, J.-K., Liang, Q. et al. (2008). A novel insensitive high explosive 3,4-bis(aminofurazano)furoxan. *Propellants Explos. Pyrotech.* 33: 347–352.

104 Wang, J., Li, J.-S., Huang, Y.-G., and Dong, H.-S. (2008). A novel high energy density material compound 3,4-bis(nitrofurazano)furoxan: synthesis characterization and properties. In: *11th Seminar New Trends in Research of Energetic Materials*, Pardubice, Czech Republic (9–11 April 2008) (ed. P.S. Zeman), 182–193. Pardubice: University of Pardubice.

105 Wang, Q.-h. (2003). Properties of DNTF-based melt cast explosives. *Chin. J. Explos. Propellants* 26 (3): 57–59.

106 Sinditskii, V.P., Burzhava, A.V., Sheremetev, A.B., and Aleksandrova, N.S. (2010). Thermal and combustion properties of 3,4-bis(3-nitrofurazan-4-yl)furoxan (DNTF). *Propellants Explos. Pyrotech.* 35: 1–6.

107 Duddu, R., Hoare, J., Sanchez, P. et al. (2017). Synthesis of nitro, dinitro and polynitroalkylamino derivatives of trifurazanoxide. *J. Heterocyclic Chem.* 54: 3087–3092.

108 Yu, Q., Chinnam, A.K., Yin, P. et al. (2020). Finding furoxan rings. *J. Mater. Chem. A* 8 (12): 5859–5864.

109 Tang, Y., He, C., Mitchell, L.A. et al. (2016). Small cation-based high-performance energetic nitraminofurazanates. *Chem. Eur. J.* 22 (33): 11846–11853.

110 Leonard, P.W., Pollard, C.J., Chavez, D.E. et al. (2011). 3,6-Bis(4-nitro-1,2,5-oxadiazol-3-yl)-1,4,2,5-dioxadiazene (BNDD): a powerful sensitive explosive. *Synlett* 14: 2097.

111 Kim, T.-K., Lee, B.W., and Chung, K.-H. (2011). Synthesis and characterization of bisnitrofurazanodioxadiazine (BNFOZ). *Bull. Korean Chem. Soc.* 32: 3802–3804.

112 Astrat'ev, A.A., Dashko, D.V., and Stepanov, A.I. (2012). Unusual reaction of 3,4-bis(3-nitrofurazan-4-yl)furoxan with ammonia, primary amines and hydrazine. *Cent. Eur. J. Chem.* 10 (4): 1087–1094.

113 Stepanov, A.I., Astrat'ev, A.A., Dashko, D.V. et al. (2012). Synthesis of linear and cyclic compounds containing the 3,4-bis(furazan-3-yl)furoxan fragment. *Russ. Chem. Bull. Int. Ed.* 61 (5): 1024–1040.

114 Zhang, Y., Zhou, C., Wang, B. et al. (2014). Synthesis and characteristics of bis(nitrofurazano)furazan (BNFF), an insensitive material with high energy-density. *Propellants Explos. Pyrotech.* 39: 809–814.

115 Astrat'ev, A.A., Stepanov, A.I., Sannikov, V.S., and Dashko, D.V. (2016). Synthesis and some chemical characteristics of 4″-nitro-3,3′:4′,3″-ter-1,2,5-oxadiazol-4-amine. *Russ. J. Org. Chem.* 52 (8): 1194–1202; Astrat'ev, A.A., Dashko, D.V., and Stepanov, A.I. (2013). Synthesis, energetic and some chemical properties of new explosive-3,4-bis(4′-nitrofurazanyl-3′-yl)furazan (BNTF). In: *16th International Seminar on New Trends in Research of Energetic Materials* (ed. P.S. Zeman), 474–488. Pardubice: University of Pardubice.

116 Sergievskii, A.V., Romanova, T.V., Mel'nikova, L.G., and Tselinsky, I.V. (2005). Reactivity of 3,5-bis-(4-amino-1,2,5-oxadiazol-3-yl)-1,2,4-triazole. *Russ. J. Org. Chem.* 41 (2): 261–267.

117 Gasco, A.M., Cena, C., Di Stlo, A. et al. (1996). Synthesis and structural characterization of the trimeric furoxan system, a new potent vasodilating moiety. *Helv. Chim. Acta* 79: 1803–1817.

118 Larin, A.A., Fershtat, L.I., and Makhova, N.N. (2020). Tandem reactions of thermolysis and [3+2] cycloaddition in the synthesis of 3-hetaryl-4-nitrofuroxans from 4-nitrofuroxannitrolic acid. *Chem. Heterocyclic Compd.* 56 (5): 607–610.

119 He, C., Gao, H., Imler, G.H. et al. (2018). Boosting energetic performance by trimerizing furoxan. *J. Mater. Chem. A* 6: 9391–9395.

120 Fershtat, L.L., Struchkova, M.I., Goloveshkin, A.S. et al. (2014). Dinitrogen trioxide-mediated domino process for the regioselective construction of 4-nitrofuroxans from acrylic acids. *Heteroat. Chem.* 25 (4): 226–237.

121 Ovchinnikov, I.V., Makhova, N.N., Khmel'mitskii, L.I. et al. (1998). Dinitrodiazenofuroxan as a new energetic explosive. *Dokl. Chem.* 359: 67–70.

122 Adolph, H.G. (1975). Oxidation reactions of 2,2,2-fluorodinitroethylamine and some *N*-alkyl-2,2,2-fluorodinitroethylamines. *J. Org. Chem.* 40 (18): 2626–2630.

123 Mehta, N., Oyler, K., Chemng, G. et al. (2014). Primary explosives. *Z. Anorg. Allg. Chem.* 640 (7): 1309–1313.

124 Fronabarger, J.W., Williams, M.D., Sanborn, W.B. et al. (2011). DBX-1 – a lead free replacement for lead azide. *Propellants Explos. Pyrotech.* 36: 541–550.

125 Smith, D.M., Manship, T.D., and Piercey, D.G. (2020). Synthesis of 5-nitrotetrazolates by the direct oxidation of 5-aminotetrazole in a single pot synthesis without isolation of explosive intermediates. *ChemPlusChem* 85: 2039–2043.

126 Fischer, D., Klapotke, T.M., and Stierstorfer, J. (2014). Potassium 1,1′-dinitramino-5,5′-bistetrazolate: a primary explosive with fast detonation and high initiation power. *Angew. Chem. Int. Ed. Engl.* 53 (31): 8172–8175.

127 Li, H., Zhao, F.-Q., Gao, H.-X. et al. (2014). Three energetic salts based on oxy-bridged bis(*gem*-dinitro) furazan: syntheses, structures and thermal behaviors. *Inorg. Chim. Acta* 423: 256–262.

128 Li, H., Zhao, F.-q., Wang, B.-z. et al. (2015). A new family of energetic salts based on oxy-bridged bis(dinitromethyl)furazan: syntheses, characterization and properties. *RSC Adv.* 5 (27): 21422–21429.

129 Tang, Y., Gao, H., Imler, G.H. et al. (2016). Energetic dinitromethyl group functionalized azofurazan and its azo furazanates. *RSC Adv.* 6: 91477–91482.

130 Tang, X.-W., He, C., Mitchell, L.A. et al. (2016). Potassium 4,4′-bis(dinitromethyl)-3,3′-azofurazanate: a highly energetic 3D metal–organic framework as a promising primary explosive. *Angew. Chem. Int. Ed. Engl.* 55: 5565–5567.

131 Zhai, L., Fan, X., Wang, B. et al. (2015). A green high-initiation-power primary explosive: synthesis, 3D structure and energetic properties of dipotassium 3,4-bis(3-dinitromethylfurazan-4-oxy)furazan. *RSC Adv.* 5 (71): 57833–57841.

132 Li, W., Wang, K., Qi, X. et al. (2018). Construction of a thermally stable and highly energetic metal–organic framework as lead-free primary explosives. *Cryst. Growth Des.* 18 (3): 1896–1902.

133 He, C. and Shreeve, J.M. (2016). Potassium 4,5-bis(dinitromethyl)furoxanate: a green primary explosive with a positive oxygen balance. *Angew. Chem. Int. Ed. Engl.* 55 (2): 772–775.

134 Li, Y., Huang, H., Yang, J. et al. (2017). Potassium nitraminofurazan derivatives: potential green primary explosives with high energy and comparable low friction sensitivities. *Chem Eur. J.* 23: 7353–7360.

135 Huang, H., Li, Y., Yang, J. et al. (2017). Materials with good energetic properties resulting from the smart combination of nitramino and dinitromethyl group with furazan. *New J. Chem.* 41 (15): 7697–7704.

136 Yu, Q., Yang, H., Ju, X. et al. (2017). 1D energetic metal–organic framework: sodium 6-nitro-5-oxidopyrazolo[3,4-*c*][1,2,5]oxadiazol-4-ide with good thermal stability. *ChemistrySelect* 2 (17): 4673–4677.

10

Insensitive High Explosives Containing Tetraazapentalene Moiety

Ernst-Christian Koch

Lutradyn Energetic Materials, Burgherrenstraße 132, D-67661 Kaiserslautern, Germany

10.1 Introduction

[1,2,3]Triazolo[1,2-*a*][1,2,3]triazol-8-ium-1-ide(1,3*a*,6,6*a*-tetraazapentalene; **1**) is a fused-ring hetero-aromatic structure containing 10 π-electrons (Figure 10.1) [1]. Although attempts to prepare compound **1** were not successful until now, its benzo and pyrido derivatives, such as 2,4,8,10-tetranitrobenzo[*d*]benzo-[4,5][1,2,3]-triazolo[1,2-*a*][1,2,3]triazol-6-ium-5-ide (y-TACOT; **3**) [2], mixed benzo-pyrido 2,4,8,10-tetranitrobenzo-[4′,5′]-[1,2,3]-tri-azolo[2′,1′:2,3][1,2,3]triazolo[4,5-*b*]pyridin-6-ium-5-ide-(2,4,8,10-tetranitrobenzopyrido-5,6,7,12-tetraazapentalene, BPTAP; **4**) [3], and dipyrido derivatives aza-TACOTs **5** [4] and **6** [5], were synthesized (Figure 10.1). It should be mentioned that preparation and spectral properties of structurally related 1,3*a*,4,6*a*-tetraazapentalene (z-TACOT; **2**) were described for the first time by Carboni in 1959 [6].

In the family of the y-TACOT analogs, compound **4** assumes a prominent position as a highly thermostable and insensitive explosive. Though the synthesis of **4** was reported by Maquestiau in 1983 [3], it was not until the mid-2000's that explosive properties of this compound were described in the open literature [7]. In general, these tetraazapentalenes are highly thermally stable, and the above and others have been proposed as insensitive explosives when annulated to a nitroaromatic. Annulation to other heterocycles with higher energetic performances such as 1,2,3,4-tetrazine-1,3-dioxides is also known.

10.2 Synthesis of TACOT Derivatives

Synthesis of z-TACOT (**2**) was achieved in four steps (Figure 10.2), starting with oxidation *ortho*-phenylene-diamine to corresponding diazo compound **8**, which was further converted into bis-azido derivative **9** *via* diazonium chemistry. These transformations were followed by thermal cyclization reactions to key precursor **11** and the synthesis was completed by nitration [6].

Nitrogen-Rich Energetic Materials, First Edition. Edited by Michael Gozin and Leonid L. Fershtat.

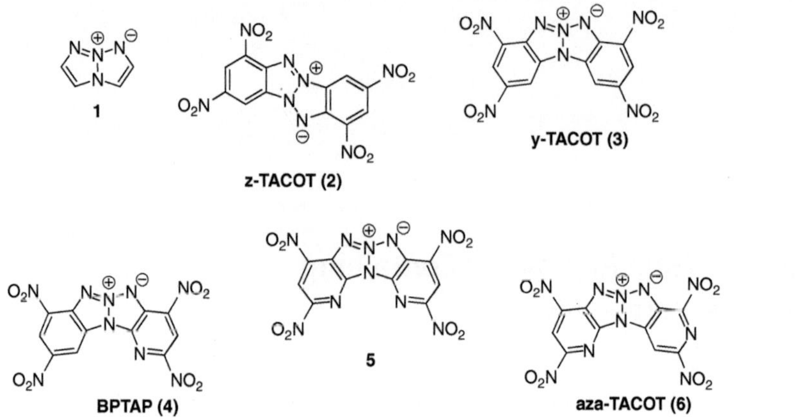

Figure 10.1 Structures of 1,3*a*,6,6*a*-tetraazapentalene (**1**), y-TACOT (**2**), BPTAP (**3**), and other related tetraazapentalenes.

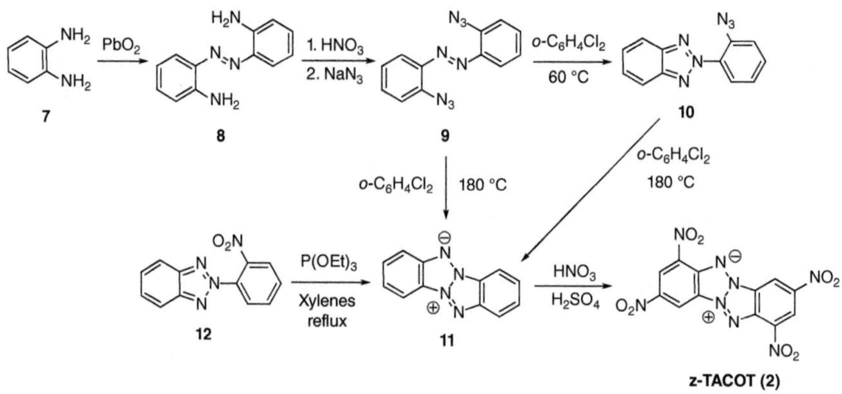

Figure 10.2 Synthesis of z-TACOT (**2**).

Synthesis of y-TACOT (**3**) was also conducted in four steps (Figure 10.3), starting with coupling of *ortho*-phenylene-diamine with *ortho*-chloronitrobenzene to yield N^1-(2-nitrophenyl)benzene-1,2-diamine (**13**). The latter product was reacted with sodium nitrite in acetic acid, leading to formation of 1-(2-nitrophenyl)-1*H*-benzo[*d*]-[1,2,3]triazole (**14**), which was further treated with triethyl phosphite in refluxing xylenes, to produce key precursor **16**. The synthesis of the target y-TACOT (**3**) was completed by nitration of **16** in a mixture of nitric and sulfuric acids [2].

Synthesis of aza-TACOT (**6**) was conducted in four steps, as well (Figure 10.4), and was shown to be accomplished in two synthetic routes, using pyridine-3,4-diamine (**17**) or pyridine-2,3-diamine (**21**), as starting materials. Each of these starting materials was converted to corresponding 1*H*-[1,2,3]triazolo[4,5-*c*]pyridine (**18**) and to 1*H*-[1,2,3]-triazolo[4,5-*b*]pyridine (**22**), respectively, *via* reaction with sodium nitrite in acetic acid. Formed triazolo-pyridine derivatives were subsequently coupled with 2-chloro-3-nitropyridine (**19**) or with 4-chloro-3-nitro-pyridine (**23**) in DMSO to yield intermediates 1-(3-nitropyridin-2-yl)-1*H*-[1,2,3]triazolo[4,5-*c*]

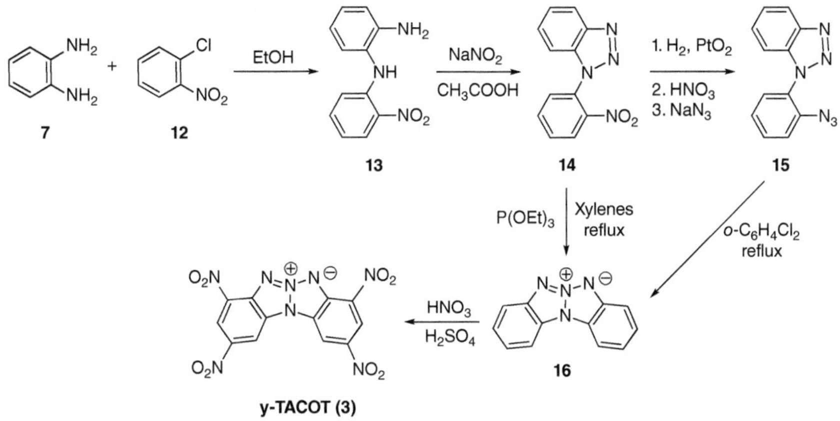

Figure 10.3 Synthesis of y-TACOT (**3**).

pyridine (**20**) and 3-(3-nitropyridin-4-yl)-3*H*-[1,2,3]triazolo[4,5-*b*]pyridine (**24**). Upon cyclization of each of these intermediates with triethylphosphite in refluxing xylenes, key precursor **20** (analogous to compound **16**, Figure 10.3) was formed. The synthesis of the target aza-TACOT (**6**) was completed by nitration of compound **20** in a mixture of nitric and sulfuric acids [6].

Currently, the best-known synthetic route to obtain BPTAP (**4**) (Figure 10.5) is similar to the methodology described in Figure 10.4. The synthesis of compound **1** was made in an analogue manner to y-TACOT [1, 6] (Figure 10.2).

A similar approach was utilized for the synthesis of amino-substituted BPTAP – amino-BPTAP. Initial 5-chloro-benzotriazole **28** was coupled with 2-chloro-3-nitropyridine **19** followed by reductive cyclization of formed derivative **29** into tetraazapentalene **30**. Nitration of the latter compounds with the subsequent nucleophilic substitution of the chlorine with ammonia afforded amino-BPTAP (Figure 10.6) [8].

Analogously, the nucleophilic substitution of one of the nitro groups in the pyridine ring of BPTAP, upon reaction with ammonia, led to formation of 4-amino-2,8,10-trinitrobenzo[4′,5′][1,2,3]triazolo[2′,1′:2,3][1,2,3]triazolo-[4,5-*b*]-pyridine-6-ium-5-ide **32** (Figure 10.7) [8].

Further structural diversification of the dibenzotetraazapentalene core was accomplished by an assembly of the furoxan ring according to a synthetic scheme depicted in Figure 10.8. Nucleophilic substitution of two nitro groups in y-TACOT with NaN_3 afforded diazido derivative **33**, which was nitrated to the tetranitrodibenzotetraazapentalene **34**. Subsequent thermal cyclization of vicinal azido and nitro groups in **34** furnished the formation of y-DBBD [9].

Reactivity of **z-TACOT (2)** was extensively studied. Unfortunately, selective reduction of the nitro groups in **z-TACOT** proved difficult, but a variant of this approach suggested itself. An interaction of **z-TACOT** with sodium azide in DMSO at 100 °C resulted in a formation of 4,10-diazido-2,8-dinitrobenzotriazolo[2,1-*a*]-benzotriazole **35** (Figure 10.9). Reaction of this material with triphenylphosphine in ethanol or benzene, at ambient temperature or under reflux conditions, gave

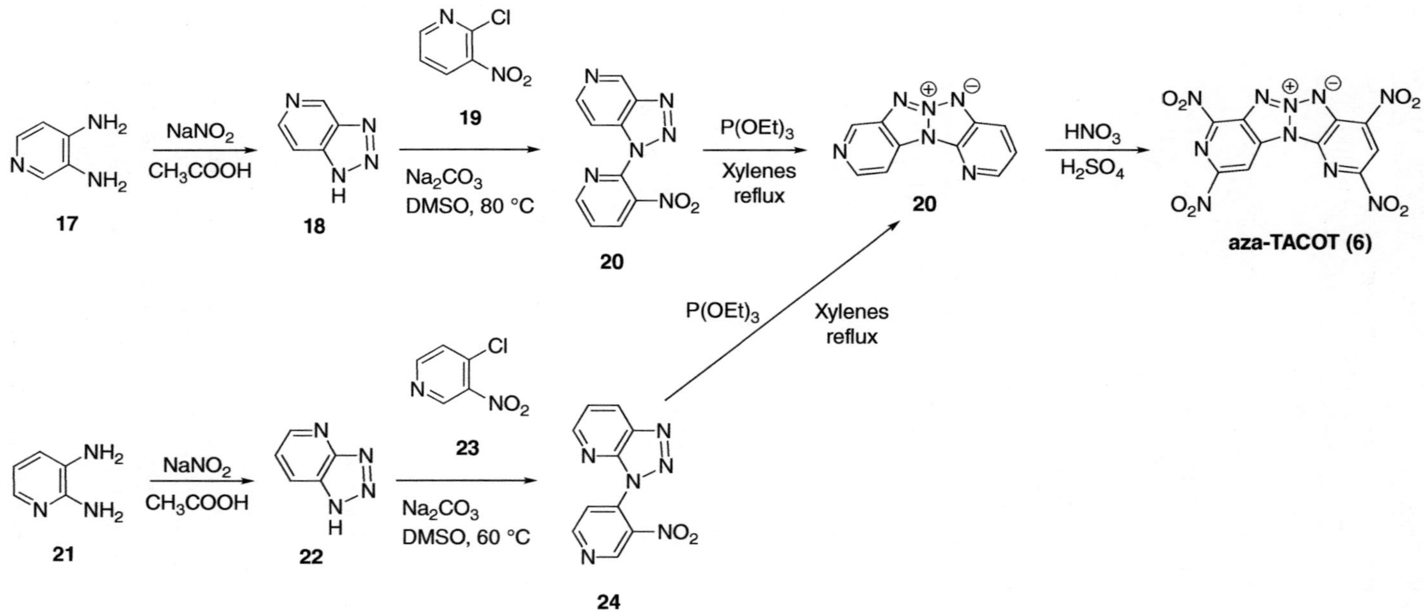

Figure 10.4 Synthetic routes for preparation of aza-TACOT (**6**).

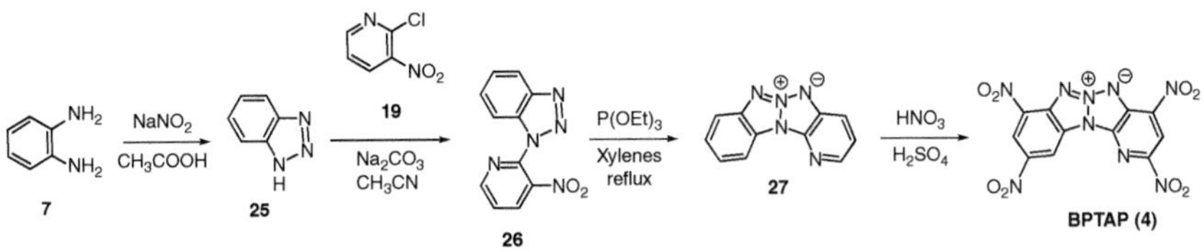

Figure 10.5 Synthesis of BPTAP (**4**).

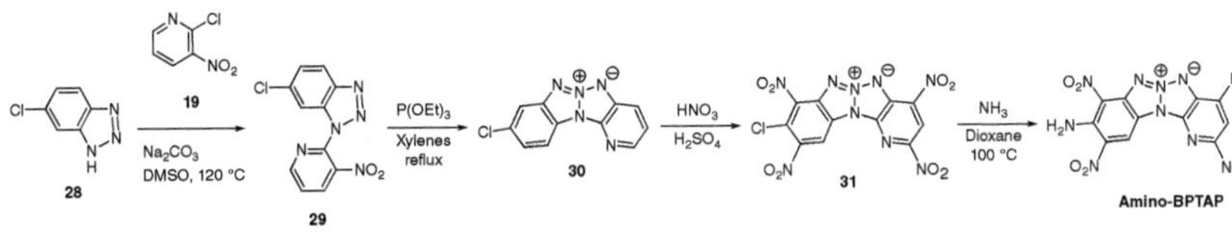

Figure 10.6 Synthesis of amino-BPTAP.

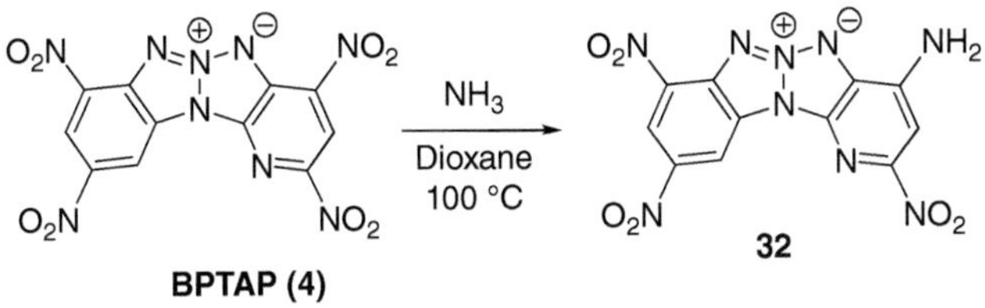

Figure 10.7 Synthesis of pentalene **32**.

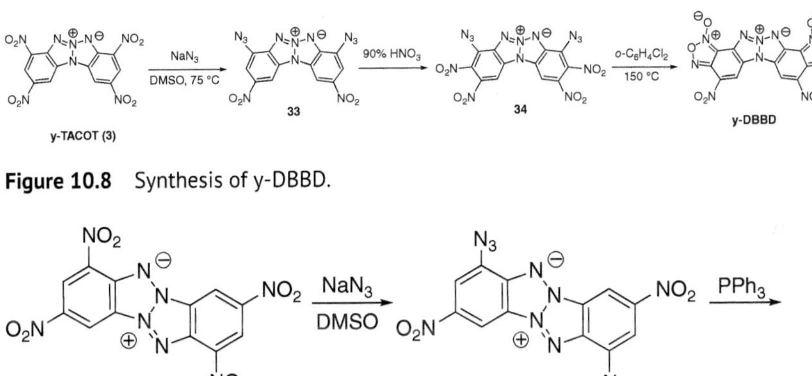

Figure 10.8 Synthesis of y-DBBD.

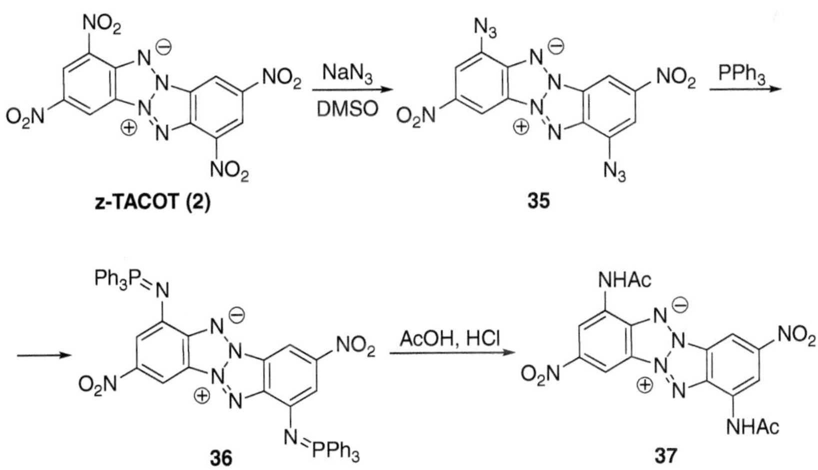

Figure 10.9 Reactivity of nitro groups in **z-TACOT (2)**.

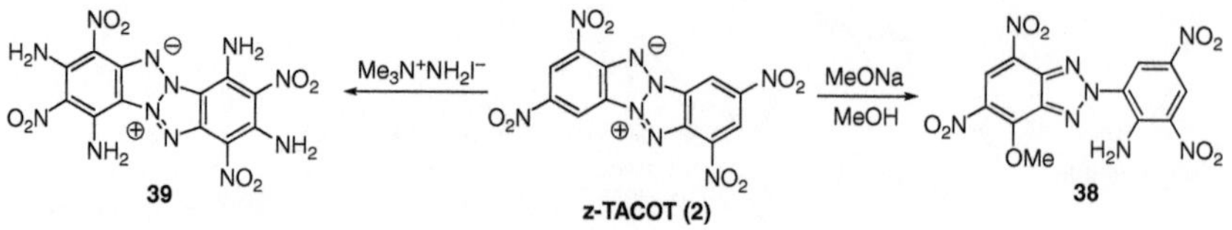

Figure 10.10 Interaction of **z-TACOT (2)** with sodium methoxide and aminating reagents.

an insoluble purple solid that on the basis of infrared and mass spectroscopy was proposed to be bis(triphenylphosphinimine) **36**. Hydrolysis of this phosphinimine required more vigorous conditions than were usually necessary, but use of concentrated hydrochloric acid in acetic acid under reflux conditions produced 4,10-bis(acet-amido)-2,8-dinitrobenzotriazolo[2,1-*a*]benzotriazole **37**. Surprisingly, the latter compound was found to be resistant to both hydrolysis to the 4,10-diamino compound and to further nitration, giving only decomposition products under harsh conditions [10].

Interestingly, the reaction of **z-TACOT (2)** with sodium methoxide took a substantially different course. Instead of substitution of a nitro group, addition of methoxy group occurred at the 1-position, followed by scission of the remote triazole ring to give 2-(2′-amino-3′,5′-dinitrophenyl)-7-methoxy-4,6-dinitrobenzotriazole **38** (Figure 10.10). It should also be noted that the methoxy group in **38** is quite labile and can undergo a facile nucleophilic substitution by ammonia, by azide ion, and even by water. Subjecting **z-TACOT (2)** to the vicarious nucleophilic substitution conditions, using either hydroxylamine or trimethylhydrazinium iodide gave an insoluble red solid that melts with decomposition at 350 °C, and was identified as 1,3,7,9-tetraamino-2,4,8,10-tetranitro-benzo-triazolo[2,1-*a*]benzotriazole **39** [10].

The thermal cyclization of formylaminotriazole **40** in 1,2-dichlorobenzene at 180 °C resulted in a formation of a mixture of (formylamino)tetraazapentalenes **41a** and **41b** in 52% and 31% yields, respectively, bearing an exploso-phoric *tert*-butylazoxy group. Deformylation of compounds **41a** and **41b** was performed in methanol in the presence of methanesulfonic acid at 50 °C, leading to a complete conversion into aminotetraazapentalenes **42a** and **42b** (Figure 10.11). Interestingly, thermal cyclization of aminotriazole **43** at 150 °C afforded aminotetraazapentalene **42a** in 70% yield with no formation of isomer **42b** [11].

The heating of tetrazine 1,3-dioxide **44a** in 1,2-dichlorobenzene at 170 °C yielded a mixture of two isomeric tetraazapentalenes **45a** and **46a** in a 5.3 : 1 ratio (according to ^{1}H NMR) in 75% total yield (Figure 10.12). The cyclization of nitrosubstituted tetrazine 1,3-dioxide **44b** required heating under the same conditions and afforded a mixture of tetraazapentalenes **45b** and **46b** in a 1.1 : 1 ratio in high yield. Due to the poor solubility in organic solvents and close R_f values, only small amounts (several milligrams) of tetraazapentalenes **45a**, **46a**, and **46b** were isolated in pure form using preparative TLC [12].

Nitration of tetraazapentalenes **45a** and **45b** in a mixture of concentrated HNO_3 and H_2SO_4 gave dinitrotetraazapentalene **45c** in good yields (Figure 10.13). It should be noted that nitration of a mixture of isomeric tetraazapentalenes **45a** and **46a**

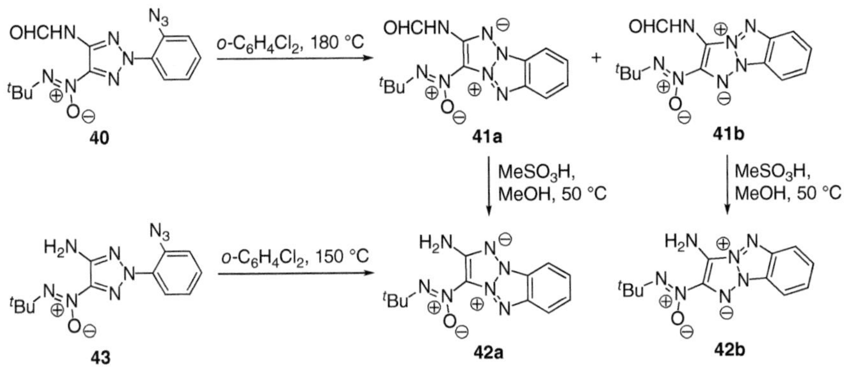

Figure 10.11 Synthesis of tetraazapentalenes bearing *tert*-butylazoxy group.

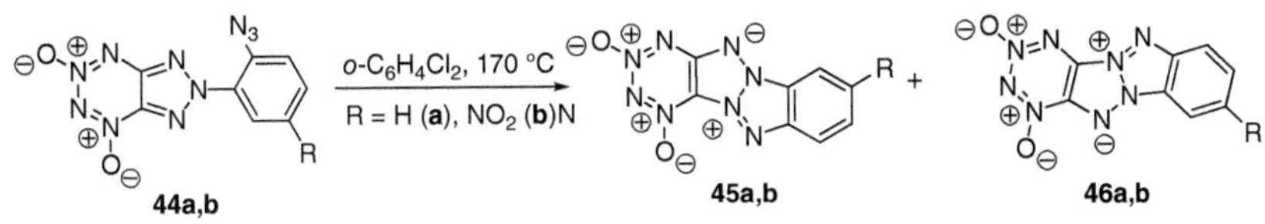

Figure 10.12 Synthesis of tetraazapentalenes annulated with a tetrazine 1,3-dioxide ring.

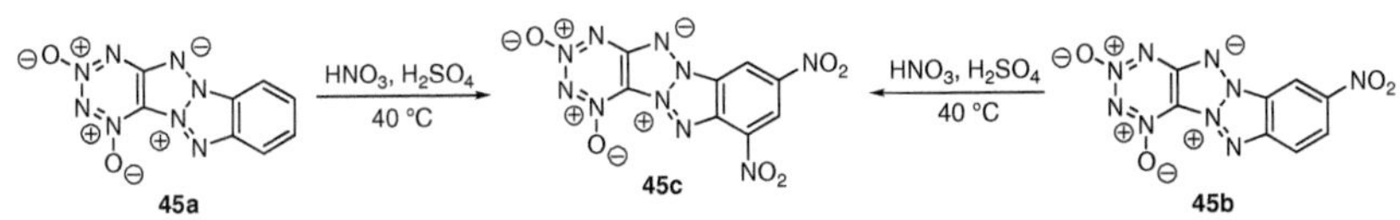

Figure 10.13 Nitration of tetraazapentalenes **45a** and **45b**.

under similar conditions afforded also only dinitrotetraazapentalene **45c**, because compound **46a** decomposed at these reaction conditions [12].

10.3 Crystal and Molecular Structure

BPTAP (**1**) forms orange platelets, crystallizing in the orthorhombic space group $P2_12_12_1$ with a ρ of 1.868 g cm^{-3}. BPTAP is a planar molecule; only the nitro groups are tilted slightly out of the plane by 10–15° (Figure 10.14 and Table 10.1) [13].

The bond lengths and angles for the pentalene unit and the aromatic substructures are generally in good agreement with calculated (B3LYP/6-311+G(2d,p); DFT-HF) and with measured values for the unsubstituted compound [14, 15]. They are also in the good agreement with other derivatives of BPTAP [16, 17], as well. However, one significant deviation can be observed in the central pentalene unit. While the pentalene core in the parent compound is symmetrical with regard to the central N1—N2 bond (C_2 axis), BPTAP is not. Instead, it shows a N1—C5 bond which is markedly shorter (135.27 pm) in comparison to N1—C11 bond (140.48 pm). The observed bond lengths indicate the significance of the resonance structures I, II, III, and IV [18] (Figure 10.15).

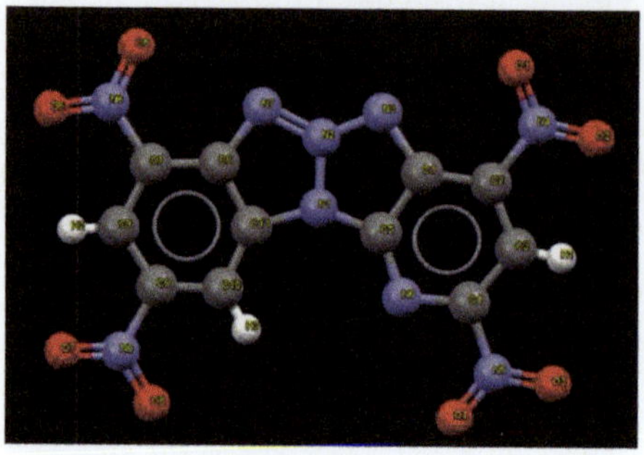

Figure 10.14 Crystal structure of BPTAP (**1**). Source: Adapted from Carboni et al. [1].

Table 10.1 Selected bond length and angles of BPTAP and its non-nitrated parent compound **6.**

Bond	Bond length (pm)	Parent molecule [11]	Bond	Bond length (pm)	Bonds	Bond angle (degree)
N1—N2	136.83	138.41	N8—O6	123.84	N1—N2—N3	114.34
N2—N3	133.50	132.33	C9—N9	146.68	N1—N2—N7	114.17
N3—C1	140.90	137.25	N9—O8	122.48	N2—N1—C5	108.36
C1—C5	142.61	140.00	N9—O7	121.72	N2—N1—C11	107.89
N1—C5	135.27	138.27	C2—N4	146.5	N3—C1—C5	111.60
N2—N7	133.47	131.88	N4—O1	121.7	N7—C6—C11	113.96
N7—C6	139.93	138.05	N4—O2	123.3	N7—N2—N1—C5	178.73
C6—C11	140.25	139.76	C7—N8	145.33	N6—C4—N5—O4	10.08
C11—N1	140.48	138.06	N8—O5	121.64	C3—C2—N4—O2	14.27
C5—N6	132.17	132.92	C4—N5	145.3	C6—C7—N8—O5	14.00
N6—C4	134.26	133.94	N5—O4	120.9	C8—C9—N9—O7	16.23
C11—C10	134.36	138.41	N5—O3	123.9		
C10—C9	134.42	138.73				

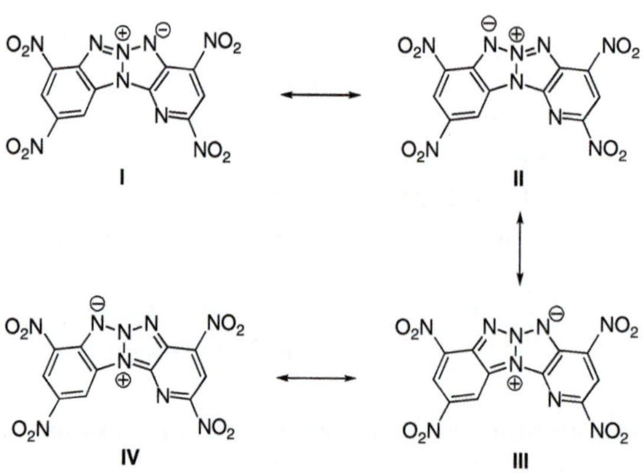

Figure 10.15 Resonance structures of BPTAP (**4**).

In general, pyridine suffers from imbalanced electron density caused by the nitrogen lone pair. Hence, the nitro-substitution on the pyridine moiety of compound **1** affects its electron density and consecutively increases its aromaticity and thermal stability of this molecule [19].

10.4 Spectroscopy

10.4.1 NMR Spectroscopy

The ^{1}H NMR spectrum of BPTAP shows a singlet and a pair of doublets that were assigned to protons H1, H2, and H3 (Figure 10.14 and Table 10.2).

Only a decoupled ^{13}C NMR spectra of BPTAP have been reported so far. Hence, the assignment of the eleven chemical resonances in Table 10.3 is tentative only.

Table 10.2 ^{1}H NMR (DMSO-d_6) chemical shifts in spectrum of BPTAP (**4**).

Chemical shift (ppm)	Assignment
9.27	H1
9.30 (d, $J = 2.0$ Hz)	H3
9.38 (d, $J = 2.0$ Hz)	H2

Table 10.3 ^{13}C NMR (100 MHz, DMSO-d_6) chemical shifts in spectra of BPTAP (**4**) and in spectra of its parent compound **6**.

Compound 1 Chemical shift (ppm)	Assignment	Compound 6 Chemical shift (ppm)
129.8	C1	137.3
150.6	C2	123.3
118.9	C3	121.4
151.6	C4	142.6
134.2	C5	136.1
143.3	C6	143.8
140.6	C7	116.6
120.5	C8	126.8
137.5	C9	123.4
113.4	C10	110.6
124.1	C11	121.9

Numbering according to the structural diagram in Figure 10.2.

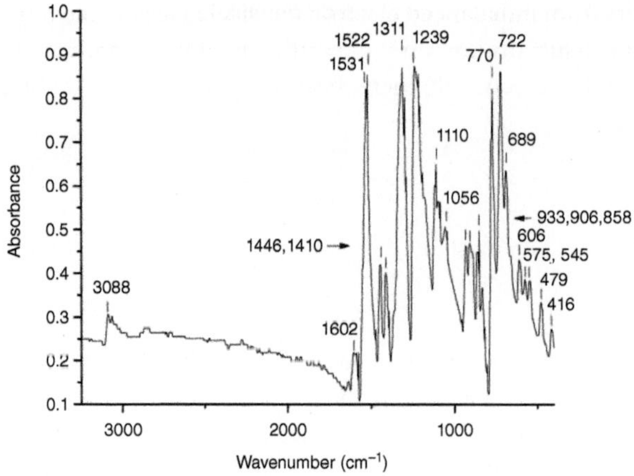

Figure 10.16 FTIR spectrum of compound **4**. Source: Adapted from Tan et al. [22].

Basis for the assignment are both the predicted and measured ^{13}C NMR data for the parent compound **6** [14, 15], as well as reported data for 2,4-dinitrobenzotriazol [20] and 2-nitro-pyridine [21].

The observed ^{13}C NMR shifts for compound **1** are in line with the BPTAP parent compound **6**. The IR spectrum of **4** shows characteristic resonances for NO_2 groups (1522 and 1531 cm^{-1}), C—N bonds (1239 and 1311 cm^{-1}), N=N bonds (1602 cm^{-1}), and the remainder characteristic fingerprint for both pyridine and benzene rings in a spectral range of 416–1110 cm^{-1} (Figure 10.16) [15, 23].

10.5 Thermochemistry

The enthalpy of formation of BPTAP has been determined by combustion calorimetry to +443.5 kJ mol^{-1} [7]. The same authors reported a lower value of +378.2 kJ mol^{-1} [23] as well. Zhang and coworkers reported experimental enthalpy of formation (ΔH_f) to be +497.8 kJ mol^{-1} [15]. In the following calculations, ΔH_f of +497.8 kJ mol^{-1} will be used as reference value. The deflagration of BPTAP (**1**) formally yields products described in Eq. (10.1).

$$C_{11}H_3N_9O_8 \rightarrow 1.5H_2O + 6.5CO + 4.5C + 4.5N_2 \tag{10.1}$$

However, the calculated isobaric deflagration at pressures between 0.1, 0.5, 1, 5, 10, 50 MPa predicts primarily CO, N_2, C(gr), H_2 , and practically no water. In addition, there is only marginal influence of pressure on the equilibrium composition explaining the nearly constant temperature (DT ~ 100 K) over this range, due to the low oxygen balance (Table 10.4).

$$C_{11}H_3N_9O_8 + 7.75O_2 \rightarrow 1.5H_2O + 11CO_2 + 4.5N_2 \tag{10.2}$$

Isobaric combustion with oxygen at pressures between 0.1, 0.5, 1, 5, 10, 50 MPa yields the equilibrium compositions and temperatures (Table 10.5).

Table 10.4 Adiabatic equilibrium composition of isobaric BPTAP deflagration.

Entity ($mol\,mol^{-1}$)	Pressure (MPa)					
	0.1	1	10	50	100	500
T (K)	2670	2712	2729	2736	2740	2762
CO	0.477 64	0.480 81	0.482 04	0.482 08	0.481 74	0.478 32
CO_2	0.000 00	0.000 01	0.000 06	0.000 29	0.000 57	0.002 68
C_2H_2	0.001 55	0.001 88	0.002 02	0.002 05	0.002 05	0.001 99
H	0.012 56	0.004 76	0.001 62	0.000 74	0.000 53	0.000 24
HCN	0.015 84	0.017 70	0.018 47	0.018 66	0.018 69	0.018 61
HNC	0.002 03	0.002 37	0.002 52	0.002 56	0.002 58	0.002 62
H_2	0.072 76	0.075 79	0.076 74	0.076 06	0.074 95	0.067 56
H_2O	0.000 00	0.000 01	0.000 06	0.000 31	0.000 61	0.002 61
N_2	0.259 59	0.260 32	0.260 67	0.260 97	0.261 23	0.262 75
C(gr)	0.157 72	0.156 15	0.155 50	0.155 49	0.155 69	0.157 66

Table 10.5 Adiabatic equilibrium composition of isobaric BPTAP combustion in air.

Entity ($mol\,mol^{-1}$)	Pressure (MPa)					
	0.1	1.0	10	50	100	500
T (K)	3184	3505	3861	4127	4244	4519
CO	0.323 55	0.309 23	0.285 99	0.263 54	0.252 25	0.222 54
CO_2	0.176 59	0.204 72	0.244 12	0.279 27	0.296 34	0.340 05
H	0.022 71	0.015 18	0.009 27	0.006 17	0.005 08	0.003 09
H_2	0.007 10	0.006 46	0.005 58	0.004 87	0.004 55	0.003 79
H_2O	0.030 06	0.035 42	0.041 67	0.046 59	0.048 83	0.054 18
NO	0.021 67	0.029 12	0.037 30	0.042 72	0.044 79	0.048 49
N_2	0.193 76	0.195 66	0.198 13	0.200 53	0.201 80	0.205 47
O	0.069 71	0.052 40	0.035 79	0.025 53	0.021 60	0.013 86
OH	0.039 34	0.041 16	0.040 58	0.038 45	0.037 07	0.032 87
O_2	0.115 45	0.110 51	0.101 19	0.091 53	0.086 56	0.073 31

Unlike the deflagration of BPTAP its combustion in air shows pronounced pressure sensitivity as is evidenced by the decreasing ratios of CO/CO_2 and H_2/H_2O and consequently a significant increase in temperature (DT ~ 1400 K) over the investigated pressure range.

10.6 Detonation Performance

Pristine BPTAP has not been investigated experimentally so far. Therefore, the performance of BPTAP at its crystal density has been calculated using EXPLO5 (6.04), Cheetah 2.0 [24] and the semi-empirical methods developed by Kamlet–Jacobs [23], Keshavarz [25, 26], and Frem [22] (Table 10.6). The data are compared with 1,3,5-triamino-2,4,6-trinitrobenzene, TATB, which is an equally thermally insensitive high explosive.

Table 10.6 shows that the calculated performance of pure BPTAP clearly exceeds pure TATB with regard to the detonation pressure and cylinder expansion energy at given density of 1.84 g cm^{-3}, while the detonation velocity is on par with TATB. Of note, the detonation performance of amino-substituted BPTAP derivatives **31** and amino-BPTAP are similar to the parent BPTAP (**31**: V_D of 7547 m s^{-1}, P_{CJ} of 21.9 GPa; amino-BPTAP: V_D of 7933 m s^{-1}, P_{CJ} of 24.6 GPa) [8]. Detonation properties of y-DBBD is on the same level (V_D of 7520 m s^{-1}, P_{CJ} of 24.5 GPa) [10]. PBX-9502 (95 wt% TATB, 5 wt% Kel-F) is an important insensitive high explosive with an extensive data basis [30, 31]. To assess BPTAP's performance it too has been compounded with Kel-F in the same proportions as in PBX-9502 (95 wt% BPTAP, 5 wt% Kel-F) [7]. Tests have been conducted with both 0.25-in. and 0.5-in. diameter cylindrical strands in a poly-ρ test. Figure 10.17 displays the data from 0.5-in. diameter cylinders versus results obtained with Cheetah 2.0 with the BKWC data set for the same density. Figure 10.17 also depicts results for the poly-ρ test with PBX-9502 (TATB/Kel-F, 95/2) conducted by the LANL (Los Alamos National Laboratory) [32].

In the same density range ($1.70 < \rho < 1.78$ g cm^{-3}) BPTAP/Kel-F detonates on average 200 m s^{-1} faster than PBX-9502. While Cheetah 2.0 tends to overestimate the performance of PBX-9502, it just fits the detonation velocity of BPTAP/Kel-F-800 at densities >90% theoretical maximum density (TMD). The failure diameter of PBX-9502 is 9 mm at ρ of 1.895 g cm^{-3} (98% TMD) [30], while the failure diameter of BPTAP/Kel-F-800 (95/5) is well below 3 mm [7, 24] at ρ of 1.78 g cm^{-3} (95%

Table 10.6 Detonative properties of pristine BPTAP and TATB at density of 1.84 g cm^{-3}.

	BPTAP				TATB	
	Kamlet–Jacobs	Frem[a)] Keshavarz[b)]	EXPLO5	Cheetah	Cheetah	Experimental
ρ (g cm^{-3})	1.84	1.84	1.84	1.84	1.84	1.84
V_D (m s^{-1})	7622	7672[b)]	7670	7743	7779	7728
P_{CJ} (GPa)	26.56	28.02[a)]	24.19	28.17	26.61	25.9
$\nu(2E_G)$ (m s^{-1})	2566	2381[b)]	—	—	—	2440[a)]
$V/V_0 = 9.0$ (kJ cm^{-3})	—	—	—	−7.19	−6.51	—

a) At $\rho = 1.854$ g cm^{-3} [27].
b) At $\rho = 1.847$ g cm^{-3} [28, 29].

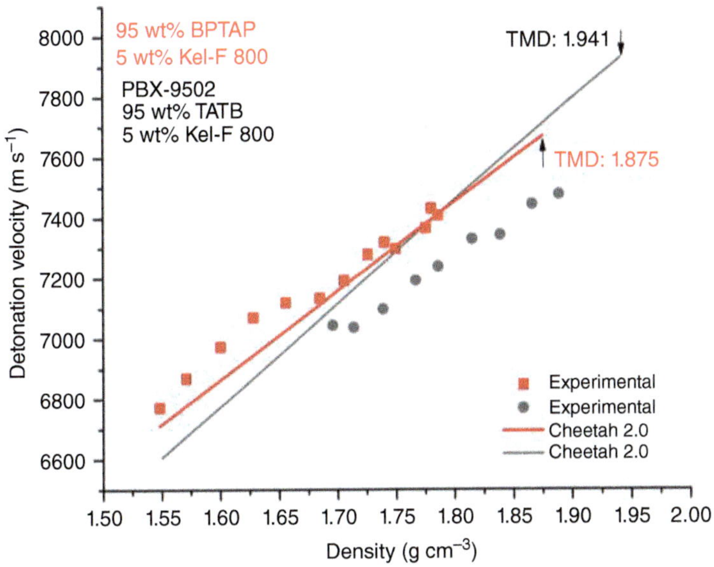

Figure 10.17 Experimental and calculated V_D for both 0.5-in. diameter charges of BPTAP/Kel-F-800 and PBX-9502 [7, 32].

TMD) emphasizing the potential of BPTAP for use in miniaturized initiation trains. Until now, the detonation pressure of BPTAP has only been sampled with the plate dent test. As plate dent tests only probe a small portion of the detonation products, this may lead to significant underestimation of performance in some cases (e.g. nitroguanidine [33]). However, in the absence of more accurate tests

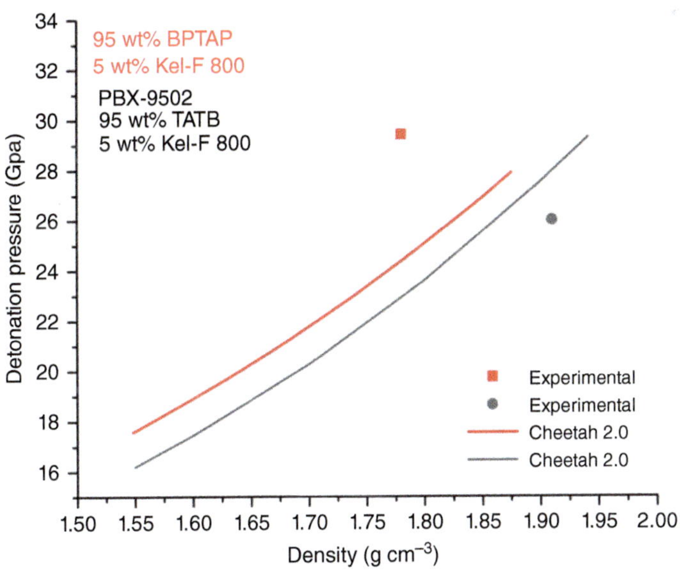

Figure 10.18 Experimental and calculated P_{CJ} for BPTAP/Kel-F-800. Source: Adapted from Huynh and Hiskey [7].

Table 10.7 Detonation properties of BPTAP/Kel-F (95/5) versus PBX-9502 at ρ of 1.84 g cm^{-3}.

	BPTAP/Kel-F (95/5)		PBX-9502		
	Experimental	Cheetah 2.0	Experimental	Experimental	Cheetah 2.0
ρ (g cm^{-3})	1.78	1.78	1.786	1.91	1.91
% TMD (%)	95	95	92	98.4	98.4
V_D (m s^{-1})	7430	7394	7239		7828
P_{CJ} (GPa)	29.4	24.36		26	27.92
$V/V_0 = 9.0$ (kJ cm^{-3})		−6.65			−6.75
$Ø_f$ (mm)	<3		9		

(e.g. copper cylinder expansion test), this will at least give a rough idea of the performance. The plate dent test was carried out with a 0.5-in. diameter charge of BPTAP/Kel-F, at ρ of 1.78 g cm^{-3}, and detonation pressure (P_{CJ}) was determined to be of 29.4 GPa. This detonation pressure was 18% higher than the value predicted with Cheetah 2.0. PBX-9502 was predicted to have a lower detonation pressure by 1.5 GPa. An experimental value, determined at ρ of 1.91 g cm^{-3}, yielded P_{CJ} of 26 GPa (Figure 10.18 and Table 10.7) [34].

10.7 Thermal Behavior

Maquestiau already observed the unusual thermal stability of BPTAP and found that its melting point exceeded the measuring capabilities of the apparatus which was limited to 300 °C [3]. Huynh found the onset of the exo-thermal decomposition at 365 °C (heating rate of 10 °C min^{-1}) and the peak decomposition temperature at 375 °C [7, 23]. The thermal stability of BPTAP has been investigated by Zhang and coworkers [15]. Under non-hermetic conditions (open vessel) the onset of the exothermal decomposition of BPTAP was found at 345.7 °C (Figure 10.19), whereas under hermetic conditions (sealed vessel), the decomposition starts at 334.8 °C (at heating rate of 5 °C min^{-1}). In the open vessel a 60%-mass loss coincides with heat flow (Figure 10.19). In the open vessel the heat of decomposition is −1729.3 J g^{-1}, whereas under confinement it is −2255.9 J g^{-1}.

The 5 second-time to ignition temperature has been determined to T_{5ex} of 400 °C. The decomposition of BPTAP is auto-catalytic in its nature [28], having an activation energy (E_a) of 182 kJ mol^{-1}. The thermal stability of BPTAP exceeds that of PYX, TACOT, and TATB [7, 15], as well as of some other benzopyridotetraazapentalene derivatives, including **32** and amino-BPTAP [8]. It can be further improved by coating with polydopamine/graphene oxide [22]. Tetraazapentalenes **41a** and **41b**, as well as **42a** and **42b** are high-melting crystalline compounds. According to differential scanning calorimetry data, isomer **42a** with the temperature of

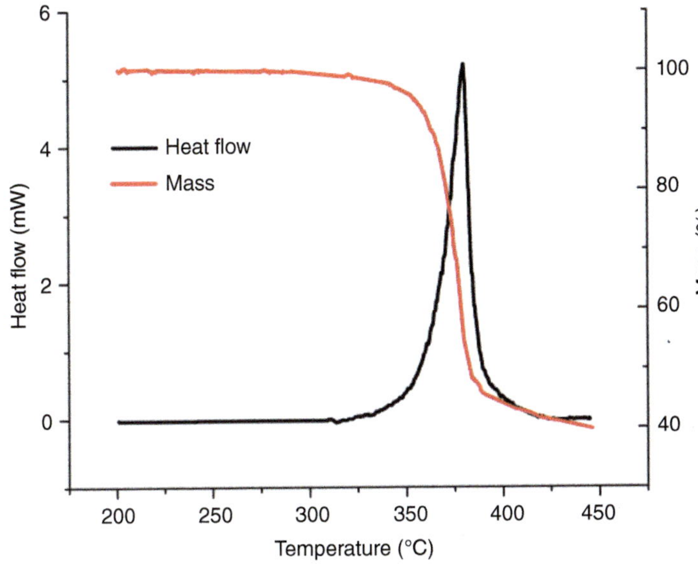

Figure 10.19 Mass loss and heat flow of a sample BPTA at 5 °C min^{-1} heating rate. Source: Adapted from Yang et al. [28].

decomposition onset equal to 240 °C is thermally more stable than isomer **42b** with the corresponding temperature value of 193 °C. The chemical resistance of these compounds probably originates from the presence of *tert*-butyl-NNO-azoxy group as a strong electron-withdrawing substituent adjacent to the amino group [11]. The thermal stabilities of tetraazapentalenes **45a–45c** were determined by differential scanning calorimetry (DSC). These compounds decompose without melting. Compound **45c** was the most stable in the series and had a thermal decomposition onset of 234 °C (10 °C min^{-1}) and 227 °C (5 °C min^{-1}). Tetraazapentalenes **45a** and **45b** had thermal decomposition onsets of 183 °C (10 °C min^{-1}) and 191 °C (10 °C min^{-1}), respectively [12].

10.8 Sensitivity

BPTAP is neither friction sensitive (BAM test) nor sensitive to electrostatic discharge [7]. Pristine BPATP has a 50%-initiation height in the ERL impact test (2.5 kg weight, Type 12 tool) of h_{50} of 59 cm, whereas HMX (class 5) typically yields 25 cm [35]. Wang reported that for BPATP the limiting impact energy was determined to be 9 J, whereas for HMX it was determined to be 7.5 J [23, 36]. For TATB the 50%-initiation height is above 320 cm and the limiting impact energy exceeds the measuring capability of the BAM machine (>50 J) [26]. BPTAP formulated with 5 wt% Kel-F has a 50% initiation height in the ERL impact test of 155 cm [24]. Mechanical sensitivities of closely related pentalenes **31** and amino-BPTAP are on the same level as of BPTAP [8].

Table 10.8 Synopsis of basic properties of BPTAP with TATB.

Property	Unit	BPTAP	TATB
CAS number		86 662-96-6	3058-38-6
Oxygen balance (Ω, CO_2)	wt%	−63.6	−55.78
Nitrogen content	wt%	32.34	32.55
MW	$g\,mol^{-1}$	389.2	258.15
ρ (at 298 K)	$g\,cm^{-3}$	1.868	1.937
ΔH_f	$kJ\,mol^{-1}$	+497.8	−154
V_D (at density)	$m\,s^{-1}$	7430 (1.78 $g\,cm^{-3}$)	7239 (1.786 $g\,cm^{-3}$)
P_{CJ} (at density)	GPa	29.4 (1.78 $g\,cm^{-3}$)	26 (1.91 $g\,cm^{-3}$)
Friction sensitivity	N	>353	>353
Impact sensitivity	cm	59	>250

10.9 Conclusions

BPTAP belongs to a family of thermally and friction insensitive high explosives. BPTAP's detonation performance (V_D, P_{CJ}) and a critical diameter below 3 mm is clearly outperforming other high-temperature insensitive explosives, such as TATB and HNS. A straightforward synthesis of BPTAP from non-expensive starting materials and its properties makes this material a viable alternative for TATB (Table 10.8).

Acknowledgments

The author thanks Mrs. Elizabeth Francois (Los Alamos National Laboratory, Los Alamos, NM, USA), Dr. Ernest Baker (NATO – Munitions Safety Information Analysis Center [MSIAC], Brussels, Belgium), Dr. Randall Simpson (Lawrence Livermore National Laboratory, Livermore, CA, USA), and Alexander Tappan (Sandia National Laboratory, Albuquerque, NM, USA) for helpful discussions on BPTAP.

Abbreviations

BPTAP	benzopyridotetraazapentalene
HNS	hexanitrostilbene, $C_{14}H_6N_6O_{12}$
LANL	Los Alamos National Laboratory
TACOT	tetranitro-2,3,5,6-dibenzo-1,3*a*,4,6*a*-tetraazapentalene
TATB	1,3,5-triamino-2,4,6-trinitrobenzene, $C_6H_6N_6O_6$
TMD	theoretical maximum density

References

1 Carboni, R.A., Kauer, J.C., Castle, J.E., and Simmons, H.E. (1967). Aromatic azapentalenes. I. Dibenzo-1,3*a*,4,6*a*-tetraazapenatlene and dibenzo-1,3*a*,6,6*a*-tetraazapentalene. *J. Am. Chem. Soc.* 89: 2618–2625.

2 Balachari, D. and Trudell, M.L. (1997). Synthesis of new dipyridotetraazapentalenes. *Tetrahedron Lett.* 38: 8607–18610.

3 Maquestiau, A., Flammang-Barbieux, M., Vilain, E., and Pfister-Guillouzo, G. (1983). Synthèse et Réactivité du [3,2-*b*] Pyridino [4,5]benzo-1,3*a*,6,6*a* tetraazapentalene. *Bull. Soc. Chim. Belg.* 92: 67–75.

4 Maquestiau, A., Flammang-Barbieux, M., and Vilain, E. (1986). Nitration de Pyridinobenzotetraazapentalenes. *Bull. Soc. Chim. Belg.* 95: 1117–1122.

5 Ramanaiahh, K.C.V., Stevens, E.D., Trudell, M.L., and Pagoria, P.F. (2000). Synthesis of 1-substituted [1,2,3]-triazolo[4,5-*d*]pyridazines as precursors for novel tetraazapentalene derivatives. *J. Heterocyclic Chem.* 37: 1597–1602.

6 Subramanian, G., Eck, G., Boyer, J.H. et al. (1996). Synthesis of benzotriazolo[1,2-*a*]benzo-triazole derivatives as new high density, insensitive energetic materials. *J. Org. Chem.* 61: 5801–5803.

7 Huynh, M.H.V. and Hiskey, M.A. (2005). Preparation and properties of 2,4,8,10-tetranitro-*5H*-pyrido[3′,2′:4,5]-[1,2,3]-triazolo[1,2-*a*]benzotriazol-6-ium inner salt as a thermally stable high explosive. *J. Energetic Mater.* 23: 27–32.

8 Geng, W.J., Ma, Q., Chen, Y. et al. (2020). Structure–performance relationship in thermally stable energetic materials: tunable physical properties of benzopyridotetraazapentalene by incorporating amino groups, hydrogen bonding, and π–π interactions. *Cryst. Growth Des.* 20: 2106–2114.

9 Subramanian, G., Eck, G., Boyer, J.H. et al. (1996). Synthesis of benzotriazolo [1,2-*a*]benzotriazole derivatives as new high density, insensitive energetic materials. *J. Org. Chem.* 61 (17): 5801–5803.

10 Altmann, K.L., Chafin, A.P., Merwin, L.H. et al. (1998). Chemistry of tetraazapentalenes. *J. Org. Chem.* 63 (10): 3352–3356.

11 Konnov, A.A., Klenov, M.S., Churakov, A.M. et al. (2020). Synthesis and crystal structure of the first amino-1,3*a*,4,6*a*-tetraazapentalenes. *Mendeleev Commun* 30: 139–141.

12 Konnov, A.A., Klenov, M.S., Churakov, A.M. et al. (2018). Synthesis of 1,2,3,4-tetrazine 1,3-dioxides annulated with 1,3*a*,4,6*a*-tetraazapentalene systems. *Asian J. Org. Chem.* 7: 2534.

13 Huynh, M.H.V., Hiskey, M.A., Chavez, D.E., and Gilardi, R.D. (2006). CCDC 277008: Experimental crystal structure determination.

14 Galasso, V., Jones, D., Modelli, A., and Trudell, M.L. (2000). A study of the molecular structure and spectroscopic properties of benzo- and pyrido-tetraazapentalenes. *Chem. Phys.* 254: 375–384.

15 Yang, X., Lin, X., Wang, Y. et al. (2019). TACOT-derived new nitrogen rich energetic compounds: synthesis, characterization and properties. *New J. Chem.* 43: 19180–19185.

16 Geng, W., Ma, Q., Chen, Y. et al. (2020). Structure–performance relationship in thermally stable energetic materials: tunable physical properties of benzopyridotetraaza-pentalene by incorporating amino groups, hydrogen bonding, and *p–p* interactions. *Cryst. Growth. Des.* 20: 2106–2114.

17 Zhang, Z., Geng, W., Yang, W. et al. (2021). Heat-resistant energetic materials deriving from benzopyridotetraazapentalene: halogen bonding effects on the outcome of crystal structure, thermal stability and sensitivity. *Propellants Explos. Pyrotech.* 46: 593–599.

18 Chia, Y.T. and Simmons, H.E. (1967). Aromatic azapentalenes. IV. Heats of combustion of monobenzo- and dibenzo-1,31,6,6*a* tetraazapentalenes and monobenzo- and dibenzo-1,3*a*,4,6*a*-tetraazapentalenes. The structure of tetraazapentalenes. *J. Am. Chem. Soc.* 89: 2638–2643.

19 Türker, L., Gümüş, S., and Atalar, T. (2010). A DFT study on nitro derivatives of pyridine. *J. Energetic Mater.* 28: 139–171.

20 Klapötke, T.M., Pflüger, C., and Reintinger, M.W. (2016). Energetic materials based on 5,7-dinitrobenzotriazole and 4,6-dinitrobenzotriazol-3-ium 1-oxide derivatives. *Eur. J. Inorg. Chem.* 138–147.

21 Rasała, D. and Gawinecki, R. (1993). Effect of the *ortho*-nitro group on the ^{13}C NMR chemical shifts of substituted pyridines. *Magn. Reson. Chem.* 31: 38–44.

22 Tan, L., Lu, Y., Liu, N., and Yan, Q. (2021). Further enhancing thermal stability of thermostable energetic derivatives of dibenzotetraazapentalene by polydopamine/graphene oxide coating. *Appl. Surf. Sci.* 543: 148825.

23 Huynh, H.V., Hiskey, M.A., and Gilardi, R.D. (2005). Tetraazapentalene chemistry: unexpected intramolecular electron rearrangement induced by highly reactive *y*-dinitroso substituents. *Angew. Chem.* 117: 7251–7256.

24 Huynh, H.V. and Hiskey, M.A. (2006). New suitable replacement for the high-temperature explosive HNS-4. *Nuclear Weapons J.* 1: 2–5.

25 Fried, L.E. and Howard, W.M. (1998). Cheetah 2.0. Lawrence Livermore National Laboratory, Livermore, CA, USA.

26 Koch, E.-C. (2021). *High Explosives, Propellants Pyrotechnics*, 219. Boston: De Gruyter.

27 Frem, D. (2017). The use of the [H_2O–CO_2] arbitrary decomposition assumption to predict the performance of condensed high explosives. *Combust. Explos. Shock Waves* 54: 704–711.

28 Yang, F., Chen, Y., Zhao, C., and Zhang, L. (2019). Studies on the thermal behavior and safety of a novel thermostable explosive BPTAP. *RSC Adv.* 9: 22198–22204.

29 Keshavarz, M.H. and Klapötke, T.M. (2017). *Energetic Compounds, Methods for Prediction of Their Performance*, 110. Berlin/Boston: De Gruyter.

30 Gibbs, T.R. and Popolato, A. (ed.) (1980). *LASL Explosive Property Data*, 120–129. Berkeley: University of California Press.

31 Mader, C.L., Shaw, M.S., and Ramsay, J.B. (1981). *PBX-9502 Performance, LA-9053-MS, UC-45*. Los Alamos, NM: Los Alamos National Laboratory. 25 pp.

32 Harry, H.H., Uher, K.J., Hagelberg, S.I., and Hartline, E.L. (2005). The poly-rho test as a screening tool for explosive performance. *Propellants Explos. Pyrotech.* 30: 430–434.

33 Koch, E.-C. (2019). Insensitive high explosives IV. Nitroguanidine, initiation & detonation. *Defence Technol.* 15: 467–487.

34 Tarver, C.M. (2020). Jones–Wilkins–Lee unreacted and reaction product equations of state for overdriven detonations in octogen- and triaminotrinitribenzene-based plastic bonded explosives. *J. Phys. Chem. A* 124: 1399–1408.

35 Koch, E.-C. and Schaffner, D. (2009). Sensitivity of nanoscale energetic materials, I. High explosives. MSIAC-Limited Report L-158. Brussels. 26 pp.

36 Geng, W., Ma, Q., Chen, Y. et al. (2020). Structure–performance relationship in thermally stable energetic materials: tunable physical properties of benzopyridotetraaza-pentalene by incorporating amino-groups, hydrogen bonding, and π–π interactions. *Cryst. Growth. Des.* 20: 2106–2114.

11

Nitrogen-Rich Pyrotechnic Materials for Light and Smoke

Thomas M. Klapötke and Magdalena Rusan

Ludwig-Maximilian University, Department of Chemistry, Energetic Materials Research, Butenandtstrasse 5 – 13 (Haus D), Munich 81377, Germany

11.1 Light-Generating Pyrotechnics

Pyrotechnic formulations which emit electromagnetic radiation in the visible region are used as parachutes or hand-held signal flares for large-area illumination, or to mark specific positions (targets, landing areas for planes or paratroopers, etc.) [1]. For example, red is often used by aircraft personnel to indicate their position in the event of an emergency landing or accident. Green and yellow are used by ground troops to mark their positions, and white ($Mg + Ba(NO_3)_2 + KNO_3$; Al, Ti, and Zr are also used as fuels for producing white light) allows large-area illumination at night. The intensity and wavelength (color) of the emitted light depend on the components present in the burning pyrotechnic mixture. Pyrotechnics which burn at approximately 2200 °C usually contain perchlorates as the oxidizer as well as organic propellants (e.g. nitrocellulose). In order to increase the flame temperature to 2500–3000 °C, metal powders such as magnesium are added. The emission of specific spectral colors occurs as a result of the addition of other metals and other metal salts. Sodium is used for yellow, strontium for red, and barium is usually used for green. The main emitters of the corresponding spectral bands are atomic Na for yellow, $Sr^{(I)}OH$ and $Sr^{(I)}Cl$ for red, and $Ba^{(I)}Cl$ and $Ba^{(I)}OH$ for green (Table 11.1). BaO is also a strong emitter in barium illuminants (Table 11.1). The hydrogen necessary for the formation of the mono-hydroxides SrOH and BaOH comes from the decomposition products of the binder or the PVC (Table 11.1). The Cl atoms are supplied by the perchlorate oxidizing agent, but also from PVC. Table 11.1 shows the composition of some important illuminants in the visible range.

The emission in pyrotechnic compositions generally originates from one of the four following sources:

- Atomic (elemental) emission (e.g. Na) → line spectrum
- Molecular emission (e.g. BaCl, SrOH) → band spectrum
- Incandescence → continuous spectrum
- Luminescence → band spectrum

Nitrogen-Rich Energetic Materials, First Edition. Edited by Michael Gozin and Leonid L. Fershtat.

Table 11.1 Spectra of the most important atoms and molecules used for pyrotechnics.

Element	Emitter	Wavelength (nm)	Color
Lithium	Atomic Li	670.8 610.4 460 413 497 427	Red Red Blue Violet Blue/green Violet
Sodium	Atomic Na	589.0, 589.6	Yellow
Copper	CuCl	420–460 510–550	Blue/violet Green
Strontium	SrCl	661.4, 662.0, 674.5, 675.6	Red
	SrCl	623.9, 636.2, 648.5	Orange
	SrCl	393.7, 396.1, 400.9	Violet
	SrOH	605.0, 646.0 659.0, 667.5, 682.0	Orange Red
	Atomic Sr	460.7	Blue
Barium	BaCl	507, 513.8, 516.2, 524.1, 532.1 649	Green Red
	BaOH	487 512	Blue/green Green
	BaO	604, 610, 617, 622, 629	Orange
	Atomic Ba	553.5 660	Green Red

Colored flames are usually produced by the excitation of species which emit metal atoms (atomic emission). Other metal atoms may combine with other radicals in the flame, forming a color-emitting molecule (molecular emission). The vapors of the emitting species are then excited by the thermal energy of the flame. The atoms or molecules in an excited state then relax to the ground state with the emission of visible light. For example, for the color red, strontium chloride is the main emitting species.

Over the past several years, interest in the development of environment-friendly pyrotechnics for military and civilian applications has grown [1]. In particular, the removal of heavy metals such as barium from green-light-emitting pyrotechnics, and the removal of perchlorate oxidizers from many colored pyrotechnic formulations lies at the forefront of this research area. Barium compounds, long associated with green-light emission, are believed to be a hazard both to human health and to the environment.

Table 11.2 M126A1 control formulation.

Components	Composition (mass %)
Strontium nitrate	39.3
Magnesium 30/50	14.7
Magnesium 50/100	14.7
Polyvinyl chloride (PVC)	14.7
Potassium perchlorate	9.8
Laminac 4116/Lupersol	6.8

Recently, the scaled-up prototype testing of perchlorate-free, hand-held, signal illuminants for the US Army's M126A1 red star parachute hand-held signal has been described [2]. Compared to the perchlorate-containing control, the new illuminants showed excellent stabilities toward various ignition stimuli, while offering superior pyrotechnic performance. These illuminants proved that the development of smaller hand-held signal devices for military applications, but which are environmentally conscious is a realistic and obtainable goal. The results are also important from the perspective of civilian fireworks, as the development of brighter, longer-burning, and environmentally compatible red-light-emitting pyrotechnics is now possible.

Perchlorate-free, red-light-emitting, HHSs (hand-held signals) illuminants have been developed and have advanced from the prototype stage to production level. These high-nitrogen containing illuminants (**1**) are environment friendly, safer to handle (Table 11.2), free of the single-point-of-failure Laminac binder and have exceptional performance. Once consolidated, formulation **A** (Tables 11.3–11.7) was 13% shorter but burned 17% longer and with a luminous intensity comparable with that of the control. Formulation **A** is an excellent candidate for smaller HHSs, because it has the potential to be 30% shorter than those currently used, but still yields comparable results. The acceptable burn time and exceptional luminous intensity means that formulation **B** can also be used as an HHS illuminant which is visible from larger distances. Since formulation **B** burns brighter, formulation **B** (or a derivative of this formulation) may also be of interest to the commercial fireworks industry, with respect to the development of environment-friendly and brighter-burning red-light-emitting fireworks. Now that formulations **A** and **B** have been proven to be successful at the prototype level, they will be incorporated into an HHS, launched into the air, and the performances will be recorded ballistically as part of a systems demonstration [2].

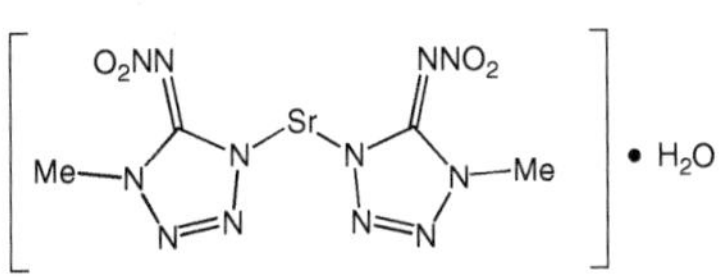

Table 11.3 Perchlorate-free formulations **A** and **B**.

Components	Formulation A (mass %)	Formulation B (mass %)
Strontium nitrate	39.3	39.3
Magnesium 30/50	29.4	35.4
Polyvinyl chloride (PVC)	14.7	14.7
1	9.8	3.8
Epon 813/Versamid 140	6.8	6.8

Table 11.4 Small-scale performance of formulations **A** and **B** in comparison with the control.

Formulation	Burn time (s)	Luminous intensity (cd)	Dominant wavelength (nm)	Spectral purity (%)
Control	32.2	423.5	623.0	86.4
A	39.4	544.3	620.6	89.1
B	35.6	812.8	618.2	90.6

Table 11.5 A comparison of the prototype performance of formulations **A** and **B** with that of the perchlorate-containing control.

Formulation	Burn time (s)	Luminous intensity (cd)	Dominant wavelength (nm)	Spectral purity (%)
Military requirements	50.0	10 000	620 ± 20	76.0
Control	54.0	17 195	613.1	88.6
A	63.3	16 285	612.5	89.9
B	55.1	24 490	612.7	91.6

The development of perchlorate-free hand-held signal illuminants for the US Army's M195 green star parachute has also recently been reported [3]. Compared with the perchlorate-containing control (Table 11.8), the optimized perchlorate-free illuminants were less sensitive toward various ignition stimuli, while still showing comparable burn times and visible-light outputs (Tables 11.9–11.14). The results are also important from the perspective of civilian fireworks, because the development of perchlorate-free illuminants remains an important objective of the commercial fireworks industry.

Environmentally friendly green-light-emitting formulations have been developed and certified for the M195 green-light-emitting HHS parachute at the prototype level. Although the high-nitrogen compounds **1–4** served as initial potassium perchlorate replacements, the barium bis-tetrazolate salt **4** was determined to be the best candidate based on its performance in small-scale M195 formulations (Table 11.8). The

Table 11.6 Packing densities of formulations **A** and **B** and the perchlorate-containing control.

Formulation	Height (cm)	Mass (g)
Control	7.06	89.9
A	6.15	90.1
B	6.22	90.1

Table 11.7 Behavior of formulations **A** and **B** and the perchlorate-containing control toward various ignition stimuli.

Formulation	Impact (IS, J)	Friction (FS, N)	ESD[a] (J)	Thermal onset (°C)
Control	8.8	80	>0.25	264.6
A	10.8	160	>0.25	235.1
B	10.8	>360	>0.25	231.2

a) ESD = electrostatic discharge.

Table 11.8 M195 control formulation.

Components	Composition (mass %)
Barium nitrate	48
Magnesium 30/50	22
Dechlorane plus	15
Potassium perchlorate	10
Laminac 4116/Lupersol	5

bis-tetrazolate salt **4** did not undergo outgassing when subjected to vacuum thermal stability, and X-ray diffraction has previously shown that the water molecule is not simply crystal water but is actually coordinated to barium metal.

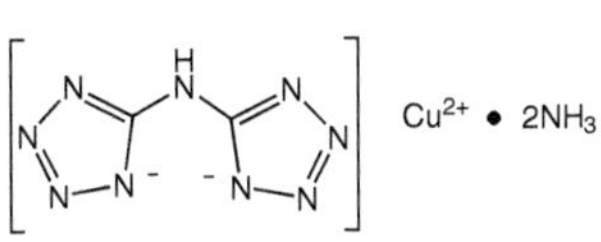

1

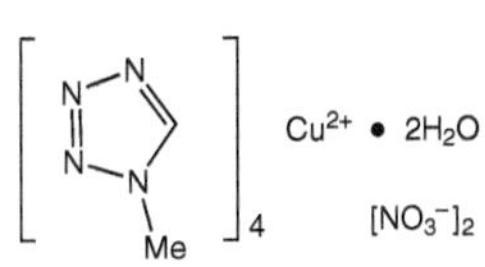

2

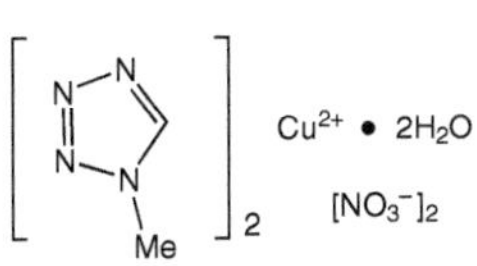

3

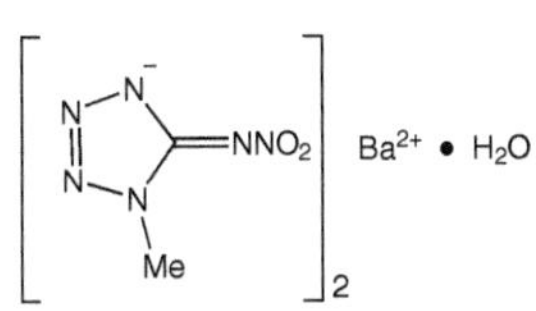

4

Table 11.9 Energetic properties of high-nitrogen salts **1–4**.

	IS (J)	FS (N)	ESD (J)	N (%)	Ω (CO_2) (%)	T_{dec} (°C)	ρ (g cm^{-3})	$-\Delta_c U$ (kJ kg^{-1})
1	100	>360	0.70	61.95	−55	281	1.99	9353
2	30	120	2.00	49.70	−54	168	1.71	10840
3	30	196	2.00	39.85	−30	142	1.95	7415
4	15	>360	0.65	30.05	−20	349	2.30	6050

Table 11.10 Small-scale high-nitrogen M195 formulations.

Formulation	Components	Composition (mass %)
A	Barium nitrate	48
	Magnesium 30/50	22
	Dechlorane plus	15
	1	10
	Epon 813/Versamid 140	5
B	Barium nitrate	48
	Magnesium 30/50	22
	Dechlorane plus	15
	2	10
	Epon 813/Versamid 140	5
C	Barium nitrate	48
	Magnesium 30/50	22
	Dechlorane plus	15
	3	10
	Epon 813/Versamid 140	5
D	Barium nitrate	48
	Magnesium 30/50	22
	Dechlorane plus	15
	4	10
	Epon 813/Versamid 140	5

The bis-tetrazolate salt **4** is stable up to its decomposition temperature of 349 °C. Although formulation **D** showed acceptable performance, was thermally stable, and had a high level of stability toward impact, friction, and ESD, a new formulation needed to be developed due to the presence of dechlorane plus and the concerns associated with its potential bioaccumulation. Consequently, formulation **F** (Tables 11.15–11.19) was developed and certified at the prototype level.

Table 11.11 Small-scale performance of formulations **A–D** in comparison with the control.

Formulation	Burn time (s)	Luminous intensity (cd)	Dominant wavelength (nm)	Spectral purity (%)
Control	44.6	998	559.4	61.2
A	36.6	1420	554.2	58.6
B	38.8	1516	552.9	56.9
C	42.0	1349	554.3	58.1
D	40.2	1773	554.0	61.0

Table 11.12 Burning properties of small-scale formulations.

Formulation	Height (cm)	Mass (g)	Burn rate ($cm\,s^{-1}$)	Mass consumption ($g\,s^{-1}$)
Control	3.92	5.46	0.088	0.122
A	4.35	6.48	0.112	0.177
B	3.92	6.44	0.101	0.166
C	4.07	6.48	0.097	0.154
D	3.98	6.51	0.099	0.162

Table 11.13 Performance of the control and formulation **D**.

Formulation	Burn time (s)	Luminous intensity (cd)	Dominant wavelength (nm)	Spectral purity (%)
Military requirements	50.0	5000	540 ± 20	50.0
Control	55.3	6973	562.3	64.8
D	56.2	6537	561.9	65.3

In contrast to the control, formulation **F** was perchlorate-free and dechlorane plus-free, and it utilized an epoxy binder system instead of a Laminac binder system, thus eliminating single-point-of-failure concerns. The performance of formulation **F** far exceeded the military requirements in terms of burn time, luminous intensity, and spectral purity (Table 11.13). Furthermore, formulation **F** also showed similar burn times and luminous intensities, and lower sensitivities toward various ignition stimuli, while maintaining reasonable thermal stabilities when compared with the perchlorate-containing control. While the dominant wavelength observed for formulation **F** was outside the 540 ± 20 nm dominant wavelength requirement outlined in the Mil-Spec, the US Army in-service control

Table 11.14 Burning properties of the control and formulation **D** prototype formulations.

Formulation	Height of composition (cm)	Mass of composition (g)	Burn rate ($cm\,s^{-1}$)	Mass consumption ($g\,s^{-1}$)
Control	6.08	98.99	0.110	1.79
D	6.11	98.86	0.109	1.76

Table 11.15 Burning properties of formulations **D** and **E**.

Formulation	Height of composition (cm)	Mass of composition (g)	Burn rate ($cm\,s^{-1}$)	Mass consumption ($g\,s^{-1}$)
D	6.11	98.86	0.109	1.76
E	7.22	98.87	0.110	1.50

Table 11.16 PVC-based formulation **F**.

Components	Composition (mass %)
Barium nitrate	48
Magnesium 30/50	27
PVC	15
4	5
Epon 813/Versamid 140	5

Table 11.17 Performance of formulation **F**.

Formulation	Burn time (s)	Luminous intensity (cd)	Dominant wavelength (nm)	Spectral purity (%)
F	59.3	6609	564.6	69.4

formulation was also observed to be slightly outside of this dominant wavelength range. It was therefore concluded that the dominant wavelength issue was not problematic in obtaining a green flame of acceptable quality. Formulation **F** will now be evaluated at the systems demonstration level. As was the case with bis-tetrazolate **4**, formulation **F** will undergo vacuum thermal stability testing as part of the energetic materials qualification process if the systems demonstration testing is successful.

Table 11.18 Burning properties of formulation **F**.

Formulation	Height of composition (cm)	Mass of composition (g)	Burn rate ($cm\,s^{-1}$)	Mass consumption ($g\,s^{-1}$)
F	7.21	98.86	0.122	1.66

Table 11.19 Behavior of formulations **D** and **F** and the perchlorate-containing control toward various ignition stimuli.

Formulation	Impact (IS, J)	Friction (FS, N)	ESD[a] (J)	Thermal onset (°C)
Control	6.9	120	>0.25	293
D	11.8	>360	>0.25	280
F	11.8	>360	>0.25	239

a) ESD = electrostatic discharge.

11.2 Smokes

In general, smokes are cooler burning pyrotechnic mixtures. A cooler burning smoke will rise gradually from the ground when deployed, which will serve as an effective screening/obscuration tool for military personnel. A smoke mixture which has a higher temperature will result in the smoke rising too rapidly. When this happens, the screening/obscuration purposes of the smoke is lost. A smoke which has a high temperature will show combustion of the smoke dye (i.e. oxidation to carbon monoxide and carbon dioxide). Smoke dyes are added into smoke formulations for sublimation purposes. Not only does the sublimation event keep smokes cool (i.e. sublimation is an endothermic process), but it is the sublimation of the smoke dye which gives the smoke its distinct color (i.e. blue, red, yellow, green, and black) [1].

As has been commonly taught in smokes for years, the temperatures at which oxidizers and fuels melt are largely responsible for the temperature which a smoke will ultimately reach. Potassium chlorate has been identified as the "only oxidizer" which is useful in smoke mixes because it has an exothermic decomposition, thus contributing energy to the pyrotechnic system, but it also has a relatively low melting point. Scientists at ARDEC are currently looking at ways to replace potassium chlorate oxidizer in smoke mixes with other oxidizing agents. The EPA is currently looking at potassium chlorate, and environmentalists are questioning whether potassium chlorate is as problematic as potassium perchlorate is. Another problem with potassium chlorate is its sensitivity. Potassium chlorate was used in many early fireworks, but it was replaced by potassium perchlorate, because potassium perchlorate was found to be inherently safer to work with.

As was pointed out above, $KClO_3$ is much less stable than $KClO_4$ and therefore more hazardous (especially in mixtures with sulfur and phosphorus). On the other

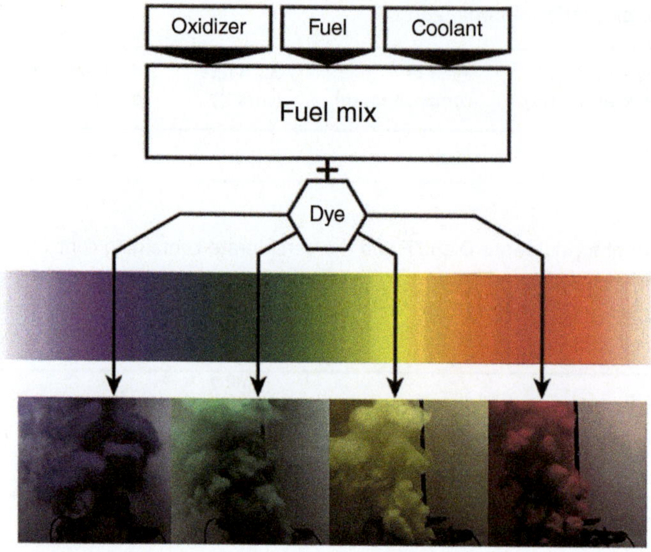

Figure 11.1 The concept of fuel mixes.

hand, it shows high burning rates and easy ignition. $KClO_3$ is slightly more hygroscopic than potassium nitrate and produces a smoke of KCl. Furthermore, it can act as a chlorine donor [4].

The mixture used for producing colored smoke is usually a cooler-burning formula based on potassium chlorate $KClO_3$ (ca. 35%) as the oxidizer, sugar as the fuel (ca. 20%), and one or more dyes, with about 40–50% of the content being the dye. About 2% sodium bicarbonate may be added as a coolant to lower the burning temperature. Its coolant properties are a consequence of the endothermic decomposition of 2 mol of sodium bicarbonate into 1 mol each of sodium carbonate, carbon dioxide, and water. In addition to reducing the flame temperatures of smokes, coolants are also used to reduce the sensitivity of smoke mixes, given the large amount of potassium chlorate used in these mixes. The presence of the coolants minimizes undesired acid formation and helps to prolong storage of the smoke mixes.

The development of sugar-free multi-colored smoke formulations, so-called fuel mixes, has recently been reported [5]. These simple four ingredient-based mixtures of dye, potassium chlorate, 5-amino-1*H*-tetrazole, and a magnesium carbonate derivative are able to produce a variety of colors by using the same basic pyrotechnic system (Figure 11.1). All of the components except the dye are pre-mixed, with addition of the dye is occurring in the final step. New colored smoke formulations were developed, based on previous results which indicated an overall higher smoke performance in terms of efficiency and persistence on utilizing 5-amino-1*H*-tetrazole as the fuel. For large producers as well as consumers, the development of fuel mixes is an effective way to reduce costs and provide a higher degree of safety. The main focus was on dyes applied in the U.S. M18 colored smoke grenades.

The most promising fuel mixes are illustrated in the ternary diagram shown in Figure 11.2. The first fuel mix, FM1, consists of 50 wt% 5-AT, 30 wt% $KClO_3$, and

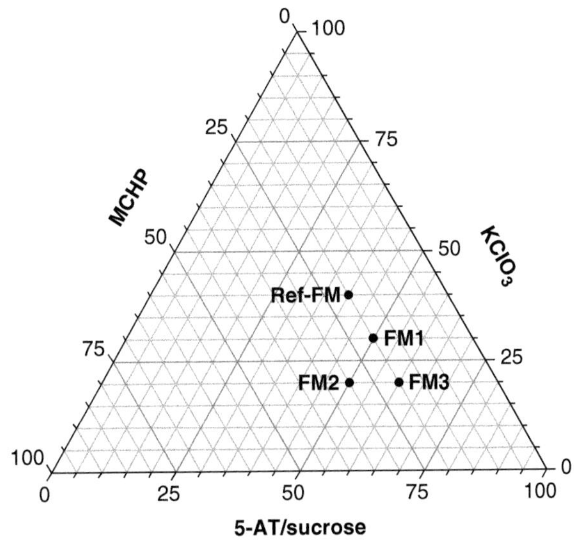

Figure 11.2 Fuel mixes which have been developed. FM1 = 5-AT (50 wt%), $KClO_3$ (30 wt%), MCHP (20 wt%); FM2 = 5-AT (50 wt%), $KClO_3$ (20 wt%), MCHP (30 wt%); FM3 = 5-AT (60 wt%), $KClO_3$ (20 wt%), MCHP (20 wt%); Ref-FM = sucrose (40 wt%), $KClO_3$ (40 wt%), MCHP (20 wt%).

20 wt% MCHP and, therefore, contained the highest amount of oxidizer. FM2 has a composition of 50 wt% 5-AT, 20 wt% $KClO_3$, and 30 wt% MCHP, and finally, fuel mix FM3 contains the highest amount of fuel with 60 wt% 5-AT and equal wt% of $KClO_3$ and MCHP (20 wt%). The reference fuel mix Ref-FM contains 40 wt% sucrose, 40 wt% $KClO_3$, and 20 wt% MCHP.

It was therefore successfully demonstrated that 5-AT-based fuel mixes can be used to produce green, yellow, red, and violet smoke. The formulations that have been developed reveal that non-traditional high-nitrogen fuels can produce smokes of high color quality. More precisely, whereas FM1-based mixtures result in rapid, strong smoke generation in a short time, FM3-based compositions are characterized by a slow, continuous smoke generation over a much longer period of time. 5-AT is only one of many potential candidates which will be investigated in the future. All of the formulations are insensitive toward friction, and in addition, all the 5-AT-based colored smoke formulations are completely insensitive toward impact. The fuel mixes were more sensitive toward mechanical stimuli. A comparison of the aerosols collected revealed similar yields for FM1-based and sugar-based formulations. The superior performance of the sugar-based reference formulations was illustrated by the measured transfer rates. Future investigations will no doubt focus on providing even more colored smoke formulations, e.g. blue and black, by applying the same fuel mixes. However, in order to achieve proper evaluation, new strategies to characterize the aerosol produced by a mixture of two dyes (e.g. green dye mix) have to be established.

In 2017, the Strategic Environmental Research and Development Program (SERDP) announced the need for so-called "next generation pyrotechnics" that show a reduction in the environmental and health impact. The field of smoke-producing pyrotechnics offers a wide variety of research possibilities. The first colored smoke signals consisted mainly of an organic dye, sulfur, potassium chlorate, sodium bicarbonate, and variable amounts of kerosene or tricalcium phosphate. Sulfur – in combination with potassium chlorate – offers a low ignition and combustion temperature, and therefore, it is the perfect candidate for low-temperature smokes. The formation of colored smoke can be obtained by a sublimation–recondensation mechanism. In more detail, the pyrotechnic mixture provides the energy necessary to sublime the dye, which can subsequently recondense as small particles. Maintaining a lower combustion temperature is vital; otherwise, the organic dye would get burned rather than sublimed. Finally, the gaseous combustion products disperse the emerging dye particles, resulting in a dense colored smoke cloud. A big step forward was taken with the substitution of sulfur by alternative fuels. During the combustion of sulfur-based smoke mixtures, hazardous SO_2 is formed, which causes a burning sensation in the lungs when inhaled. Fortunately, sugar-based fuels like sucrose or lactose act in a similar manner to sulfur when paired with potassium chlorate. Sugar is considered to be a less toxic alternative, since the resulting combustion products are mainly limited to harmless water and carbon dioxide.

Another important goal in current research is the elimination of halogens and halogen-containing molecules. In the case of smokes, potassium chlorate still seems to be the only suitable oxidizing agent to ensure the optimal temperature range for dye sublimation is reached. However, several serious issues arise from the use of potassium chlorate. It is highly reactive and tends to undergo accidental ignition, particularly when combined with combustible fuels with low melting points. Furthermore, since it is soluble in water and persistent, it can cause problems for aquatic life due to its toxicity [1]. In addition, its combustion products are an underestimated risk in pyrotechnics. Chlorates in combination with organic materials are known to form toxic and carcinogenic chlorinated organic compounds such as polychlorinated dibenzo-*p*-dioxins (PCDD) or polychlorinated dibenzofurans (PCDF). These gaseous side-products can be inhaled very easily during combustion. Nevertheless, due to the lack of sufficient alternatives, potassium chlorate is still accepted as the common oxidizer in smoke-producing mixtures.

In 2015, the very first chlorine-free red flare was investigated by Sabatini et al. [6]. Their strategy was to add molecules like hexamine or 5-amino-1*H*-tetrazole to a common flare formulation, instead of undesirable potassium perchlorate. They concluded that these nitrogen-rich fuels have a deoxidizing effect on the flame, thus resulting in a lower combustion temperature. This concept was successfully transferred to smoke-generating pyrotechnics by Glück et al., for both colored and white smoke formulations [7]. It was possible to replace sugar as main fuel with hexamine or 5-amino-1*H*-tetrazole, which led to an overall more persistent, and thick, smoke cloud. Another positive side-effect is the release of a high volume of gaseous products such as N_2, which further disperse the dye particles. Nevertheless, until now, it has not been possible to eliminate potassium chlorate in smokes, not even through the use of nitrogen-rich molecules.

The next logical step was to find a suitable nitrogen-rich molecule for smokes, which in the best case can be used without any involving potassium chlorate or other halogen sources. An extensive literature research revealed an interesting

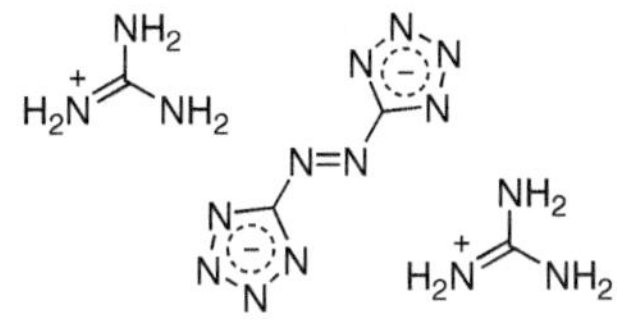

Figure 11.3 Guanidinium 5,5′-azotetrazolate (GZT).

molecule with an overall nitrogen content of 78.9%, namely, guanidinium 5,5′-azotetrazolate (GZT, Figure 11.3) [8]. GZT is a bright yellow powder and burns generating white smoke. Due to its excellent properties such as its high thermal stability, insensitivity toward external stimuli, and desirable low combustion temperatures, it is expected to be an ideal candidate for application in clean propulsion systems, new gas generators, and low signature propellants. GZT derives its power from a high heat of formation in combination with a large gas volume release, which is mainly cool and inert due to the formation of N_2. As a consequence of its excellent properties, GZT was also considered as a gas-generating compound in smokes.

Firstly, the energetic properties of GZT were examined in more detail (Table 11.20) [8]. GZT burns with moderate white smoke generation, and shows a burn time of 19 seconds, which correlates with a burn rate of 11.8 mm s^{-1}. As was reported in the literature, the impact sensitivity of GZT is 35 J, and it is insensitive toward friction and electrostatic discharge. GZT decomposes at 239 °C. The combination

Table 11.20 Selected properties as well as sensitivities of GZT and GZT-based smokes in comparison with chlorate-based reference materials [5]. (For the exact compositions for the smoke formulations, see Table 11.21).

	BT (s)	BR (mm s^{-1})	m_a (mg)	Y (%)	m_d (mg)	$T\%$ (%)	T_{dec} (°C)
GZT	19	11.8	654	32	—	—	239
W3	26	6.9	582	29	—	—	215
Y3	26	5.3	783	38	164	55	239
R3	25	5.6	745	36	169	56	239
V3	29	5.6	779	38	—	—	237
B3	24	6.8	726	35	—	—	239
G3	19	7.3	715	35	—	—	238
Ref-W	38	—	693	35	—	—	203
Ref-Y	13	—	670	33	435	73	178
Ref-R	21	—	729	36	514	86	172
Ref-V	15	—	652	32	—	—	178
Ref-G	19	—	642	32	—	—	172

BT = burn time (2.0 g scale); BR = linear burn rate (5.0–6.0 g scale); m_a = mass of aerosol produced (2.0 g scale); Y = yield factor; m_d = dye content present in aerosol produced; $T\%$ = transfer rate; T_{dec} = temperature of decomposition.
The exact composition of the formulations is summarized in Table 12.21; Table 12.22 shows the composition of the reference formulations. The structures of the dyes are shown in Figure 12.4.

Table 11.21 Exact composition of the smoke formulations from Table 12.20.

Formulation	Dye 1	Dye 1 (mass %)	Dye 2	Dye 2 (mass %)	GZT (mass %)
W3	Terephthalic acid	15			85
Y3	Solvent yellow 33	15			85
R3	Solvent red 1	15			85
V3	Solvent violet 47	15			85
B3	Solvent green 3	15			85
G3	Solvent green 3	10	Solvent yellow 33	5	85

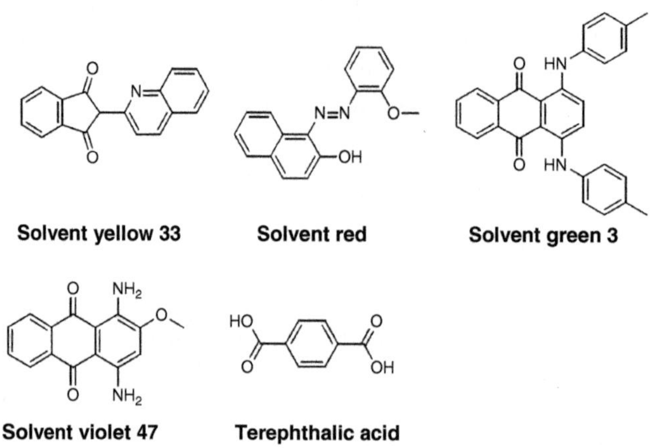

Figure 11.4 Structures of the dyes used for smoke formulations.

of its insensitive energetic properties and high decomposition temperature is advantageous for its use in modern pyrotechnics, since it guarantees safe preparation, storing, and handling (Figure 11.4).

The newly developed smoke formulations should combine acceptable smoke performance, economical manufacturing process, as well as safe application (also by untrained people). For this reason, the smoke system must be as simple and efficient as possible. As a starting point, an easy, quick-mix of GZT with various proportions of dye (5–35%) to develop two-component smokes was tried out. The combustion of GZT was assumed to provide enough heat for sublimation of the dye, and the white smoke produced should appear colored due to the presence of the dye. All GZT-based smoke mixtures were characterized according to their burning behavior (burn time: *BT*, burn rate: *BR*), smoke quality (mass of aerosol m_a, yield factor Y, mass of dye present in aerosol m_d, transfer rate $T\%$), as well as energetic properties. Furthermore, a comparison was made of the results with those of classic chlorate/sugar-based smokes. The reference formulations consisted of different dyes (Terephthalic Acid, Disperse Red 9, Violet Dye Mix, Solvent Yellow 33,

Table 11.22 Compositions of the reference smoke formulations (values given in mass %).

Reference	Solvent yellow 33	Solvent green 3	Disperse red 9	Violet dye mix	Terephthalic acid	$KClO_3$	Hexamine	Sucrose	MCHP[a)]
W					45	27.5	22		5.5
Y	30					28		28	14
R			30			28		28	14
V				30		28		28	14
G	10	20				28		28	14

a) MCHP = magnesium carbonate hydroxide pentahydrate.

(a) White (b) Yellow (c) Red (d) Violet (e) Blue (f) Green

Figure 11.5 Burning of GZT-based colored smoke formulations. (a) White, (b) Yellow, (c) Red, (d) Violet, (e) Blue, (f) Green.

Solvent Green 3), potassium chlorate, sucrose or hexamine and magnesium carbonate hydroxide pentahydrate (Figure 11.4).

11.2.1 White Smoke

It is obvious to assume GZT could be suitable as a white smoke generator (Figure 11.5a). For this reason, the mass of aerosol produced during combustion (654 mg) as well as the yield factor (32%) were determined. These results are very similar to those for the white smoke generating reference compound (693 mg, 35%). GZT may be advantageous due to its superior impact sensitivity (35 J compared to 10 J); however, for application, the spectral properties must also be considered in detail.

In common, less toxic, white smoke signals, the dye of choice is terephthalic acid (TA), since it is easily available at moderate prices and shows acceptable properties [1, 9, 10]. To improve the performance of only GZT, terephthalic acid was added stepwise. The upper limit was found to be 15% TA content; otherwise, it was not possible to ignite the GZT/TA mixture [8]. The sublimation temperature of TA is 402 °C, and therefore slightly higher than those of anthraquinone or quinoline dyes (approximately 300–350 °C) [11]. As a result, the sublimation of TA may consume too much of the system's energy, since it was observed that the smoke properties dropped significantly to 582 mg aerosol and 29% yield. Consequently, the combination of GZT and TA was excluded from further investigations.

11.2.2 Colored Smoke

Environmentally benign, multi-colored smoke signals are currently a highly relevant research topic, since the number of possible areas of application, as well as the demand is increasing. GZT was simply mixed with various dyes to obtain red (Solvent Red 1), violet (Solvent Violet 47), yellow (Solvent Yellow 33), blue (Solvent Green 3), and green (Solvent Yellow 33 + Solvent Green 3) colored smoke. The same procedure as that used for white smoke was applied, in which the dyes were added stepwise to GZT. Since GZT burns with white smoke, it is advantageous to use as much dye content as possible; otherwise, the smoke may lack color intensity. As was observed for TA, the upper limit for colored dyes was also 15% with respect to reliable igniting and burning behavior. Initial tests with the two-component dye/GZT mixtures showed strong and thick smoke generation resulted (Figure 11.5). The color intensity was slightly paler than expected due to GZT, however, it is still clearly recognizable. Finally, the composition of the new smoke system was fixed to 15% dye in combination with 85% GZT for all colors. It is noteworthy, that the dye content of the GZT system is half as much as that for the reference formulations, which needs to be considered during characterization and comparison. Unlike the simple two-component GZT system, the reference formulations were based on sucrose and potassium chlorate.

The burn times of the colored smokes were in the range of 24–29 seconds, with **G3** as the only exception (19 seconds). In contrast to this, the corresponding redox reaction of sucrose/chlorate mixtures is much more violent, resulting in faster smoke generation and shorter burn times (13–21 seconds). The yield factor of pure GZT was 32%, which could be further increased by the addition of dye (35–38%). In comparison with the reference formulations, the yield factor, as well as the mass of aerosol $m_{\underline{a}}$ are in a similar range. These results indicated that the dye particles have been dispersed by GZT. However, it is also possible that the organic dyes could be transformed or destroyed during combustion. Therefore, it was necessary to perform HPLC analysis in order to determine the exact amount of dye present in the aerosol [5, 7b]. The resulting transfer rate, $T\%$, is a measure of the effectiveness of smoke mixtures to disperse the dye, rather than to combust it. In comparison with the red and yellow reference formulations (73–86%), the two-component GZT system reached up to 55–56% transfer rate. This means that about 55% of the dye particles in the pellet are sublimed and present in the smoke cloud which is produced. It should be noted that only half of the dye content was used in the GZT system compared to that used in the reference mixtures (15% instead of 30%). Therefore, it was suggested that the probability of sublimation might be influenced by the significantly different particle content. Nonetheless, these novel and simple "nitrogen-rich-only" smokes already show promise of reaching similar smoke performance as that of presently used formulations.

Furthermore, a particularly important aspect was the overall non-sensitive energetic properties of GZT, in combination with its high decomposition temperature (239 °C). The addition of dye resulted in formulations that are not sensitive to impact, friction, or electrostatic discharge. The temperature of decomposition of all of the colored smoke mixtures was in the same range as that of GZT. This is in stark contrast to the sugar/chlorate mixture in particular, since it is known for the occurrence of spontaneous reactions, as well as its unpredictability. This causes considerable risks during production and storage [12]. By eliminating potassium chlorate, not only is a controversial compound removed, but a much safer way of generating smoke is provided at the same time. This is also advantageous for untrained people, since they can use pyrotechnics without exposing themselves to the danger of unexpected reactions occurring. As potassium chlorate and sugar are essential materials in global industry, they are available in adequate quantities at reasonable prices. In contrast, the nitrogen-rich salt GZT is more complex to produce and therefore more expensive. However, commonly used smokes consist not only of an oxidizer and fuel, but also a coolant and other additives are required. In contrast, GZT can be applied as single compound, which greatly simplifies the manufacturing process, purchasing and storage, as well as improving the previously mentioned safety issues.

Guanidinium 5,5′-azotetrazolate is only one representative of nitrogen-rich, gas-generating compounds. There may be other possible candidates which could further improve the effectiveness of smoke mixtures. For future compounds, the dye content should be increased to ensure unique and brilliant color intensity. One of the most challenging tasks in pyrotechnics is the toxicity of combustion products. Potassium chlorate in combination with organic material leads to the formation carcinogenic polychlorinated compounds, the formation of which can now be prevented by using "nitrogen-rich-only" smokes. However, the toxicity of these combustion products should also be further evaluated to satisfactorily answer questions regarding aspects of health.

Acknowledgments

Financial support of this work by the Ludwig-Maximilian University (LMU), the Office of Naval Research (ONR), and the Strategic Environmental Research and Development Program (SERDP) under contract no. W912HQ19C0033 is gratefully acknowledged.

References

1 Klapötke, T.M. (2019). *Chemistry of High Energy Materials*, 5th edn. Berlin/Boston: Walter de Gruyter.

2 Sabatini, J.J., Nagori, A.V., Chem, G. et al. (2012). *Chem. Eur. J.* 18: 628–631.

3 Sabatini, J.J., Raab, J., Hann, R.K. Jr., et al. (2012). *Chem. Asian J.* 7: 1657–1663.

4 Klapötke, T.M. (ed.) (2021). *Energetic Materials Encyclopedia*, 2e, vol. 1–3. Berlin/Boston: De Gruyter.

5 Glück, J., Klapötke, T.M., and Küblböck, T. (2018). *New J. Chem.* 42: 10670–10675.

6 (a) Sabatini, J.J., Koch, E.-C., Poret, J.C. et al. (2015). *Angew. Chem. Int. Ed.* 54: 10968–10970. (b) Sabatini, J.J., Koch, E.-C., Poret, J.C. et al. (2015). *Angew. Chem.* 127: 11118–11120.

7 (a) Glück, J., Klapötke, T., and Shaw, A. (2017). *Cent. Eur. J. Energetic Mater.* 14: 489–500. (b) Glück, J., Klapötke, T.M., Rusan, M., and Shaw, A.P. (2017). *Propellants Explos. Pyrotech.* 42: 131–141.

8 Küblböck, T., Angé, G., Bikelytė, G. et al. (2020). *Angew. Chem. Int. Ed.* 59: 12326–12330.

9 Sabatini, J. and Shaw, A. (2014). *Green Energetic Materials*, 63–102. Chichester: Wiley.

10 Domanico, J.A. (2008). *Proceed. 35th International Pyrotechnics Seminar, International Pyrotechnics Seminar* (ed. F.J. Schelling), 75–81. Fort Collins, CO, USA: International Pyrotechnics Society.

11 IFA, GESTIS-Substance Database. (2020). Institut für Arbeitsschutz der Deutschen Gesetzlichen Unfallversicherung, Sankt Augustin, Germany.

12 Conkling, J.A. and Mocella, C. (2010). *Chemistry of Pyrotechnics: Basic Principles and Theory*. Boca Raton, FL: CRC Press.

Index

r

s

t